ACID DEPOSITION AND ECOSYSTEM SENSITIVITY IN EAST ASIA

ACID DEPOSITION AND ECOSYSTEM SENSITIVITY IN EAST ASIA

Proceedings of International Symposium
"ACID DEPOSITION AND ECOSYSTEM SENSITIVITY IN EAST ASIA"
held during 7th INTECOL Congress, 19-25 July, 1998,
Florence, Italy

VLADIMIR BASHKIN
AND
SOON-UNG PARK
EDITORS

Nova Science Publishers, Inc.
Commack, New York

Editorial Production: Susan Boriotti
Office Manager: Annette Hellinger
Graphics: Frank Grucci and John T'Lustachowski
Information Editor: Tatiana Shohov
Book Production: Donna Dennis, Patrick Davin, Christine Mathosian and Tammy Sauter
Circulation: Maryanne Schmidt
Marketing/Sales: Cathy DeGregory

Library of Congress Cataloging-in-Publication Data

Acid deposition and ecosystem sensitivity in East Asia/Vladimir Bashkin and Soon-Ung Park, eds..
   p.   cm.
Includes bibliographical references and index.
ISBN 1-56072-611-3
1. Acid deposition--Environmental aspects--East Asia. 2. Air-- Pollution-- Environmental aspects--East Asia. I. Bashkin, V.N.(Vladimir Nikolaevich). II. Park, Soon-Ung.
TD195.54.E18 A27   1998   98-38394
363.738'62'095—dc21   CIP

Copyright © 1998 by Nova Science Publishers, Inc.
   6080 Jericho Turnpike, Suite 207
   Commack, New York 11725
   Tele. 516-499-3103   Fax 516-499-3146
   e-mail: Novascience@earthlink.net
   e-mail: Novascil@aol.com
   Web Site: http://www.nexusworld.com/nova

All rights reserved. No part of this book may be reproduced, stored in a retrieval system or transmitted in any form or by any means: electronic, electrostatic, magnetic, tape, mechanical photocopying, recording or otherwise without permission from the publishers.

The authors and publisher have taken care in preparation of this book, but make no expressed or implied warranty of any kind and assume no responsibility for any errors or omissions. No liability is assumed for incidental or consequential damages in connection with or arising out of information contained in this book.

This publication is designed to provide accurate and authoritative information with regard to the subject matter covered herein. It is sold with the clear understanding that the publisher is not engaged in rendering legal or any other professional services. If legal or any other expert assistance is required, the services of a competent person should be sought. FROM A DECLARATION OF PARTICIPANTS JOINTLY ADOPTED BY A COMMITTEE OF THE AMERICAN BAR ASSOCIATION AND A COMMITTEE OF PUBLISHERS.

Printed in the United States of America

CONTENTS

INTRODUCTION   1

PART 1   MODELING OF ACID DEPOSITION IN EAST ASIA   3

CHAPTER 1   NUMERICAL MODELING OF ACID DEPOSITION ON EURASIAN
              CONTINENT   5
              MICHAIL A. SOFIEV

CHAPTER 2   ESTIMATIONS OF DRY DEPOSITION OF SO_2 AND NO_2 OVER
              SOUTH KOREA   49
              SOON-UNG PARK AND SI-WAN KIM

PART 2   MONITORING OF PRECIPITATION IN EAST ASIA   93

CHAPTER 3   ACID RAIN MONITORING IN EAST AND SOUTH-EAST ASIA   95
              MIROSLAV RADOJEVIC

PART 3   EXPERIMENTAL STUDIES OF ACIDITY INFLUENCE ON
              TERRESTRIAL ECOSYSTEMS IN EAST ASIA   123

CHAPTER 4   CHANGES IN DISTRIBUTION OF ALUMINUM SPECIES IN SOIL
              SOLUTION DUE TO ACIDIFICATION   125
              KAZUO SATO, TAKASHI WAKAMATSU AND AKIRA TAKAHASHI

CHAPTER 5	DIFFERENTIAL SENSITIVITY OF TREES TO SIMULATED ACID RAIN OR OZONE IN COMBINATION WITH SULFUR DIOXIDE	143
	YOSHIHISA KOHNO, HIDEYUKI MATSUMURA, AND TAKUYA KOBAYASHI	
CHAPTER 6	ACID DEPOSITION EFFECTS ON THE DYNAMIC OF HEAVY METALS IN SOILS AND THEIR BIOLOGICAL ACCUMULATION IN THE CROPS AND VEGETABLES IN TAIWAN	189
	ZUENG-SANG CHEN, JEN-CHYI LIU, AND CHING-YI CHENG	

PART 4 EAST ASIAN ECOSYSTEM SENSITIVITY TO ACID DEPOSITION

CHAPTER 7	ACID DEPOSITION AND ECOSYSTEM SENSITIVITY IN EAST ASIA	229
	VLADIMIR N. BASHKIN	
CHAPTER 8	ACID DEPOSITION AND ECOSYSTEM SENSITIVITY IN CHINA	267
	JIMING HAO, SHAODONG XIE, LEI DUAN, XUEMEI YE	
CHAPTER 9	MODEL APPLICATION FOR ASSESSING THE ECOSYSTEM SENSITIVITY TO ACIDIC DEPOSITION BASED ON SOIL CHEMISTRY CHANGES AND NUTRIENT BUDGETS	313
	JUNKO SHINDO	
CHAPTER 10	SENSITIVITY OF THAILAND'S ECOSYSTEMS TO ACIDIC DEPOSITION	335
	MICHAEL KOZLOV AND SIRINTORNTHEP TOWPRAYOON	
CHAPTER 11	ACID DEPOSITION AND FOREST ECOSYSTEM SENSITIVITY IN TAIWAN	379
	LIN TENG-CHIU	

CONCLUSIONS	413
INDEX	423

INTRODUCTION

The rapid growth of industrialization and urbanization and a significant increase in agricultural production in East Asia have caused a remarkable increase in SO_2 and NO_x emissions during past several decades. The consequences of these growing emissions have been closely connected with enforced acidification loading on ecosystems and that results in actual and potential risk on many sensitive ecosystems.

Air pollution is an emerging important environmental problem in Asia as a whole and especially in East Asia. Ever growing emissions of sulfur dioxide, nitrogen oxides, ammonia and ground-level ozone concentration in many big cities in this region make the people worry on the synergetic effects of various pollutants.

Emission projections indicate that potentially large increases in emissions may occur during the next 25-50 years in Asia, especially in China, Thailand, India, etc, in accordance with the planned development programs. If this occurs (the probability is very high), the negative impacts that have been experienced in Europe and North America (USA and Canada) during this century will become increasingly apparent in large parts of Asia within the next 50-100 years. According to the present estimates SO_2 emission in Asia will surpass the combined emissions of North America and Europe by the year 2000 if the current growth rate of energy consumption keeps to remain. The primary man-made source of sulfur and nitrogen acidifying compounds in East Asia is low quality fossil fuel with high content of sulfur (up to 7% in Thai lignite, Chinese brown coal etc.) and nitrogen (heavy oil etc).

There is a growing concern on the long-range transboundary air pollution problem that is not limited to the geographical domains. Consequently the assessment of acidification loading on the ecosystems, monitoring of the acid rains, experimental assessment of terrestrial ecosystem response to acid input, calculation and mapping of critical loads as the indicators of ecosystem sensitivity to acid deposition in East Asia are of great scientific and political interest.

It is well known that during the winter the major weather patterns in East Asia facilitate the transboundary transport of air pollutants from west to east, from land to sea and the reverse in summer. Pollutants can thus be transported from country to country in the whole region of East Asia. It is therefore impossible for the individual country to solve the problem of air pollution and its impact on the ecosystems alone. It requires a regional intergovernmental cooperation. Currently, regional/sub-regional agreements on the pollutant emission abatement strategy do not exist at all in this region. However, some urgent concerns are emerging. In order to reduce the actual and potential risk of environmental degradation the general idea should to be set up, for example, the possible expansion of UN/ECE Convention on Long-Range Transboundary Air Pollution to cover the whole Eurasian continent

Until now only some preliminary attempts have been made to calculate the acidification loading in Asia. Thus, the main goal of this international project is to gather the existing information on the acid deposition and ecosystem sensitivity in East Asia from the national research teams. The participants of the project are from different countries in East Asian including Japan, China, Russia, South Korea, Thailand, Taiwan and Brunei. All of them mainly concern about the investigations on the acid deposition and ecosystem sensitivity in East Asia with different research backgrounds. An important feature of this volume is to give a comprehensive knowledge on acidification problem in East Asia covering the topics from monitoring of acid rain to the ecological assessment modeling.

We hope that this volume will provide useful information in developing more profound insight into the complexity of air pollution and its interaction with ecosystems in East Asia.

Vladimir Bashkin, Professor
Soon-Ung Park, Professor
Pushchino State University, Russia,
Seoul National University, Korea

Part 1

MODELING OF ACID DEPOSITION IN EAST ASIA

Chapter 1
NUMERICAL MODELING OF ACID DEPOSITION ON EURASIAN CONTINENT

MICHAIL A. SOFIEV
Home address: 4-1-18, Stroiteley str., Moscow, 117311, Russia, Tel/fax +7-095-930-0961, e-mail mas@glasnet.ru

CONTENT

Abstract
1. Introduction
2. Numerical models for evaluation of atmospheric pollution
    2.1. Model structure, input and output data
    2.2. Lagransian models
    2.3. Eulerian models
    2.4. Chemical transformations and depositions
    2.5. Vertical structure of model
3. Model validation with measurements
    3.1. Measurement representativeness and device precision
    3.2. Statistical techniques for the comparison
4. Emission estimates for the Asian region
    4.1. Temporal Variation of the Emission Intensity
5. Preliminary calculations of acidification loading in Eurasia
    5.1. Main units: advection scheme, vertical structures, chemistry transformations
        5.1.1. Advection scheme
        5.1.2. Vertical schemes
        5.1.3. Chemistry scheme
    5.2. Source – receptor calculations
    5.3. Model validation with measurement data
    5.4. Calculation results
        5.4.1. Deposition maps
        5.4.2. Source-receptor mass budget matrices
6. Conclusions
References

ABSTRACT

The problem of acid rains is known for several decades. One of the first steps in discovering the acidification as a large-scale environment damage was the Programme of Long-Range Transport of Atmospheric Pollution executed in Europe in 1970's. Since that time the investigations have been being concentrated in two main areas - field measurements and model simulations. Several studies have shown that transport distance of oxidised sulphur in the atmosphere can be hundreds and thousands of kilometres. It was also shown that powerful emission sources (first of all, power plants) can produce very considerable load to far-located environment systems.

Sulphur oxides were the first species whose transport and transformation were investigated. Since the physical and chemical processes taken place in the atmosphere with these substances are almost linear, the first models were based on trajectory approach. They considered the pollution puff as a heavy point following the wind flows with consideration of chemical transformations and sequential removing of species because of dry and wet scavenging. Providing acceptable estimates for sulphur compounds, these models can not reproduce non-linear chemistry reactions, and also not adapted to multi-layer structure of the lower troposphere. These limitations reduce the applicability of such models to more complicated simulations of atmospheric transport of oxidised and reduced nitrogen.

More sophisticated and precise but also more resource consuming instruments are Eulerian models. They are based on algorithms for numerical solution of systems of differential equations describing both chemical and physical processes affecting the pollution distribution.

Several numerical simulations have shown that in Northern Hemisphere there are three almost self-polluted regions - Europe, South East Asia and North America. The effects of cross-boarding influence of these regions to each other are detectable only close to the boundaries between them. Thus it is worth mentioning that prevailing transport direction in Eurasia is from West to East. As a result, the impact from European pollution sources can be significant in Central Asia, Asian territory of Russia, and some CIS countries.

The specific question is the precision of the obtained results. In industrial regions it is often dependent upon the quality of the emission data, while in remote areas the quality of the model definitions and algorithms takes the first priority. The most direct way to check the accuracy of the models is to compare them with available measurement data, but it raises several problems. First of all, long-range models have comparably coarse resolution - at least several tens of kilometres - while the measurement data are of point character. In this connection spatial representativeness of the measured data becomes crucial. The second difficulty is a long-term effect of acidification, which requires a characteristic time scale of several years. Consequently, it implies that measurements should also be temporally representative (to have regular character and similar features for a long period of time). To resolve these main difficulties several methodologies of the monitoring station evaluation and comparison of their results with model outputs were created.

1. INTRODUCTION

The problem of distribution of atmospheric pollutants over large distances was formulated at the beginning of 1970's for substances emitted from industrial sources and power plants. Sulphur oxides were considered as species of primary importance. Special investigations (see e.g. (Bolling & Persson, 1975), (The OECD Programme, 1977), (Eliassen, 1978)) have shown that these compounds can be distributed over the thousands of kilometres, and that the distribution process is accompanied by the number of chemical and physical transformation of the initial substances.

Large distance of the pollution transport results in considerable redistribution of the substances between regions. Thus, powerful emission source in one region can make a considerable impact to far-located remote areas. As a result, consideration of any region should include the evaluation of budget of pollutant masses emitted in this region and transported outside, and the masses come from outside sources. This approach requires calculation of a pollution budget table (see e.g. (Barrett *et al.*, 1995), (Galperin & Sofiev, 1998)) for each substance, which is useful for the development of the emission reduction measures (Klaassen *et al.*, 1992).

Specific feature of the mass budget tables is that they can not be obtained from observations. Generally monitoring station can detect the source of pollution only if it is located in close neighbourhood or has very specific speciation of the emission plume. Both these cases are not characteristic for the large-scale problem. In order to be representative for the regional scale, the monitoring station should be located far from domestic sources, which also results in mixture of their plumes and makes direct source detection practically impossible. To the opposite, atmospheric transport models can follow the distribution pattern of substances from each emission source and finally calculate the required budget table.

In addition to construction of the mass budget tables, the model enables to create maps of the pollution, which gives more regular information about the spatial pollution pattern than it can be obtained from measurements. Finally, the models can be used for the prognostic and investigation purposes. Variation of the input information like meteorological and emission data introduced into the model enables to simulate different scenarios of industry development and estimate the reply from the environment.

Current paper is dedicated to the general description of the types of long-range models and outline of currently available information about the acidification processes occurred in Eurasian regions. The second part of the paper contains a description of numerical model applied to calculations over the continent, and presents some results of these simulations.

2. NUMERICAL MODELS FOR EVALUATION OF ATMOSPHERIC POLLUTION

2.1. MODEL STRUCTURE, INPUT AND OUTPUT DATA

Classic mathematical model of the atmospheric pollution contains the following main parts (Fig.1).

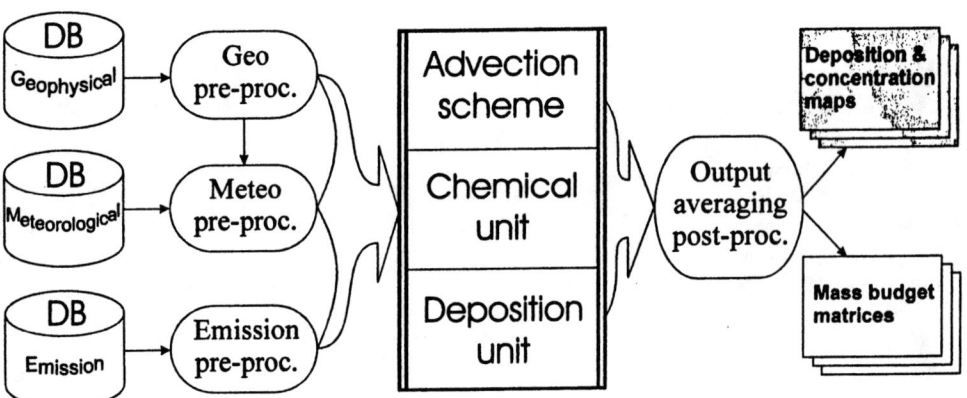

Figure 1. **General scheme of atmospheric pollution model**

The central part of any model is its advection scheme. Choice of the scheme type determines many key features of the model. This module works very closely with chemistry scheme and deposition unit, and they altogether represent the main model processor.

There are three types of input information required for calculations. They are – emission, meteorological and geographical data. These data are stored separately and loaded into the model via pre-processing units, dealing with unpacking of data and calculating on their basis some intermediate variables consumed by the model (e.g. provide the spatial / temporal interpolation of data in order to match to the internal calculation environment). In special cases such pre-processors can be very complicated. Thus, if the emission spatial / temporal resolution becomes the matter of the first priority, the calculations may require a separate emission model (Lenhart & Friedrich, 1995). In other cases the meteorological pre-processor can be presented by rather complicated prognostic model, which is also driven by necessity to create a data of adequate characteristics (Moussiopoulos, 1994).

Output data of the model consists of the maps of concentrations and depositions averaged over some time period, and budget matrices reflecting the pollution exchange between various regions. Both averaging procedures and calculation of the matrices are performed by the output post-processor.

The specific question is the choice of the meteorological data. One possible way is to apply climatological information with corresponding adaptation of the advection –

diffusion scheme of the model. It provides acceptable global budgets of substances, and also can be used for some small-scale tasks. For regional scale such approach shows insufficient precision of the final estimates. This is primarily connected with intensive (and timely irregular) mixture of the pollutants by cyclone activity, which can hardly be reproduced with climatological data. Regional-scale calculations require operational meteorological data with appropriate time and space resolution, which enable to resolve the daily variations of meteorological parameters and cyclone movement. The largest time step is normally 6 hours (corresponding to standard meteorological period), and the spatial resolution is equalled to that of the model itself.

2.2. LAGRANSIAN MODELS

As it was mentioned above, the type of advection scheme is the most important characteristic of the atmospheric model.

The first air transport models were based on so-called "trajectory approach", sometimes called as Lagransian one. In this type of routine the advection caused by wind is considered as a motion of heavy particles with wind flows. It is taken that at each model time step each emission source ejects one heavy point with prescribed chemical composition and with mass proportional to the emission rate. Then this point flies in the atmosphere following the actual 2-D (rarely - 3-D) wind flows - so, the trajectory of this mass was calculated. This mass is fully isolated from the others - both from those emitted by different sources and also from those emitted by the same source but at different moments of time. As the initial speciation of this puff is prescribed, chemical reactions and processes of dry deposition and scavenging with precipitation can be considered.

Some of such models have proved their reliability and used also now. One of the best examples is a model developed in Meteorological Synthesising Centre – West (MSC-W) of Co-operative Programme for Monitoring and Evaluation of the Long-Range Transmission of Air Pollutants in Europe (EMEP) (Eliassen, 1976; Eliassen & Satbones, 1975, 1983; Barrett *et al.*, 1995).

The main advantage of these models is an intuitive simplicity of the algorithms. This approach also enables to consider each source of pollution separately, and thus automatically provides the budget of pollution exchange between different regions.

The main disadvantage is that heavy points have to be treated separately, which prevents the consideration of the non-linear chemical reactions. In addition, the amount of required resources grows proportionally to the number of considered trajectories, which is very large (every emission source at every time step of the model ejects one separate trajectory). Algorithms of horizontal diffusion are also very expensive.

2.3. EULERIAN MODELS

Limitations of trajectory schemes resulted in intensive investigations in the field of direct numerical solution of the diffusion equations. Currently the most sophisticated models are based on various Eulerian advection schemes. This approach resolves all of the above mentioned difficulties but raises different problems.

Classic Eulerian approach implies direct numerical solution of 3-D advection – diffusion equation:

$$(1) \qquad \frac{\partial c}{\partial t} = (\nabla, K\nabla c) + (v, \nabla c) + E + D + T,$$

where $c(x,y,z,t)$ is instantaneous concentration of a particular substance, $K(x,y,z,t) = \|k_{i,j}\|_{i,j=\overline{1,3}}$ - is matrix of diffusion coefficients, $v(.)$ is vector of wind velocity, E, D, T represent the increase / decrease of the amount of the substance because of emission, deposition and chemistry transformations correspondingly ($E>0, D<0, T><=0$).

Taking the reasonable assumption that matrix K is diagonal (which means that there is no inter-action between diffusion in different directions), it is possible to split the calculations in each direction and perform them sequentially in three steps. At each step a simplified 1-D advection – diffusion equation is to be solved:

$$(2) \qquad \frac{\partial c}{\partial t} = \frac{\partial}{\partial l} K_l \frac{\partial c}{\partial l} + v_l \frac{\partial c}{\partial l} + E + D + T, \quad l \in \{x, y, z\}$$

Additional simplifications and different discretization algorithms give various advection schemes – see (Bott, 1989), (Egan & Mahoney, 1972), (Pedersen & Prahm, 1974), (Pepper & Long, 1978), (Smolarkiewich, 1983).

The main problem of Eulerian schemes is numerical viscosity originated from interaction of finite temporal and spatial resolution of the models. This effect causes a disturbing of the form of pollution plume in a certain way, which looks like appearing of additional non-symmetrical diffusion processes. Often this effect is also called as pseudodiffusion. Various attempts to reduce it resulted in creation of set of advection schemes of different complexity and quality. The most widely used one is so-called Bott scheme originally presented in (Bott, 1989). Currently this type of schemes is considered as a good compromise between the required computer resources and general precision of the scheme. At the same time continuos efforts in creation of better numerical algorithms result in appearance of some new schemes (often based on Bott algorithm) like those described in (Syrakov & Galperin, 1998). The last paper also contains some results of the scheme comparison.

A review of Eulerian modelling and its prospects can be found in (Peters et al., 1995).

2.4. CHEMICAL TRANSFORMATIONS AND DEPOSITIONS

Currently it is accepted that chemical transformations of acidic compounds include the following main processes (Barrett et al., 1995), (Moussipoulos, 1994), (Pressman et al., 1991), (Galperin & Sofiev, 1998).

Oxidised sulphur. Emission of sulphur occurs primarily in form of SO_2 with small fraction of sulphates (the percent value is about 5% but can considerably vary depending of the source type). The primary process in the atmosphere is sequential irreversible oxidation to sulphates in the aerosol form. In addition, both substances are subjects to dry deposition and scavenging with precipitation (more intensive for SO_2 than for sulphates). Most of reactions can be assumed to be linear, with exception of the scavenging of sulphur dioxide near powerful emission sources, where the effect of saturation of the precipitation elements with molecules of sulphates produced from SO_2 may take place. It is taken into account in several models (Moussipoulos, 1994), (Pressman et al., 1991). Numerical characteristics of this effect can be derived directly from observations as it is done in (Galperin, 1989). The analysis of significance of the saturation leads to a conclusion that it affects heavily polluted regions with very high concentration of sulphur dioxide while in remote regions it is negligible (Hass et al., 1996).

Nitrogen chemistry transformations are much more complicated. It is accepted now that emission takes place in a form of mixture NO and NO_2 with various proportions dependent upon the type of the source. These two oxides play the primary role in the ozone-producing circle, which also involves volatile organic compounds (VOC). The complexity of this part of chemical scheme is very high. Currently even the number of chemical reactions included into this circle is not finally determined. Fortunately acid calculations themselves can be carried out with simplified consideration of ozone-related reactions. If ozone concentration is not the goal of the model application, but just an intermediate parameter required for more accurate evaluation of proportion between nitrogen oxides, then the ozone circle can be reduced down to 2-3 reactions. VOCs are considered either in a form of ozone creation potential on the basis of (Sympson, 1995), or skipped completely. In the last case the ozone production is calculated from meteorological and geophysical parameters like insolation, cloudiness, etc. Deposition of both NO and NO_2 is rather slow and sometimes not detectable.

In addition to ozone cycle, there is comparably slow oxidation of NO_2 to NO_3^-, which then comes to a set of reactions and equilibria between nitric acid and various nitrate salts (first of all, nitrate of ammonium). There is an equilibrium between $NH_3 + NHO_3$ and NH_4NO_3, which strongly depends upon the concentrations of all reagents. More details can be found in (Finlayson – Pitts & Pitts, 1986). All these compounds can be deposited with precipitation and have considerable dry deposition velocity. The most intensive deposition is attributed to nitric acid, smaller rates characterise the processes with nitrates.

As it is seen from the above outline, the full-range chemistry simulations require proper reflection of non-linear effects and thus can not be fully implemented in Lagransian schemes.

2.5. VERTICAL STRUCTURE OF A MODEL

It is well known that troposphere is well structured to several layers with different thickness and physical parameters. Its simplified structure is presented in Figure 2.

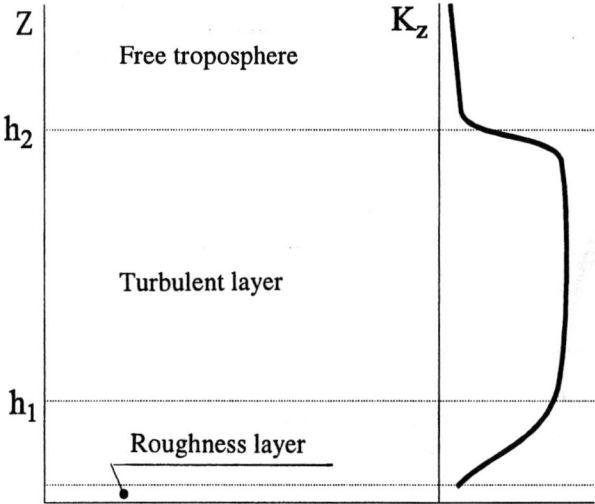

Figure 2. Simplified scheme of vertical structure of the troposphere. (Stable and neutral conditions)

The lowest thin layer (so-called roughness layer) is of several millimetres or, at most, centimetre height. The next one is the most important layer for regional transport - boundary layer (or turbulent layer). Its thickness varies from hundreds of meters (in winter conditions, above cold sea surface, etc.) up to several thousand of meters (summer conditions, well-developed convection, etc.). Finally, above the boundary layer (free troposphere) it is assumed that turbulent mixing is rather small, and the prevailing movement is close to laminar.

There are two types of motion of pollution masses along the vertical. The first one is caused by turbulent mixing (or by diffusion in layers with small turbulence); the second one - by vertical component of wind.

Local vertical winds can be quite strong, but for large grid cells (more than 100 km × 100 km) the mean value of vertical wind speed inside cyclone or anti-cyclone is about centimetre per second, which is in good agreement with results of direct observations (Dulac et al., 1992).

The specific of the description of turbulent mixing in terms of classic diffusion is that corresponding equations can not be closed (Nieuwstadt & van Dop, 1982). To reduce the number of variables down to number of equations some additional assumptions are required. One of the simplest ways is so-called first-order closure, which involves semi-empirical vertical mixing coefficient K_z, which enables to transform the initial set of equations to one widely used semi-empirical formula:

$$\text{(3)} \qquad \frac{\partial c}{\partial t} = \frac{\partial}{\partial z} K_z \frac{\partial c}{\partial z}$$

where c is concentration at a height z, t is time.

Typical profile of K_z is presented in Figure 2. Starting from roughness layer it grows almost linearly with the z up to a height h_1. Between h_1 and boundary layer height h_2 it is nearly constant, while at height of boundary layer it rapidly decreases by the factor of times (or even order) and then slowly decreases with the growth of the height up to the top of troposphere. This scheme is very rough (in particular, well-developed convection can not be described in terms of K_z) but acceptable for many real atmospheric conditions and can be used in large-scale models.

Above outlined vertical structure of the troposphere requires adequate mechanisms in transport models. The simplest approach widely used in Lagransian schemes considers the only layer up to boundary layer height h_2 as well-mixed box, and takes into account only horizontal motions. Giving qualitative picture this mechanism is not applicable without corrections describing the local depositions near the emission sources and exchange with free troposphere. For example, one of Lagransian models used for regular simulations of acidification over Europe – model ROOT (*R*eceptor *O*riented *O*ne-layer *T*rajectory model) (Barrett *et al.*, 1995) involves empirical α-factor for representing the fraction of local depositions in grid cell where the emission source is located. This approach was originally presented in (Janssen & Asman, 1988). Additional measurement information is also involved for description of exchange between boundary layer and free troposphere.

Slightly better results can be obtained from the consideration of continuous profiles of concentrations with variable parameters like that presented in (Galperin & Sofiev, 1998). In that model the concentration distribution along the vertical is taken to be truncated gaussian function with dynamic parameters. This method is a compromise between resource saving and necessity to reproduce crucial features of the vertical motion – growing of thin pollution plume near the source while the transport, modification of the concentration profile with time depending on the source and boundary layer heights. It also enables to describe the inter-action of two plumes with different vertical characteristics (although the final distribution function is forced to be uni-modal). Finally this profile enables to utilise the multi-layer meteorological parameters (first of all wind direction and velocity) and calculates the mean value on the basis of convolution:

$$\text{(4)} \qquad \overline{p} = \int_0^\infty p(z) g(z) dz,$$

where g is vertical distribution function of the concentrations, p is meteorological parameter (e.g., wind vector), z is vertical axis.

The main disadvantage of this approach is that it gives precise concentration distribution only for unrealistic profile of K_z. In other cases it gives only qualitative simulation of the actual concentration distribution. Uni-modal requirement to the distribution function results in high disturbance of a pollution plume if both low and high emission sources are existing in the grid cell. Results are also become uncertain if multi-layer wind has large deviations with height.

The best currently existing mechanism for description of the vertical concentration profile is fully 3-dimentional grid with several layers along the vertical. One of possible realisation of this approach was used for the calculations presented in this paper for Eurasia. The outline of the scheme can be found below.

3. MODEL VALIDATION WITH MEASUREMENTS

All numerical models currently used for calculations are constructed on the basis of simplifications and assumptions, which are not always fulfilled in reality. As a result, the model output reflects the actual process with certain error. Hence, the verification of the model results becomes mandatory for quantitative evaluation of the precision of the model results.

The most widely used method of the model verification is a comparison of the output results with available measurements. More complex methods are applied rarely in special cases, first of all, under conditions of lack of measurements – see (Sofiev, 1996). Simple comparison of mean modelled and measured concentrations and depositions, and calculation of correlation coefficients create no technical difficulties, although there are important specifics which should be taken into account. Two mostly important ones are – limited representativity of point measurements in a model grid cell, and limited applicability of the standard statistics (like arithmetic averaging, correlation coefficient or linear regression restored by method of least squares) to the real data sets.

3.1. MEASUREMENT REPRESENTATIVENESS AND DEVICE PRECISION

The measurement data suitable for the validation of the large-scale atmospheric transport models are provided by the stationary monitoring stations. Such stations give regular data for a long periods with stable quality. Data produced by such stations can easily be averaged over months and years. These features are in good agreement with long-term character of acidification and seasonal variability of the processes.

The main disadvantage of these stations is that they can not represent the general pollution pattern in large area like grid cell of regional model. Samples collected at a location of a site can deviate from the mean concentrations in the region. Corresponding error can be reduced by proper location of the stations oriented to large-scale pollution monitoring – far from powerful emission sources, in homogeneous terrain, etc. The remained uncertainty of monitoring data should be taken into account during the model-measurement comparison.

The numerical estimations of representativeness of point measurements were studied in several investigations (Ebel *et al.*, 1994), (Galperin & Sofiev 1994), (Sirois & Vet, 1994). Summarising these findings presented on the EMEP Workshop on the Accuracy of measurements it can be stated that standard deviation of the mean concentration in a square grid cell depends on the size of a cell, averaging period, and type of a substance. This deviation should be attributed to measured values as a measure of their uncertainty.

Below the results of comparison of closely located stations are presented. The first table contains the values taken from (Galperin & Sofiev, 1994). The Table 1 shows the sum of squares of regular and chaotic differences between monthly values measured by two independent closely located monitoring sites:

$$(5) \qquad D = \sqrt{\overline{\Delta}^2 + \sigma_\Delta^2}$$

Here, $\overline{\Delta}$ is the mean value of relative differences Δ between station data, and σ_Δ is its standard deviation.

The distance between stations was about 100 km, several pairs of stations were considered for each compound. This enables to calculate maximum and minimum of D and its mean value.

Table 1. **An estimate of uncertainty of monthly measurements for a large area (about 100 km linear size).**

Substances	Maximum D	Minimum D	Mean value
SO_2 air concentrations	28 %	20%	23%
NO_2 air concentrations	34%	27%	31%
$SO_4^=$, NO_3^-, NH_4^+ concentration in precipitation	30%	14%	25%

In addition to the representativeness of the point measurements the precision of the monitoring technique is also important. Large-scale models often deal with small concentrations in remote regions, which sometimes close to the device detection limit and / or to the background level of concentrations. It raises the problem of calibration of the monitoring devices and numerical estimation of their precision. Now virtually each large-scale monitoring campaign includes special part devoted to the quality assurance (QA) of the measurement technology and devices. Thus a regular QA of the monitoring sites and their intercalibration are performed within the scope of EMEP. The results are published each year in the annual reports of Chemical Coordination Centre (e.g. (Semb et al., 1994)). The investigations have shown that precision of sulphur oxide measurements is higher than that of nitrogen (both oxidised and reduced).

3.2. STATISTICAL TECHNIQUES FOR THE COMPARISON

The most popular statistics used for the model-measurement comparison are mean modelled and measured values for each monitoring site or calculation region; the correlation coefficient between measured and calculated data; various types of confidential ranges and percentiles; linear regression analysis commonly based on the method of least squares (MLS). A very popular consideration is a qualitative analysis of various graphs of temporal variations of the concentrations at the stations with further qualitative conclusion about the model ability to reproduce the peaks and lowest values as well as main observed trends.

The main advantage of the above methods is their simplicity and availability. They are also quite easy for interpretation of the obtained results. The main drawback is that they sometimes lead to inaccurate results of the validation.

Almost all of the above mentioned methods are based on assumptions of "normality" of the considered data sets. First of all, they are supposed to be normally distributed $N(x, \alpha, \sigma)$ (well known Gaussian distribution function):

$$(6) \quad N(x, \alpha, \sigma) = \frac{1}{\sqrt{2\pi}\sigma} e^{-\frac{(x-\alpha)^2}{2\sigma^2}},$$

where α is mathematical expectation of the stochastic variable x, and σ is standard deviation of x.

It is easy to see that this function permits the negative values, which is completely useless for the concentration and many other physical parameters. Changing of this assumption may lead to necessity to change statistical instruments used for the analysis.

The second problem is the robustness of the applied analytic methods. Robustness of a method means that small deviations in the input data result in small variations of the output results. It is essential that small deviations in the input data imply both small distortions of many values as well as large distortions of a small number of values. So, a robust method is not sensitive to positions of one-two-three points in the scatter plot. They can be completely out of the main trend/cloud as still do not affect considerably the results of analysis. One of the best examples of non-robust statistics is the arithmetic mean. Another one - is the correlation coefficient. It is often possible to make some modifications of these methods and create some robust statistics on their basis [Huber, 1981] or use different methodology (e.g. median mean instead of arithmetic one).

Another approach to this task was suggested in several papers, e.g. [Pedersen, 1994], [Journel & Huijbregts, 1978] where the measurement network was used for the creation of smoothed map of monitored parameters which then can be compared with smoothed model pattern. Being very promising this method is based on spatial correlograms and consequently requires rather dense network of the monitoring stations.

Since the compared data sets are finite-size, the obtained results have to be considered as sample estimates of corresponding parameters. The standard deviations of these sample estimates can be approximately taken as follows:

For arithmetic mean $E[x]$ (M is size of sample $\{x\}$, σ_x is standard deviation of $\{x\}$):

(7) $$\sigma_E = \frac{1}{M}\sigma_x$$

For correlation coefficient r_x:

(8) $$\sigma_r \sim \frac{1-r_x^2}{\sqrt{M}}$$

For regression slope A:

(9) $$\sigma_A = \frac{\sigma_\varepsilon}{\sqrt{M}}\frac{1}{\sqrt{\overline{x^2}-\overline{x}^2}} = \frac{\sigma_\varepsilon}{\sqrt{M}\sigma_x}$$

For regression bias:

(10) $$\sigma_B = \frac{\sigma_\varepsilon}{\sqrt{M}}\frac{\sqrt{\overline{x^2}}}{\sqrt{\overline{x^2}+\overline{x}^2}}$$

Sometimes application of the MLS for the regression analysis might give uncertain results. Some reasons for that and possible solutions are discussed in a set of papers (Sofiev, 1994), (Sofiev & Galpein, 1994), [Sofiev et al., 1994]. The last paper contains the description of more complicated but more robust methodology of linear regression.

4. EMISSION ESTIMATES FOR THE ASIAN REGION

The emission information is one of key input data for the atmospheric models. Its quality has a direct influence to the quality of final modelling results. Thus for substances like aerosol particles the relationship between emission rates entered into the model and output values of concentrations and depositions is strictly linear. For acidifying compounds non-linear chemistry transformations and deposition processes make the situation more complicated but the sensitivity of the model output to variation of the emission input data is still high (Sofiev et al., 1996).

Contrary to Europe and North America, there are no regular emission databases for Asia compiled at official level. Only expert estimates for several substances are available. The most important source for such information is Global Emission Inventory Activity – GEIA – (Benkovitz et al., 1996). GEIA project was started in 1990 and aimed at creation of global emission inventories of gases and aerosols emitted into the atmosphere from natural and anthropogenic sources. A list of first-order inventories was compiled in 1995. GEIA is a component of the International Global Atmospheric Chemistry (IGAC) Project, a core project of the International Geosphere-Biosphere Program (IGBP).

Currently available emissions of primary acidifying pollutants in GEIA database are: low and high sources of both oxidised sulphur and nitrogen (low sources are less than 100 m height, high ones – above 100 m). In addition, there is a population map, some geographical and land-use information. All the maps are compiled for the globe with spatial resolution of $1^0 \times 1^0$ latitude-longitude. The maps with sulphur and nitrogen emission for Eurasia recalculated to the polar stereographic projection are presented on Figure 3, Figure 4.

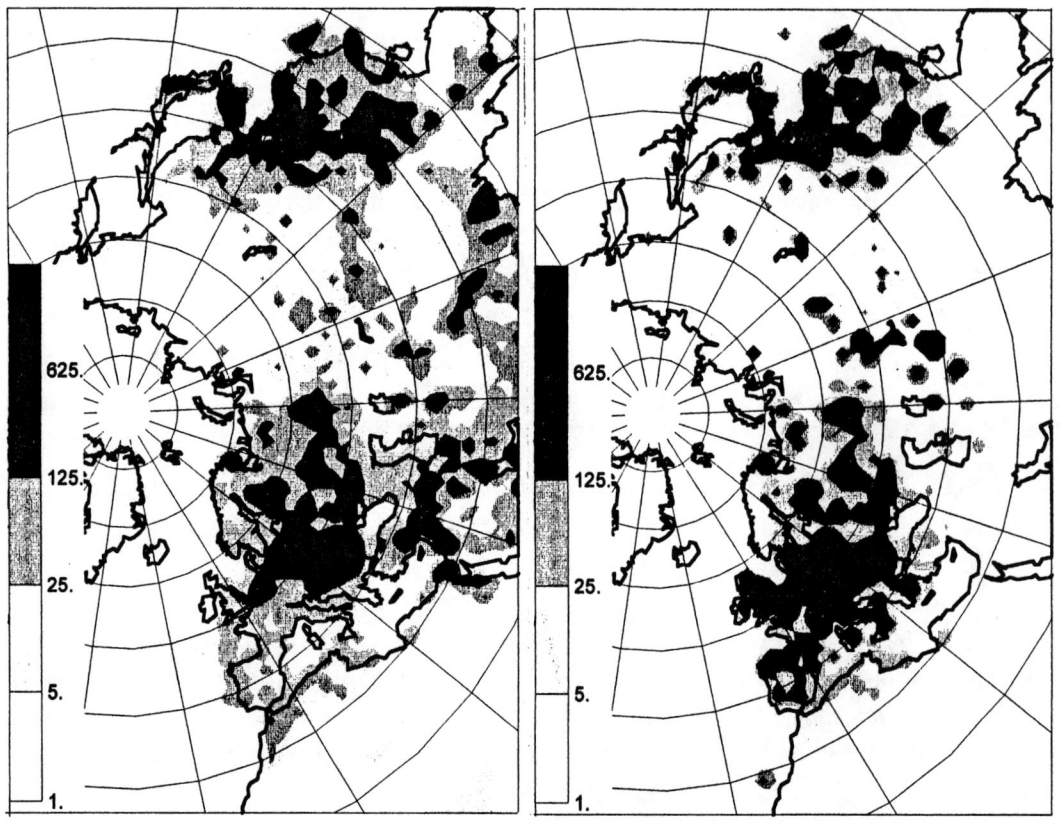

Low sources (below 100 height)　　　High sources (above 100 m height)

Figure 3. **Emission of sulphur oxides. Unit = 100 tones S year^{-1}.**
(Benkovitz *et al*, 1996).

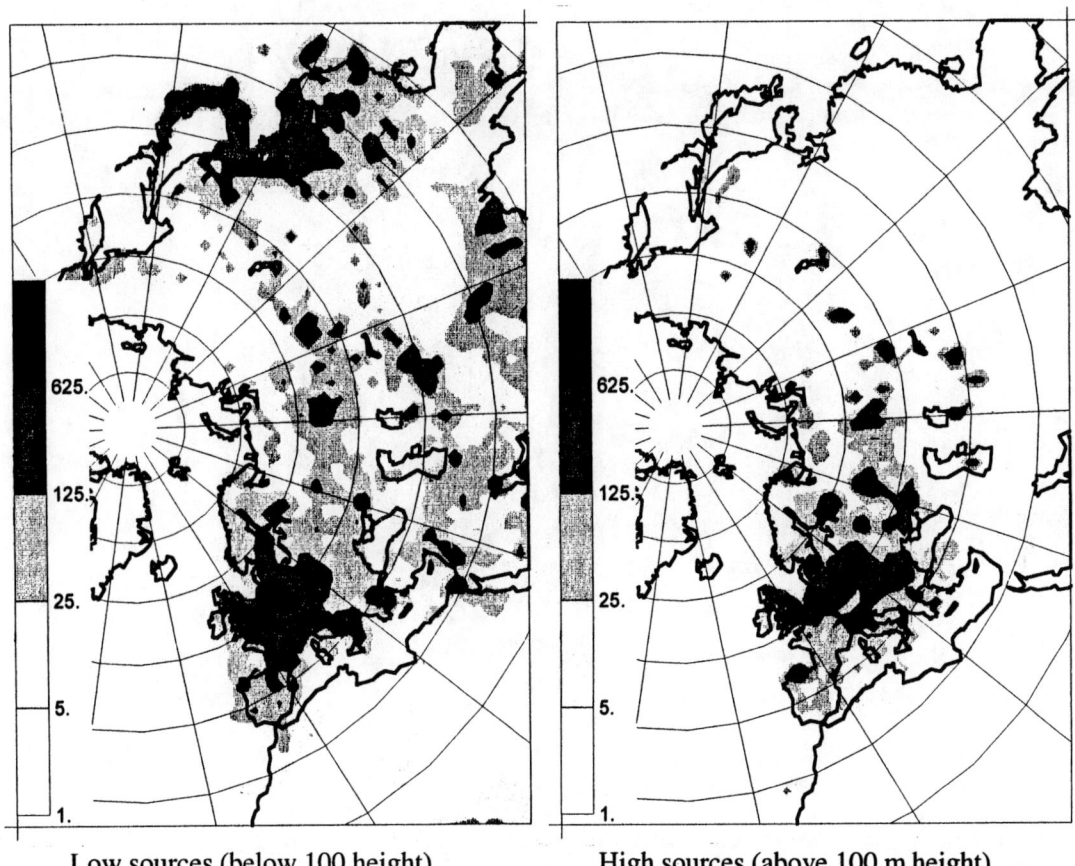

Low sources (below 100 height)　　　High sources (above 100 m height)

Figure 4. **Emission of nitrogen oxides. Unit = 100 tones N year^{-1}. (Benkovitz** *et al,* **1996).**

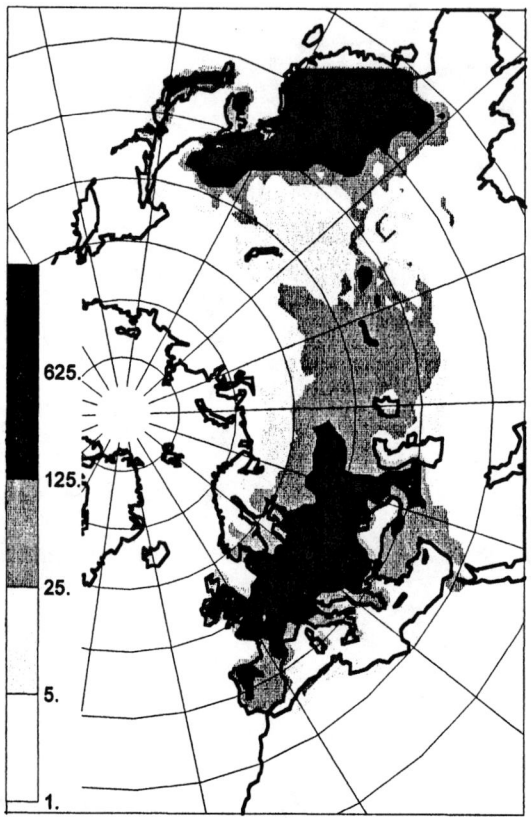

Figure 5. **Emission of ammonia. Unit = 100 t N year^{-1}. (Dianwu & Anpu, 1994).**

As it was mentioned above, proper simulation of nitrogen chemistry requires the consideration of ammonia, although its emission is not available in GEIA database. Still some other estimates are available (Dianwu & Anpu, 1994). On this basis a map of ammonia emission was compiled for preliminary hemispheric calculations (Sofiev et al., 1995), and also used for simulations (Galperin & Sofiev, 1998). There are several peculiarities of this ammonia emission map. (i) The bulk of ammonia is emitted from agricultural sources, so it can be considered as low emission (less that 100 m source height). (ii) Calculation domain do not cover the complete Northern Hemisphere (see Figure 5). (iii) There was practically no information about the space distribution of the agricultural activity in China. In these conditions the uniform distribution of the total emission values from (Dianwu & Anpu, 1994) was used.

4.1. TEMPORAL VARIATION OF THE EMISSION INTENSITY

In additional to absolute emission rates and spatial distribution of the sources, temporal variability of the emission intensity also considerably affects the pollution distribution pattern. The importance of this effect was estimated in (Sofiev et al., 1996). It can be stated that the most crucial parameter is the monthly variation of the emission intensity, while the weekly and daily deviations are not so important. For current preliminary calculations the monthly variability of the emission intensity was calculated following the approach presented in (Afinogenova & Galperin, 1985).

Corresponding monthly correction coefficients are shown in Figure 6. Similar estimates are also available in GEIA database, where they are calculated separately for each grid cell for 4 seasons.

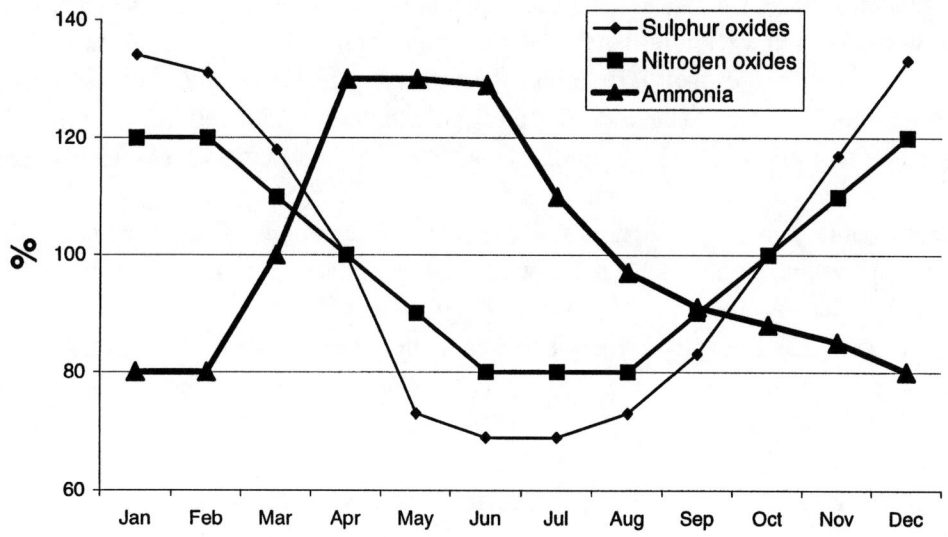

Figure 6. **Relative correction coefficients to the monthly emission rates.**

5. PRELIMINARY CALCULATIONS OF ACIDIFICATION LOADING IN EURASIA

This chapter contains some assessments of the acid deposition in Asian region. Since Western part of Asia is affected by the European sources, the simulations were made for the whole Eurasian continent.

Calculations were split to two parts. The first one included general assessment of the acid deposition, while the second part was dedicated to the mass budget and source-receptor pollution exchange between countries and regions. For these purposes two versions of acid model were used. Calculations of total deposition and concentration maps were performed with fully 3-D Eulerian model with 6 vertical layers. For source-receptor relationship evaluation a simplified version of this model was utilised. It considers a dynamic continuous vertical profile of mass distribution along the height. Such a compromise was forced by available computer resources.

The development of the model was started in scientific team of Prof. Y.Pressman and Dr. Sci. M.Galperin in middle of 1980-s, and initial version of the model applied to the calculations of the sulphur and oxidised nitrogen depositions over Europe (Galperin, 1989),

(Pressman *et al.*, 1991). The model is intended for long-term (from 1 month to decades) calculations of airborne pollution transport and source-receptor relationship on the basis of actual meteorological and emission data. The calculation results are concentration and deposition maps and emitter-receptor matrices. The model chemistry unit includes 17 reactions (5 of second order) for SO_2, H_2SO_4, sulphates, NO, NO_2, HNO_3, nitrates, NH_3, ammonium, PAN, O_3.

The model input consists of geographical, meteorological and emission data.

Geographical information includes geographical co-ordinates, surface type (sea, large fresh water basins, coastal line, land) and the height above the sea level. The state of underlying surface (dry or wet) is determined on the basis of precipitation intensity taken from meteorological data. At latitude 60° the grid cell size is 150×150 km^2 in the polar stereographic projection. At other latitudes the size is corrected to the actual one. The area correction can be as much as 30% in tropics.

The meteorological information used in the current calculations (Table 2) was prepared in Russian Hydrometeorological Centre within the framework of numerical short-term weather prediction (Frolov *et al.* 1994). For 1985 calculations the information was taken from (Grid point data set, v.II) and processed following the algorithm (Rubinshtein *et al.*, 1998). The averaging interval is 6 hours, spatial resolution is 150×150 km^2 for the whole region.

Table 2 *Meteorological information used for the model calculations*

Parameter	Level, mbar	Measurement units
Wind speed u, v	1000, 925, 850, 700	km·h^{-1}
Geopotential trend	925, 850 (only for 1995)	mbar h^{-1}
Temperature	2m above surface, 1000, 925, 850, 700	4°C (1°C for 1995)
Precipitation	Surface	0.1 mm (6 h)$^{-1}$
Cloudiness	Inferior and middle clouds (only for 1995)	Amount

The emission information contains the total maps of low and high emission (see Figures 3 –5); and the separate maps with emissions of each country / region considered in source-receptor calculations. Each country, region, water basin etc. can be considered as an emitter, receptor or both. A source can also have no area (point sources).

In order to create more realistic initial conditions in the model by the beginning of a considered period the 3-14 days advance pre-computation is used. The length is determined to be close to pollutants lifetime in a calculated domain.

5.1. MAIN UNITS: ADVECTION SCHEME, VERTICAL STRUCTURES, CHEMISTRY TRANSFORMATIONS

5.1.1. Advection scheme

The advection scheme used in the model is based on combination of Lagransian trajectories within the grid cell limits with Eulerian transport routine for mass exchange between cells. The presented scheme has common features with numerical advection schemes (Egan & Mahoney, 1972; Pepper & Long, 1978, Pedersen & Prahm, 1974). Its main idea is presented in Figure 7, Figure 8. Full description can be found in (Sofiev et al., 1996), (Sofiev, 1998).

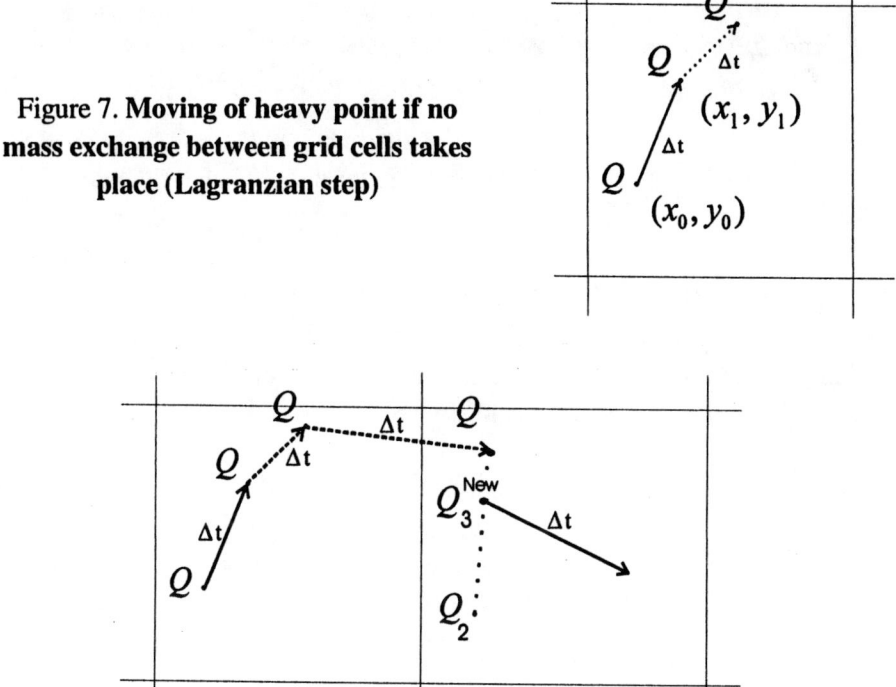

Figure 7. **Moving of heavy point if no mass exchange between grid cells takes place (Lagranzian step)**

Figure 8. **Conservation of the first moments during the mass exchange between cells (Eulerian step). Example for two masses in a cell**

The pollution mass in a grid cell is considered as heavy point representing the centre of mass in a cell. Calculations are split to two parts. The first Lagransian part (Figure 7) is executing while the heavy point is moving inside a cell and no other masses come into it. During this period heavy point moves along the trajectory driven by wind velocity vector.

If a heavy point leaves a cell and comes to a new cell where another mass is located (Figure 8) than Eulerian step takes place. It results in creation a new heavy point, which mass is equalled to sum of all masses in a cell, and position is calculated from the equation for first moment concervativity. Finally, we have:

$$(11) \qquad \mathbf{r}^{new} = \frac{1}{Q^{total}} \sum Q^i \mathbf{r}^i,$$

where $\mathbf{r}^{new}$ is a co-ordinate vector of new heavy point (Q_3^{New} in Figure 8), and $\mathbf{r}^i$ are co-ordinates of initial heavy points before the Eulerian step (Q and Q_2 in Fugure 8).

The advantage of the algorithm is zero numerical viscosity, which is the main problem of most of Eulerian schemes. Its high efficiency enables to realise this algorithm on INTEL – based computers with extended configurations.

The main disadvantage of the scheme is a considerable non-monotonicity of instantaneous concentrations and depositions. These amplitude oscillations at constant wind V are resulted from interference of quantization motion with period $|V \Delta t|$ and the grid with a constant cell size Δs. These fluctuations are of finite amplitude and do not break down the positive definiteness of the scheme. Randomization of the process, i.e. random fluctuations of real wind and other meteorological parameters, suppress the oscillations, which was proved during the multiannual experience of model application within the scope of EMEP.

The computational scheme described above does not contain terms of horizontal diffusion. It is explained by the fact that at the model spatial scale (cells 150×150 km² and distribution over continents) with averaging time over a day the horizontal diffusion is mainly connected with wind variations in strength and direction at various heights (Pekar, 1989).

The scheme keeps track of the pollution export outside the grid limits. If a certain pollution amount has left the grid across one of its four boundaries it is supposed that it has left it for good. The total amount left is summed up for each boundary.

If mass in a cell is lower than the threshold value, the cell drops out of the calculations for a given time step and this mass is added to background container. The mass in this container is considered to be uniformly distributed over the whole grid and as origin of unattributable deposition. The threshold value is chosen to be lower than natural background concentrations.

These two supplementary procedures assure that total mass balance for the whole calculation period and for the whole model domain is kept.

5.1.2. Vertical schemes

Current model can be used with two vertical schemes. More precise (but also more expensive in exploitation) one is based on multi-layer approach and can consider up to 10 layers of different thickness. A simplified scheme considers a continuous profile of concentrations with dynamic parameters, which are recalculated at each model time step.

concentrations with dynamic parameters, which are recalculated at each model time step. As it was mentioned above, for current calculations the multi-layer scheme was utilised for full-chemistry calculations, while simplified Gaussian profile was used for creation source-receptor budget matrix.

For consideration of multi-layer scheme let introduce the following notations (Figure 9).

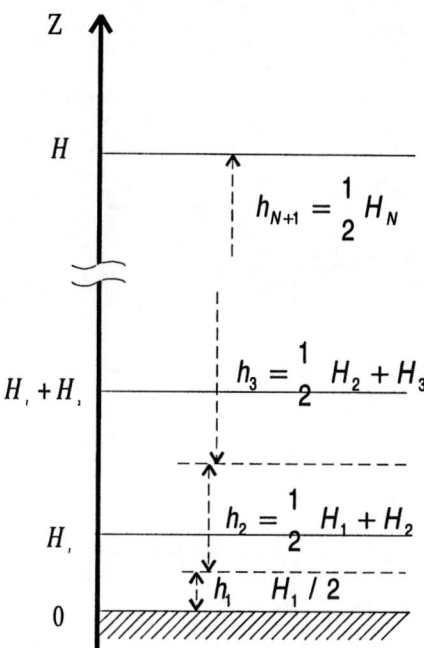

Figure 9. **Scheme of model vertical structure.**

Let H be a total height of the transport layer. It is assumed that there is no mass moving above its upper limit. Transport downwards from this layer is considered as a dry deposition process, which velocity is designated as V_d.

H_i, $(i=1..N)$ is the thickness of a sublayer. $\sum_{i=1}^{N} H_i = H$.

h_i, $(i=1,N+1)$ is the distance between central points of the two adjacent layers (but h_1, h_{N+1}):

(12)
$$h_1 = H_1/2;$$
$$h_i = H_{i-1}/2 + H_i/2, \quad i = \overline{2, N};$$
$$h_{N+1} = H_N/2$$

$K_z(z)$ is the coefficient of vertical turbulent diffusion, z is vertical co-ordinate, $z \in [0,H]$.

Multi-layer scheme of vertical diffusion is based on the expanded resistance-capacity analogy.

In 'classical' resistance analogy it is assumed that the process of the dry deposition can be described as the motion of the electric charge (pollution mass amount) through three sequential resistors [Wesely (1989)]. Following the same idea it is possible to consider the process of the transport of the pollution between different horizontal layers as the process of the electric charge motion between capacitors through resistors. The electric scheme is shown in Figure 10.

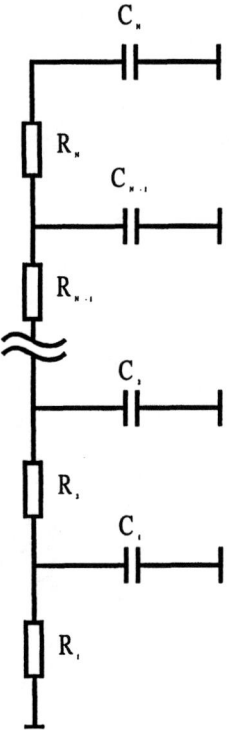

Figure 10. **Resistive analogy for the multi-layer pollution exchange**

Capacitors C_i correspond to the heights of the layers H_i (it is assumed that all vertical processes are considered in the column above the square with unit area). The resistors (but R_1) could be taken as the ratio of the distance between corresponding layer centres and mean value of K_z for the layer h_i:

$$(13) \quad R_i = \frac{h_i}{K_z\big|_i}$$

It is possible to show that equation for this electric scheme will be similar to the result of gridded discretisation of the standard diffusion equation:

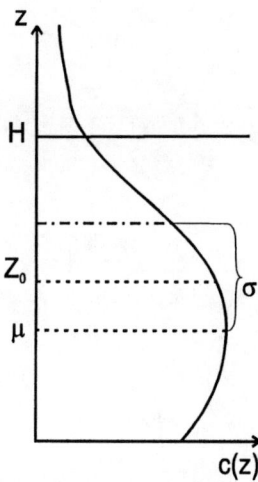

Figure 11. **Gussian distribution of concentration along the height**

(14) $\quad \dfrac{dc}{dt} = \dfrac{\partial}{\partial z} K_z \dfrac{\partial c}{\partial z}$

following the implicit scheme:

(15) $\quad \left.\dfrac{dc}{dt}\right|_{l,\tau} = \dfrac{c_{l,\tau+1} - c_{l,\tau}}{\Delta t}$, etc.

Here l,τ are the co-ordinates in discrete grid, Δt is the time discretisation step.

The resistor R_I reflects dry deposition from the centre of the first layer and in accordance with the 'classical' resistance analogy it is equal to V_d^{-1} (or $R_I = R_a + R_b + R_c$ in the commonly used notations, see (Wesely, 1989), (Erisman et al., 1994)). Hence, the scheme shown in Figure 11 meets the following boundary conditions:

(16) $\quad \left.\dfrac{\partial c}{\partial z}\right|_{z=H} = 0 \,, \quad \left. K_z \dfrac{\partial c}{\partial z}\right|_{z=0} = V_d\, c$

Dry deposition is resulted from solution of (14) - (15) as a difference between total charge (mass) of capacitors before and after time step.

Discretisation of the system (14), (16) using the implicit numerical scheme leads to three-diagonal matrix, which then can be treated by a diagonalisation method (e.g. so-called sweep method).

Less expensive **one-layer Gaussian dynamic profile** is based on the following distribution function (see Figure 11):

$$(17) \quad c(z) = Q \frac{\varphi(z)}{P(\mu/\sigma)},$$

where c is the concentration at a height z, Q is a total mass of a given pollutant in a cell and:

$$(18) \quad \varphi(z) = \frac{1}{\sigma\sqrt{2\pi}} \exp\left[\frac{-(z-\mu)^2}{2\sigma^2}\right]$$

$$(19) \quad P(\mu/\sigma) = \int_0^\infty \varphi(z)dz = \frac{1}{\sqrt{2\pi}} \int_{-\infty}^{\mu/\sigma} \exp(-\lambda^2/2)$$

Here $\varphi(z)$ describes the shape of the vertical concentration profile, $P(\mu/\sigma)$ - normalizing divisor, μ - vertical co-ordinate of maximum concentration and σ - measure of the variance along the vertical.

It is known (Turner, 1969; Stern, 1968; Bowne, 1974) that far from a source the dependence $\sigma(t)$ is on the average linear. In this case K_z should be considered as a linear function of the diffusion scale $\sigma(t)$, i.e. the dependence of $\sigma \sim t$ requires the consideration of K_z growth proportional to σ. (Theoretically $K_z \sim L_D^\alpha$, where L_D^α is a spatial scale of diffusion and $\alpha \approx 1$ (Leihtman, 1970)). In the first approximation:

$$(20) \quad K_z \cong k_z \sigma / L_k$$

where k_z - mean coefficient of turbulence, obtained from meteodata, L_k - spatial scale of averaging proportional to the boundary layer depth: $L_k \sim H_{BL}$. Here the proportionality factor is within the limit of 1-1.5 (Nieuwstadt & van Dop, 1982; Kaimal et al., 1976).

For the evaluation of μ the boundary condition

$$(21) \quad V_d c(0) = -K_z \frac{\partial c}{\partial z}\bigg|_{z \to +0}$$

can be used. Substituting (17)-(19) to (21) and taking into account (20) we obtain:

$$(22) \quad \frac{\mu}{\sigma} = \frac{V_d L_k}{k_z}$$

The value of μ is defined not only by the pre-history of the process (recorded in σ) but also by local conditions on the surface (V_d) and in the boundary layer (k_z, L_k). Taking into account the relationship of σ and μ the value of Z_0 can be re-written as follows:

$$(23) \quad Z_0 = \frac{1}{Q}\int_0^\infty zc(z)dz = \frac{1}{P(\mu/\sigma)}\int_0^\infty z\varphi(z)dz = \sigma\left[\frac{\varphi_0(\sigma/\mu)}{P(\sigma/\mu)} + \frac{\sigma}{\mu}\right]$$

This system (22), (23) for three variables Z_0, μ, σ enables to determine any two values if one is known. Initially μ can be equal to effective source height. Functions $\varphi_0(x)$ and $P(x)$ can be taken from tables.

The procedure of vertical profile changing caused by inter-cell mass transport is based on the vertical moment conservation during the pollution advection. If the Eulerian step occurs then new height of the mass centre Z_0^{new} is calculated from the moment conservation equation:

$$(24) \quad Z_0^{new} = \frac{1}{Q_N}\sum_{m=1}^{N} Q_m Z_{0m},$$

where Q_m are masses appeared to be in a cell after time step realisation, Z_{0m} are the heights of their centres of gravity, $Q_N = \sum_m Q_m$.

At known $Q(t)$ and $Z_0(t)$ the complete distribution $c(z)$ may be calculated following (22), (23).

5.1.3. Chemistry Scheme

The model chemical scheme is an extended version of that described in (Pressman *et al.*, 1991). The main addition is connected with ammonia transformations described in (Galperin & Sofiev, 1998).

The scheme contains the descriptions of the following main processes:

sequential oxidation of sulphur dioxide to sulphur acid and sulphates;
ozone producing cycle considered as a regulator of the NO – NO_2 – PAN proportion;
oxidation of NO_x to nitric acid and nitrates;
equilibrium between ammonia + nitric acid and ammonium nitrate

It is assumed that dry deposition velocity V_d depends on the surface type and state, and V_d is taken at the height of roughness layer (i.e. for a level of a millimetre to several tens of centimetre.

For the estimations of V_d of gases and aerosols with particle sizes less-than 0.1 μm the data available in the literature: (Sehmel, 1980), (Rodhe & Soderlund, 1980) and (Nowicki, 1987) were used. The data provide a general image of the underlying surface impact on dry deposition velocity. In particular these data indicate that the extent of surface moistening inflects drastic impact on V_d variation for SO_2. It was shown that mean V_d is abruptly changed at the temperature about 0^0C i.e. when water freezes.

The specific problem is the parameterization of the dry deposition velocity onto sea surface and its dependence upon wind speed. The approach applied in the model is based on the effect of growth of the marine surface roughness with the wind increase, wave creation and crashing.

In the first approximation this effect could be described by the Charnock formula

$$z_0 \propto \frac{u_*^2}{g} \quad (25)$$

which leads to square dependence of the roughness on the wind speed. The correction factor for dry deposition velocity is assumed to be proportional to z_0.

This formula is suitable for moderate wind speeds (approximately up to 3 balls of the Bofort scale). In case of stronger winds the wave crashing process leads to further increase of the deposition velocity because the sedimentation to the sea surface is coming to be practically wet scavenging process. There are a few data suitable for the quantitative parameterization of the process. Some of them have been presented in one of the EMEP reports (Ukrainian contribution, 1995).

For **wet deposition** calculations the non-linear washout model developed in (Galperin, 1989) is used. As it was shown the washout intensity depends upon several parameters, in particular upon precipitation intensity and air concentration of the pollutants. The analysis of experimental data for sulphur enabled to suggest analytical function, which is capable to reproduce the main characteristics of the observed values:

$$(26) \quad C_{Pr} = C_{Pr}^0 \left[1 - \exp\left(\frac{C_{Air}}{C_{Air}^0}\right)\right], \quad C_{Pr}^0 = C_{Pr}^0(I), \quad C_{Air}^0 = C_{Air}^0(I),$$

where C_{Pr}, C_{Air} are concentrations of sulphur oxides in the air and precipitation respectively. C_{Pr}^0 and C_{Air}^0 are normalising values depended on precipitation intensity (in current calculations they are taken as $C_{Pr}^0 \sim 3$ mg S l^{-1}, and $C_{Air}^0 \sim 2$ μg S m^{-3}).

The same equation can be derived from the hypothesis of sub-cloud scavenging with saturation effect. This effect was registered by several investigators, and in (Galperin, 1989) it is shown that correct numerical values can be obtained if accept that pollution molecules do not come into the rain droplet but make a uni-molecular film on its surface. In this case the saturation takes place if pollutant molecules occupy all free space on the cloud droplet surface. It is noted that this effect could be explained by electrical effects which may be significant for dissolved ions like $SO_4^=$ or NO_3^-.

The saturation effect for pollutants producing cations in solution (for example, NH_4^+) is remained open and is not taken into account in the current model version.

5.2. Source - Receptor Calculations

The problem of source-receptor calculations is quite simple for passive pollutant like aerosol carried species (i.e. the pollutant which concentration does not affect the transport process). In this case a model should only meet the standard mass conservation requirements and calculation routine is reduced to a sequential simulation of the pollution distribution from each of considered sources. Total concentration and deposition levels are the sum of individual impacts.

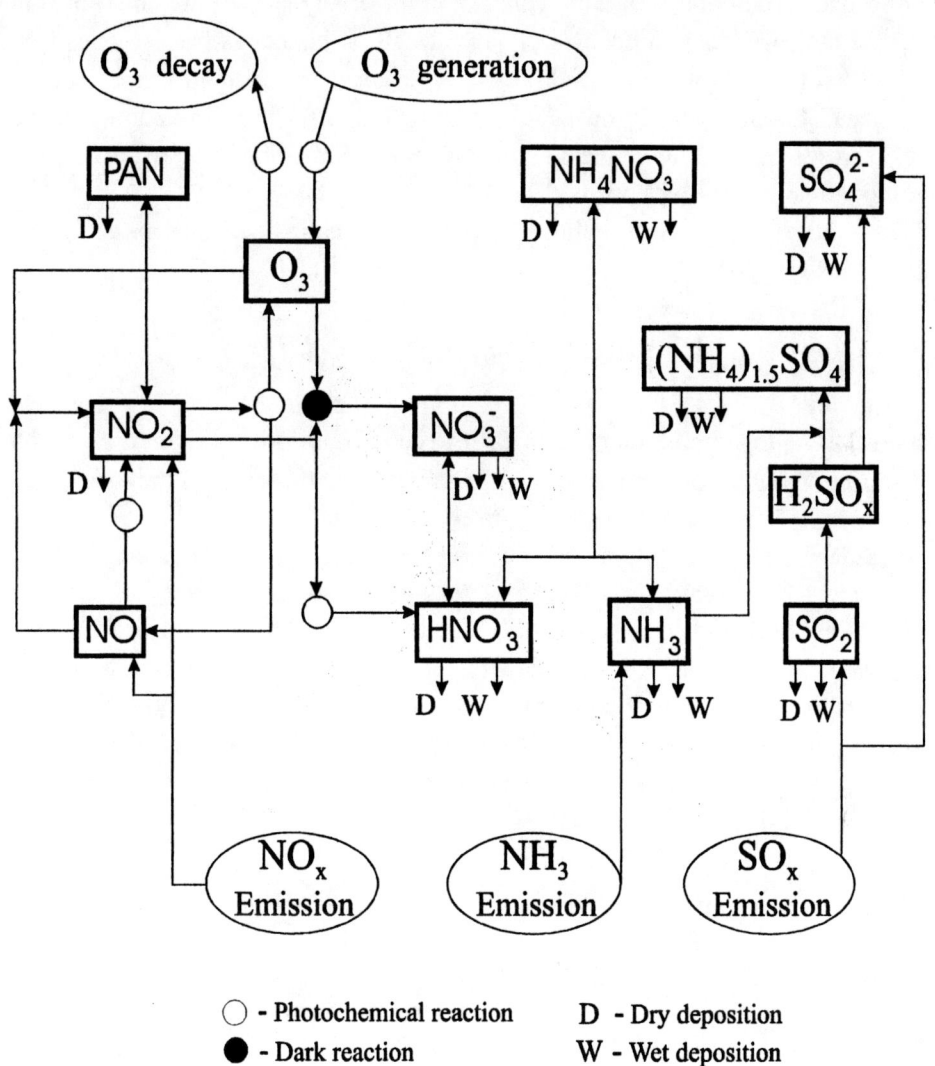

Figure 12 **Model chemical scheme**

For acid compounds this approach is hardly acceptable because of several second-order chemical reactions. They result in dependence between the substance concentrations and the reaction speeds and directions. Since different species have different transport features (first of all, different removal characteristics), correct evaluation of the pollution distribution requires consideration of all emission sources without exclusions.

In above described conditions source-receptor relationship evaluation requires some simplifications which can be described by the following two assumptions.

It is assumed that within a 3-D grid cell pollutants are well mixed. This assumption of "well mixing" is fulfilled in remote regions and lead to limited errors in industrial regions.

The second assumption is that the concentrations of any species are small in comparison with main air fractions. It enables to consider all pollutants as passive ones from the point of view of meteorological conditions and physical transport processes (both wind advection and turbulent diffusion). This assumption also has limited accuracy. For example, aerosol concentration may affect the cloud formation process. Anyway, it is common for all pollution transport models.

In view of this idea the total pollution mass in a grid cell is a sum of masses originated from different sources:

$$(27) \qquad Q(i,j) = \sum_{\xi=1}^{S} q_\xi(i,j) \;, \quad q_\xi(i,j) = \gamma_\xi(i,j)\, Q(i,j)$$

where Q is the total pollution mass in the grid cell (i,j), q_ξ is a pollution mass in this cell originated from the ξth emission source, γ_ξ is the fraction of this mass in the total value, S is the total number of sources considered in current model run.

The model considers two types of substances for each of three pollutants (see Figure 12) - primary (SO_2, NO&NO_2, NH_3) and secondary (sulphates, nitrates, ammonium). Concentrations of other species (like PAN, etc.) are assumed to be comparably small and corresponding masses are added to one of the above 6 pollutants. This simplification enables to reduce the whole chemistry to the process of sequential transformation of "primary" pollutants to the "secondary" ones with step-by-step dry and wet removing of both types of substances. The coefficients of these transformations and scavenging are taken from chemical unit of the model.

The calculations are carried out in two steps.

During the **first step** of "full-mass" chemical and advection calculations the coefficients of transformations and removals are stored for each grid cell and for each time step together with the advection and diffusion information.

After recording of above values the **second step** is started for each individual emitters, which follows the same algorithm, but both the advection and chemistry schemes are substituted by previously stored coefficients. All sources are addressed separately by the same procedure.

This approach is quite time-effective. The experience of its application shows that for about 60 sources the second step is only 3 times longer than the first one. Such a speed is reached because of fast algorithm of implementation of the coefficient sets to a large number of countries. This algorithm is also a good subject for parallel computation.

A disadvantage of this method is a high requirement to operational memory of a computer. In addition, the intermediate files (which have to be processed many times) with the coefficients are about one-two orders bigger than output data volume.

5.3. MODEL VALIDATION WITH MEASUREMENT DATA

The initial version of the model with one-layer gaussian vertical scheme was validated against measurement data collected in Europe within the scope of EMEP (Pedersen *et al.*, 1988; Schaug *et al.*, 1989, 1990, 1991, 1992; Lovblad *et al.*, 1995; Hjellbrekke *et al.*, 1996, 1997). The results for 5-years validation 1987-1992 can be found in (Sofiev et al., 1994). The adjusted ammonia chemical unit was re-calibrated in (Galperin & Sofiev, 1998).

Currently the measurements in Asia are rather scarce and performed mainly during some campaigns rather than produced by stationary stations. Hence, calibration of the model used in current calculations was split to two parts – multi-annual juxtaposition of calculated values with European data, and simple comparison of available Asian measurements with corresponding modelled values.

The results of validation of multi-layer model for European continent are outlined in Table 3. It contains the following parameters (columns) – substance acronym, unit, mean observed value over 1987-1995 and over all stations, corresponding mean calculated value, correlation coefficient for monthly values (mean over 1987-1995), correlation coefficient for annual values (mean over 1987-1995), slope and bias of linear regression, number of participated stations (mean over 1987-1995), and number of monthly values produced by these stations (mean 1987-1995).

Table 3. **Aggregated results for model-measurement comparison for Europe for 1987-1995.**

Substance	Unit	Mean observed	Mean calculated	Correl. monthly	Correl. annual	Regr. slope	Regr. bias	Number of stations	Monthly sample volume
SO_2 in air	0.1 µg S m^{-3}	29.3	31.2	0.75	0.83	0.98	4.5	45	515
$SO_4^=$ in aerosol	0.01 µg S m^{-3}	134.6	102.1	0.58	0.75	0.73	5.6	42	489
$SO_4^=$ in precip.	0.01 mg S l^{-1}	97.7	86.9	0.54	0.75	0.78	14.1	38	434
NO_2 in air	0.1 µg N m^{-3}	26.4	25.6	0.69	0.76	0.98	3.8	25	281
NO_3^- + HNO_3 in air	0.01 µg N m^{-3}	52.6	48.7	0.72	0.90	0.75	12.0	16	185
NO_3^- in precip	0.01 mg N l^{-1}	51.6	56.8	0.49	0.70	0.79	15.8	39	450
NH_3 + NH_4^+ in air	0.01 µg N m^{-3}	134.4	143.4	0.78	0.90	0.99	11.4	14	166
NH_4^+ in precip.	0.01 mg N l^{-1}	66.3	62.9	0.57	0.76	0.68	17.5	39	448

Asian data were taken from (Rodhe et al., 1995), (Okita et al., 1994), (Hatakeyama[1] et al., 1995), (Hatakeyama[2] et al., 1995), (Hatakeyama et al., 1997). They are aggregated in Table 4. Almost all measurement data in this table taken from monitoring campaign and thus have certain specifics. They are not strongly linked to a geographical location, and often obtained during aircraft flight. Temporal coverage of these data is at most couple of days, so they are hardly representative in a monthly scale. It is explicitly mentioned in (Hatakeyama[1,2] et al., 1995) that variation of concentrations in space was very high and to a large extent driven by atmospheric conditions (cyclone fronts, typhoons).

Table 4. **Model comparison with available data from Asian measurement campaigns.**

Data description	Measured value (range)	Modelled value (range)
Northern part of Russia (Norilsk area). Wet deposition of sulphur. Unit = mg S m^{-2}.	~ 400	200 – 1000
Japan sea, aircraft measurements at 850 mbar. SO$_4^=$ aerosol concentrations. Unit = µg S m^{-3}.	0.6 – 1.2	0.3 – 1
Northern part of Russia (Norilsk area). Wet deposition of nitrogen. Unit = mg N m^{-2}.	~ 60	30 – 80
Japan sea, aircraft measurements at 500 m (data from several campaigns). SO$_2$ air concentrations. Unit = 1 µg S m^{-3}.	0.45 – 0.65 0.45 – 4.5 0.35 – 0.7 0.45 - 1	0.8 – 1.5 0.4 – 2.6 0.1 – 0.4 0.2 – 3.9
Japan sea, aircraft measurements at 500 m (data from several campaigns). NO$_x$ air concentrations. Unit = 1 µg N m^{-3}.	1.8 – 2.3 0 – 1.2 1 – 2.5	0.1 – 0.4 0.2 – 0.4 0.3 – 0.4

Consideration of the comparison results (first of all, more detailed European data Table 3) leads to several general conclusions.

· Precision of reproducing of the primary pollutants (like SO$_2$, NO$_2$, NH$_x$ emitted directly from the emission sources) is much higher than that of secondary ones (sulphates and nitrates). This effect is caused by the simplifications in the model chemistry scheme and meteorological data. It also confirms that the emission inventory in Europe is rather good.

Accuracy of the precipitation concentration is the worst in the total list, which is originated from the stochastic features of the precipitation pattern and its high variability in space and time. The model spatial resolution (150 km) is insufficient to reproduce this thin structure, which leads to significant stochastic deviation from measurement data even for annual values.

Monthly estimates of the model are less accurate than annual ones. There are two evident reasons for that. The first one is possible errors in meteorological information, in particular wrong reflection of seasonal trends of some of meteoparameters. Annual averaging enables to smooth their influence, but monthly one does not. The second reason is small representativness (see Table 1) of monthly measurements in comparison with annual ones. Switching from monthly averaging to yearly one reduces the corresponding standard deviation approximately by the factor of 2 or 3.

Table 4 does not enable to derive any quantitative conclusions about the model precision in Asia. It can only be stated that calculated and observed data are within the factor of 2 – 5.

Model validation only on the basis of European monitoring information is clearly insufficient and can not be automatically extended to Eurasia. So, currently available experimental information does not enable to execute a complete validation. The results of model-measurement comparison can give just a general picture of the quality of the calculation results in Asia.

Another limitation factor of the model precision in Asia is poor quality of the emission data in this region. For some substances (first of all, for ammonia) it makes the biggest impact into errors in model output.

5.4. CALCULATIONS RESULTS

The results of the preliminary calculations consist of two main parts – deposition maps and mass budget tables. Some examples of them are presented below. Concentration maps produced by the model were used only during the model validation with measurements (see previous chapter) and not presented in the paper.

5.4.1. Deposition maps

Deposition maps of oxidised sulphur, oxidised nitrogen and ammonia are shown in Figure 13 – Figure 15. It is well seen that there are three main sources of pollution in Eurasian continent – Europe, Ural industrial region and Southeast Asia. The main direction of the atmospheric transport is from West to East. As a result, the influence of Europe is detectable practically up to Baikal lake area, especially for nitrogen compounds. The highest deposition of sulphur compounds is in "Black triangle" region in the centre of Europe and on East coast of China – up to 7 grams of sulphur per square metre per annum. Nitrogen deposition is more homogeneous with less pronounceable peaks. According to current results, in 1995 the highest deposition was on East Coast of China and accounted to 2 grams of nitrogen per square metre per annum. In central Europe corresponding value is as much as half of this amount. Ammonia deposition is driven by spatial emission distribution, which is almost artificial in current calculations.

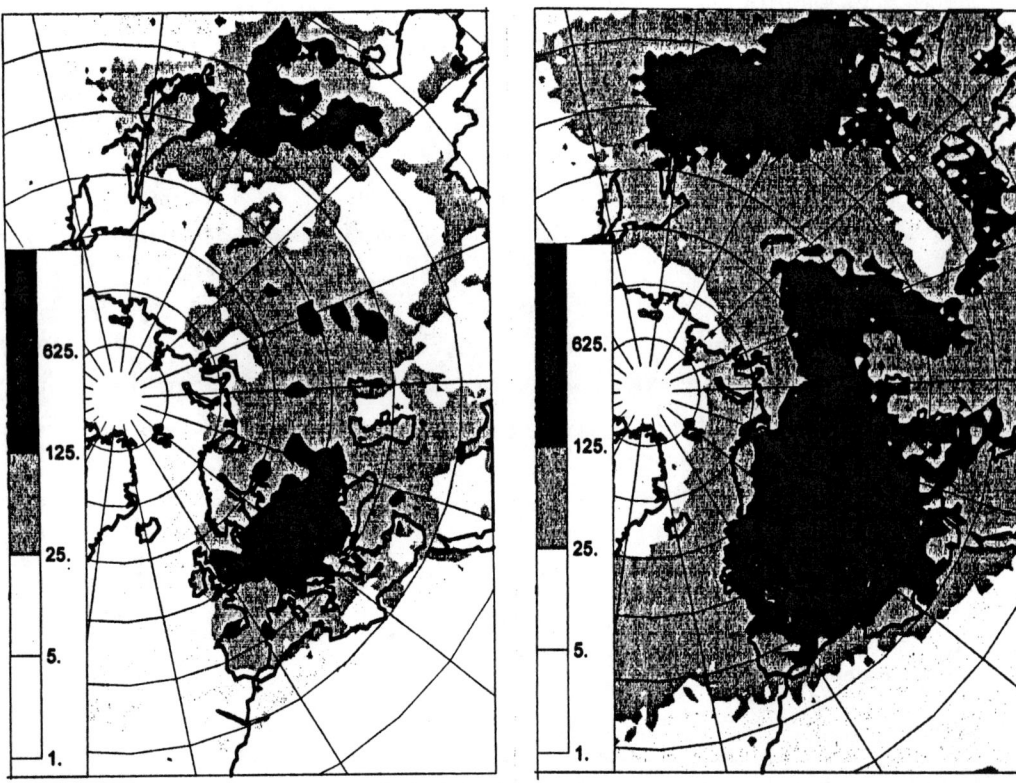

Figure 13. **Annual Deposition of Oxidised Sulphur in 1995. Unit=10 mg S m^{-2}**

Figure 14. **Annual Deposition of Oxidised Nitrogen in 1995. Unit=1 mg N m^{-2}**

South-east Asia is practically self-polluted region. The bulk of emission sources are located close to coast, while the prevailing wind direction leads to further transport of these masses to the ocean. This effect is well seen in all deposition maps. Rather short transport distance of all compounds over the marine surface is probably explained by fast increase of the dry deposition velocity onto water surface with growing of wind speed and creation and crashing of waves. Corresponding feature of the dry deposition unit of the model was mentioned in previous chapter.

5.4.2. Source-Receptor Mass Budget Matrices

As it was mentioned above, the most important characteristic of the acidification is deposition of sulphur and nitrogen oxides. Deposition of bound nitrogen is crucial for eutrophication. Because of the atmospheric redistribution of pollutants a considerable part of deposition onto some region can be originated from remote sources. This pollution exchange should be taken into account during the development of the emission reduction measures. Necessary information can be presented in the form of mass budget matrices. Corresponding tables for Eurasia are presented in

Table 6 - Table 8. Notations of emitters and receptors are shown in Table 5. Data in tables are averaged over 4 years – 1991 – 1994 – in order to reduce the influence of meteorological variability.

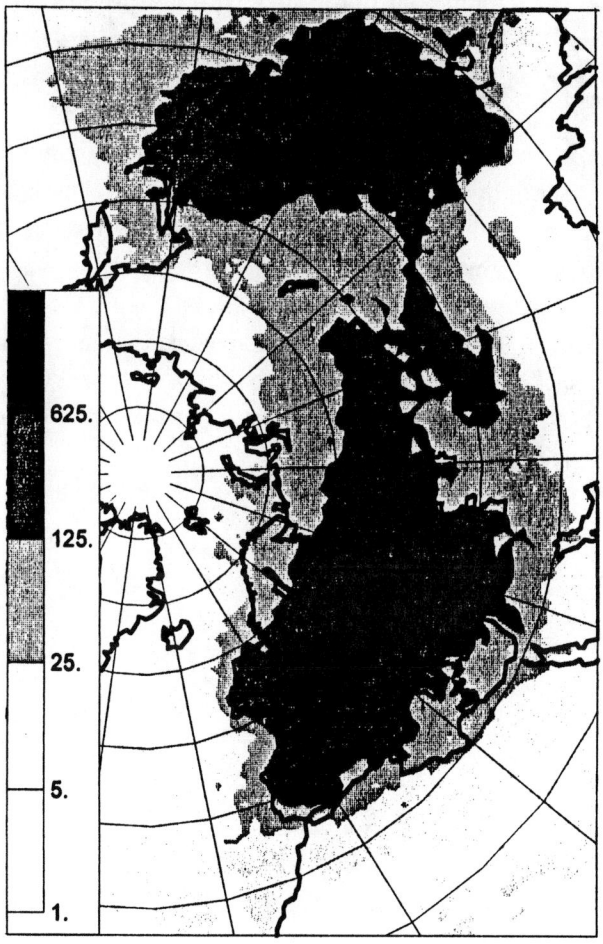

Figure 15. **Annual Deposition of Reduced Nitrogen. Unit=1 mg N m^{-2}**

Columns in these tables reflect the distribution of the emission from a particular source. Total annual emission of the source is shown in the bottom cell of the column (row "qa").

Rows show the amount of pollution deposited onto the territory of a particular receptor. Hence, each cell contains the amount of the deposition onto the specific receptor originated from the specific emitter.

Accumulation of artificial errors during the long chain of computation results in some small difference between actual annual emission (row "qa") and sum of depositions originated from the source (row "sum" plus transport outside the calculation domain "bsm"). Such a deviation never exceeds 1 % of the source emission, which confirms the stability of the applied scheme.

Brief analysis of the tables confirms that three above mentioned main regions of pollution (Europe, Ural industrial region, and Southeast Asia) have limited influence to each other. There is a considerable transport from Europe to the east. In particular, up to 40% of the oxidised nitrogen deposition onto European territory of Russia come from Europe. For Asian part of Russia this fraction is about 10%. Corresponding figures for sulphur and ammonia are not so big, which is explained by comparably short transport distance of these species.

Southeast Asia is mainly self-polluted region. Only few countries are affected by remote sources (first of all, by emitters located in Asian part of Russia and in Central Asia). The mostly affected territory belongs to Mongolia, where the fraction of sulphur deposition from remote sources is accounted to 30 %.

Sources in Asia affect eastern Russia, shelf seas and Pacific Ocean. The main source of all types of pollutants is China. Its annual emission exceeds values of other countries taken together by the factor of times (for sulphur and oxidised nitrogen) or by one decimal order (for ammonia). Consequently, this country is the most important donor of the acidifying and euthrophic substances in Asia.

Table 5. Notations for mass budget matrices

Notation	Description
Ner	Denmark, Finland, Norway, Sweden, Baltic Sea
Wer	Belgium, France, Luxembourg, Netherlands, Portugal, Spain, Ireland, Great Britain
Cer	Austria, Czech Republic, Germany, Hungary, Poland, Slovakia, Switzerland
Ser	Armenia, Azerbaijan, Bosnia & Hercegovina, Bulgaria, Croatia, Cyprus, Georgia, Greece, Italy, Moldavia, Romania, Slovenia, Turkey, Yugoslavia, Albania, Macedonia
Eer	Belarus, Estonia, Latvia, Lithuania, Ukraine
Rue	European part of Russia: Arkhangelsk, Karelian, Komi, S.Peterburg, Murmansk, Ladoga lake, White Sea, Astrakhan, Bashkortostan, Belgorod, Bryansk, Chuvashiya, Ivanovo, Jaroslavl, Kaliningrad, Kalmykia, Kaluga, Kavkaz, Kirov, Kostroma, Krasnodar, Kursk, Lipetsk, Marial, Mordovia, Moscow, Nihznii Novgorod, Novgorod, Orel, Orenburg, Penza, Perm, Pskov, Rostov, Ryasan, Samara, Saratov, Smolensk, Stavropol, Tambov, Tatarstan, Tula, Tver, Udmurtiya, Ulyanovsk, Vladimir, Volgograd, Vologda, Voronehz
rua	Asian part of Russia: Altai, Chelyabinsk, Ekaterinburg, Kemerovo, Kurgan, Novosibirsk, Omsk, Tomsk, Tuva, Tyumen, Krasnoyarsk, Amur, Buryatiya, Chita, Kamchatka, Khabarovsk, Magadan, Primorie, Sakhalin, Yakutiya, Yrkutsk, Baikal lake
cas	Iran, Iraq, Israel, Jordan, Lebanon, Syria, Kazakhstan, Kirghizia, Tajikistan, Turkmenistan, Uzbekistan, Aral Sea, Caspian Sea

Notation	Description	Notation	Description
chn	China	atl	Atlantic
jpn	Japan	arc	Arctic, oceanic part
nkr	North Korea	pas	Pacific ocean
skr	South Korea	med	Mediterranean sea
mgl	Mongolia	sum	Sum of column/row
asi	Asia other	bsm	Transport outside the grid
		qa	Annual emission

Table 6. Mass budget matrix for oxidised sulphur. Mean over 1991-1994. Unit = kt S per annum

Rc/Em	ner	wer	cer	ser	eer	rue	rua	cas	chn	Jpn	nkr	skr	mgl	sum
ner	136	144	323	19	51	65	2	0	0	0	0	0	0	739
wer	29	2991	509	72	20	12	5	10	0	0	0	0	0	3648
cer	21	282	2163	149	28	8	0	1	0	0	0	0	0	2650
ser	3	79	421	1377	78	16	1	25	0	0	0	0	0	1999
eer	16	42	492	128	597	47	1	2	0	0	0	0	0	1324
rue	34	61	401	108	426	1057	59	21	0	0	0	0	0	2165
rua	9	24	124	25	103	318	1516	249	106	1	4	5	5	2486
cas	4	96	112	142	108	198	91	864	16	0	0	0	0	1630
chn	0	1	8	2	6	13	52	35	7152	0	11	9	10	7299
jpn	0	0	1	0	1	1	9	1	79	196	8	43	0	338
nkr	0	0	0	0	0	1	2	1	111	0	42	56	0	213
skr	0	0	0	0	0	0	1	0	59	1	9	139	0	208
mgl	0	0	4	1	3	7	33	20	44	0	0	0	30	141
asi	0	0	0	0	0	0	0	10	46	0	0	0	0	56
arc	12	46	89	7	26	135	216	7	1	0	0	0	0	539
pas	2	15	13	2	8	18	143	14	1235	175	83	314	4	2024
med	4	354	340	963	127	21	1	109	0	0	0	0	0	1919
sum	269	4134	4998	2992	1581	1914	2131	1368	8848	372	155	565	48	29375
bsm	1	115	63	177	15	7	10	221	671	100	12	84	1	1829
qa	270	4250	5069	3172	1598	1928	2161	1590	9531	472	166	647	50	30904

Table 7. **Mass budget matrix for oxidised nitrogen. Mean over 1991-1994.**
Unit = 100t N per annum

Rc/Em	ner	wer	cer	ser	eer	rue	rua	cas	chn	Jpn	nkr	skr	mgl	sum
ner	152	107	122	10	14	10	0	0	0	0	0	0	0	415
wer	44	1403	263	35	5	3	0	1	0	0	0	0	0	1756
cer	24	220	643	68	8	2	0	0	0	0	0	0	0	965
ser	5	63	143	468	18	11	0	4	0	0	0	0	0	713
eer	25	44	161	52	125	24	0	0	0	0	0	0	0	433
rue	63	57	123	46	111	383	19	6	0	0	0	0	0	806
rua	14	14	24	6	18	119	335	49	43	1	4	2	2	633
cas	6	47	27	59	25	88	30	165	5	0	0	0	0	454
chn	0	0	1	0	1	3	18	7	1470	1	9	3	5	1519
jpn	0	0	0	0	0	0	2	0	20	182	7	13	0	225
nkr	0	0	0	0	0	0	1	0	38	0	29	11	0	80
skr	0	0	0	0	0	0	0	0	17	1	9	28	0	55
mgl	0	0	0	0	0	2	12	3	11	0	0	0	11	40
asi	0	0	0	0	0	0	0	3	7	0	0	0	0	10
arc	28	35	19	1	5	19	13	1	0	0	0	0	0	121
pas	2	5	1	0	1	4	50	5	373	212	77	90	2	821
med	7	204	125	397	31	18	0	16	0	0	0	0	0	797
sum	370	2200	1653	1143	362	686	483	263	1984	398	134	146	22	9844
bsm	1	54	14	61	3	3	3	52	132	171	9	23	0	911
qa	372	2255	1667	1205	365	690	487	315	2119	569	143	169	22	10378

Table 8. Mass budget matrix for reduced nitrogen. Mean over 1991-1994. Unit = 100t N per annum

Rc/Em	ner	wer	cer	ser	eer	rue	rua	cas	chn	Jpn	nkr	skr	mgl	sum
ner	129	30	92	8	38	14	0	0	0	0	0	0	0	312
wer	37	1120	140	20	14	5	0	2	0	0	0	0	0	1336
cer	21	131	678	54	30	5	0	0	0	0	0	0	0	919
ser	2	26	101	793	59	27	0	15	0	0	0	0	0	1022
eer	11	10	117	84	564	56	0	1	0	0	0	0	0	843
rue	20	10	57	69	219	835	8	26	0	0	0	0	0	1245
rua	3	2	7	8	26	142	298	70	194	2	1	1	10	762
cas	2	23	10	91	31	151	26	433	48	0	0	0	1	816
chn	0	0	0	0	1	4	21	20	6403	0	4	2	33	6489
jpn	0	0	0	0	0	0	2	0	56	139	1	5	1	204
nkr	0	0	0	0	0	0	1	0	80	0	12	5	1	99
skr	0	0	0	0	0	0	0	0	28	1	2	21	0	52
mgl	0	0	0	0	0	2	13	5	35	0	0	0	46	101
asi	0	0	0	0	0	0	0	10	82	0	0	0	0	93
arc	7	6	6	1	6	14	3	1	1	0	0	0	0	44
pas	2	1	0	1	1	7	43	7	762	132	20	43	9	1027
med	2	94	58	332	85	29	0	21	0	0	0	0	0	621
sum	236	1451	1268	1461	1073	1289	415	613	7687	273	40	76	102	15983
bsm	0	36	8	95	8	5	2	107	309	57	1	7	2	696
qa	236	1486	1275	1557	1079	1295	417	720	8007	330	41	82	104	16629

6. CONCLUSIONS

Current paper presents a short outline of the modelling activities dedicated to the investigation of the acidification processes in Eurasia. The bulk of these studies were performed in Europe where the development of scientific instruments enables to start operational numerical simulations of the acid pollution distribution. The modelling efforts are accompanied by the regular measurements on stationary monitoring stations distributed all around Europe.

Accumulated experience in this field enables to detect the main features of the numerical models applicable to regional tasks. They are based on algorithms of direct solution of the advection – diffusion equations together with chemical and physical transformations. The input data include emission inventory, geophysical and operational meteorological data. Climatological models are applicable only in some specific cases.

Output data of the models contain three types of information – deposition maps, concentration maps and source-receptor budget matrices. Deposition fields reflect the direct anthropogenic load onto ecosystems and can be used for the estimation of the damage in natural systems caused by industrial and agriculture activity. Concentration fields are used for the model validation against monitoring data. Finally, the source-receptor budget matrices reflect the pollution exchange between different regions. They are used during the development of the strategy of the emission control and reduction.

Since all models contain sets of simplifications and assumptions, their results always deviate from the actual situation. Quantitative estimate of the model precision is available from the quality assurance process. The most widely used method of validation is the comparison with measurements. Despite the intuitive simplicity of the comparison technique, there are certain limitations in the applicability of some methods. One of the most ones is problem of representatively of point measurements in the surrounding region.

The second part of the paper is devoted to the preliminary calculations performed on a 3-D Eulerian model for the Eurasian continent. The description of this model is presented. It is mentioned that the model considers the non-linear transformations of some of pollutants and is also capable to perform the source-receptor calculations for these substances.

The preliminary calculations were based on the emission inventory produced by the Global Emission Inventory Activity project (GEIA), which was combined with extra data for ammonium. The meteorological information was prepared in Russian Hydrometeorological Centre.

The results of calculations show that there are three main sources of the acidifying pollutants in Eurasia – Europe, Ural industrial region and Southeast Asia. The main transport direction is from West to East, which results in considerable impact from European sources into depositions onto Russian territory. Consequently, some Southeast Asian countries are affected by remote sources located in Russia and in Central Asia.

REFERENCES

Afinogenova O.G., Galperin M.V. (1985) Temporal non-uniformity of SO2 emissions to the atmosphere in European countries with moderate climate. In Institute of Applied Geophysics, **62**, *Huydrometeoizdat*, 1985, Moscow, 52-62.

Barrett K., Seland O., Foss A., Mylona S., Sandnes H., Styve H., Tarrason L. (1995) European transboundary acidifying air pollution. Ten years calculated fields and budgets to the end of the first Sulphur Protocol. In *EMEP / MSC-W Report 1/95*, Oslo, 290p.

Benkovitz C.M., Scholtz T., Pacyna J., Tarrason L., Dignon J., Voldner E.C., Spiro P.A., Logan J.A. and Graedel T.E. (1996) Global gridded inventories of anthropogenic emissions of sulphur and nitrogen, *J. Geophysical Research*, **101 (22)**, 29239 - 29253.

Bolling B., Persson G. (1975) Regional dispersion and deposition of atmospheric pollutants with particular application to sulphur pollution of Western Europe, In *Tellus*, **3**, 281-310

Bott A. (1989) A positive definite advection scheme obtained by non-linear renormalisation of the advective fluxes. In *Mon.Wea.Rev.*, **117**, 1006-1015.

Bowne N.E. (1974) Diffusion Rates, *J. Air Pollution Control Assoc.*, **24**, No 9, 832.

Dianwu Z., Anpu W. (1994) Estimation of anthropogenic ammonia emissions in Asia. In *Atm.Env.*, **28**, 689-694.

Dulac F., Bergametti G., Losno R., Remoudaki E., Gomes L., Ezat U., Buat-Menard P (1992) *Dry deposition of mineral aerosol particles in the marine atmosphere: Significance of the large particle fraction, in precipitation scavenging and atmosphere-surface exchange.* Ed. by Schwartz S.E. and Slinn W.G.N. **2** pp841-854, Hemisphere Publishing Corp., Richland, WA.

Ebel, A., H. Hass H. and H. Petry (1994) Correlation distances for air pollutants and implications for mesoscale modelling, *Proceedings of EMEP workshop on the Accuracy of Measurements, EMEP/CCC Rep.2/94,* Co-Operative Programme for monitoring and evaluation of the long-range transmission of air pollutants in Europe, Chemical Co-ordinating Centre, Lillestrom, pp.201-207.

Egan A.B. and Mahoney J.R. (1972) Numerical modelling of advection and diffusion of urban area source pollutants, *J.Appl.Met.*, **11**, 312-322.

Eliassen A. (1976) The trajectory model: a technical description. In *Norwegian Institute of Air Research*, Oslo, 1976, 14p.

Eliassen A. (1978) The OECD study of long-range transport of air pollution long-range transport modelling. In *Atmos.Env.* **12**, N 1, 28-40.

Eliassen A., Satbones J. (1975) Decay and transformation rates of SO_2, as estimated from emission data, trajectories and measured air concentrations. In *Atm.Env.* **9**, 425-429.

Eliassen A., Satbones J. (1983) Modelling of long-range transport of sulphur over Europe: a two-year model run and some model experiments. In *Atm.Env.* **17**, 1457-1473.

Erisman J.W., van Pul A., Wyers P. (1994) Parametrisation of surface resistance for the quantification of atmospheric deposition of acidifying pollutants and ozone. In *Atm. Env.* **28**, No 16, 2595 – 2607.

Finlayson-Pitts B.J., Pitts J.N. (1986) *Atmospheric chemistry fundamental and experimental techniques.* Wiley & Sons, New York.

Frolov A., Vazhnik A., Astakhova E., Alferov Yu., Kiktev D., Rosinkina I., Rubinshtein K. (1994) System for diagnosis of the lower atmosphere state (SDA) for pollution transport models, *EMEP / MSC-E Technical report 6/94*, Moscow.

Galperin M.V. (1989) Adsorption-kinetic nonlinear washout model of sulphur and nitrogen compounds from the atmosphere. In *Air Pollution Modeling and Its Application VII*, N.Y. & London, Plenum Press, 475-484.

Galperin, M., M. Sofiev (1994) Errors in the validation of models for long-range transport and critical loads stipulated by stochastic properties of pollution fields. *Proc. of EMEP workshop on the Accuracy of Measurements, EMEP/CCC Rep.2/94*, Co-Operative

Programme for monitoring and evaluation of the long-range transmission of air pollutants in Europe, Chemical Co-ordinating Centre, Lillestrom, pp. 162-179.

Galperin M.V., Sofiev M.A. (1998) The long-range transport of ammonia and ammonium in the northern hemisphere, In *Atm.Env.* **32**, No 3, 373-380.

Grid point data set, v.II. *Gridded MNC analysis for the Northern Hemisphere*. National Meteorological Centre , Atmospheric Sci.Univ.of Washington, data support section NCAR.

Hass H., Berge E., Ackerman I., Jakobs H.J., Memmesheimer M., Tuovinen J.P. (1996) A diagnostic comparison of EMEP and EURAD model results for a wet deposition episode in July 1990. In *EMEP / MSC-W Note 4/96*, The Norwegian Meteorological Institute, Oslo, Norway.

Hatakeyama[1] S., Murano K., Bandow H., Mukai H., Akimoto H. (1995) High concentration of SO_2 observed over the Sea of Japan. In *J. of terrestrial, atmospheric and oceanic sciences*, **6**, No.3, 403-408.

Hatakeyama[2] S., Murano K., Bandow H., Sakamaki F., Yamato M., Tanaka S., Akimoto H. (1995) The 1991 PEACAMPOT aircraft observation of ozone, NO_x and SO_2 over the east China Sea, the Yellow Sea, and the Sea of Japan. In *J. of Geophysical Res.*, **100**, No.D11, 23143-23151.

Hatakeyama S., Murano K., Mukai H., Sakamaki F., Bandow H, Watanabe I., Yamato M., Tanaka., Akimoto H. (1997) SO_2 and sulphate aerosols over the seas between Japan and Asian continent. In *Japan Atmospheric Journal* **12**, No. 2, 91-95.

Hjellbrekke A.-G., Schaug J., Hanssen J.E., Skjelmoen J.E. (1997) Data report 1995. Part 2: monthly and seasonal summaries. In *EMEP / CCC Report 5/97,* NILU, Norway.

Hjellbrekke A.-G., Schaug J., Skjelmoen J.E. (1996) Data report 1994. Part 2: monthly and seasonal summaries. In *EMEP / CCC Report 5/96,* NILU, Norway.

Huber, P.J. (1981) *Robust Statistics*. Wiley series in probability and mathematical statistics. Wiley and Sons, 304 p.

Janssen A.J., Asman W.A.H. (1988) Effective removal parameters in long-range air pollution transport models. In *Atm.Env.*, **22**, 359-367.

Journel, A.G. and C.J. Huijbregts (1978) *Mining geostatistics*. Academic Publishing Inc., London, 215 p.

Kaimal J.C., Wingaard J.C., Haugen D.A., Gote O.R., Izumi Y., Coughey S.J., Readings C.J. (1976) Turbulence structure in the convective boundary layer, *J.Atmos. Sci.*, **33**, 2152-2169.

Klaassen G., Amann M., Schopp W. (1992) Strategies for reducing sulphur dioxide emissions in Europe based on critical sulphur deposition values. Background paper for the UN/ECE Task Force on Integrated Assessment Modelling, 30 Nov – 2 Dec 1992, Geneva. International Institute for Applied System Analysis (IIASA), Viena, Austria.

Leihtman D.L. (1970) *Physics of the atmospheric boundary layer*, Leningrad, Hidrometeoizdat. (Russian edition).

Lenhart L., Friedrich R. (1995) European emission data with high temporal and spatial reoslution, In *Air pollution III*, Computational mechanism publication, Southhampton, UK, **2**, 285-292.

Lovblad G., Hjellbrekke A.-G., Sjoberg K., Skjelmoen J.E., Schaug J. (1995) Data report 1993. Part 2: monthly and seasonal summaries. In *EMEP / CCC Report 8/95*, NILU, Norway.

Moussipoulos N. (editor) (1994) The EUMAC Zooming Model. Model structure and applications. In *EUROTRAC Special Publications*, Garmish-Partenkirchen, 266 p.

Nieuwstadt F.T.M., van Dop H. (editors) (1982) *Atmospheric turbulence and air pollution modelling*. Royal Meteorological Institute, de Bilt, D.Reidel Publishing Company Dordtechet/Boston/London. (Russian edition: Yaglom A.M. editor, Hydrometeoizdat, S.Peterburg, 351p.).

Nowicki M. (1987) *Measurements of sulphur dioxide dry deposition velocity on the Earth's surface*, EMEP/PL/01/87, Warsaw.

The OECD Programme on LRTAP. Measurements and findings. – OECD, Paris, 1977, 320p.

Okita T., Hara H., Fukuzaki N. (1994) Determination of SO_2 and SO_4 concentrations and wet scavenging coefficient of SO_4 in winter monsoon. Presentation on 4th *Int. conf. On Atmospheric Sci. and Applications to Air Quality* & private communication. Korea National Univ. of Education, Chongwon, Chongbook 363-792, South Korea.

Pekar M.I. (1989) *The effect of wind shear on pollution dispersion*, News (Izvestija) of the USSR Academy of Sciences, 24, 1 (Russian edition).

Pedersen U. (1994) Improvement of spatial correlation structure by use of anisotropic variogram analysis. *Proc. of EMEP workshop on the Accuracy of Measurements*, EMEP/CCC Rep.2/94, p.190-200.

Pedersen L.B. and Prahm L.P. (1974) A method for numerical solution of the advection equation, *Tellus*, **XXVI**, 5, 594-602.

Pedersen U., Schaug J., Skjelmoen J., Hansen J. (1988) Data report 1988. Part 2: Monthly and seasonal summaries. *EMEP/CCC, NILU, Norway*. 267 p.

Pepper D.W., Long P.E. (1978) A comparison of results using second-order moments with and without wind correction to solve the advection equation, In *J.Appl.Met.*, 17, 228-233.

Peters L.K., Berkowitz C.M., Carmichael G.R., Easter R.C., Fairweather G., Ghan S.J., Hales G.M., Leung L.R., Pennell W.R., Potra F.A., Saylor R.D., Tsang T.T. (1995) The current state and future direction of Eulerian modelings in simulating the tropospheric chemistry and transport of trace species: a review. In *Atmos. Environ.* 29, N 2, 189-222.

Pressman A.Ya., Galperin M.V. Popov V.A., Afinogenova O.G., Subbotin S.R., Grigoryan S.A., Dedkova I.S. (1991) A Routine Model of Chemical Transformation and Transport of nitrogen compounds, ozone and PAN within a regional scale. In *Atm. Env.*, 25A. No 9, 1851-1862.

Rodhe H., Langer J., Gallardo L., Kjellstrom E. (1995) Global scale transport of acidifying pollutants. In *Water, Air and Soil pollution*, 85 37-50.

Rodhe H., Soderlund R. (1980) Deposition of airborne pollutants on the Baltic Sea, *Ambio*, 9, 3-4, 168-173.

Rubinshtein K., Kiktev D. et al. (1998) Specialised Multiannual Archive of Daily Meteorological Data (SMAMD) 1967-1988, Russian edition, Rushsian Hydrometcentre, in prep.

Schaug J., Pedersen U., Skjelmoen J. (1989) Data report 1989. Part 2: Monthly and seasonal summaries. *EMEP/CCC,* NILU, Norway. 291 p.

Schaug J., Pedersen U., Skjelmoen J., Arnesen K. (1992) Data report 1992. Part 2: Monthly and seasonal summaries. *EMEP/CCC,* NILU, Norway. 324 p.

Schaug J., Pedersen U., Skjelmoen J., Kvalvagnes I. (1990, 1991) Data report 1990(1991). Part 2: Monthly and seasonal summaries. *EMEP/CCC,* NILU, Norway. 327 p., 334 p.

Schaug J. Skjelmoen J.E., Walker S.E., Pedersen U., Harstad A. (1987) Data report 1987. Part 2: Monthly and seasonal summaries. *EMEP/CCC,* NILU, Norway. 267 p.

Sehmel G.A. (1980) Particle and gaz dry deposition: a review, *Atmos.Environ.*, 14, 9, pp.983-1011.

Semb, A., J. Schaug J. and J.E. Hansen (1994) Accuracy and precision in the EMEP data. A first evaluation on the basis of available information. *Proc. of EMEP workshop on the Accuracy of Measurements, EMEP/CCC Rep.2/94,* Co-Operative Programme for monitoring and evaluation of the long-range transmission of air pollutants in Europe, Chemical Co-ordinating Centre, Lillestrom, pp. 53-66.

Sirois, A. and R.J., Vet (1994) Comparability of precipitation chemistry measurements between the Canadian air and precipitation monitoring network (CAPMoN) and 3 other north American networks *Proc. of EMEP workshop on the Accuracy of Measurements, EMEP/CCC Rep.2/94,* Co-Operative Programme for monitoring and evaluation of the long-range transmission of air pollutants in Europe, Chemical Co-ordinating Centre, Lillestrom, pp. 88-114.

Smolarkievich P.M. (1983) A simple positive definite advection scheme with small implicit diffusion. In *Monthly Weather Review*, 111, 479-486.

Sofiev, M. (1994) Statistical Properties of the Model Verification Problem and Special Methods for Comparison of Measured and Calculated Data. In: P. M. Borrell, P. Borrell, T. Cvitas, W. Seiler (Eds.) *Proceedings of Eurotrac Symposium'94.* SPB Academic Publishing, The Hague, pp.869-873.

Sofiev M.A. (1998) *The model for the evaluation of airborne pollution transport at regional and continental scales.* (In prep.)

Sofiev, M. and M. Galperin (1994), Robustness of Methods for Comparison of Measured and Calculated Data, *Proc. of EMEP workshop on the Accuracy of Measurements, EMEP/CCC Rep.2/94*, Co-Operative Programme for monitoring and evaluation of the long-range transmission of air pollutants in Europe, Chemical Co-ordinating Centre, Lillestrom, pp.315-341.

Sofiev M.A., Gusev A.V., Afinogenova O.G. (1995) Atmospheric transport of acid compounds in the Northern Hemisphere in 1991 – 1994. *EMEP / MSC-E report 8/95.* Co-Operative Programme for monitoring and evaluation of the long-range transmission of air pollutants in Europe, Meteorological Synthesising Centre – East, Moscow.

Sofiev M.A., Gusev A.V., Strijkina I.E. (1994) Results of MSC-E current model calibration with measurements for SO_x, NO_x, NH_x. 1987-1993. In *EMEP / MSC-E report 4/94*. Moscow, 124p.

Sofiev M.A., Maslyaev A. and Gusev A.V. (1996) Heavy metal model intercomparison. Methodology and results for Pb in 1990. In *EMEP / MSC-E Report 2/96*, Moscow, 103p.

Stern A.C., ed. (1968) Air Pollution, **1**, Second ed., NY, Academic Press.

Sympson D. (1995) Hydrocarbon reactivity and ozone formation in Europe. In *J.of Atm.Chem.*, **20**, 163-177.

Syrakov D., Galperin M.V. (1998) On some effective Bott-type advection schemes. In *EUROTRAC Symposium 1998 Proceedings*, in press.

Turner D.B. (1969) Workbook of atmospheric dispersion estimates, Washington D.C., HEW.

Ukrainian contribution (1995), Annual report of Ukrainian Scientific Centre of the Ecology of Sea, Ministry for Environment Protection and Nuclear Safety of Ukraine and Meteorological Synthesizing Centre - East of EMEP, Odessa-Kiev-Moscow.

Wesely M.L. (1989) Parametrisation of surface resistances to gaseous dry deposition in regional-scale numerical models. In *Atm.Env.* **23**, 1293-1304.

Chapter 2
ESTIMATIONS OF DRY DEPOSITION OF SO_2 AND NO_2 OVER SOUTH KOREA

SOON-UNG PARK AND SI-WAN KIM
Department of Atmospheric Sciences, Seoul National University, Seoul, 151, Korea

CONTENTS

Abstract
1. Introduction
2. Data
    2.1. Meteorological data
    2.2. SO_2 and NO_2 concentrations
3. Methodology
    3.1. K-mean divisive clustering for the synoptic meteorological classification
    3.2 Estimation of dry deposition velocity
        3.2.1 Estimation of aerodynamic resistance
        3.2.2 Estimation of quasi-laminar sublayer resistance
        3.2.3 Estimation of the surface or canopy resistance
4. Results
    4.1. Synoptic meteorological classification
    4.2. Seasonal mean SO_2 and NO_2 concentrations
5. Relationship between the SO_2 and NO_2 concentrations and the synoptic meteorological classification
6. Dry deposition of SO_2 and NO_2
    6.1. Dry deposition velocities
    6.2. Dry deposition fluxes of SO_2 and NO_2
7. Conclusions
Reference

ABSTRACT

Synoptic meteorological conditions are categorized into several similar types using a divisive Q-clustering technique based on the K-means principle with the 0000 UTC 500 hPa grid pointed geopotential height in East Asia for five years from 1989 to 1993. The surface observation data from 64 stations over South Korea are averaged hourly for the non-precipitating days belonging to each cluster to get diurnally varying micrometeorological data that are required for the estimation of the dry deposition velocity. With the use of the monitored SO_2 and NO_2 concentrations at 31 sites in South Korea, the estimated deposition velocity belonging to each cluster and the supplement of some empirical constants used in the Regional Acid Deposition Model (RADM), the dry deposition fluxes of SO_2 and NO_2 are estimated using the inferential method over South Korea. The classified meteorological clusters can lead to identify the SO_2 and NO_2 concentration levels without any further requirement other than to know the synoptic class for the next day. The classified synoptic cluster that is related with the high pollution level is not necessarily coincide with the cluster that has a high deposition velocity. The estimated annual mean dry deposition flux over South Korea is about 8.07 t km^{-2} yr^{-1} for SO_2 and 3.8 t km^{-2} yr^{-1} for NO_2 which are corresponding to 45% and 40% of the annual total emission rate respectively. However, in the eastern parts of Korea except for the regions where high emission sources are located the dry deposition fluxes exceed the emission rates. This suggests a significant horizontal transport of pollutants from west to east due to prevailing westerly winds and local circulations.

1. INTRODUCTION

Anthropogenic acid precursor emissions (SO_2 and NO_X) in South Korea are increasing at significant rates (Park and Kim, 1995), in line with increases in emissions throughout the wider region of East Asia as a whole. Accordingly, acid deposition must be considered to be a potentially significant environmental issue in South Korea, as else where in East Asia.

An awareness of the phenomena of acid deposition, commonly known as "acid rain" entered the public consciousness in the early 1970s, particularly in Stockholm in 1972 at the UN Conference on the Human Environment. Oden (1968) and Engstrom (1972) concluded that there had been a significant acidification of European precipitation over the period of 1956 to 1966. Similar conclusions were soon reached for the north eastern parts of the North American continent (Likens and Bormann, 1974). These conclusions were that growing anthropogenic emissions of sulfur dioxide and oxides of nitrogen in heavily industrialized regions were the cause of the observed acidity, produced by chemical conversion of these gases in the atmosphere to the respective strong acids, and that the levels of acidity had reached the point where they posed serious threat to sensitive environments. It is now generally believed that resultant increases in atmospheric depositions of acidic materials have a number of quite adverse environmental consequences including deteriorative effects upon sensitive soils, plants, aquatic ecosystems, and building materials.

Deposition processes are usually considered under two mechanisms: Wet deposition and dry deposition (Park, 1995). The wet deposition comprises removal by falling precipitation (washout) and rainout in the cloud, and the dry deposition denotes the direct collection of gaseous and particulates on lands or water surfaces.

Widely used techniques that allow the collection of wet-only deposition including analysis of precipitation chemistry have been developed, however, there is no reliable and widely accepted direct measurement technology for dry deposition (Lindberg and Lovett, 1985; Davison and Wu, 1990). Nevertheless, dry deposition could be as effective as wet deposition for removing atmospheric pollutants in many ecosystems.

Recently, Indirect means of estimating dry deposition, so called the inferential method have been developed using the surface layer micrometeorological conditions, including land-use types, surface roughness and vegetation (Baldocchi et al., 1987; Wesely, 1989; Hicks et al., 1991; Park, 1995). This method yields a reasonably good agreement with the direct method provided that micrometerological conditions are well described (Hicks and Matt,1986).

For the estimation of dry deposition using the inferential method, the micrometeorological conditions are required. These conditions are greatly influenced by synoptic meteorological situations. A wide variety of meteorological conditions directly and indirectly influence not only air quality but also micrometeorological conditions, which, in turn, influences the dry deposition velocity. Therefore, it is impossible to make any definitive statements regarding the dry deposition of a region without accounting for the underlying synoptic meteorology or climatology.

The synoptic meteorological conditions are usually categorized into several similar types having similar meteorological conditions by using the principal components analysis (e.g., Davis and Gay, 1993), and/or the clustering technique (Sanchez et al., 1990), and the geostrophic wind speed and direction in the free atmosphere (Park and Yoon, 1991). In this study, a synoptic meteorological classification has been performed by means of a divisive Q-clustering algorithm based on the K-means principle with the 0000 UTC 500hPa grid point values in East Asia obtained from Japan Meteorological Agency. The surface observation data from 64 stations over South Korea are averaged hourly for the non-precipitating days belonging to each cluster to get the diurnally varying micrometeorological data that are required for the estimation of the dry deposition velocity. With the use of the monitored SO_2 and NO_2 concentrations at 31 sites in South Korea and the estimated dry deposition velocity using micrometeorological data belonging to each cluster with some empirical constants used in the Regional Acid Deposition Model (RADM), the horizontal distribution of the dry deposition fluxes of SO_2 and NO_2 are estimated over South Korea.

In this study we will describe the method to estimate the seasonally averaged dry deposition fluxes of SO_2 and NO_2 using the inferential method by means of a Q-mode clustering technique of a synoptic meteorological classification.

2. DATA

2.1 METEOROLOGICAL DATA

The 2.5°latitude×2.5°longitude 500 hPa GPV data obtained from Japan Meteorological Agency from 1989 to 1993 in the domain of 20°N ⃞ 60°N and 90°E ⃞ 150°E are used for synoptic meteorological classifications by means of a divisive Q-clustering algorithm based on the K-mean principle.

The characteristics of the categorized each cluster are identified with the averaged meteorological variables for the days belonging to each cluster. The selected meteorological variables are given in Table 1.

Surface meteorological data observed at 64 stations for the estimation of the dry deposition velocity at each station, and the 9 stations for the identification of the character of each cluster are given in Fig. 1.

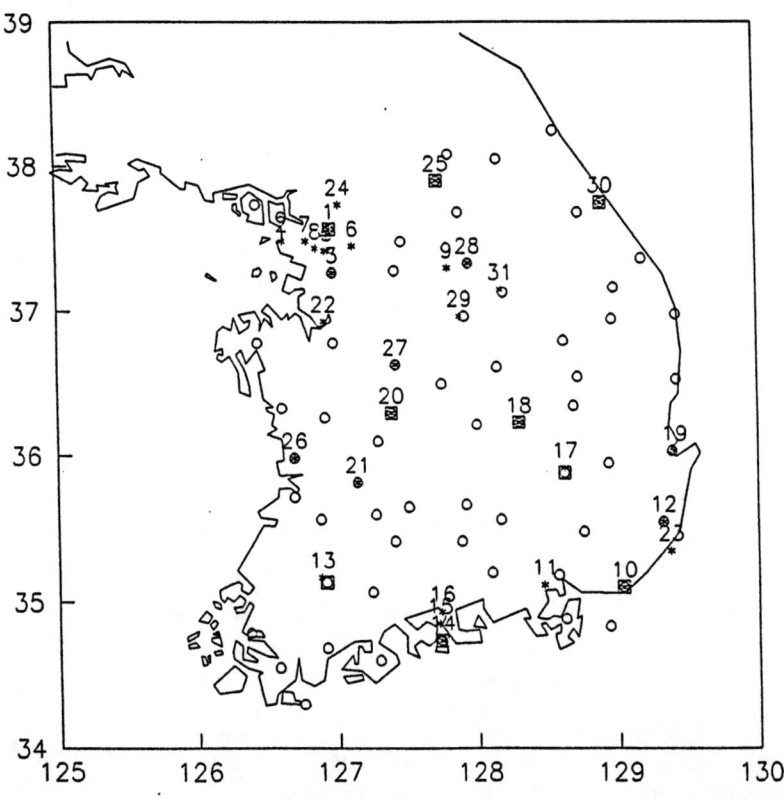

Figure 1. Locations of surface meteorological observation sites (O), surface meteorological observation sites used for the identification of the cluster character (□) and air pollution monitoring sites (*) over South Korea.

2.2 SO$_2$ AND NO$_2$ CONCENTRATIONS

The SO$_2$ and NO$_2$ concentrations data comprise hourly averaged concentrations for the non-precipitating days belonging to each cluster recorded at 31 monitored sites over South Korea (Fig. 1) for 5 years from 1989 to 1993. Each city has a various number of air pollution monitoring sites. Therefore, the concentrations monitored at all sites belonging to each city are averaged with a proper weighting factor (Jung et al., 1996) at each site to get the representative concentration of that city.

Table 1. The selected meteorological variables to identify the characteristics of each cluster. h represents the topographical height of the station, C$_p$ the specific heat at constant pressure and g the gravity.

H$_{500}$	500 hPa geopotential height at 00 UTC from Osan sounding
T$_{500}$	500 hPa air temperature at 00 UTC from Osan sounding
RH	Surface relative humidity averaged at nine observation stations at 00 UTC
T$_{min}$	Surface minimum temperature averaged at nine observation stations
Ws	Surface wind speed averaged at nine observation stations at 00 UTC
Ps	Surface pressure
T$_{dew}$	Surface dew point temperature
ED	Stability parameter defined as (T$_{500}$-T$_{min}$)/(H$_{500}$-h)·100
EDI	Isoin index defined as T$_{500}$-T$_{dew}$
MP	Stability parameter defined as C$_p$T$_{500}$+gH$_{500}$
W$_{850}$	850 hPa geostrophic wind speed estimated by the 850 hPa Geopotential height at 00 UTC
WD$_{850}$	850 hPa geostrophic wind direction calculated by W$_{850}$ at 00 UTC

3. METHODOLOGY

3.1. K-MEAN DIVISIVE CLUSTERING TECHNIQUE FOR THE SYNOPTIC METEOROLOGICAL CLASSIFICATION

K-mean algorithm starts from a given partition specified by the assignment of vector Q (K,ϕ) with the geopotential height ϕ allocated to K clusters. In a stepwise manner, the algorithm separates the ϕ trying to find their optima location, transferring the geopotential height between clusters and looking for the final cluster assignment. The division of the geopotential height ϕ for a given day into the different clusters is based on the function error, E, defined as :

$$E[Q(K,\phi)] = \sum_{1} D[I, L(I)]^2 \quad (1)$$

where $L(I)$ is the cluster containing the Ith case. $D[I,L(I)]^2$ represents the sum of the squared absolute deviations from the cluster centroid (mean of geopotential height) over all the grid values of the geopotential height in the cluster. The number of clusters to be retained has been chosen by analyzing the values of the error function (E) for an increasing number of clusters.

3.2. ESTIMATION OF DRY DEPOSITION VELOCITY

It is usually assumed that the flux of any pollutant to the ground, F is linearly proportional to the pollutant concentration, c, in the lower boundary layer,

$$F = V_d \cdot c \quad (2)$$

where V_d is the proportionality which has the unit of length/time, thus it is called a deposition velocity. This deposition velocity varies highly in time and space for a given pollutant (NCAR, 1985; Park, 1995).

The deposition velocity is given by (Park, 1995),

$$V_d = \frac{1}{R_a + R_b + R_c} \quad (3)$$

where $R_a = \dfrac{\ln\left(\dfrac{z}{z_0}\right) - \Psi_h\left(\dfrac{z}{L}\right)}{ku_*}$ is the aerodynamic resistance, $R_b = \dfrac{\ln\left(\dfrac{z_0}{z_s}\right)}{ku_*} + \dfrac{z_s}{D_g}$ the quasi-laminar sublayer resistance, and R_c the surface resistance. Here z_0 is the surface roughness length, k the von Karman constant, $\Psi_h\left(\dfrac{z}{L}\right)$ the integrated diabatic function, L the Moin-Obukhov length scale, u_* the friction velocity, z_s the height of the quasi-laminar sublayer and D_g the molecular diffusivity of the pollutant.

3.2.1 Estimation of aerodynamic resistance

The analytical method for the estimation of the aerodynamic resistance rather than using the profile method (Chang et al., 1990) is to utilize the bulk Richardson number, R_{iB} (Garratt, 1992),

$$R_{iB} = \frac{g}{\theta_0} \frac{Z_u^2(\theta_a - \theta_g)}{Z_T u_a^{-2}} \quad (4)$$

where g is gravity, θ_0 the mean surface layer temperature, θ_a the observed air temperature at the height of Z_T, $\overline{u}_a$ the observed mean wind at the height of Z_u, and θ_g the ground surface temperature.

For the unstable stratification ($R_{iB}<0$), the aerodynamic resistance (Park, 1994) is given by

$$R_a = \frac{1}{C_{HN}\left[1 - \frac{10R_{iB}}{1 + b_H|R_{iB}|^{1/2}}\right]\overline{u}_a} \quad (5)$$

where

$$C_{HN} = \frac{k^2}{\ln\left(\frac{Z_u}{Z_0}\right)\ln\left(\frac{Z_T}{Z_{0T}}\right)}$$

is the drag coefficient of the heat in the neutral stability, and z_{0T} is the thermal surface roughness length, and

$$b_H = 10\left(6.3 - 0.18C_{DN}^{-1/2}\right)C_{HN}\left(\frac{Z_T}{Z_{0T}}\right)^{1/2}$$

with $C_{DN} = \dfrac{k^2}{\left[\ln\left(\dfrac{Z_u}{Z_0}\right)\right]^2}$

the drag coefficient of momentum in neutral stability.

For the stable stratification ($R_{iB}>0$), R_a is estimated as,

$$R_a = \frac{1}{C_{HN}\left[(1 + 5R_{iB})^{-2}\right]\overline{u}_a} \quad (6)$$

The ground surface temperature θ_g is estimated by the soil-canopy energy budget model (Park, 1994).

3.2.2 Estimation Of Quasi-Laminar Sublayer Resistance

Following the result of Wesely and Hicks (1977), the sublayer resistance is given by,

$$R_b = \frac{2}{ku_*}\left[\frac{v_h}{D_g}\right]^{2/3} \quad (7)$$

where v_h is the thermal diffusivity of air. It is believed that R_b is smaller than R_a.

3.2.3 Estimation of the surface or canopy resistance.

The surface resistance (or canopy resistance), R_c, is broken into several components. Series and parallel resistances are usually identified for the various parts of the canopy (Baldocchi et al., 1987 ; Hicks et al., 1991).

Analogously to Ohm's law in the electrical circuits, R_c can be expressed as (e.g., Wesely, 1989)

$$\frac{1}{R_c} = \frac{1}{r_s + r_m} + \frac{1}{r_{lu}} + \frac{1}{r_{dc} + r_{cl}} + \frac{1}{r_{ac} + r_{gs}} \quad (8)$$

where r_s, r_m, r_{lu}, r_{dc}, r_{cl}, r_{ac} and r_{gs} represent resistances of the stomata, plant mesophyll, leaf tissue in the upper canopy, buoyant convection, leaf tissue in the lower canopy, canopy and ground surface uptake, respectively. Eq.(8) is derived under assumptions that concentrations representative of the plant mesophyll, plant tissue of leaves in the upper canopy, substrates in the lower canopy, and substrates at the ground surface are in equilibrium with the concentration in air immediately above these surface elements.

The stomatal resistance (r_s) depends upon the leaf area index, the photosynthetically active radiation, the soil moisture content, the vapour-pressure deficit of the atmosphere, and air temperature (Noilhan and Planton, 1989; Milhailovic and Rajikovic, 1992; Park, 1994).

But we use the following generalized simple function to estimate the bulk canopy stomatal resistance as in Wesely (1989);

$$r_s = r_{min}\left[1 + \left(\frac{200}{S^\downarrow + 0.1}\right)^2\right]\left[\frac{400}{T_a(40 - T_a)}\right] \quad (9)$$

where r_{min} is the minimum bulk canopy stomatal resistance for the water vapor, $S^{\downarrow}$ the incoming solar radiation, T_a the surface air temperature in degree Celsius between 0 and $40°C$. Outside the range, r_s is set to a very large value so that the transfer through stomata is stopped. The combined minimum stomatal and mesophyll resistance for substance X can be found as

$$r_{smx} = r_s \frac{D_{H_2O}}{D_x} + r_{mx} \quad (10)$$

where D_x is the molecular diffusivity of gas X in air, D_{H_2O} the molecular diffusivity for water vapor, and r_{mx} the resistance of the mesophyll for the gas of interest. The mesophyll resistance is computed for any substance X as,

$$r_{mx} = \frac{1}{\left(\frac{H^*}{3000} + 100 f_0\right)} \quad (11)$$

where H^* is the effective Henry's law constant for the substance X and f_0 the reactivity factor for substance X.

The resistance of the outer surfaces of leaves in the upper canopy is estimated as

$$r_{lux} = \frac{r_{lu}}{\left(\frac{H^*}{10^5} + f_0\right)} \quad (12)$$

The resistance of the plant tissue of leaves in the upper canopy (r_{lu}) varies widely with vegetation. The resistance to SO_2 uptake by a dew-covered surface is

$$r_{lus} = 100 \text{ s m}^{-1} \quad (13)$$

on the other hand for O_3, the dew acts as a barrier to surface removal:

$$r_{lux} = \left(\frac{1}{3000} + \frac{1}{3 r_{luo}}\right)^{-1} \quad (14)$$

For substances other than SO_2 and O_3, the following equation may be used if either dew or rain wets the surface:

$$r_{lux} = \left[\frac{1}{3r_{lu}} + 10^{-7} H^* + \frac{f_0}{r_{luo}}\right]^{-1} \quad (15)$$

where reduction of dry area, solubility in water, and chemical reactivity are taken into account by the first, second, and third terms.

The resistance r_{dc} is the effects of mixing forced by buoyant convection when sunlight heats the ground or lower canopy and by penetration of wind into canopies on the sides of hills. This is estimated by

$$r_{dc} = \frac{100\left[1 + \frac{1000}{(S^{\downarrow} + 10)}\right]}{1 + 1000 S_{ter}} \quad (16)$$

where S_{ter} is the slope in radians of the local terrain.

The values of r_{ac} are estimated solely by land use and season, mostly on the basis of the depth of the structure, which is usually a plant canopy.

The resistance of the exposed surfaces in the lower portions of structure above the ground (r_{cl}) is estimated as,

$$r_{clx} = \frac{1}{\frac{H^*}{10^5 r_{cls}} + \frac{f_0}{r_{clo}}} \quad (17)$$

The subscripted s in r_{cls} and o in r_{clo} represent the values of r_{cl} for SO_2 and O_3, respectively.

The resistance to uptake at the ground surface (r_{gs}) is computed as

$$r_{gsx} = \frac{1}{\frac{H^*}{10^5 r_{gss}} + \frac{f_0}{r_{gso}}} \quad (18)$$

For surfaces with a pH significantly less than seven, it would be desirable to use approximately adjusted values of H^*.

Table 2 shows resistance components for the various land-use types that are used in the RADM model and Table 3 is properties of gaseous species relevant to estimate the surface resistance to dry deposition.

4. RESULTS

4.1. SYNOPTIC METEOROLOGICAL CLASSIFICATIONS

Based on the analysis of the error function, 6 meteorological clusters in spring, 5 clusters in summer, 4 clusters in autumn and in winter are retained. Table 4 shows the number of non-precipitating days belonging to each cluster in each season. The number of precipitating days are largest in summer due to the summer monsoon in East Asia.

The obtained mean values of each variables in each cluster, or centroids are shown in Table 5.

Cluster 1 in spring (Table 5) is characterized by very low 500 hPa geopotential height and temperature with very low surface minimum temperature and dew point temperature at the surface. However, it has relatively high surface pressure with relatively strong surface wind and northerly geostrophic wind at the 850 hPa level.

The composite synoptic maps at the 1000 hPa and 500 hPa level belonging to cluster 1 in spring are given in Figs. 2 and 3. The Korean peninsula is influenced by the high pressure system located at the Yellow Sea at the 1000 hPa level. However, an upper level trough is passing through the central part of Korea from northeast to southeast, resulting in predominant northerly flows over the Korean peninsula.

Cluster 2 in spring (Table 5) is characterized by very low 500 hPa level temperature with very low surface relative humidity and surface dew point temperature. The surface pressure is the highest among all clusters in spring due to the influence of a high pressure system centered in the East China Sea (Fig. 2). The southerly surface wind is relatively strong but the upper level northerly geostrophic wind is weak with wind veering with height as seen in the pressure pattern at the 500 hPa level (Fig. 3).

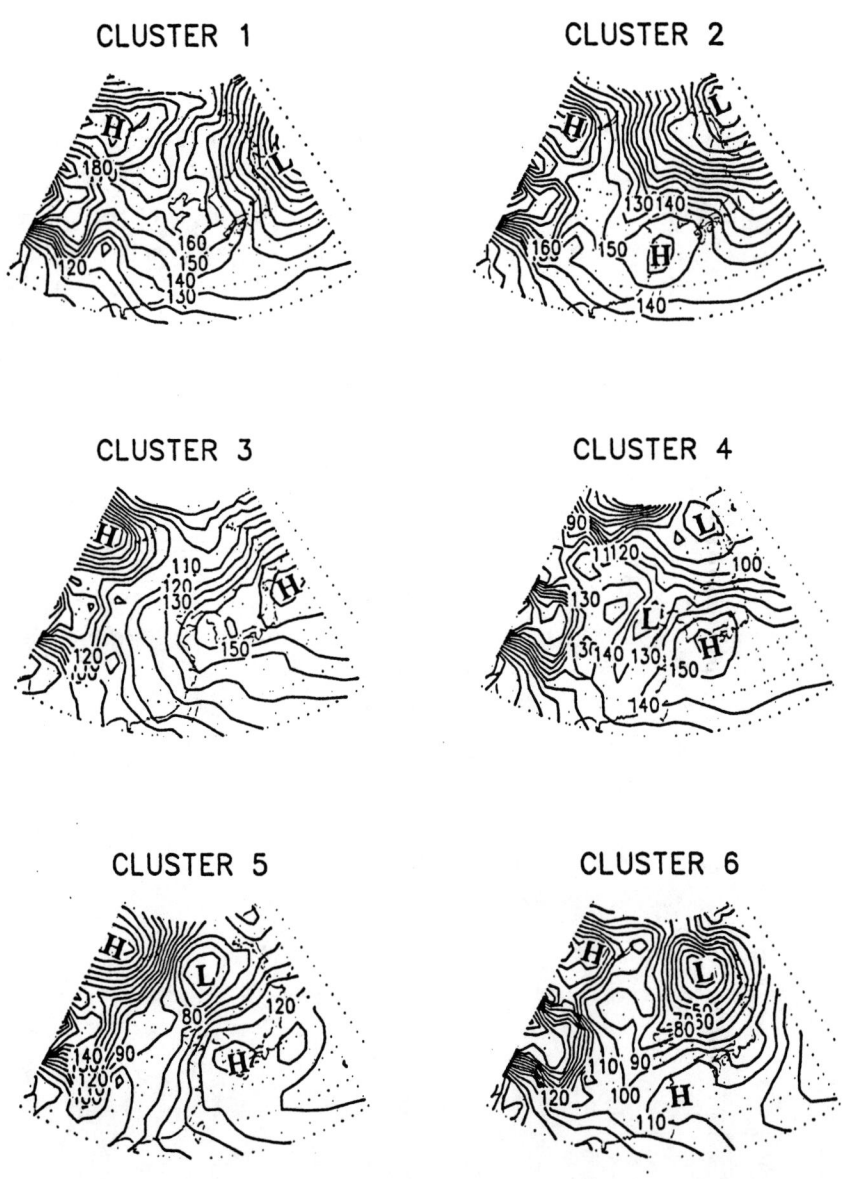

Figure 2. Composite 1000 hPa maps for each cluster in spring.

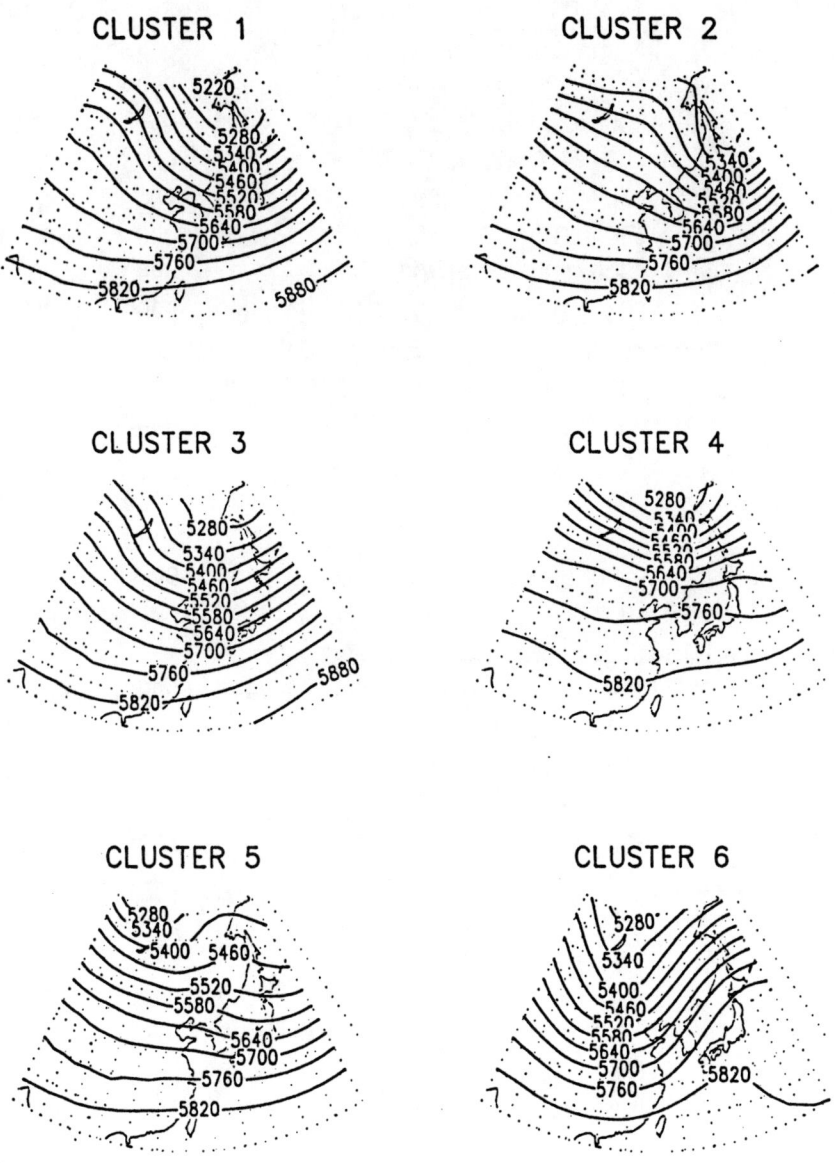

Figure 3. Composite 500 hPa maps for each cluster in spring.

Table 2. Input Resistances (s m^{-1}) to module for Computations of Surface Resistances (r_c). Entries of 9999 indicate that there is no air-surface exchange via that resistance pathway (From RADM).

Resistance Component	Land-use Types										
	1a	2b	3c	4d	5e	6f	7g	8h	9I	10j	11k
Seasonal Category 1: midsummer with lush vegetation											
r_{min}	9999	60	120	70	130	100	9999	9999	80	100	.150
r_{lu}	9999	2000	2000	2000	2000	2000	9999	9999	2500	2000	4000
r_{ac}	100	200	100	2000	2000	2000	0	0	300	150	200
r_{gsS}	400	150	350	500	500	100	0	1000	0	220	400
r_{gsO}	300	150	200	200	200	300	2000	400	1000	180	200
r_{clS}	9999	2000	2000	2000	2000	2000	9999	9999	2000	2000	4000
r_{clO}	9999	1000	1000	1000	1000	1000	9999	9999	1000	1000	1000
Seasonal category 2: autumn with unharvested cropland											
r_{min}	9999	9999	9999	9999	250	500	9999	9999	9999	9999	9999
r_{lu}	9999	9000	9000	9000	4000	8000	9999	9999	9000	9000	9000
r_{ac}	100	150	100	1500	2000	1700	0	0	200	120	140
r_{gsS}	400	200	350	500	500	100	0	1000	0	300	400
r_{gsO}	300	150	200	200	200	300	2000	400	800	180	200
r_{clS}	9999	9000	9000	9000	2000	4000	9999	9999	9000	9000	9000
r_{clO}	9999	400	400	400	1000	600	9999	9999	400	400	400
Seasonal category 3: late autumn after frost, no snow											
r_{min}	9999	9999	9999	9999	250	500	9999	9999	9999	9999	9999
r_{lu}	9999	9999	9000	9000	4000	8000	9999	9999	9000	9000	9000
r_{ac}	100	10	100	1000	2000	1500	0	0	100	50	120
r_{gsS}	400	150	350	500	500	200	0	1000	0	200	400
r_{gsO}	300	150	200	200	200	300	2000	400	1000	180	200
r_{clS}	9999	9999	9000	9000	3000	6000	9999	9999	9000	9000	9000
r_{clo}	9999	1000	400	400	1000	600	9999	9999	800	600	600
Seasonal category 4: winter, snow on ground and subfreezing											
r_{min}	9999	9999	9999	9999	400	800	9999	9999	9999	9999	9999
r_{lu}	9999	9999	9999	9999	6000	9000	9999	9999	9000	9000	9000
r_{ac}	100	10	10	1000	2000	1500	0	0	50	10	50
r_{gsS}	100	100	100	100	100	100	0	1000	100	100	50
r_{gsO}	600	3500	3500	3500	3500	3500	2000	400	3500	3500	3500
r_{clS}	9999	9999	9999	9000	200	400	9999	9999	9000	9999	9000
r_{clo}	9999	1000	400	400	1000	600	9999	9999	800	600	600
Seasonal category 5: transitional spring with partially green short annuals											
r_{min}	9999	120	240	140	250	190	9999	9999	160	200	300
r_{lu}	9999	4000	4000	4000	2000	3000	9999	9999	4000	400	8000

	a	b	c	d	e	f	g	h	I	j	k
r_{gsS}	100	50	80	1200	2000	1500	0	0	200	60	120
r_{gsO}	300	150	200	200	200	300	2000	400	1000	180	200
r_{clS}	9999	4000	4000	4000	2000	3000	9999	9999	4000	4000	8000
r_{clO}	9999	1000	500	500	1500	700	9999	9999	600	800	800

a Urban land, almost no vegetation   g Water, including both salt and fresh water
b Agricultural land, usually wet watered   h Barren land, mostly desert
c Range land, usually with low soil moisture   I Non-forested wetland
d Deciduous forest   j Mixed agricultural and range land
e Coniferous forest   k Rocky open areas occupied by low-growing shrubs
f Mixed forest including wet land

Table 3. Gases and their properties relevant to estimating resistances to dry deposition (from RADM).

Gaseous species	Symbol	D_{H2O}/D_x	H^*	f_o
Sulfur dioxide	SO_2	1.9	1×10^5	0
Ozone	O_3	1.6	0.01	1
Nitrogen dioxide	NO_2	1.6	0.01	0.1
Nitric oxide	NO	1.3	3×10^{-3}	0
Nitric acid vapor	HNO_3	1.9	1×10^{14}	0.1
Ammonia	NH_3	1.0	2×10^4	0

Table 4. Number of non-precipitating days belonging to each cluster in each season.

	Cluster	Spring	Summer	Autumn	Winter
	1	37	22	36	34
	2	34	44	42	39
	3	39	17	43	29
	4	23	16	71	25
	5	43	11	-	-
	6	19	-	-	-
Number of non-precipitating days Used for clustering (N_c)		195	110	192	127
Available total number Of days in 5 years (N_T)		397	438	412	304
N_C/N_T (%)		49	25	47	42

Table 5. Composition of the mean variables obtained by means of clustering technique in each season.

	CL	H_{500} (m)	T_{500} (°C)	RH (%)	T_{min} (°C)	W_s (ms^{-1})	P_s (hPa)	T_{dew} (°C)	ED (°C/100m)	EDI (°C)	MP (m^{-2}s^{-2})	W_{850} (ms^{-1})	WD_{850} (°)
SPRING	1	5566	-21.9	59	5.6	1.9	1010.5	2.8	-0.50	-24.6	32605	5.5	308
	2	5589	-21.8	56	5.8	2.0	1013.1	2.5	-0.50	-24.3	32880	4.7	305
	3	5579	-21.7	58	7.0	2.1	1011.2	3.6	-0.52	-25.3	32892	6.2	263
	4	5667	-16.6	58	10.6	1.9	1007.4	7.8	-0.49	-24.4	38870	3.8	263
	5	5640	-18.5	64	8.4	1.5	1011.5	6.6	-0.49	-25.1	36670	3.3	265
	6	5600	-20.0	59	7.3	1.6	1010.6	4.1	-0.50	-24.1	34795	3.8	281
SUMMER	1	5762	-9.5	74	17.5	1.5	1003.4	16.4	-0.48	-25.9	46910	2.5	278
	2	5852	-5.1	75	22.4	1.6	1004.0	21.3	-0.48	-26.4	52207	2.5	228
	3	5747	-11.2	70	16.8	1.1	1004.2	15.4	-0.49	-26.6	45082	2.5	258
	4	5778	-7.8	74	18.0	1.4	1003.4	17.1	-0.45	-24.9	48832	2.3	270
	5	5808	-8.0	69	18.6	1.7	1004.1	17.2	-0.47	-25.2	48934	0.7	254
AUTUMN	1	5588	-21.4	58	4.8	2.1	1016.5	0.2	-0.47	-21.7	33249	6.6	308
	2	5720	-15.3	71	11.5	1.5	1011.9	10.2	-0.48	-25.4	40733	1.6	313
	3	5697	-15.4	68	11.1	1.5	1012.1	9.6	-0.47	-25.0	40376	4.2	287
	4	5707	-17.6	67	8.2	1.4	1017.7	5.9	-0.46	-23.5	38225	3.1	283
WINTER	1	5498	-26.8	57	-4.0	1.9	1021.3	-8.4	-0.42	-18.4	26970	7.3	321
	2	5473	-26.8	58	-2.1	2.1	1017.5	-6.1	-0.46	-20.7	26690	7.1	307
	3	5477	-27.1	58	-3.6	1.7	1018.3	-7.8	-0.44	-19.3	26483	5.7	306
	4	5564	-24.3	64	-0.2	1.8	1019.7	-3.5	-0.44	-20.7	30158	6.4	271

Cluster 3 in spring (Table 5) is characterized by the relatively low 500 hPa geopotential height due to the upper level trough passing through the west of the peninsula (Fig. 3) and low 500 hPa temperature. However, the surface wind is the strongest among all clusters in spring. The upper-level southwesterly geostrophic wind (Fig. 3) is also strong. The vertical stability of the cluster is the weakest of all clusters in spring. The composite surface map shows the high pressure center over the Korean peninsula.

Cluster 4 in spring (Table 5) is characterized by the highest 500 hPa geopotential height and temperature at the 500 hPa level. The surface minimum temperature and dew point temperature are also the highest of all clusters in spring. However, the surface pressure is the lowest of all clusters in association with the surface trough extending from northeast to southwest to the west of the Korean peninsula (Fig. 2). The atmospheric stability expressed by ED values shows moderate stable stratification. The southwesterly geostrophic wind at the 850 hPa level is weak. The composite 500 hPa chart (Fig. 3) shows a relatively weak zonal flow over the Korean peninsula.

Cluster 5 in spring (Table 5) reveals the highest relative humidity. But both the surface wind speed and the 850 hPa geostrophic wind speed are the lowest among all clusters in spring. The surface pressure is relatively high under the influence of the high pressure center located in the Korean strait (Fig. 2). The composite 500 hPa map shows a weak geopotential height ridge over Korea (Fig. 3).

Cluster 6 in spring is characterized by a relatively low 500 hPa geopotential height, low temperature at the 500 hPa level, and low surface wind speed. The westerly geostrophic wind at 850 hPa is also relative weak.

Cluster 2 in summer (Table 5) is associated with high geopotential height and temperature at the 500 hPa level due to the influence of the subtropical high pressure system (Fig. 5). This cluster is characterized by the high values of relative humidity, minimum temperature, dew point temperature, and the wind speed at the surface in association with the influence of the subtropical high (Fig. 4). On the other hand, cluster 3 in summer (Table 5) is characterized by low geopotential height and temperature at the 500 hPa level influenced by the upper-level pressure trough extending southeastward from the Lake Baykal (Fig. 5). This cluster is also characterized by low value of relative humidity, minimum temperature, and wind speed at the surface. But the surface pressure is high in association with the surface high pressure system located over Korea (Fig. 4). A relatively weak stable stratification prevails throughout the mid-troposphere.

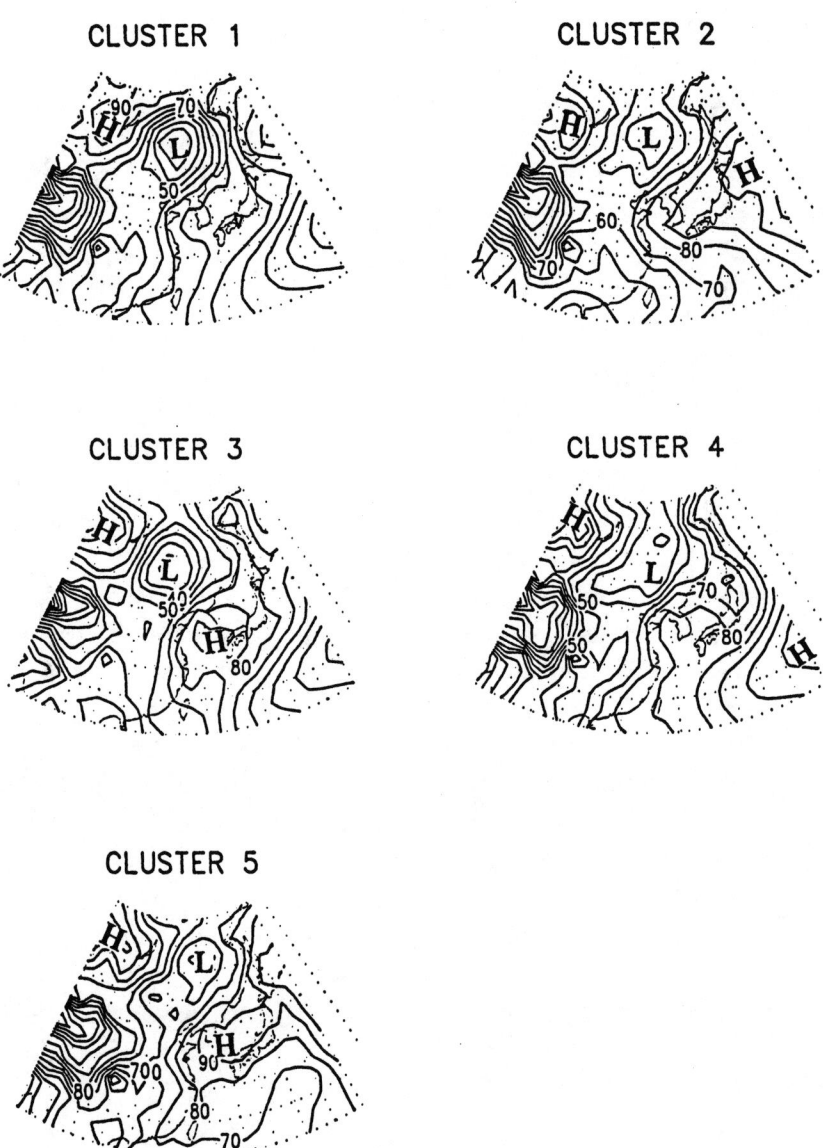

Figure 4. Composite 1000 hPa maps for each cluster in summer.

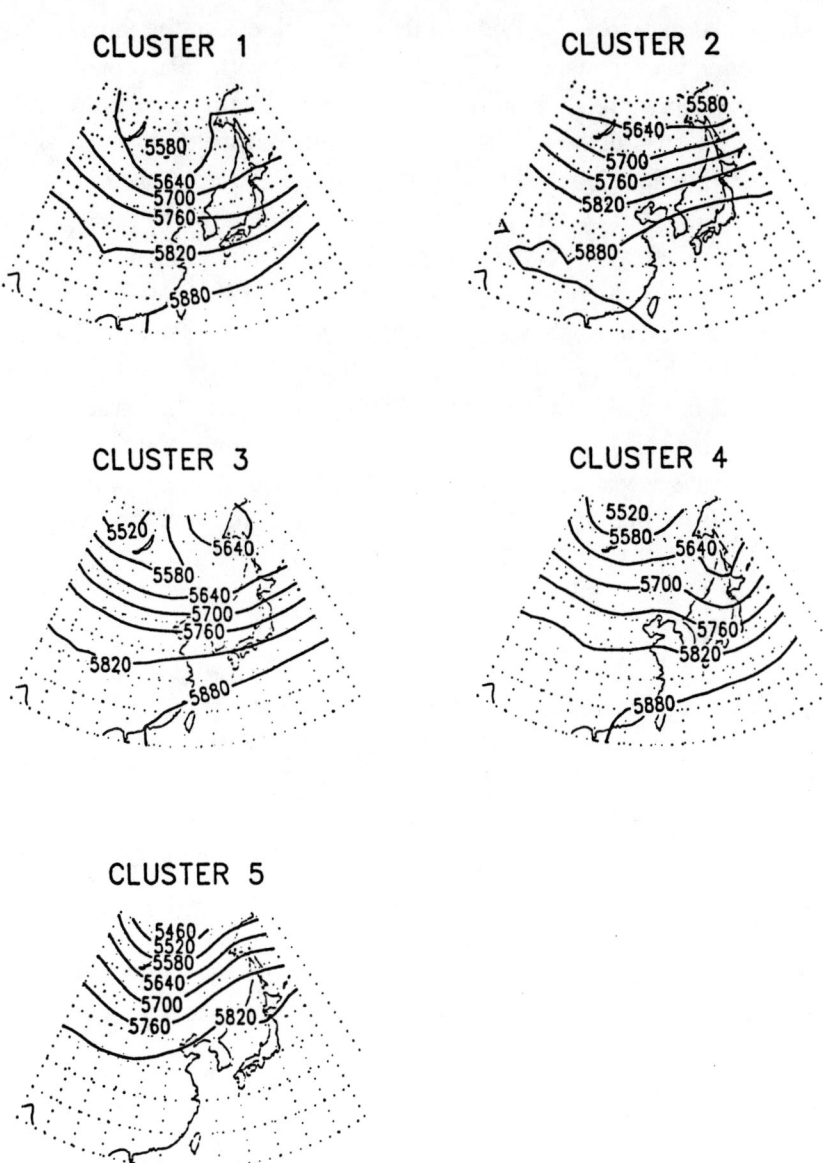

Figure 5. Composite 500 hPa maps for each cluster in summer.

Cluster 1 in autumn (Table 5) shows low mean values of 500 hPa geopotential height, 500 hPa temperature, surface relative humidity, surface minimum temperature, surface dew point temperature and montgomery potential energy, but high values of the surface pressure and surface wind speed in association with the frequent outbreak of the Siberian high (Fig. 6). Cluster 2 in autumn is characterized by the high values of 500 hPa geopotential height, 500 hPa temperature, surface relative humidity, surface minimum temperature and surface dew point temperature, while the low values of surface pressure, surface wind speed, 850hPa geostrophic wind speed and ED (weak stable stratification). Clusters 3 and 4 show in-between values of clusters 1 and 2 except for the values of surface pressure and montgomery potential energy in cluster 4. The highest surface pressure in cluster 4 is associated with the migratory high pressure system located over Korea (Fig. 6).

Cluster 2 in winter (Table 5) shows low mean values of 500 hPa geopotential height, surface pressure, but high surface wind speed and strong northerly 1000 hPa geostrophic wind due to the cold outbreak of the Siberian high (Figs. 8 and 9). Cluster 4 in winter shows high values of 500 hPa geopotential height and temperature, and surface dew point temperature with relative weak surface wind speed and 1000 hPa-level westerly geostrophic wind in association with the migratory high pressure system over Korea (Fig.8).

4.2. SEASONAL MEAN SO_2 AND NO_2 CONCENTRATIONS

Seasonally averaged SO_2 and NO_2 concentrations for 5 years from 1989 to 1993 at each city in Korea are presented in Table 6. Table 6 clearly indicates that most of big cities including Seoul, Inchon, Suwon, Pusan, Kwangju, Taegu, Taejon and Chunchon have large seasonal variations of the SO_2 concentration with maxima in winter and much reduced minima in summer. However, most of industrial cities including Changwon, Ulsan, Yochon, and Dongkwangyang, Gume and Pohwang have relatively small seasonal variations of the SO_2 concentration. On the other hand the mean NO_2 concentration has different seasonal variation compared with SO_2 concentration. Even though the maximum concentration of NO_2 occurs in winter, some of cities have their maximum values in spring and autumn and the seasonal range is much smaller than that of SO_2.

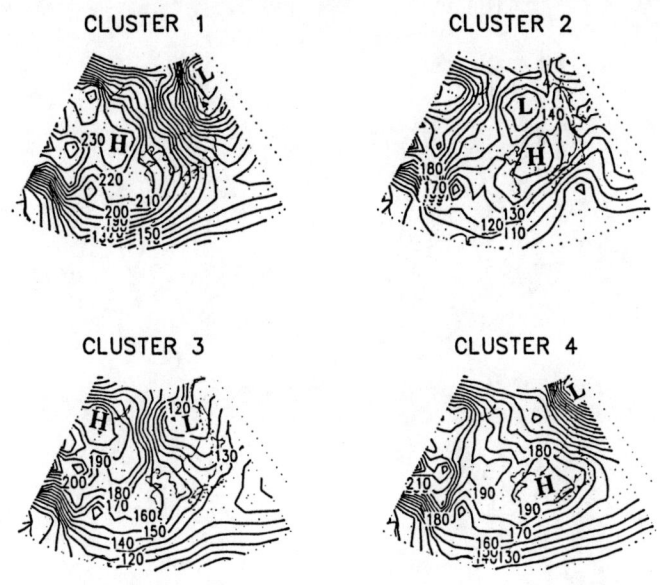

Figure 6. Composite 1000 hPa maps for each cluster in autumn.

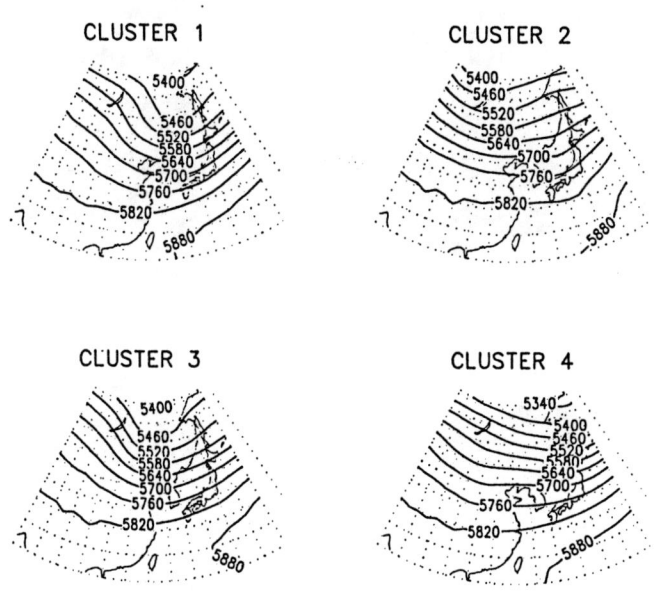

Figure 7. Composite 500 hPa maps for each cluster in autumn.

1000hPa GPH (WINTER)

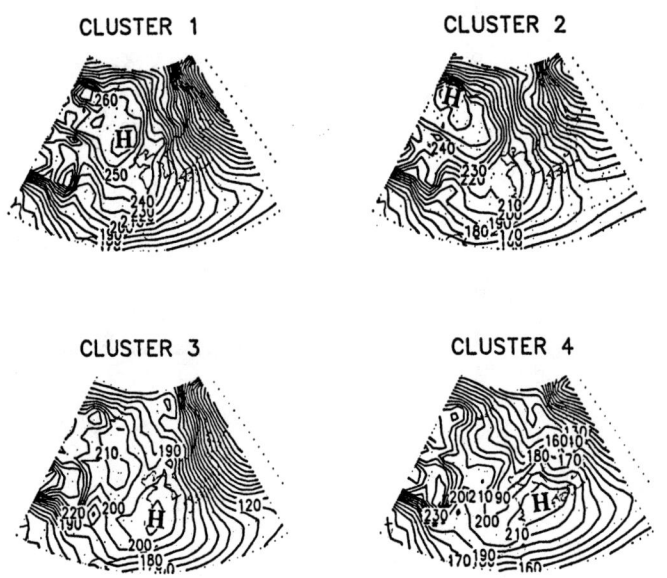

Figure 8. Composite 1000 hPa maps for each cluster in winter.

500hPa GPH (WINTER)

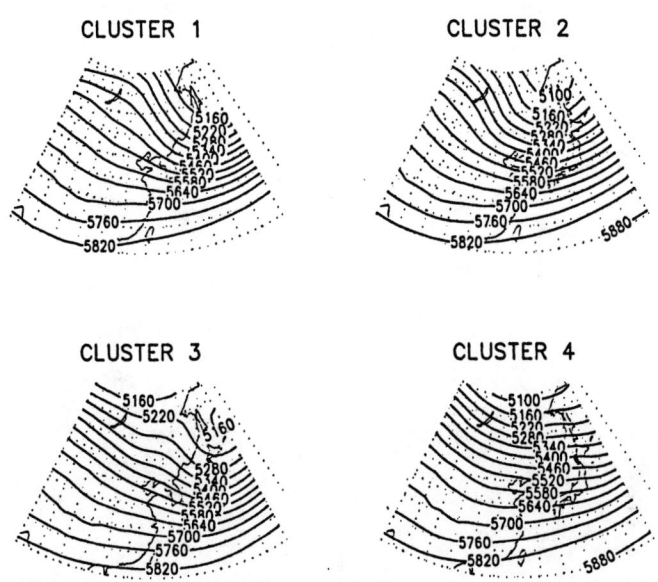

Figure 9. Composite 1000 hPa maps for each cluster in winter

5. RELATIONSHIP BETWEEN THE SO_2 AND NO_2 CONCENTRATIONS AND THE SYNOPTIC METEREOLOGICAL CLASSIFICATIONS

The seasonally averaged SO_2 and NO_2 concentrations for all cities belonging to each cluster normalized by the seasonally overall averaged concentrations for all cities (Table 6) are presented in Fig. 10. In spring (Fig. 10a) the maximum concentration of SO_2 is associated with the synoptic meteorological classification of Cluster 3 while the minimum concentration is related with Cluster 4. However, the maximum concentration of NO_2 is associated with Cluster 6 while the minimum is related with Cluster 2, which implies that the synoptic patterns that are associated with the maximum SO_2 concentration may not coincide with that of NO_2 concentration in spring.

In summer (Fig. 10b) Cluster 5 is associated with the maximum SO_2 and NO_2 concentrations while the minimum SO_2 concentration occurs in Cluster 2 but the minimum NO_2 concentration is related with Cluster 3, where a relatively large SO_2 concentration occurs. This discrepancy is probably due to the varying degree of photochemical reaction of these two pollutants.

In autumn (Fig. 10c), the overall averaged maximum concentrations of both SO_2 and NO_2 occur in Cluster 4 with the minimum concentrations in Cluster 3. This suggests that the synoptic patterns that control the concentrations of SO_2 and NO_2 coincide each other.

In winter (Fig. 10d) as is the case in autumn, the overall averaged maximum concentrations of SO_2 and NO_2 occurs in Cluster 4, while the minimum concentrations occur in Cluster 2.

Above cluster analyses indicate that the air pollution potential can be identified with the classification of synoptic patterns. Throughout the whole seasons, it can be seen that the surface anticyclonic center located in Korea (Figs. 2,4,6, and 8) with the westerly zonal type flow at the 500 hPa level (Figs. 3,5,7, and 9) gives rise to high average pollution levels in Korea.

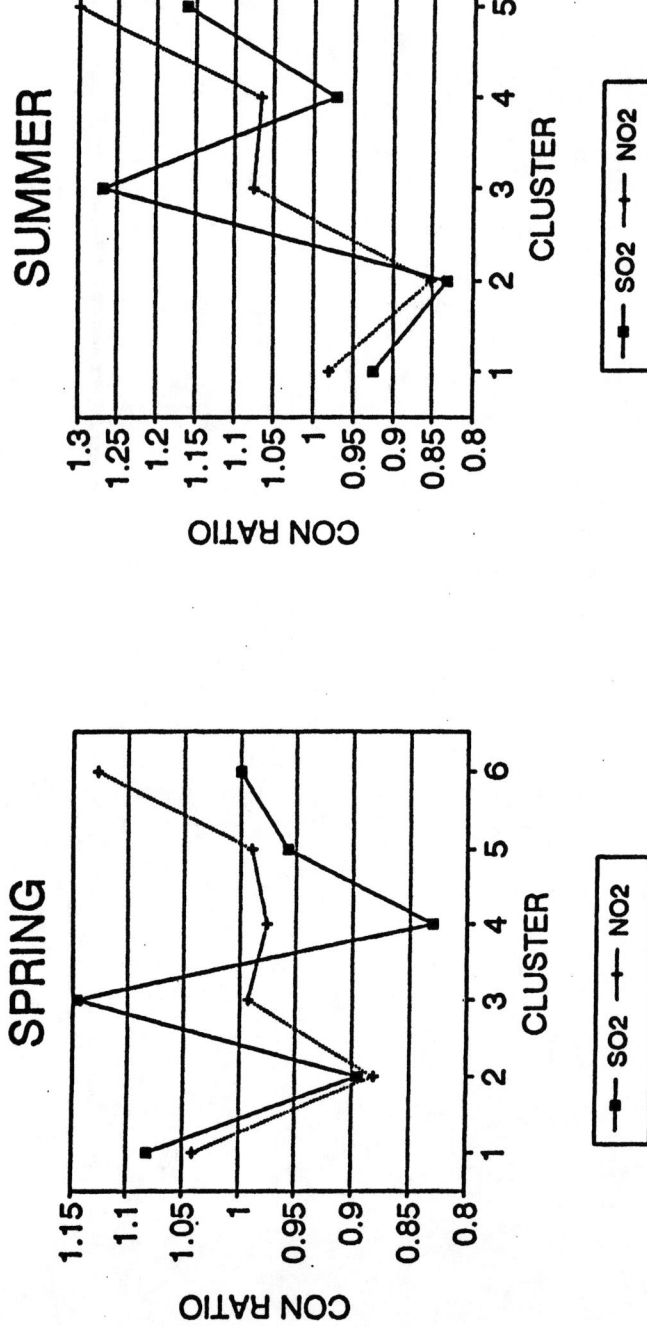

Figure 10. Variations of SO_2 and NO_2 concentration normalized by seasonal mean concentration in each retained cluster in (a) spring, (b) summer.

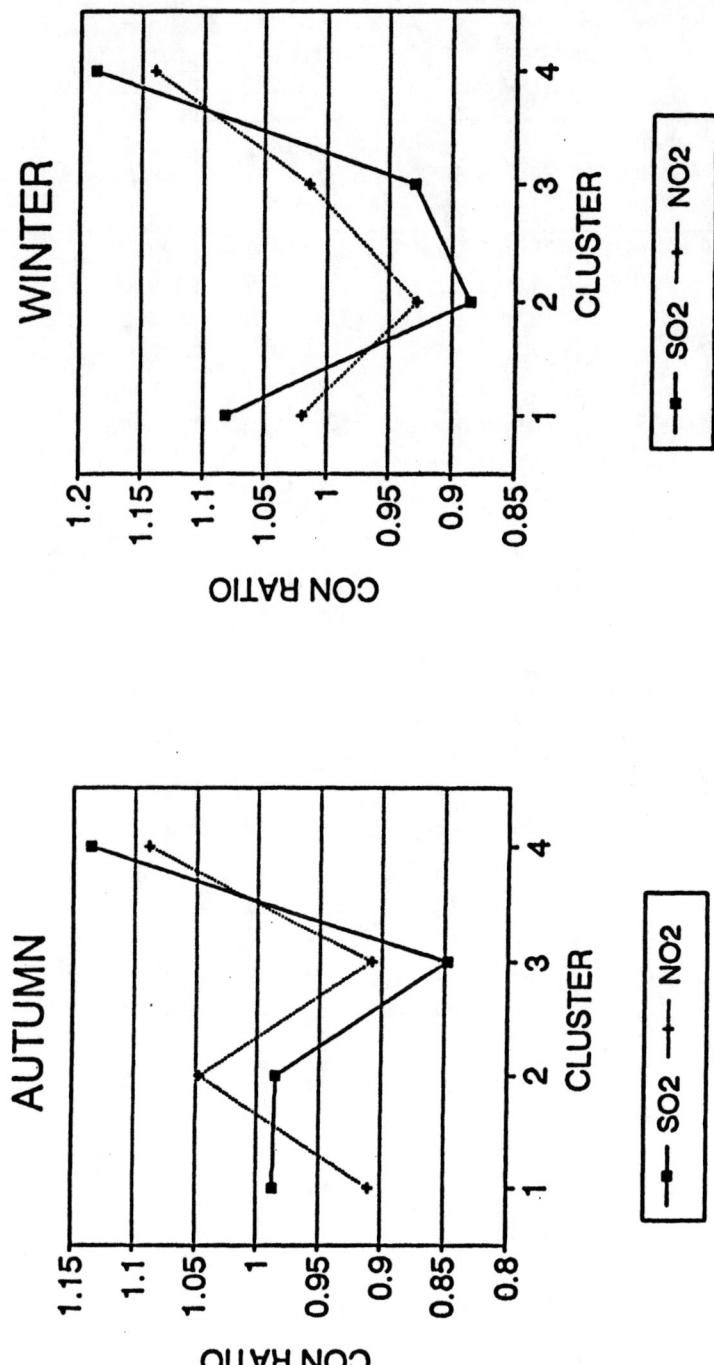

Figure 10 (cont'd) Variations of SO_2 and NO_2 concentration normalized by seasonal mean concentration in each retained cluster in (c) autumn, and (d) winter.

Table 6. Five years averaged seasonal mean SO_2 and NO_2 concentrations (ppb) at each city in South Korea

Season City / Pollutant	SPRING		SUMMER		AUTUMN		WINTER	
	SO_2	NO_2	SO_2	NO_2	SO_2	NO_2	SO_2	NO_2
Seoul	41.8	33.4	16.7	27.4	44.8	32.0	99.0	37.2
Inchon	39.5	29.8	19.3	21.8	41.3	29.9	106.6	32.8
Suwon	44.0	29.3	18.0	21.3	42.4	27.1	106.6	31.9
Anyang	44.7	32.4	17.3	25.8	59.8	21.2	82.7	47.4
Kwachon	29.8	28.3	13.4	21.4	25.7	22.6	61.3	36.8
Songnam	41.3	34.5	16.9	28.4	38.1	34.6	79.9	39.3
Buchon	41.4	28.5	17.3	22.1	46.1	35.8	108.1	39.5
Kwangmyong	35.6	32.7	18.4	24.8	50.5	40.2	88.4	30.4
Ansan	26.5	23.5	15.5	18.2	26.1	23.6	39.9	29.3
Pusan	41.1	26.6	26.5	20.3	36.3	23.3	64.8	26.9
Changwon	19.8	21.9	13.5	17.3	20.2	21.2	30.0	26.2
Ulsan	38.0	29.2	31.6	26.8	28.8	26.4	44.3	25.6
Kwangju	18.8	16.0	5.8	10.8	18.2	12.8	36.5	14.4
Yosu	22.6	27.6	12.0	21.2	26.8	19.9	35.1	23.5
Yochon	17.6	15.2	16.2	12.7	17.7	13.7	15.9	15.5
DongKwangyang	16.0	14.2	15.1	12.1	10.7	15.1	11.7	17.4
Taegu	41.2	28.1	22.7	23.6	45.9	17.6	72.0	26.7
Gumi	34.8	22.1	20.5	16.0	42.6	14.0	45.2	12.7
Pohwang	21.1	14.3	15.4	13.2	19.5	16.2	27.6	14.5
Taejon	26.0	18.2	10.9	18.5	29.4	21.6	60.5	31.5
Chonju	19.9	21.0	10.6	15.5	19.2	21.4	29.3	21.2
Iri	24.2	19.2	14.5	21.1	35.0	17.7	40.3	20.2
Ulsan city	29.8	25.8	27.4	23.7	26.7	28.8	26.9	26.8
Ujungbu	37.6	23.0	12.0	22.9	46.0	26.5	85.0	39.0
Chunchon	32.1	16.3	6.6	10.9	39.1	13.2	93.7	21.1
Kunsan	33.8	20.3	13.9	14.8	22.1	16.6	44.1	22.1
Chungju	19.7	15.9	13.6	12.8	18.2	16.0	25.5	25.0
Wonju	23.5	20.9	7.0	13.0	32.2	16.9	76.1	25.0
Chongju	27.0	26.2	9.6	17.7	30.4	30.9	71.2	30.4
Kangnung	25.3	14.7	15.5	12.1	16.7	11.7	30.8	10.5
Jechon	14.3	20.3	4.7	20.5	17.5	18.0	37.5	23.6
Mean	30.0	23.8	15.4	18.7	31.4	22.2	57.3	26.6

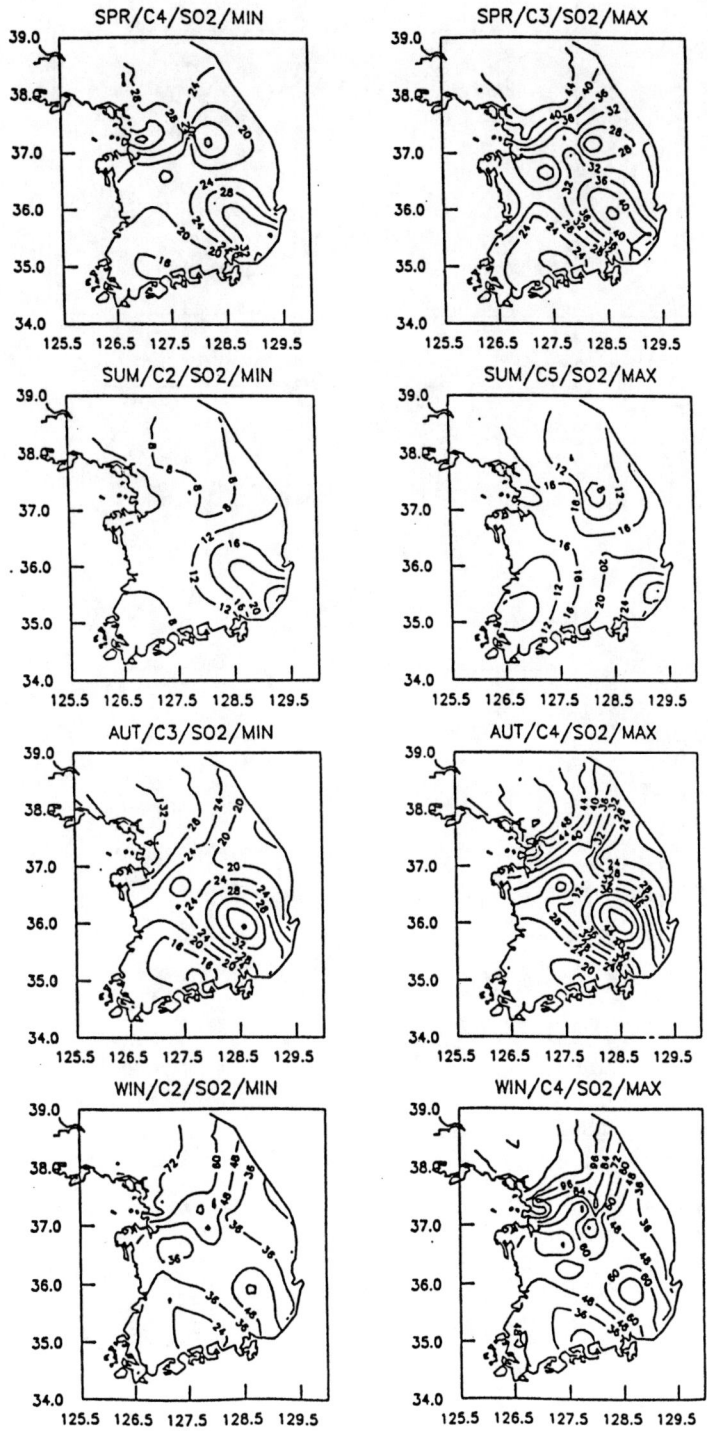

Figure 11. Horizontal distributions of seasonal mean minimum (left panel) and maximum (right panel) SO_2 concentrations (ppb) on each retained cluster for each season.

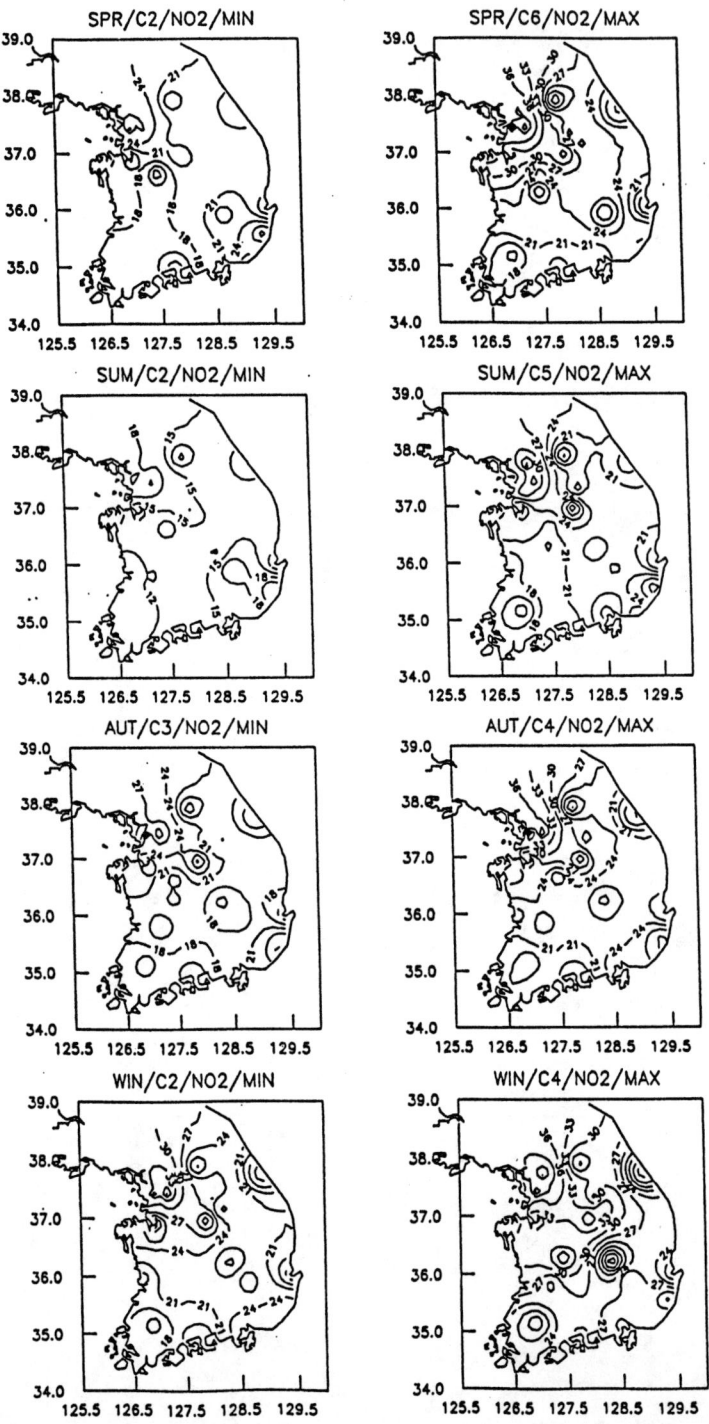

Figure 12. Horizontal distributions of seasonal mean minimum (left panel) and maximum (right panel) NO$_2$ concentrations (ppb) on each retained cluster for each season.

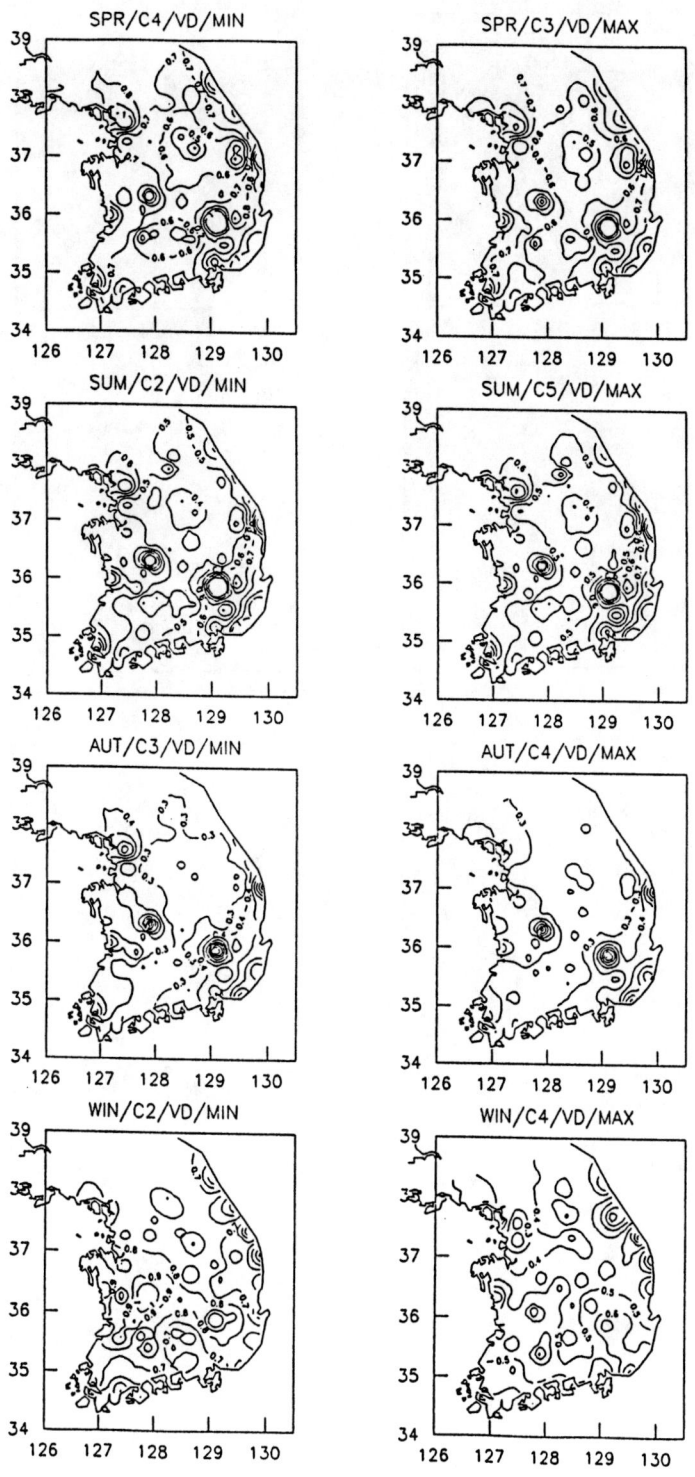

Figure 13. The same as in Figure 11 except for dry deposition velocity (cm s^{-1}).

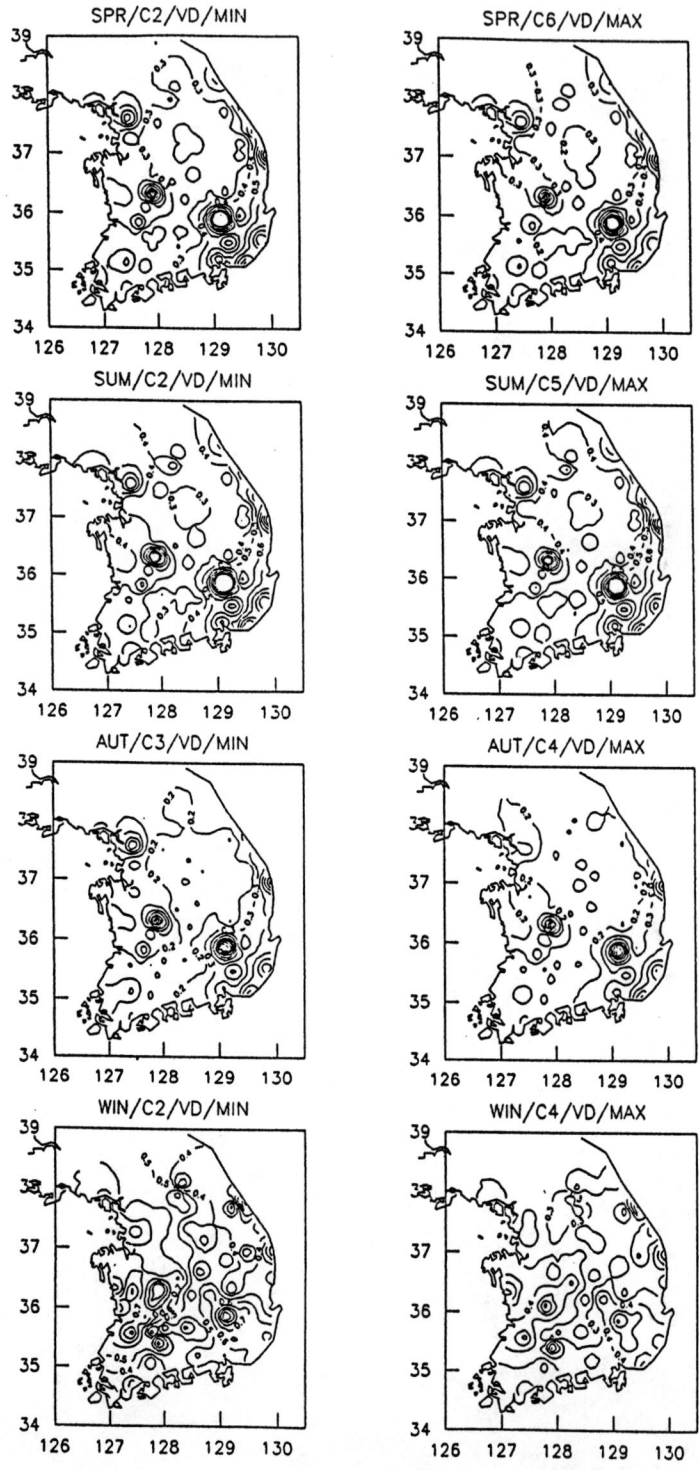

Figure 14. The same as in Figure 13 except for NO_2.

The horizontal distributions of daily averaged SO_2 concentrations belonging to the clusters that are associated with the maximum concentration (right panel in Fig. 11) are quite similar in all seasons with relatively high concentrations in the regions of Kyungsangdo, Daegu, Pusan, and Kyongin (Inchon and Seoul). However, the magnitude of concentration changes with season with a maximum in winter and a minimum in summer. Very similar horizontal distribution patterns are also seen in clusters that are associated with the minimum SO_2 concentration (left panel in Fig. 11) with much reduced SO_2 concentration.

Seasonal variations of the horizontal distributions of mean NO_2 concentrations belonging to the clusters that are associated with both the maximum and minimum concentrations (Fig. 12) are quite similar to those of SO_2 concentrations in Fig. 11.

6. Dry Depositions of SO_2 and NO_2

6.1. Dry Deposition Velocities

The estimated diurnal mean dry deposition velocities of SO_2 and NO_2 belonging to the clusters that are associated with the maximum and minimum concentrations in each season are shown in Figs. 13 and 14 respectively. The horizontal distributions of the dry deposition velocities of both SO_2 and NO_2 show significant horizontal variations with maxima in the regions of Taegu, Taejon and Kyongin and relatively large values along the coastal region for all clusters. However, the magnitude and the horizontal distribution pattern of the diurnal mean dry deposition velocity do not change significantly with cluster, even though hourly deposition velocity varies with cluster (not shown here). It is worthwhile to note that presently estimated magnitude of the dry deposition velocities of SO_2 and NO_2 range 0.3-1.4 cm s^{-1} all over the peninsula. This value is quite reasonable compared to the estimates of Sehmal (1980). He indicates that the dry deposition velocities of SO_2 and NO_2 vary from 0.04 to 7.5 cm sec^{-1}. The maximum deposition velocities of SO_2 and NO_2 in winter occur for the case of Cluster 2 in which case the minimum concentrations of both SO_2 and NO_2 occur. This implies that the cluster that is associated with maximum pollutant concentration may not coincide with that of maximum deposition velocity.

6.2. Dry Deposition Fluxes of SO_2 and NO_2

Figs. 15 and 16 show horizontal distributions of diurnal mean dry deposition fluxes of SO_2 and NO_2 for the clusters related to the maximum and minimum concentrations of SO_2 and NO_2 in each season respectively. The deposition flux depends on the product of the concentration and deposition velocity. The cluster related to the maximum concentration may not necessarily coincide with the cluster that is related to the maximum deposition velocity. Therefore, the maximum deposition flux may not occur in the cluster that is associated with the maximum concentration. This is the case in winter. The dry deposition fluxes in Cluster 4 in winter are rather smaller than those in Cluster 2 that is related with the minimum SO_2 and NO_2 concentrations.

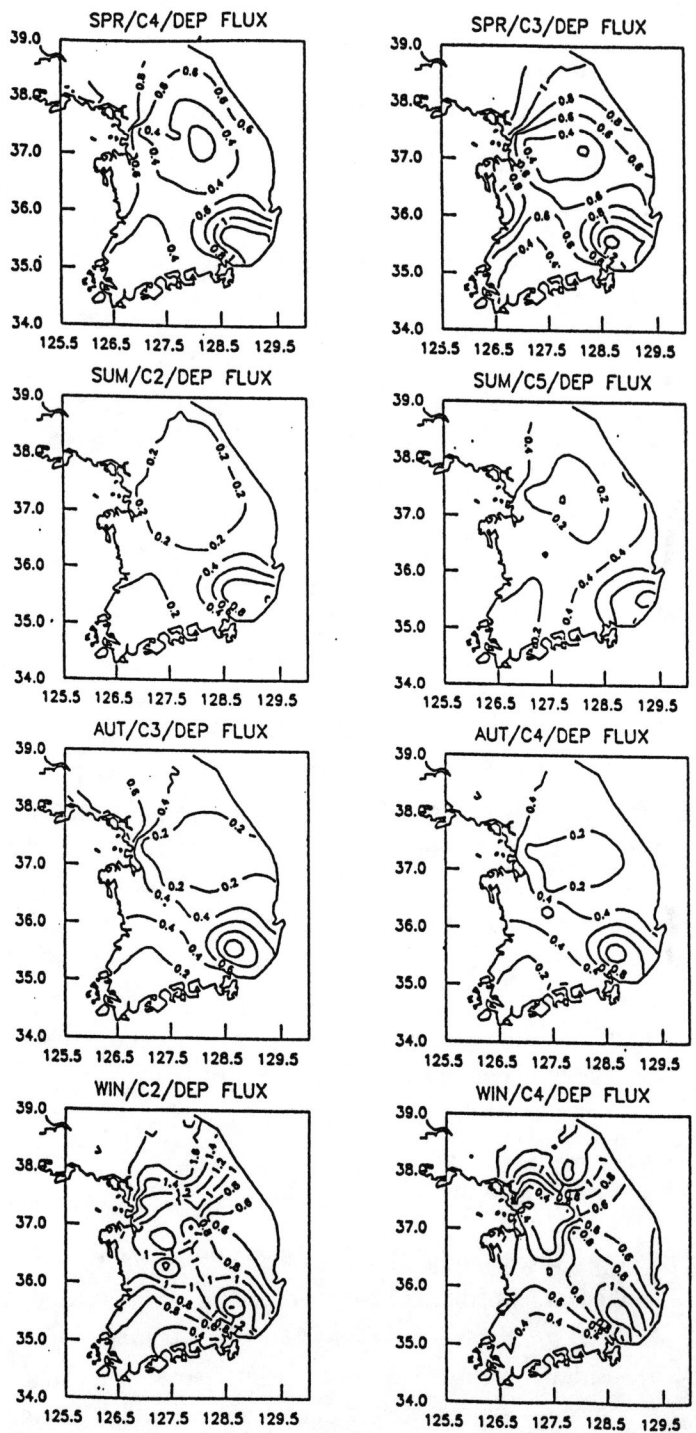

Figure 15. The same as in Figure 11 except for dry deposition flux of SO_2 (µg m^{-2}s^{-1}).

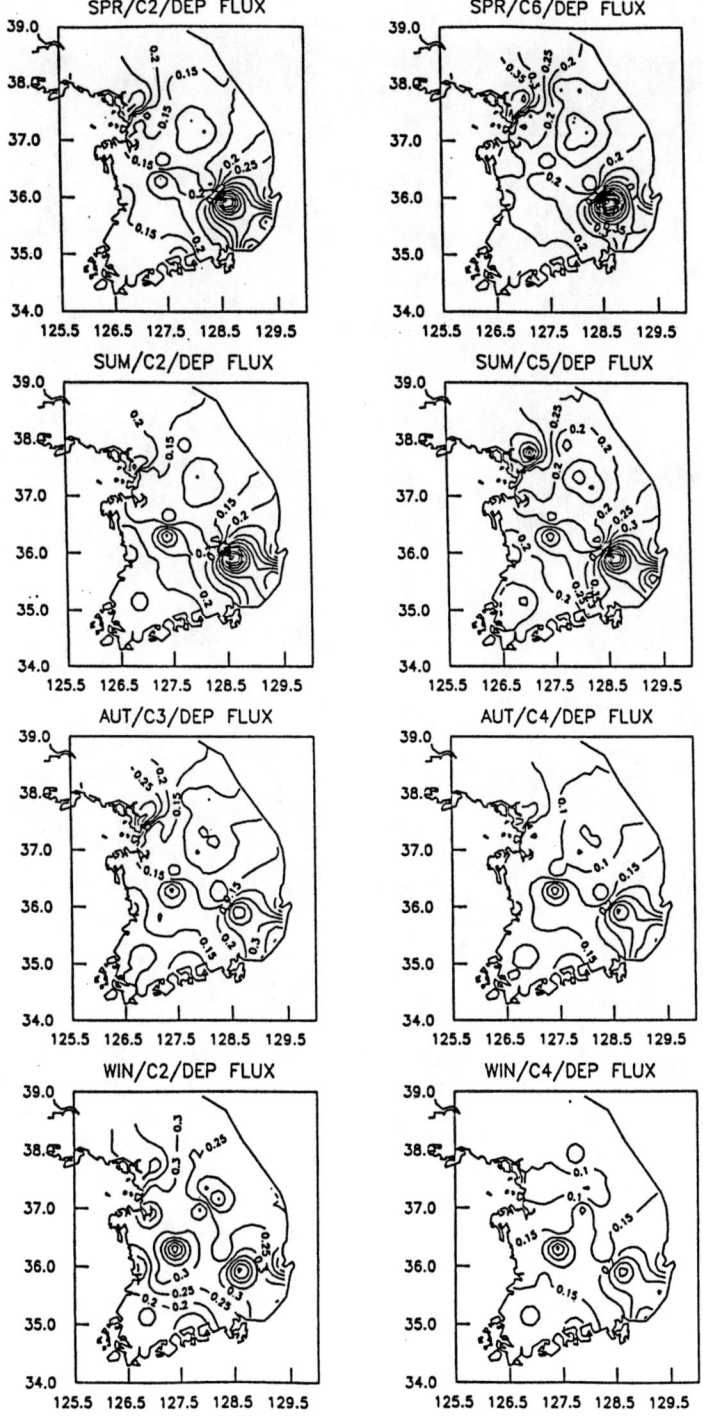

Figure 16. The same as in Figure 11 except for dry deposition flux of NO_2 ($\mu g\ m^{-2} s^{-1}$).

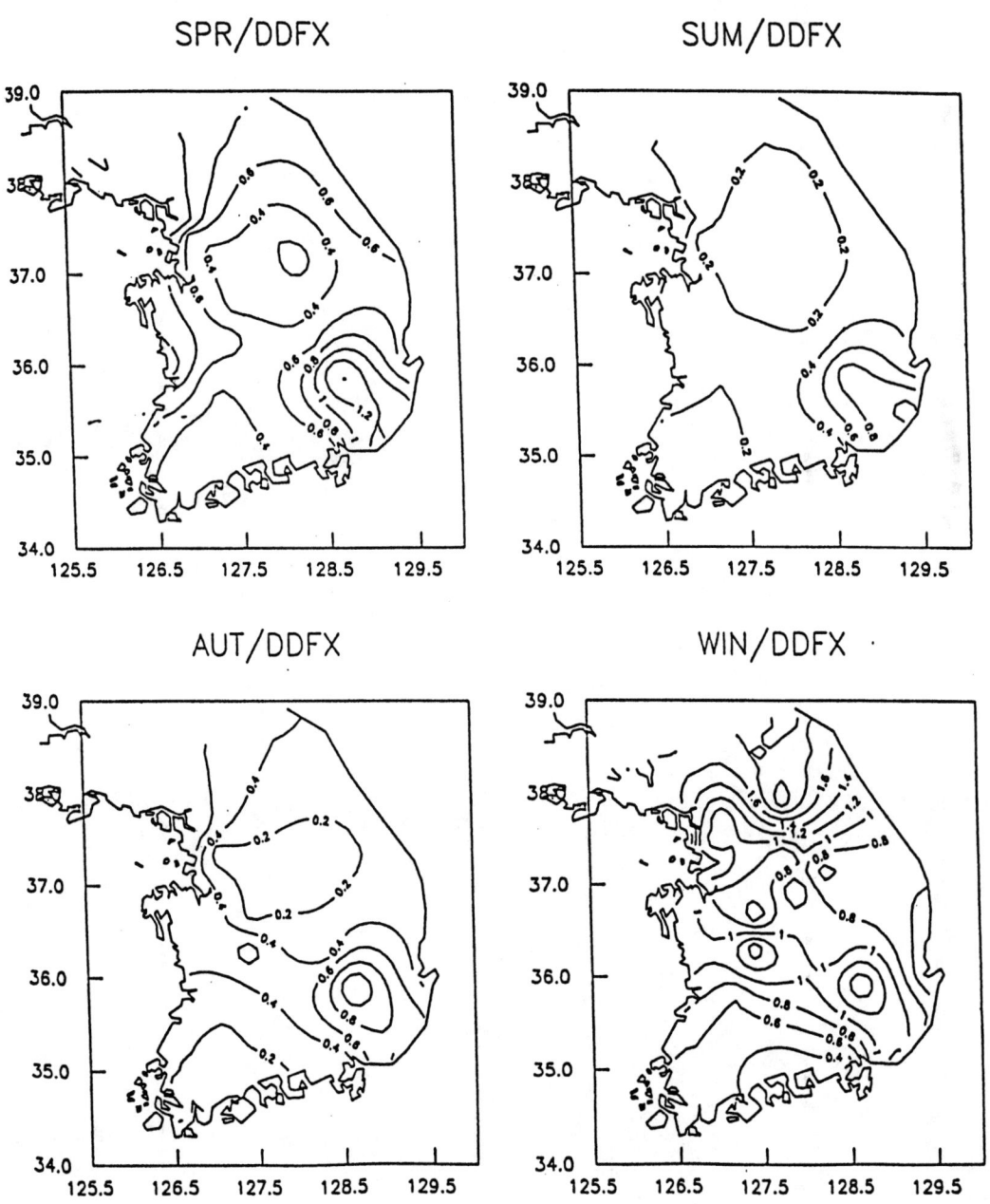

Figure 17. Horizontal distribution of seasonal mean dry deposition flux of SO_2 ($\mu g\ m^{-2} s^{-1}$)

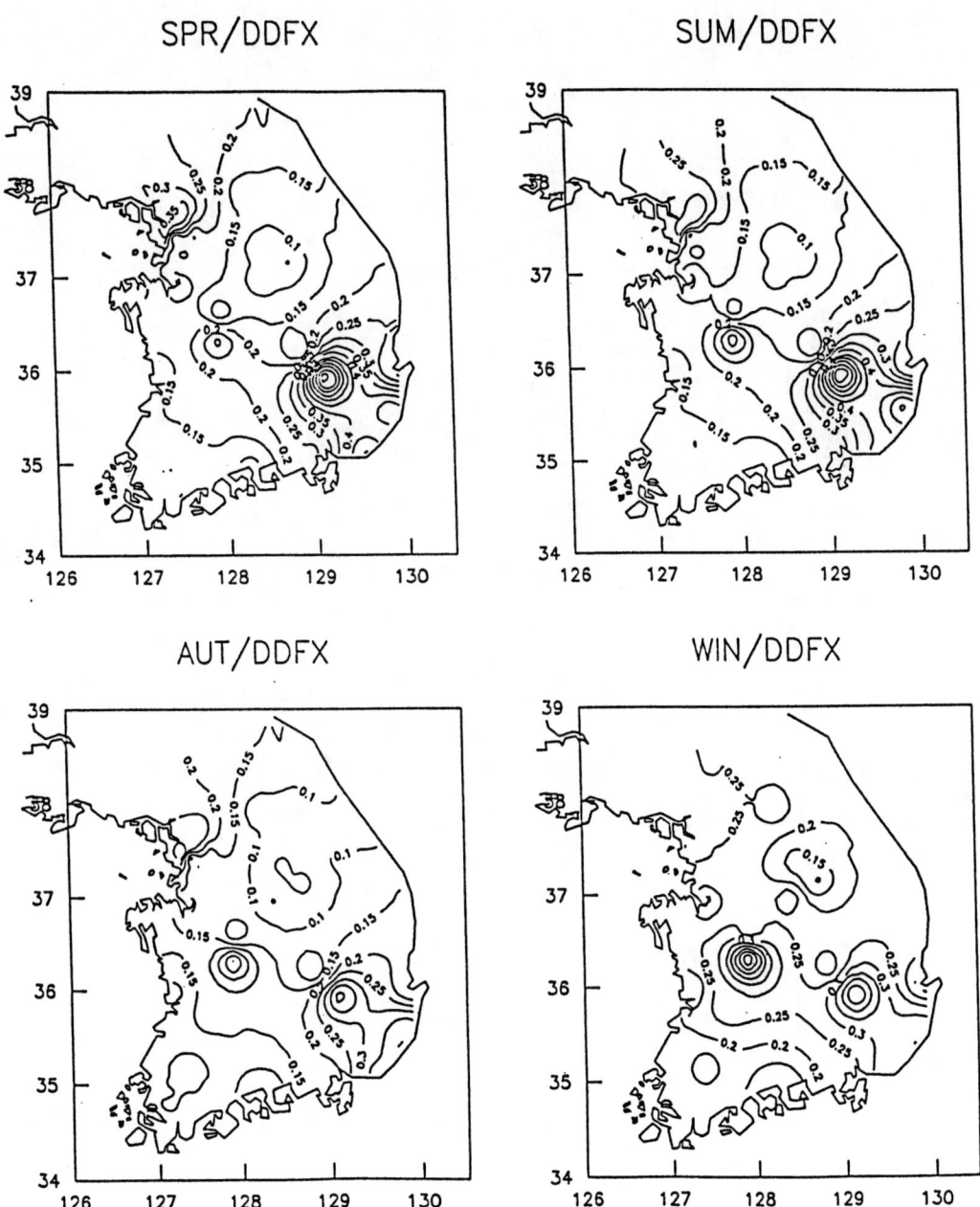

Figure 18. Horizontal distribution of annual mean dry deposition flux of NO_2 ($\mu g\ m^{-2} s^{-1}$).

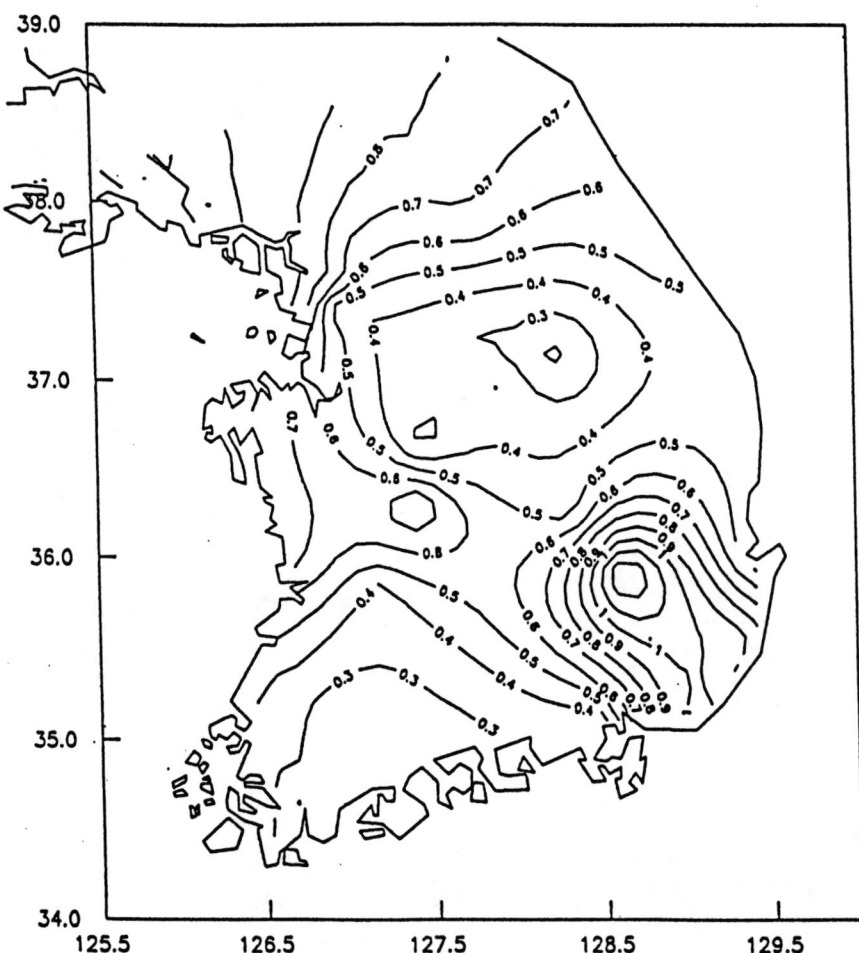

Figure 19. Horizontal distribution of annual mean dry deposition flux of SO_2 (ug $m^{-2}s^{-1}$).

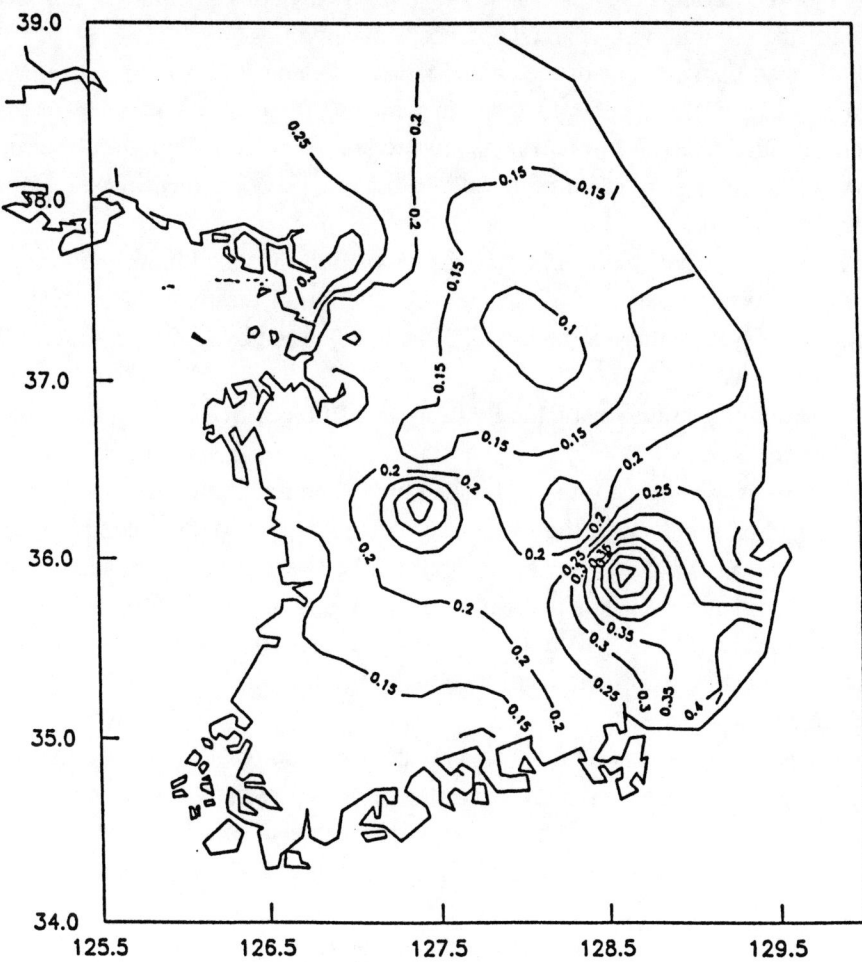

Figure 20. Horizontal distribution of annual mean dry deposition flux of NO_2 (ug $m^{-2}s^{-1}$).

Seasonal distributions of averaged dry deposition fluxes of SO_2 and NO_2 are shown in Figs. 17 and 18 respectively. The highest dry deposition fluxes occur in winter due to the high SO_2 and NO_2 concentrations while the minimum dry deposition fluxes occur in autumn due to the minimum SO_2 and NO_2 deposition velocities.

The horizontal distributions of the annual mean dry deposition fluxes of SO_2 (Fig. 19) and NO_2 (Fig. 20) quite resemble to those in winter (Fig. 16) with maxima in the zone extending from Taejon to Taegu and in Inchon region. However, the Seoul metropolitan area shows a relatively low value of the deposition flux even though the SO_2 and NO_2 concentrations are high.

It is worthwhile to make comparison of the deposition fluxes with the emission rates to understand transport processes of pollutants associated with the regional circulation. The estimated anthropogenic emission rates of SO_2 and NO_2 made by Park and Kim (1995) are used for this purpose.

Fig. 21a shows the annual total emission and deposition of SO_2 per unit area in the $1° \times 1°$ gridded area in Korea while Fig. 21b indicates the ratio of the annual total deposition to the annual total emission of SO_2 in the gridded area. The maximum emission rate of SO_2 occurs in the Inchon region with the value of 65.5 t km^{-2} yr^{-1} and the second maximum of 44.6 t km^{-2} yr^{-1} occurs in the Keochedo region, while the maximum dry deposition flux occurs in the Ulsan and Pusan regions with the maximum value of 11.4 t km^{-2} yr^{-1} and the second maximum of 9.7 t km^{-2} yr^{-1} in the region of Taegu. The dry deposition flux greater than 9 t km^{-2} yr^{-1} occurs in the regions of Inchon, Taegu, Pusan, Ulsan and the northeastern parts of Kyungbuk.

Fig. 22a shows the annual total emission and deposition of NO_2 for the same grid that is used for the SO_2. The maximum emission rate of NO_2 occurs in the same region where the SO_2 emission is maximum with the maximum value of 26.1 t km^{-2} yr^{-1}, while the maximum dry deposition flux occures at Ulsan and Pusan with the maximum value of 4.8 t km^{-2} yr^{-1}, and the second maximum of 3.8 t km^{-2} yr^{-1} at Taegu.

It is interesting to note that most of the eastern parts of Korea except in the regions of big cities and industrial areas including Taegu, Pusan and Ulsan experience net excess deposition of SO_2 and NO_2 compared to the local emission rate. The highest excess deposition of a factor of 2.8 for SO_2 and 3.2 for NO_2 occurs in the northeastern parts of Kyungbuk and the next maximum in the southwestern tip of Korea where the local emission rates are small. Fig. 22b also clearly indicates that high emissions of NO_2 in the western parts of Korea including the Kyungin region are transported eastward due to prevailing westerly winds and contribute to the high deposition flux in the eastern coastal region.

The average total dry deposition flux in South Korea is about 8.09 t km^{-2} yr^{-1} for SO_2 and 2.81 t km^{-2} yr^{-1} for NO_2 which are corresponding to 45% and 40% of the total emission rate respectively over South Korea. The total emission rates of SO_2 and NO_2 are respectively 18.09 t km^{-2} yr^{-1} and 7.09 t km^{-2} yr^{-1} over South Korea.

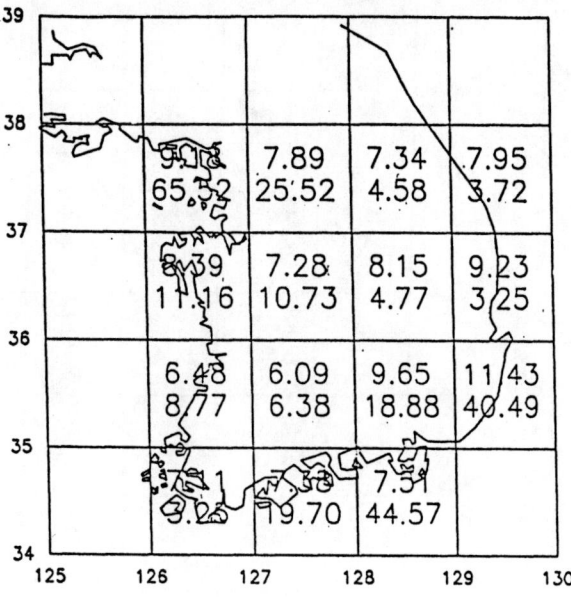

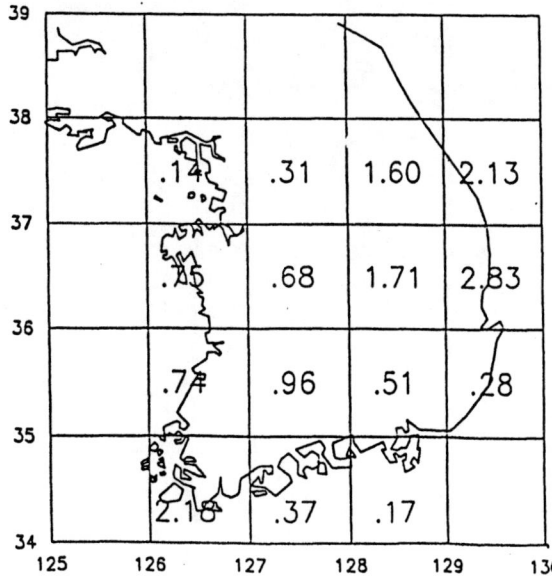

Figure 21. Geographical distributions of mean (a) dry deposition flux (upper number, t km^{-2} yr^{-1}) and emission rate (lower number, t km^{-2} yr^{-1}) of SO$_2$ in a 1°x1° gridded area and (b) the ratio of dry deposition flux to the emission rate.

a. NO2 DRY DEPOSITION VS EMISSION(TON/KM2/YEAR)

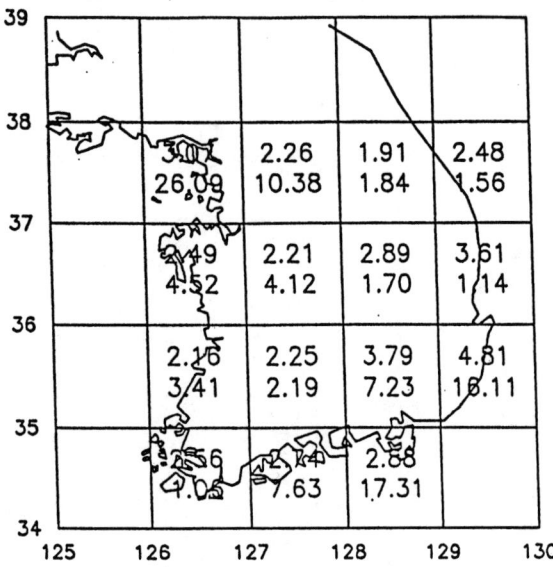

b. NO2 DRY DEPOSITION/EMISSION RATIO

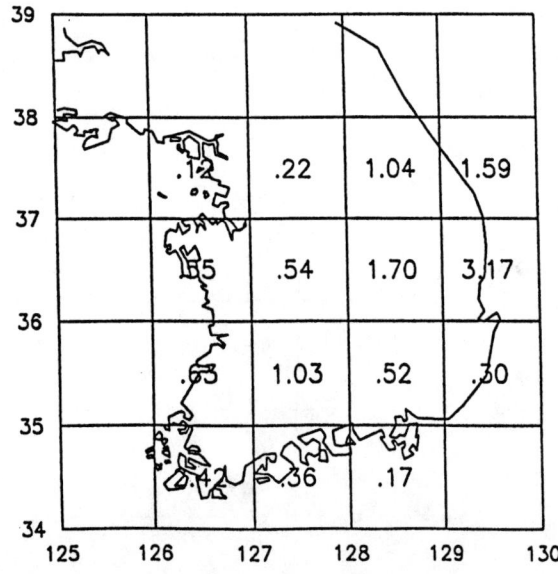

Figure 22. Geographical distributions of mean (a) dry deposition flux (upper number, t km^{-2} yr^{-1}) and emission rate (lower number, t km^{-2} yr^{-1}) of NO$_2$ in a 1°×1° gridded area and (b) the ratio of dry deposition flux to the emission rate.

7. Conclusions

A classification of synoptic meteorological conditions based on the 500 hPa GPV data for the period of five years from 1989 to 1993 and its relationship with the SO_2 and NO_2 concentrations observed at 31 monitoring sites over South Korea has been established using a K-mean clustering technique.

The application of a K-mean cluster algorithm to the 500 hPa geopotential height has allowed to identify 6 synoptic meteorological classes in spring, 5 in summer, and 4 each in autumn and winter. These classified meteorological clusters can lead to identify the SO_2 and NO_2 concentration levels, which imply that the concentration levels can be inferred without any further requirement other than to know the synoptic class for the next day. However, the classified synoptic cluster that is related with the high pollution level is not coincide with the cluster that is associated with the high deposition velocity. This implies that the deposition velocity is mainly controlled by the local scale atmospheric conditions together with the land use types rather than synoptic scale meteorological conditions. Consequently, the maximum dry deposition flux may not occur in the same cluster that is related to the high pollutant level. However, the clustering technique used here has not only the advantage of permitting us to visualize easily the different synoptic classes but also to describe the main features of each one of them in local areas in a quantitative manner so that the estimated dry deposition flux associated with a given cluster can be easily understood by knowing underlying physical processes that give rise to the resulting flux.

The estimated dry deposition fluxes of SO_2 and NO_2 show that about 45% and 40% of the total corresponding emission are deposited on the ground by the dry deposition processes over the Korean peninsula. However, the eastern parts of Korea except for the regions where high emission sources located the dry deposition fluxes exceed the emission rates. This suggest a significant horizontal transport of pollutants from west to east due to prevailing westerly winds and local circulations.

The present clustering technique would meet an important need in the field of acid rain modeling. It would alleviate the heavy costs and computing requirements of the full-scale chemical model of the acid rain.

Acknowledgements

This research was partially supported by the Ministry of Environment and the Ministry of Science and Technology under the grant of the G-7 project.

REFERENCES

Baldocchi, D. D., B. B. Hicks, and P. Camara, 1987: A canopy stomatal resistance model for gaseous deposition to vegetated surface. *Atmos. Environ.*, 21, 91-101.

Chang, J. S., and Co authors, 1990 : The regional acid deposition model and engineering model. *NAPP Report 4*, 13088.

Davis, R. E., and D. A. Gay, 1993: A synoptic climatological analysis of air quality in the Grand Canyon National Park. Atmos. *Environ.*, 27A (5), 713-727.

Davison, J. I., and Y.-L. Wu, 1990: *Acidic precipitation III: Sources, Deposition, and Canopy Interactions*. Springer-Verlag, New York, 103-216.

Engstrom, A., 1972: *The impact on the environment of sulphur in air and precipitation*. Swedish Roy. Ministry of Foreign Affairs and Roy. Ministry of Agriculture. Stockholm.

Garratt, J. P., 1992 : *The atmospheric boundary layer*. Cambridge Univ. Press, 316pp.

Hicks, B. B., and D. Matt, 1986: Combining biology, chemistry and meteorology in modeling and measuring dry deposition. *2nd Inter. Symposium on Biosphere Atmosphere Exchange*. Mainz, Germany.

Hicks, B. B., R. P. Hosker, Jr, T. P. Meyers, and J. D. Womack, 1991: Dry deposition inferential measurement techniques I. Design and Tests of a prototype meteorological and chemical system for determining dry deposition. *Atmos. Environ.*, 12A, 2345-2359.

Jung, Y.-S., S.-U. Park and I.-H. Yoon, 1996 : Characteristic features of local air quality associated with meteorological conditions. *J. Kor. Meteor. Soc.*, 32 (2), 271-290.

Likens, G. E., and F. H. Bormann, 1974: Acid rain: a serious regional environmental problem. *Science*, 184, 1176-1179.

Lindberg, S. E., and G. M. Lovett, 1985: Field measurements of particle dry deposition rates to foliage and inert surfaces in a forest canopy. *Environ. Sci. Tech.*, 19, 238-244.

Mihailovic, D. T., and B. Rajikovic, 1992: Surface vegetation parameterization in atmospheric model : A numerical study. *Meteor. Z.*, 41, 29-33.

Ncar, 1985 : The NCAR Eulerian regional acid deposition model. *ADMP-85-3, NCAR/TN-256+STR*. National Center for Atmospheric Research, Boulder, Co. 178pp.

Noilhan, J. M., and S. Planton, 1989: A simple parameterization of land surface processes for meteorological models. *Mon. Wea. Rev.*, 117, 536-549.

Oden, S., 1968: *The acidification of air and precipitations and its consequences on the natural environment*. Swedish Nat. Sci. Res. Council Ecol. Committee Bull., 1, 1-86.

Park, S.-U., and I.-H. Yoon, 1991: The characteristic features of local weather phenomena under the various synoptic conditions over South Korea. *J. Kor. Meteor. Soc.*, 27 (2), 87-118.

Park, S.-U., 1994: The effect of surface physical conditions on the growth of the atmospheric boundary layer. *J. Kor. Meteor. Soc.*, 30, 119-134.

Park, S.-U., and C.-H. Kim, 1995: Interannual trend of the anthropogenic emissions of SO_2 and NO_2 in South Korea. *J. Kor. Meteor. Soc.*, 31 (1), 65-78.

Park, S.-U., 1995: The effect of dry deposition on the ground-level concentration. *J. Kor. Meteor. Soc.*, 31 (2), 97-115.

Sanchez, M. L, D. Pascual, C. Ramos, and I. Perez, 1990: Forecasting particulate pollutant concentrations in acidity from meteorological variables and regional weather patterns. *Atmos. Environ.*, 24A (6), 1509-1519.

Sehmal, G. A., 1980: Particle and gas dry deposition: A Review. *Atmos. Environ.*, 14, 983-1011.

Wesely, M. L., 1989: Parameterization of surface resistances to gaseous dry deposition in regional scale numerical models. *Atmos. Environ.*, 23 (6), 1293-1304.

Wesely, M. L., and B. B. Hicks, 1977: Some factors that affect the dispersion rates of sulfur dioxide and similar gases on vegetation. *J. Air Pollu. Control. Assoc.*, 27, 1110-1116.

Part 2
MONITORING OF PRECIPITATION IN EAST ASIA

Chapter 3
ACID RAIN MONITORING IN EAST AND SOUTH-EAST ASIA

MIROSLAV RADOJEVIC
Department of Chemistry, University of Brunei Darussalam, Bandar Seri Begawan
Brunei Darussalam, miro@ubd.edu.bn

CONTENT

Abstract
1. Introduction
2. Experimental results of acid rain monitoring
    2.1 Japan
    2.2 China
    2.3 Vietnam
    2.4 Indonesia
    2.5 Philippines
    2.6 Singapore
    2.7 Malaysia
    2.8. Brunei Darussalam
    2.9. Thailand
    2.10. Hong Kong
    2.11. South Korea
    2.12. Cambodia
    2.13. Laos
    2.14. Myanmar
    2.14. Taiwan
    2.16. An acid rain monitoring network for East Asia
3. Conclusions
References

ABSTRACT

The experimental monitoring data has been collected for all the countries of East Asia. It was shown that acid rain is observed regularly at many sites in East Asia, especially in China, South Korea, Taiwan, Hong Kong and Japan where the numbers of rain events with pH less than 5.6 are in the limits of 50-70%. Lower pHs tend to be observed in industrialized countries and regions while higher pH values tend to be observed in areas with little or no industries. Among Asian countries Japan has the most sophisticated and longest running precipitation chemistry and air quality monitoring program in East Asia. With the exception of Japan, other countries in East and South-East Asia have started monitoring acidic precipitation and air quality relatively recently. Monitoring programs vary from being relatively comprehensive (e.g. South Korea, Hong Kong, China) to rudimentary (e.g. Indonesia, Vietnam). Some countries have yet to set up air quality and acid rain monitoring programs of any kind (e.g. Myanmar, Laos, Cambodia). The quality of data produced by many of the monitoring networks in the region, with the possible exception of Japan, leaves much to be desired. There is an absence of quality control procedures in many of the networks. The proposed international acid rain survey in East Asia would help by standardizing methods and procedures. From the measurements available it should be concluded that acid rain is the major problem in China, South Korea, Hong Kong, Taiwan, and Japan. In less industrially developed countries of South-East Asia such as Laos, Cambodia, Myanmar, most rainwater pH measurements tend to be around 5.6, the pH of "natural" rainwater. Instances of pH <5 are encountered at some sites, and these are mainly due to localized industrial pollution. There is some evidence that pH values below 5 at unpolluted sites may be due to the contribution of weak organic acids, such as formic and acetic acids.

1. INTRODUCTION

Acid rain and its harmful effects have been extensively studied in the industrialized countries of Europe and North America over the last thirty years (Radojevic and Harrison, 1992). Acid rain was first reported in Europe in the last century (Ducros,1854). Concerns about the multifarious and far-reaching effects of acid rain have led to the setting up of detailed rainwater monitoring networks since the late 1960s/early 1970s in industrialized countries. These networks are producing a long-term record of rainfall acidity and composition, which could be useful in assessing the effects of control measures and policies. Industrialized countries have also enacted legislation relating to the emissions of acidic precursors and have implemented measures to control their emissions.

Table 1. Composition of rainwater in East and South-East Asia. NSS: non-sea salt. Where single values are given these represent means.

Site	Country	Reference	Sampling period	pH	EC µS/cm	SO_4^{2-} mg/L	NO_3^- mg/L	Cl^- mg/L	NH_4^+ mg/L	Na^+ mg/L	K^+ mg/L	Ca^{2+} mg/L	Mg^{2+} mg/L
Beijing	China	Zhao & Sun(1986)	1981	6.8	-	13.1	3.11	5.58	2.54	3.24	1.57	3.69	-
Beijing (urban)	China	Zhao et al (1988)	1982	6.74	-	16.2	5.02	2.10	4.04	1.78	1.49	15.2	-
Beijing (suburban)	China	Zhao et al (1988)	1982	6.54	-	7.8	2.10	1.39	2.88	1.00	0.82	9.22	-
Beijing	China	Galloway et al (1987)	1984	6.22	-	4.39	1.74	-	2.11	-	-	2.06	-
Beijing	China	Wang & Wang (1996)	1986	6.14	-	7.42	1.22	-	2.93	-	-	3.04	-
Beijing	China	Wang & Wang (1996)	1993	5.98	-	7.29	3.09	-	2.04	-	-	2.87	-
Chongqing	China	Zhao & Sun(1986)	1982	4.14	-	13.6	1.33	0.54	1.46	0.39	0.58	2.02	-
Chongqing (rural)	China	Zhao et al (1988)	1982/84	4.44	-	7.92	1.12	0.85	1.15	1.04	0.91	0.84	0.22
Chongqing (urban)	China	Zhao et al (1988)	1982/1984	4.14	-	14.7	1.96	0.53	1.91	1.18	0.29	2.21	0.59
Guiyang	China	Zhao & Sun(1986)	1982	4.02	-	16.56	0.59	0.32	1.15	0.22	0.37	2.99	-
Guiyang (urban)	China	Zhao et al (1988)	1982/84	4.07	-	19.7	1.30	0.29	1.42	0.23	1.03	4.64	0.68
Guiyang (urban)	China	Zhao et al (1988)	1982/84	4.58	-	8.0	0.99	0.75	0.91	0.14	0.27	3.52	0.71
Guiyang (suburban)	China	Zhao et al (1988)	1982/84	4.42	-	13.5	1.57	0.42	0.89	0.26	0.41	3.97	0.54
Guiyang	China	Galloway et al (1987)	1984	4.02	-	21.5	0.62	-	1.09	-	-	4.97	-

Tianjin (urban)	China	Zhao & Sun(1986)	1981	6.26	-	15.2	1.81	6.49	2.26	4.03	2.3	5.75	-
Yibin (urban)	China	Zhao et al (1988)	1982/84	4.87	-	5.37	0.91	0.88	1.28	1.10	1.14	0.07	-
Shisun	China	Galloway et al (1987)	1984	4.73	-	3.69	0.48	-	0.34	-	-	1.25	-
Kaiyang	China	Galloway et al (1987)	1984	4.76	-	7.06	0.94	-	0.97	-	-	1.50	-
Luizhang	China	Galloway et al (1987)	1984	4.91	-	5.71	0.45	-	0.39	-	-	1.99	-
Danum Valley, Sabah	Malaysia	Burghouts et al (1993)	October 1989-January 1990	-	5.3	0.342	0.307	0.427	0.085	0.246	0.160	0.093	0.033
Gunung Silam, Sabah	Malaysia	Bruijnzeel et al (1993)	1987	5.9	14	1.5	1.4	1.5	0.20	1.0	0.30	0.19	0.04
Gunung Silam, Sabah	Malaysia	Bruijnzeel et al (1993)	1989	5.4	4	<0.5	0.4	<0.5	0.08	<0.1	<0.05	0.12	0.03
Pasoh Forest Reserve, Negri Sembilan	Malaysia	Manokaran (1980)	1973	5.5	-	-	0.75	-	0.13	0.98	0.29	0.18	0.03
Serpong	Indonesia	Saefudin (1997)	June-December 1996	5.8	25	2.9	2.5	1.9	-	-	-	-	-
Watubelah, Java	Indonesia	Bruijnzeel (1989)	1976-1983	6.4	-	0.93	0.27	0.87	0.18	0.28	0.21	0.21	0.09
Saigon (Ho Chi Min City)	Vietnam	Vialard-Goudou & Richard (1956)	1950-1954	5.9	-	-	0.49	0.31	0.48	-	-	-	-
Nam Pung Dam	Thailand	Granat et al (1996)	1991-1995	5.0	-	0.53	0.43	0.18	0.22	0.07	0.08	0.14	0.02
Srinakarin Dam	Thailand	Granat et al	1991-1995	5.2	-	0.62	0.25	0.21	0.16	0.12	0.07	0.18	0.02

Sob pad Village	Thailand	(1996)	1996	3.3-8	2.9-332.1	0.7-19.6	0.9-8.4	-	-	-	-	-
Ta Si Village	Thailand	Siriswasdi (1997)	1996	3.6-7	1.3-82.3	0.7-5.1	0.4-3.1	-	-	-	-	-
Bangkok	Thailand	Siriswasdi (1997)	1996	4.4-7.9	4.2-73.7	0.8-11.4	0.3-3	-	-	-	-	-
Pi-Lu-Chi	Taiwan	King & Yang (1984)	January 1981-December 1982	6.0	-	0.78	1.06	0.10	0.18	0.34	0.28	0.68	0.15
Bandar Seri Begawan	Brunei Darussalam	Radojevic & Lim (1995)	26 April-3 November 1994	4.85	10.9	0.86	0.72	0.78	-	0.34	0.14	0.44	0.08
average at 29 sites	Japan	Murano (1997b)	1986-1988	4.7	-	2.64	0.96	3.82	0.39	1.97	0.18	0.52	0.26
average at 29 sites	Japan	Hara (1997)	1989-1993	4.76	-	1.85 (NSS)	0.87	-	0.33	-	-	0.28	-
average at 48 sites	Japan	Iijima (1997)	1989-1993	4.8	1.85 (NSS)	-	0.87	--	0.33	-	-	0.28 (NSS)	-

Over the last ten years, the presence of acid rain has been identified at sites in Asia, Africa, Australia, and South America, and the phenomenon of acid rain is increasingly being recognized as a major global problem. Acid rain has even been reported at sites far removed from industrial sources, such as the Arctic ice caps and tropical rainforests. Recent industrial development in East and South-East Asia have raised concerns about potential acidification in the region (Rodhe et al, 1992). In this paper, studies of acid rain and related issues in East and South-East Asia are reviewed. Results of selected rainwater chemistry studies are summarized in Table 1. Studies of rainwater chemistry in eastern Russia are not discussed here; these have been reviewed elsewhere (Ryaboshapko et al, 1994).

2. EXPERIMENTAL RESULTS OF ACID RAIN MONITORING

2.1. JAPAN

Of all the countries in East Asia Japan has the longest tradition of environmental research and environmental awareness. Studies of rainwater composition were reported as long ago as 1894 (Kellner et al, 1894). Annual mean rainwater pH values were reported to be 4.1 in Tokyo, 5.2 in Kobe and 5.6 in Hamamatsu in the period 1936-1937 (Miyake, 1939). Several instances of rainwater pH values less than 3 were observed in the 1970s (Hashimoto, 1989; Hara, 1993). The incidence of acid rain seems to have declined in Japan due to the control of SO_2 and NO_x pollution (Hashimoto, 1989). Today there are numerous rainwater surveys operating in Japan, and these are carried out by government agencies, universities, and even concerned citizen groups (Hara, 1993). The World Meteorological Orginisation (WMO) operates a station in Ryori, Iwate, as part of the Background Air Pollution Monitoring Network (BAPMoN), and precipitation chemistry measurements were initiated at this station in 1976. Another WMO station in Minami Torishima started measuring precipitation chemistry in 1994.

The Japan Environment Agency has been investigating acid rain since the early 1970s, and in 1983 it established the National Acid Deposition Monitoring network. The aim of this network is to accurately assess the state of acid deposition in Japan, its spatial distribution, long-term trends, and long range transport (Iijima, 1997). The network comprises 48 urban, rural and remote stations throughout Japan. Wet-only deposition is collected using automatic samplers on a daily, weekly and forthnightly basis. Samples are refrigerated on-site and sent to the laboratory every two weeks for analysis of pH, conductivity and major ions (SO_4^{2-}, NO_3^-, Cl^-, NH_4^+, Na^+, K^+, Ca^{2+}, and Mg^{2+}). Three five year surveys have been conducted so far; 1983-1988, 1988-1993, and 1993-1998. The range of pH values for all stations was 4.5 to 5.8 for the period 1988-1993, and this is similar to pH values observed in Europe and North America.

A survey into the effects of acidic precipitation on inland waters and terrestrial eco-systems has also been initiated by JEA. Thirty-three lakes hav been surveyed for pH, major ions, and other indicators of water quality. Soil analysis is carried out at 88 sites every 3 years. Visual investigation of vegetation in the surrounding area is carried out to look for any evidence of damage or decline. The pH of all lakes, with a few exceptions, was found to be in the range of 7 to 8. So far there is no clear evidence of adverse effects on aquatic or terrestrial ecosystems. There is some evidence of tree damage, but in most cases this was attributed to causes other than acid rain: weather, insect damage, magnesium deficiency, ozone, and water deficiency. In a separate study, a 17 year record of the pH or rivers and lakes in a mountainous region of central Japan was analyzed (Kurita et al, 1991). Decreases in pH of between 0.2 to 0.3 pH units were noted in some lakes and rivers with a high acid neutralizing capacity. In some lakes and rivers with a low acid neutralizing capacity the pH decreased by as much as 0.6 pH units over a 10 year period. However, it is suspected that factors other than acid precipitation are responsible for the observed acidification.

Japan has an extensive air quality monitoring network.The concentration of acid rain precursor gases, SO_2 and NO_x, and other air quality parameters (CO, O_3, particulates) is monitored at more than 1,000 stations (Yanagisawa, 1989). In the area of air pollution control Japan has no rival; it has been a world leader in air pollution control technologies for quite some time, and it has the most stringent environmental legislation in the world. Japan started introducing SO_2 and NO_x control technologies at its power plants and other industries as far back as the early 1970s (Radojevic, 1996,1998). More than 1,000 power plants are equipped with flue gas desulfurization scrubbers, and all power plants are required to employ NO_x reduction technologies. The massive program of desulfurization and NO_x control at industrial plants initiated in the late 1960s and early 1970s, switches to low-sulfur fuels, and other air pollution control policies resulted in a significant decrease in mean annual SO_2 concentrations, from around 60 ppbv in 1967 to about 15 ppbv in 1981 (Hashimoto, 1989). Furthermore, Japan was the first to mass produce low emission motor vehicles, and it introduced strict vehicle emission standards well before the US. For example, the Japanese standard for NO_x was 0.4 g/mile in 1978, while the Californian and federal US standards in that year were 1.5 and 2.0 g/mile respectively (Nishimura and Sadakata, 1989). California only introduced the 0.4 g/mile standard in 1983, five years after Japan. Although the NO_x emissions per vehicle have been declining the number of vehicles on Japanese roads has been growing steadily throughout the 1970s and 1980s. The average annual NO_2 concentration has shown almost no change between 1973 and 1990; it has remained between 25 and 30 ppb throughout this period (Matsushita, 1996).

2.2. CHINA

After Japan, the most extensive chemistry data set in East and South-East Asia is for China. The Institute of Environmental Chemistry initiated a survey of precipitation chemistry in some cities in the late 1970s, and nationwide surveys have been reported since 1982. The results of these studies show that acid rain occurs in many parts of China, especially in the southwest (Zhao and Sun, 1986; Galloway *et al*, 1987; Zhao and Xiong, 1988, Narita *et al*, 1997).

The main cause of acid rain in China is the industrial and domestic combustion of coal. Many of the furnaces have short chimney stacks and have minimal controls. The content of sulfur varies between different regions. In Beijing the sulfur content of coal is circa 1%, whereas in Guizhou province in the south it is between 3 and 5 %.

A comparison of the survey results with precipitation chemistry results from the US and a remote area in Australia showed that the rainwater concentrations and wet deposition rates of SO_4^{2-} in China were generally higher than in the US and this was attributed to the high sulfur content of coal and absence of controls (Galloway et al, 1987). The same applied to Ca^{2+} and NH_4^+ concentrations in rainwater and their deposition rates; they were higher in China than in the US. The high content of Ca^{2+} was ascribed to calcareous soils, uncontrolled emissions of dust particles from furnaces, and the extensive use of calcareous building materials. The high NH_4^+ content was attributed to the widespread agricultural use of excretory wastes and the release of NH_3 from high pH soils in northern areas. However, for NO_3^-, concentrations and deposition rates were lower in China than in the US, due to the lower number of motor vehicles. Of the Chinese cities investigated, Beijing had the highest NO_3^- concentrations and deposition rates because of the greatest density of motor vehicles. Cities in the south had higher SO_4^{2-} concentrations and deposition rates than Beijing and this was explained in terms of the lower sulfur content of coal in Beijing, although neutralization by alkaline particles and NH_3 emitted from the alkaline soils in the north may also play a role (Zhao and Sun, 1986). A detailed field and laboratory study of the effect of wind blown dust on rainwater chemistry in Northeast China concluded that dust particles containing $CaCO_3$ are carried by air masses from Siberia/Mongolia and Northwest China and dissolve in rainwater leading to acid neutralization over Northeast China (Zhang et al, 1993). A similar conclusion was reached in a modeling study by Mo et al (1988). They used a chemical equilibrium model that included concentrations of the major precursor gases. Including $CaCO_3$ in the aerosol resulted in significantly increased precipitation pH in their model output.

Wang and Wang (1996) compared precipitation chemistry from north and south areas of China for the years 1986 and 1993 and concluded that acid precipitation is observed in the south but not in the north. Zhao and Xiong (1988) and Zhao et al (1988) have also surveyed the precipitation chemistry in several cities and concluded that about 90% of sampling sites with an average pH<5.6 are situated south of the Yangtze River.

Damage to metal, concrete, trees and crops has been reported in areas seriously affected by acid rain, however, it is not known to what extent acid rain contributed to the observed effects (Zhao and Sun, 1986; Zhao and Xiong, 1988).

China is currently undertaking a massive pollution control program that will involve the fitting of desulfurization equipment to industrial plants. Also, the number of air quality monitoring stations in urban areas is growing.

2.3. VIETNAM

Vietnam has been experiencing rapid industrial development in recent years. In response to the concerns over the potential environmental effects of this industrialization, the National Environment Agency (NEA) was established in 1993. One of the activities of NEA is the setting up and running of air quality and acid rain monitoring stations. The NEA operates several environmental monitoring stations in conjunction with other organizations, including a background air quality station located at the Cuc Phuong national reserve forest and a rainwater monitoring station located in Lao Cai province, near to the border with China. The rainwater monitoring stations has been in operation since 1996 monitoring the following parameters: pH, electrical conductivity, SO_4^{2-}, NO_2^-, NO_3^-, Cl^-, NH_4^+, Na^+, K^+, Ca^{2+}, Mg^{2+}, and PO_4^{3-}. In addition, the Hydrometeorological Service operates 22 air monitoring stations as part of a larger network of environmental monitoring stations including river water quality monitoring stations (Anh, 1996).

In 1990 the Hydrometeorological service conducted a survey of rainfall acidity at seven sites in the north of the country comprising of three in-land sites, three coastal sites, and one background site (Khahn, 1993). Rainwater was collected using a wet-only collector, and samples were analyzed using a color indicator method. The survey yielded the following results:

- The pH of rainwater was generally between 5 and 7.
- The pH values tend to be lower in winter, during the dry season with little rainfall, and higher in the summer, during the wet season when there is a large amount of rainfall, *at all* stations.
- Lowest pH values (<4 to 5) were observed at a site located in an industrialized area.

Acid rains were also recently reported in Ho Chi Minh city (Thuc and Yen, 1997). In conclusion, acid rain does not seem to be a regional problem in Vietnam, other than in industrialized areas.

Air quality monitoring was initiated in 1995 at several sites throughout Vietnam. In Hanoi SO_2 concentrations varied between 0.001 and 0.04 (mg/m^3) (Lan, 1996). Higher concentrations were observed close to industrial sources.

2.4. INDONESIA

Indonesia, the most populous country in S.E. Asia, has experienced considerable industrial development over the last twenty years. Emissions of acidic precursors have doubled between 1975 and 1987, as shown in Table 2.

Table 2 Emissions of SO_2 and NO_x in Indonesia (adapted from Kato and Akimoto, 1992).

Year	SO_2 (Tg/year)	NO_x (Tg/year)
1975	0.20	0.33
1987	0.49	0.64

An automatic air quality monitoring station was set up at the Environmental Management Center (EMC), Serpong, Tangerang, about 40 km from Jakarta, in August 1993. This station monitors SO_2, NO_x, CO, non-methane HC, total-HC, O_3, PM_{10} and meteorological data. Bulk precipitation sampling was started in June 1996, and an automatic acid rain analyzer was installed in November 1996. This measures: rainfall amount, temperature, pH, conductivity, SO_4^{2-}, NO_3^- and Cl^- automatically. Operational problems were encountered with the automatic analyzer and only data from the manual collector were available. The results obtained in 1996 are given in Table 3.

Table 3. Analytical results of bulk precipitation samples collected at Serpong, Tangerang, Indonesia in 1996 (Adapted from Saefudin, 1997)

Month	rainfall (mm)	pH	Conductivity (μS/cm)	SO_4^{2-} (mg/L)	NO_3^- (mg/L)	Cl^- (mg/L)
June	29	6.4	43	2.5	2.2	1.9
July	12	5.9	16	5.3	2.8	1.7
August	41	6.2	20	3.3	3.1	1.3
September	81	5.3	23	3.3	2.9	3.5
October	98	5.0	26	3.6	3.4	1.8
November	4	6.1	32	1.9	2.6	1.9
December	146	5.8	13	0.4	0.2	1.4
Average	59	5.8	25	2.9	2.5	1.9

In a separate study, wet-only samples were collected at four sites in West Java (Jakarta, Serang, Cilegon, and Merak) over a 12 month period from June 1991 to June 1993, however, the samples were analyzed only for SO_4^{2-} and NO_3^- (Ayers et al, 1995). Various problems were encountered in this study including incomplete sample collection *at all* sites, lack of field laboratory facilities, lack of appropriately trained staff, etc. Although it is claimed that the pH was measured both on-site and in the laboratory, no pH values are reported. The authors conclude that these deposition rates of SO_4^{2-} and NO_3^- are significantly higher than natural fluxes, indicating significant anthropogenic acidification.

Some earlier data on precipitation pH determined in Jakarta at a BAPMoN station has been reported by Ayers (1991). Samples were collected from January 1981 to July 1984 on a monthly basis and from December 1984 to October 1987 on a weekly basis. Volume weighted mean pH values were 4.79 for the monthly data and 5.33 for the weekly data.

Very few surveys of air pollution at other sites in Indonesia have been reported, due mainly to a lack of adequate instrumentation. The only nationwide surveys reported so far were based on antiquated methods such as the PbO_2 candle method for SO_2 and the triethanolamine (TEA) plate method. These methods give monthly averages and are subject to considerable errors compared to modern instrumental techniques. They do not give short-term averages and peak concentrations which are more important from a public health point of view. Nevertheless, in the absence of other measurements they give some idea on the distribution of acidic precursors throughout Indonesia. Since April 1994, the SO_2 concentration was determined in 24 cities and the concentration of NO_2 was determined in 23 cities throughout Indonesia using these methods. The results of these studies are summarized in Table 4. Jakarta was the most polluted city, having highest concentrations of both SO_2 and NO_2.

Table 4. Concentrations of SO_2 and NO_2 in air in Indonesia measured using the PbO_2 candle (Rinda,1996) and TEA plate method (Rachmawati, 1996) respectively. Results are in ppbv calculated by converting from original data in mass/100 cm^2/day using appropriate conversion factors. n=number of samples for SO_2 determination. The detection limit for SO_2 is 0.376 ppb.

City	Island	n	SO_2(ppbv) Range	SO_2 (ppbv) Area average	NO_2 (ppbv) Area average
Jakarta	Java	12	4.7-9.4	6.58	39.0
Ciputat	Java	3	0.94-1.88	0.94	5.42
Serpong	Java	10	0.94-1.88	0.94	6.95
Bandung	Java	23	<0.376-4.7	2.82	6.44
Tangerang	Java	14	<0.376-6.58	2.82	-
Yogyakarta	Java	16	<0.376-4.7	0.94	4.07
Semarang	Java	2	<0.376-0.94	0.94	4.58
Surabaya	Java	14	1.88-8.46	5.64	8.14
Palembang	Sumatra	26	<0.376-4.7	1.88	6.95
Bengkulu	Sumatra	9	<0.376-1.88	0.94	4.75
Lampung	Sumatra	9	0.94-4.7	2.82	4.07
Jambi	Sumatra	9	<0.376-4.7	0.94	0.68
Medan	Sumatra	3	0.94-1.88	0.94	8.81
Riau	Sumatra	-	-	-	4.92
Pontianak	Kalimantan	3	0.94-2.82	1.88	5.59
Samarinda	Kalimantan	6	<0.376-2.82	1.88	5.59
Bunjarmasin	Kalimantan	20	0.94-2.82	1.88	7.12
Ujung Pandang	Sulawesi	10	<0.376-2.82	0.94	5.76
Palu	Sulawesi	14	<0.376-1.88	0.94	1.02
Kendari	Sulawesi	7	<0.376-1.88	0.376	-
Mataram	Lombok	10	<0.376-2.82	0.658	0.85
Kupang	Kupang	9	<0.376-3.76	0.94	0.51
Jayapura	Irian Jaya	3	<0.376-2.82	0.94	-
Tembagapura	Irian Jaya	7	0.94-3.76	2.82	0.85

In a separate study SO_2 and NO_2 concentrations were determined in Jakarta and several other cities using a filter pack method (Ayers et al, 1995). The concentration of SO_2 in Jakarta was 7.1 ppbv in the period 1st September 1994 to 2nd January 1995. The concentration of NO_2 in Jakarta was 28.18 ppbv in the same period and 18.86 from June 1991 to July 1992. Concentrations of NO_2 were 11.51, 1.81, and 4.9 ppbv in Cilegon, Serang and Merak respectively. The concentration of HNO_3 gas in Jakarta, in the period 1st September 1994 to 2nd January 1995 was 2.38 ppbv. The results for SO_2 and NO_2 obtained using the filter pack in Jakarta are in reasonable agreement with the results given in Table 4. Concentrations of SO_2 and NO_2 have also been determined by means of bubblers followed by colorimetric analysis in a suburban area of Jakarta. In 1995, SO_2 concentrations between 1 and 4 ppbv (annual average = 2 ppbv)and NO_x concentrations between 6 and 12 ppbv (annual average= 10 ppbv), were measured at the suburban site using these methods (Lubis and Aprishanty, 1996).

Few industries in Indonesia employ air pollution control technologies. A seawater based desulfurisation plant is planned for 2x600 MW coal fired boilers in East Java (Radojevic, 1996).

2.5. PHILIPPINES

The Philippines has a population of 65 million. Its economy is based largely on agriculture, and the country is going through a phase of industrialization. There is no specific policy on acid deposition. However, laws that regulate the emission of acidic precursors, SO_2 and NO_x, have been promulgated. The administration of air pollution has undergone several changes over the years. Controls for SO_2 and NO_x were implemented by the National Air and Water Pollution Control Commission created in 1964. The agency was renamed to the national Pollution Control Commission (NPCC) in 1976. This was abolished in 1986 and the Department of Environment and Natural resources (DENR) and the Environmental Management Bureau (EMB) were created to take over the role of NPCC. Various other agencies have also been involved in projects relating to acid rain and precursor gases.

A detailed source inventory of air pollutants, including SO_2 and NO_x, was prepared for Metro Manila in 1990 (Siador, Jr. and Calderon, 1997). A study of acid rain was conducted between April and November of 1986 and 1987 with 94 event samples collected at 15 sampling stations in Metro Manila and the provinces of Bulacan, Cavite, Laguna and Bataan. The pH of rain was found to vary between 3.7 and 7.7. The lowest value of 3.7 was reported in the central district of Metro Manila. Although measurements of other parameters (alkalininty, acidity, sulfate, nitrate, chloride, calcium, magnesium, potassium and sodium) were made, these are not considered as reliable due to the use of obsolescent methods. A reactivation of the acid precipitation program, which was suspended ten years ago, is planned in the near future.

The monitoring of acid rain precursor gases, SO_2 and NO_x, and other air quality parameters (O_3, CO, particulates) was initiated in Metro Manila in 1971 using manual methods to determine concentrations on a weekly basis at several stations (Siador,C.S.,Jr., 1996). Manual methods were replaced with automatic instruments in 1974. Monitoring activity declined during the 1980s. In 1987 monitoring of SO_2 resumed at 8 stations using bubbler techniques. At present several automatic and mobile stations monitor pollutant levels in Metro Manila. The SO_2 concentration seems to have declined at most stations in Metro Manila between 1975 and 1993. Annual averages of 24 hour measurements of SO_2 ranged between 25 and 55 ppbv in 1975 at four sites. In 1993, annual averages of SO_2 ranged from 7 to 21 ppbv at four sites. NO_x concentrations tend to be below 25 ppbv.

2.6. SINGAPORE

Major sources of acidic precursors (SO_2 and NO_x) in Singapore include power stations, petroleum refineries, other industries and motor vehicles. Control of these pollutants is regulated by the Clean Air Act of December 1971. Emissions of SO_2 are controlled by restricting the sulfur content of automotive diesel and industrial fuel oil. The maximum permitted sulfur content of automotive diesel is 0.3%. As of July 1994 all new vehicles are required to be fitted with catalytic converters in order to comply with vehicular emission standards. Before 1994 air quality monitoring was performed using semi-automatic methods. Since 1994, continuous monitoring of SO_2, NO_x, O_3, CO and PM_{10} is carried out at 15 monitoring stations. Daily rainwater pH measurements have been carried out at some of the air quality monitoring stations since 1982. Between 1982 and 1994 rainwater was collected using bulk samplers. Since 1994 rainwater has been collected at four stations using an automatic precipitation collector. It is reported that the pH of rainwater is typically between 5.1 and 5.5 (Yong and Eng, 1997). Other components have not yet been measured in rainwater collected at the air quality stations. An ion chromatograph has been purchased and method development is currently under way. In the future, major ion concentrations will be reported.

A detailed analysis of the rainwater pH data collected between 1981 and 1984 at different sites in Singapore has been reported (Rahman and Chin, 1984). It has been found that sampling bottles affected the pH of rain. Samples collected in metal rain gauges had the highest pH values; most of them >5.6. Samples collected in glass bottles by the Antipollution Unit (APU) were close to 5.6. Lowest pH values were observed in samples collected using plastic containers. It was suggested that metal collectors could react with acids in rainwater to neutralize them.

Precursor gases, SO_2 and NO_x, are routinely monitored at the air quality stations (Huan and Boo, 1996). Average annual SO_2 concentrations have been in the range of 10 and 35 ppbv since 1985. There appears to have been a slight upward trend in SO_2 concentrations at most sites since 1991. Annual average NO_2 concentrations have been in the range of 5.3 to 21 ppbv since 1985.

2.7. MALAYSIA

Malaysia experienced rapid development of various economic sectors, including agriculture, industry and forestry from the early 1960s. However, the impact of human activities on the environment has only become evident over the last 10 years. The Environmental Quality Act was passed in 1974, and the Department of the Environment has taken steps to control air pollution through the introduction of regulations and policies. Sources of acidic precursor gases, SO_2 and NO_x, include: motor vehicles, oil, coal and gas fired power plants, open burning of refuse, waste incineration, oil refineries, chemical industries, steel plants, and forest fires.

The Malaysian Meteorological Service (MMS) started monitoring air pollution in 1976 at Petaling Jaya, an urban station. At the same time it started monitoring air pollution at Tanah Rata, a "background" station some distance from urban and industrial centres, as part of the WMO's BAPMoN program. In 1987 MMS operated a total of 10 air quality stations, 7 in Peninsular Malaysia, 2 in Sabah and 1 in Sarawak. At present Malaysia has 23 air quality stations measuring acid rain precursor gases such as SO_2, NO_x, as well as CO, O_3 and PM_{10}, and more are planned in the state of Sarawak in response to the problem of regional haze (Radojevic et al, 1997). Eventually, Malaysia plans to operate 50 fully automatic air quality stations (Abdullah, 1996).

In October 1984 the Klang Valley Acid Rain Monitoring network was set up by MMS in collaboration with other organisations to monitor rainfall acidity in the most rapidly growing urban area in the country. In 1985 the network of stations monitoring rainwater acidity was expanded nation-wide to include an additional 10 stations as part of the National Acid Rain Monitoring Network. This was later extended to include stations in Sarawak and Sabah on Borneo island. Both manual and automatic precipitation collectors are used in this network to collect wet-only precipitation. Samples are collected on a daily basis and accumulated in a storage container for a week to give a weekly compound sample. A portion of this sample is then sent to the laboratory for pH analysis (Ishak and Hamzah, 1997).

A detailed statistical analysis of the pH data obtained from the National Acid Rain Monitoring network prior to 1988 has been carried out (Leong et al, 1988). A seasonal variation in rainwater pH was observed. It was found that stations on the east coast of Peninsular Malaysia had significantly higher pH values during the North East monsoon (November-March) than during the South West monsoon (May-September) or the intermonsoon periods (April and October). Some of the stations on the west coast had higher pH values during the South West monsoon. The explanation for these observations could be the direction of the wind. During the SW monsoon, the west coast experiences offshore winds which bring generally cleaner air, while during the NE monsoon the air mass traverses the Malayan peninsula bringing with it pollutants that lower the pH at these stations. The situation is reversed for the stations on the east coast; these experience offshore winds during the NE monsoon and hence the higher pH. These stations experience winds which traverse the peninsula during the SW monsoon bringing with them polluted air and lowering the pH.

The annual volume weighted rainwater pH values at various stations between 1985 and 1993 are given in Table 5. It is apparent that Petaling Jaya, Perai and Senai consistently recorded the lowest pH values. The high acidity could be due to a high density of motor vehicles and industries that emit SO_2 and NO_x in these areas. Mersing in eastern Johore and Malacca also recorded relatively low pH values especially in later years, due to the expansion of industrial activity in these areas, and the transport of pollutants from other industrial areas. Central and eastern parts of Malaysia recorded relatively high pH values, due to the low level of industrial activity in these areas. *At all* stations there was a decrease in rainwater pH between 1985 and 1988. The situation seems to have stabilized after 1988, but low pH values are still being recorded at many stations. The greatest increase in acidity was observed at Petaling Jaya and Senai. In terms of H^+ concentration, the acidity increased fourfold between 1985 and 1988. SO_4^{2-} and NO_3^- were also measured in rainwater *at all* sites. Mean annual SO_4^{2-} concentrations varied between 0.08 and 0.9 mg/L while the mean annual NO_3^- concentrations varied between 0.01 and 0.25 mg/L. Petaling Jaya, Senai and Perai had the highest SO_4^{2-} concentrations. In general, these three sites also had the highest NO_3^- concentrations. A seasonal variation in the concentrations of SO_4^{2-} and NO_3^- was observed. Highest concentrations *at all* sites were generally observed during the months of May to August, which are the driest periods of the year. Also, the concentrations of SO_4^{2-} and NO_3^- increased significantly at Petaling Jaya and Senai between 1985 and 1987. The increasing acidity observed at the same sites during this period could be due to the increase in SO_4^{2-} and NO_3^- in rainwater.

Table 5. Annual volume weighted mean pH of rainwater in Malaysia

Site	1985	1986	1987	1988	1989	1990	1991	1992	1993
Petaling Jaya	5.05	4.54	4.46	4.44	4.51	4.50	4.63	4.41	4.47
Bayan Lepas	5.31	5.01	5.03	4.99	4.99	4.92	4.92	4.79	4.93
Senai	4.99	4.54	4.40	4.39	4.69	4.62	4.55	4.22	4.74
Alor Setar	5.58	5.49	5.53	5.38	5.42	5.67	5.74	5.08	5.19
Malacca	5.46	4.97	4.58	4.76	4.64	4.61	4.74	4.68	4.93
Mersing	5.35	4.89	4.84	4.65	4.78	4.90	5.05	4.70	4.93
Kota Bharu	5.51	5.47	5.47	5.18	5.37	5.37	5.28	5.14	5.31
Tanah Rata	5.64	5.23	5.21	5.08	5.22	5.30	5.11	4.95	5.33
Ipoh	5.43	5.23	5.21	5.10	5.16	4.97	5.13	4.97	5.09
Perai	4.71	4.51	4.61	4.52	4.66	4.53	4.63	4.60	4.69
Kuantan	5.35	5.22	5.13	5.10	5.09	5.09	5.23	4.85	5.10
Kuala Terengganu	5.44	5.33	5.27	5.16	5.19	5.17	5.26	5.12	5.24
Kuching	NR	NR	NR	5.25	5.34	4.84	5.50	5.35	5.56
Tawau	NR	NR	NR	5.54	5.64	5.75	5.95	5.92	5.94

It can be concluded that there is sufficient evidence to suggest increasing acidification of precipitation in industrial areas of Malaysia. On the other hand, regions without significant industrial sources, especially Sabah and Sarawak in Borneo, seem to be largely unaffected by acid rain.

It is debatable how representative the pH measurements reported in the Malaysian National Acid Rain Monitoring Network are of actual rainfall. One of the shortcomings of the survey is that the pH measurements are not taken immediately after rainwater collection but up to one week after precipitation. Clearly, many chemical changes can take place in samples stored for a week leading to significant changes in the pH. Some of the more important chemical reactions with regard to rainwater pH are the dissolution of Mg^{2+} and Ca^{2+} ions from dust particles washed out by falling rain, and the decomposition of weak organic acids such as formic and acetic acid (Radojevic, 1995). Both of these processes would tend to increase the rainwater pH during storage. A rainwater study in Brunei on Borneo island indicated significant increases in most rainwater samples during storage (Radojevic and Lim, 1995a). Therefore, actual pH values of precipitation in Malaysia could be considerably lower than reported in the National Acid Rain Monitoring Network.

Manokoran (1980) reports an earlier study of rainwater chemistry carried out at Pasih Forest Reserve, Negri Sembilan during a one-year period from January to December 1973. Samples of precipitation, throughfall and stemfall were analyzed for pH, NH_4^+, NO_3^-, albuminoid-nitrogen, K^+, Na^+, Ca^{2+} and Mg^{2+}. The average monthly pH values of precipitation varied between about 5.9 and 5.1. The study identified the following sources of ions in precipitation. Oceanic and terrestrial sources contributed to Na^+, while K^+, Ca^{2+} and Mg^{2+} were derived in the main from wind blown dust. Several sources contributed to NO_3^- including lightning discharges and atmospheric reaction of N_2O emitted from soil. Soil and atmospheric aerosol were identified as sources of NH_4^+. Dust particles, microorganisms, insects and pollen were the likely source of albuminoid nitrogen. Concentrations of ions increased when precipitation passed through the canopy in the form of stemfall and throughfall. This enrichment was attributed to leaching from within the plants and the washout of material adsorbed or settled on plant foliage.

Data on precursor gas (SO_2, NO_x) concentrations are available from the air quality stations (Abdullah, 1996). In Petaling Jaya the average annual SO_2 concentration increased from 12 to 21 ppbv between 1992 and 1995. In nearby Kuala Lumpur the average annual concentration of SO_2 varied between 2 and 7 ppbv, with no consistent trend during this period. At both sites there appears to have been an upward trend in NO_2 concentrations during the same period. In Petaling Jaya the annual average concentration of NO_2 increased from 17 ppbv to 26 ppbv, while at Kuala Lumpur it increased from 16 to 24 ppbv in 1994, thereafter declining to 21 ppbv in 1995.

2.8. BRUNEI DARUSSALAM

Brunei Darussalam, a small country with a land area of 5765 km^2 and a population of 261×10^3, is located on Borneo island. The country has few anthropogenic sources of air pollution; circa 100×10^3 on-road motor vehicles, a small refinery, and one medical incinerator. Natural and man-made forest fires also contribute to local pollution, but most air pollution originates from transboundary sources. The first ever rainwater study was conducted in the capital city, Bandar Seri Begawan, in the period between 26 April and 3 November 1994 (Radojevic and Lim, 1995a). Bulk precipitation was analyzed for pH, conductivity, Cl^-, NO_3^-, SO_4^{2-}, Na^+, K^+, Ca^{2+} and Mg^{2+}. A statistical study of the interrelationship of various parameters was also carried out. The authors also conducted a study into the short-term variation of the concentration of selected ions within individual rainstorms (Radojevic and Lim, 1995b). This revealed that most of the ionic content of precipitation was deposited within the initial stages of the rainstorms. Although organic acids were not analysed directly in this study, there was some evidence of the possible major contribution of organic acidity (formic and acetic) to the pH measured in these samples (Radojevic and Lim, 1995a).

An automatic, wet-only, collector was set up at the Brunei international Meteorological Station in Bandar Seri Begawan and this started sampling precipitation in July 1995 in conformity with the GAW protocol of the WMO (Radojevic et al, 1997). In the 185 samples analysed between July 1995 and June 1996 it was observed that 91% of the samples had pH values less than 5.6 and could therefore be classified as "acid rain". Some 42% of the samples had pH values less than 5.

Prior to 1997 there was no monitoring of precursor gases in the country. A fully-equipped air quality monitoring station was set up in Bandar Seri Begawan in October 1997 to measure routinely SO_2, NO_2, O_3, CO and PM_{10} concentrations. A further 6 PM_{10} monitoring instruments were set up throughout the country. A survey of the distribution of average monthly SO_2, NO_2 and O_3 concentrations throughout the country using diffusion tubes was also carried out in October 1997 and February 1998 (Radojevic et al, 1998). Concentrations of SO_2 varied between 0.48 and 5.71 ppbv while concentrations of NO_2 varied between 0.3 and 19.68 ppbv. Concentrations of O_3 were between 5.2 and 57.2 ppbv. Legislation with regard to air quality in Brunei is still relatively rudimentary. Air quality standards similar to those adopted by Singapore have been suggested (Goh and Ling, 1996) but they have still not been enacted.

2.9. THAILAND

At present, rainwater is routinely sampled at 5 sites in Thailand including urban, industrial and remote sites by the Pollution Control Department (Siriswasdi, 1997). Three of the sites measure pH, conductivity, SO_4^{2-} and NO_3^-, while the other sites measure only pH and conductivity. Samples are collected using wet-only automatic collectors.

In addition, precipitation has been collected on a daily basis at two rural sites in Thailand since mid-1991 in rural locations (Granat et al, 1996). Samples were analyzed for conductivity, pH, NH_4^+, Na^+, K^+, Mg^{2+}, Ca^{2+}, SO_4^{2-}, NO_3^- and Cl^-. A high correlation was observed between H^+ and SO_4^{2-} suggesting that sulfur played an important role in acidification. Also, a strong interrelationship between Ca^{2+}, Mg^{2+}, Na^+ and Cl^- was found.

The authors found a high correlation between NH_4^+ and NO_3^- but the reason for this relationship was not obvious. Bulk collectors gave results 10-30% higher than wet-only collectors.

Some earlier data on rainwater acidity in Thailand were summarized by Ayers (1991). These include 12 samples collected between November 1983 and July 1985 at a BAPMoN site on the island of Ko Sichang. The volume-weighted mean pH was 6.54, and NO_3^-, SO_4^{2-}, Ca^{2+} and K^+ were also measured. Some more data are available from four sites, Chiang Mai, Khon Kaen, Bangkok and Songkla, during 1987 (Ayers, 1991). The lowest pH measured was 4.5 and most samples had pH values >5.6. Only <10% of the samples having pH<5.

Precursor gases are measured at 52 automatic air quality monitoring stations throughout Thailand, each of which measure CO, SO_2, NO_2, O_3, TSP, PM_{10}, non-methane hydrocarbons as well as relevant meteorological data as part of the Ambient Air Monitoring Network set up by the Pollution Control Department in 1992 (Charasaiya, 1996; Siriswasdi, 1997).

2.10. HONG KONG

Results for 1988 from the BAPMoN station of Yuen Ng Fan at Sia Kung are reported by Ayers (1991). Wet-only samples were collected on a weekly basis and the volume-weighted mean pH was 4.68. Data obtained between 1988 and 1993 are discussed by Ayers and Yeung (1996). The analysis suggested elevated levels of SO_4^{2-} and NO_3^-.

Results of daily precipitation samples collected from 1989 have been reported by Peart (1995). Results are summarised in Table 6. No apparent trend in pH was observed over the years. There was evidence that rainfall was less acidic during November, December and January than for the other months.

Table 6. pH of rainwater samples collected in Hong Kong (adapted from Peart, 1995).

Year	Minimum pH	Maximum pH	Median pH
1989	3.5	6.25	4.57
1990	3.45	6.78	4.69
1991	3.45	6.09	4.48
1992	3.81	6.56	4.54
1993	3.71	6.76	4.62
1994	3.56	6.92	4.59

No significant difference was observed between pH values measured in filtered and unfiltered samples in a rural site in Hong Kong (Sequeira et al, 1996). Hong Kong has got a satisfactory air quality monitoring and management programme in place.

2.11. SOUTH KOREA

A number of separate groups monitor precipitation chemistry in Korea. Most of these measure the electrical conductivity, pH, SO_4^{2-}, NO_3^-, Cl^-, NH_4^+, Na^+, Ca^{2+}, Mg^{2+} and K^+. At most monitoring stations the volume weighted pH is generally around 5.0, although pH values as high as 6.8 have been reported during yellow sand period in springtime (Shim et al, 1997).

Measurements of wet-only precipitation between August and November 1985 at 10 sites in the Seoul area were reported by Shin et al (1989). The volume weighted mean pH values ranged from 4.39 to 4.64. Concentrations of SO_4^{2-} varied between 1.4 and 3.3 mg/L, while NO_3^- was present at low levels < 0.37 mg/L. Measured levels of Ca^{2+} suggested that alkaline aerosol was important in the ion balance.

A study of over four hundred rainwater samples collected during 1996 has been reported (Park, 1997). The most abundant anion was SO_4^{2-}, present at concentrations between 1.4 and 5.76 mg/L, and accounting between 30 and 68% of total anions. Most of this was of non-sea salt origin. Sea salt contribution to SO_4^{2-} was <10%. The NO_3^- concentration ranged between 1.0 and 1.7 mg/L, while the Cl^- concentration varied from 0.28 to 8.86 mg/L. Comparison with results from previous years showed that the NO_3^- contribution rose between 1994 and 1996 due to increasing NO_2 emissions in the country. The volume weighted mean concentrations ranged from 4.5 to 5.2 with an annual mean of 4.8. Similar values were recorded in previous years. Some 2% of the individual precipitation events had pH values <4. Deposition rates of SO_4^{2-} and NO_3^- were 2.0-6.9 g/m^2 and 1.0-2.0 g/m^2 respectively on the basis of studies over several years.

Chung et al (1996) report measurements at five stations over the period 1990-1993. The pH measurements are summarized in Table 7 indicating that acidity increased over the years at most sites. Long range transport of pollution from China was suggested as a possible factor in acidification of precipitation in Korea.

Table 7. Annual volume-weighted mean pH values at five sites in South Korea (adapted from Chung et al, 1996).

Year	Koong-Hyun Ree	Chongwoon	Tae-Ahn Peninsula	Seokeepo	Hahn-Ra Mountain
1990	5.22	4.73	NR	NR	NR
1991	4.82	4.67	4.80	5.04	4.61
1992	4.41	4.47	4.68	4.80	5.55
1993	4.39	3.99	4.42	4.51	6.08

2.12. CAMBODIA

Cambodia's population of 9 million is mainly (ca.85%) engaged in agriculture. There are few industrial sources of air pollution in Cambodia. These include three power stations in Phnom Phen and approximately 70 factories (Sophy, 1997). In addition, there are some 47,000 motor vehicles and 170,000 motorcycles. Emissions of pollutant gases are summarised in Table 8.

Table 8. Sources of air pollutants in Cambodia (in tons per annum)

Source	Particulates	SO_2	NO_x	HC	CO
3 power plants in Phnom Penh	NR	20	87	NR	22
Other mobile and stationary sources	1,000	7,700	4,400	2,300	52,000

Environmental management, control, and monitoring are still in heir infancy in Cambodia. Air quality management is undertaken by the Department of Pollution Control of the Ministry of the Environment, whose tasks include: preparation of draft air quality and emission standards, monitoring of environmental pollution, preparation of reports on pollution, provision of environmental education programs, etc. At present there is no air quality of acid rain monitoring program in the country. One of the main reasons is lack of technical expertise and absence of instrumentation.

2.13. LAOS

Laos is one of the poorest and least developed counties of the world. Its population of 4.5 million is mainly rural (80%); only 20% of the population live in urban centres. The major source of air pollution is slash and burn agricultural practices. Industry is still a minor contributor to air pollution. The number of motor vehicles is rapidly growing. Laos ranks as the number one country in the Asia/Pacific region in terms of the vehicle growth rate with a growth rate of 25% per annum (Vongphosy, 1997). Environmental management, control, legislation, and monitoring are still at a non-existent or rudimentary stage in Laos. Overall coordination of environmental activities is the responsibility of the Science, Technology and Environment Organization (STENO) within the Prime Minister's Office. This organization was set up in 1993 and charged with developing environmental planning and management procedures. A lignite power station will be built in Laos and it is expected to be equipped with desulfurization technology. At present there are no air quality or rainwater monitoring programs in the country. One of the major problems is the lack of technical expertise and appropriate instrumentation. The development of an adequate monitoring program is presently a high priority.

2.14. MYANMAR

Myanmar, with a population of 43 million, is mainly an agricultural country undergoing rapid industrialization. The extent of air pollution and acid rain in Myanmar has not been quantified. It has been reported that Myanmar has only a few minor air pollution problems (Joseph, 1997). Most problems are localized, such as around industries, or air pollution from motor vehicles during peak hours in major cities. Acid precipitation has not yet been monitored. However, air pollution is expected to rise with increasing industrialization.

2.15. TAIWAN

Although Taiwan is a highly industrialized country only a few studies of rainwater chemistry have been reported. Ayers (1991) report the results from nine stations spread across Taiwan. Between 1985 and 1988, 300 rainwater samples were analyzed. Some 56% of the samples had pH values<5.6. The latest monitoring showed that already > 70% of the precipitation in Taiwan exhibits a pH lower 5.6 and has to be considered as the acid rains (Lin, this book).

2.16. AN ACID RAIN MONITORING NETWORK FOR EAST ASIA

It has been suggested that an acid deposition monitoring network should be set up in East Asia (Murano, 1997a, b). A uniform sampling and analysis protocol together with quality assurance/quality control (QA/QC) procedures has been proposed. The draft manual is remarkably similar to that recommended by the WMO for stations participating in the Global Atmosphere Watch (GAW). It is hoped that this network would produce more reliable data which could be used to make more meaningful spatial and temporal comparisons.

3. CONCLUSIONS

• Acid rain is observed regularly at many sites in East Asia, especially in China, South Korea, Hong Kong and Japan. Lower pHs tend to be observed in industrialized countries and regions while higher pH values tend to be observed in areas with little or no industries.

• Japan has the most sophisticated and longest running precipitation chemistry and air quality monitoring program in East Asia. Also, Japan is a world leader in air quality legislation, management and control. In many areas of air quality management it is on a par with the industrialized countries of Europe and North America, and in some areas, notably air pollution control technologies, it is more advanced.

- With the exception of Japan, other countries in East and South-East Asia have started monitoring acidic precipitation and air quality relatively recently. Monitoring programs vary from being relatively comprehensive (e.g. South Korea, Hong Kong, China) to rudimentary (e.g. Indonesia, Vietnam). Some countries have yet to set up air quality and acid rain monitoring programs of any kind (e.g. Myanmar, Laos, Cambodia). General air quality management, legislation, and control measures vary between countries of East Asia in the same proportion as air quality and acid precipitation monitoring.

- The quality of data produced by many of the monitoring networks in the region, with the possible exception of Japan, leaves much to be desired. There is an absence of quality control procedures in many of the networks. Much improvement could be achieved by the introduction of QA/QC schemes in acid rain and air quality monitoring networks. The use of certified reference materials (CRM) in rainwater analysis is to be encouraged and the setting up of international monitoring programs with a centralized QC/QA scheme would be desirable. This would greatly improve the quality of the data generated and allow for more meaningful comparisons to be made between different countries. Obviously, many of the earlier conclusions will have to be revised when more reliable data become available. The proposed international acid rain survey in East Asia would help by standardizing methods and procedures.

- Many of the poorer countries in East and South-East Asia lack the technical expertise and instrumentation to set up adequate acid rain and air quality monitoring networks. Clearly, there is a need for international assistance in terms of providing training and financial resources to these countries.

- It is difficult to make reliable comparisons in pH and other measurements reported in the literature, in the absence of a uniform methodology strictly adhered to by all parties involved. Measurements of pH are especially sensitive to a variety of factors including: sampling bottle material, sampling period (event, daily, weekly, or monthly), type of sample (bulk or wet-only), and especially the time of pH analysis. The pH is not a static parameter, but is continually changing, during and after sampling, due to chemical reactions in the sample.

- From the measurements available so far it could be concluded that acid rain is still not a major problem in South-East Asia. In South-East Asia most rainwater pH measurements tend to be around 5.6, the pH of "natural" rainwater. Instances of pH <5 are encountered at some sites, and these are mainly due to localized industrial pollution. There is some evidence that pH values below 5 at unpolluted sites may be due to the contribution of weak organic acids, such as formic and acetic acids.

- It is difficult to make predictions about the future incidence of acid rain in East and South-East Asia due to the unpredictable economic situation. Many countries in the region are experiencing a slowing down of their industrial development due to the recent currency crisis, which could in turn slow down the growth in pollutant emissions. On the other hand, the result could be fewer, but more polluting, industries, as there may be less financial resources available for introducing pollution control equipment. If the economic growth recovers to the rates experienced during the late 1980s and early 1990s then we could expect to see a greater incidence of acid rain over a wider region.

•The haze phenomenon in South-East Asia is a major air pollution problem and its impact on the chemistry of precipitation is still unknown. It has been speculated that forest fires could raise the acidity of precipitation. While it is known that forest fires produce acid precursor gases such as SO_2, NO and NO_2, the non-carbonaceous matter of particles produced by these fires contains Ca and K, which give rise to alkalinity in rainwater. There is a need for a research project to study both the emissions from the forest fires in South-East Asia and their impact on precipitation chemistry and rainfall acidity.

•Most of the precipitation networks in East and South-East Asia do not monitor weak organic acids, such as formic and acetic acids, in rainwater. At many tropical sites in Africa and South America it has been demonstrated that organic acids can contribute between 40 and 80 % of the total acidity. In view of this it is important that precipitation surveys in East and South-East Asia begin routine monitoring of organic acids at the earliest stage.

REFERENCES

Abdullah, M. H. (1996) Malaysia country report, in Proceedings of the ASEAN Network on Environmental Monitoring (ASNEM) on the *3rd Workshop on Air Quality Monitoring and Analysis with Emphasis on Polycyclic Aromatic Hydrocarbons*, Environmental Research and Training Center (ERTC), Pathumthani, Thailand, pp. 177-188.

Anh, P.T.V.(1996) Vietnam country report, in Proceedings of the ASEAN Network on Environmental Monitoring (ASNEM) on the 3rd Workshop on Air Quality Monitoring and Analysis with Emphasis on Polycyclic Aromatic Hydrocarbons, Environmental Research and Training Center (ERTC), Pathumthani, Thailand, pp. 177-188.

Ayers, G.P. (1991) Atmospheric acidification in the Asian region. *Environmental Monitoring and Assessment*, Vol. 19, 225-250.

Ayers, G.P, Gillet, R.W., Ginting, N., Hooper, M., Selleck, P.W. and Tapper, N. (1995) Atmospheric sulfur and nitrogen in West Java. *Water, Air and Soil Pollution*, Vol. 85, 2083-2088.

Ayers, G.P. and Yeung, K.K. (1996) Acid deposition in Hong Kong. *Atmospheric Environment*, vol. 30, 1581-1587.

Bruijnzeel, L.A. (1989) Nutrient content of bulk precipitation in south-central Java, Indonesia, *J. Tropical Ecology*, Vol. 5, 187-202.

Bruijnzeel, L.A., Waterloo, M.J., Proctor, J., Kuiters, A.T. and Kotterink, B. (1993) Hydrological observations in montane rain forests on Gunung Silam, Sabah, Malaysia, with special reference to the "Massenerhebung" effect, *J. Ecology*, Vol. 81, 145-167.

Burghouts, T. van Straalen and Bruijnzeel, L.A. (1993) Contributions of throughfall, litterfall and litter decomposition to nutrient cycling in dipterocarp forest in the upper Segama area, Sabah, Malaysia, in *Spatial Heterogeneity of Nutrient Cycling in Bornean Rain Forest*, Vrije Universiteit, Amsterdam, pp. 47-79.

Charasaiya, T. (1996) Air quality monitoring in Pathum Thani, Thailand, in *Proceedings of the ASEAN Network on Environmental Monitoring (ASNEM)* on the 3rd Workshop on Air Quality Monitoring and Analysis with Emphasis on Polycyclic Aromatic Hydrocarbons, Environmental Research and Training Center (ERTC), Pathumthani, Thailand, pp. 245-254..

Chung, Y.-S., Kim, T.-K., and Kim, K.-H. (1996) Temporal variation and cause of acidic precipitation from a monitoring network in Korea. *Atmospheric Environment*, Vol. 30, 2429-2435.

Ducros,M (1854) Observation d'une pluie acide. *J. Pharm. Chim.*, Vol. 3, 273-277.

Galloway, J.N., Zhao, D., Jiling, X., and Likens, G.E. (1987) Acid rain: China, United States, and a remote area, *Science*, Vol. 236, 1559-1562.

Goh, K.J. and Voon, O.L. (1996) Brunei Darussalam country report, in Proceedings of the ASEAN Network on Environmental Monitoring (ASNEM) on the *3rd Workshop on Air Quality Monitoring and Analysis with Emphasis on Polycyclic Aromatic Hydrocarbons*, Environmental Research and Training Center (ERTC), Pathumthani, Thailand, 143-152.

Granat, L., Suksomsankh, K., Simachaya, S., Tabucanon, M. and Rodhe, H. (1996) Regional acidity and chemical composition of precipitation in Thailand, *Atmospheric Environment*, Vol. 30, 1589-1596.

Hara, H. (1993) Acid deposition chemistry in Japan, *Bull. Inst. Public Health*, Vol. 43, 426-437.

Hara, H. (1997) Acid deposition chemistry, paper B8010, presented at the 7th Asian Chemical Congress, 16-20 May 1997, Hiroshima, Japan.

Hashimoto, M. (1989) History of air pollution control in Japan, in *How to Conquer Air Pollution: A Japanese Experience*, Studies in Environmental Science 38, edited by H. Nishimura, Elsevier, Amsterdam, pp. 1-94.

Huan, T.H. and Boo, L.S. (1996) Singapore country report, in Proceedings of the ASEAN Network on Environmental Monitoring (ASNEM) on the *3rd Workshop on Air Quality Monitoring and Analysis with Emphasis on Polycyclic Aromatic Hydrocarbons*, Environmental Research and Training Center (ERTC), Pathumthani, Thailand, pp. 206-223.

Iijima, T. (1997) Overview of Acid Deposition in Japan and Acid deposition Monitoring Network in East Asia, paper presented at the *4th ASEAN Workshop on Air Quality Monitoring and Analysis with Emphasis on Acid Deposition*, Technopolis, Pathumthani, Thailand, 25 February - 4 March 1997.

Ishak, A. and Hamzah, W.N.W. (1997) Air quality monitoring and acid deposition in Malaysia, paper presented at the *4th ASEAN Workshop on Air Quality Monitoring and Analysis with Emphasis on Acid Deposition*, Technopolis, Pathumthani, Thailand, 25 February - 4 March 1997.

Joseph, U. (1997) Country report (Myanmar), paper presented at the *4th ASEAN Workshop on Air Quality Monitoring and Analysis with Emphasis on Acid Deposition*, Technopolis, Pathumthani, Thailand, 25 February - 4 March 1997.

Kato, N. and Akimoto, H. (1992) Anthropogenic emissions of SO_2 and NO_x in Asia: emission inventories. *Atmospheric Environment*, Vol. 26a, 2997-3017.

Kellner, O, Sawano, J., Yoshii, T. and Oku, K. (1894) Amounts of ammonium and nitrate in rainwater. *Nokadaigaku Gakujyutsushiken Iho*, Vol. 1, 28-42.

Khanh, N.H. (1993) Investigation. Evaluation of the state of pH value of rainfall. Research Report. *Hydrometeorological Service*, Hanoi, Vietnam.

King, H.B. and Yang, B.Y. (1984) Precipitation and stream water chemistry in Pi-Lu-Chi watersheds. January 1981-December 1982. *Bulletin of the Taiwan Forestry Research Institute*, Taipei, no.427.

Kurita, H, Hori, J., Hamada, Y. and Ueda, H. (1991) Decrease of pH of river and lake water in mountainous region of central Japan and its relation to acid rain. *J. Jpn. Assoc. Air Pollut.* Vol. 28, 308-315.

Lan, N.T. (1996) Vietnam country report, in Proceedings of the ASEAN Network on Environmental Monitoring (ASNEM) on the *3rd Workshop on Air Quality Monitoring and Analysis with Emphasis on Polycyclic Aromatic Hydrocarbons*, Environmental Research and Training Center (ERTC), Pathumthani, Thailand, 258-262.

Leong, C.P., Lim, S.F., and Lim, J.T. (1988) Report on Rain Acidity Analysis Based on Data from the National Acid Rain Monitoring Network, Malaysian Meteorological Service.

Lubis, S.M. and Aprishanty, R. (1996) Indonesia country report, in Proceedings of the ASEAN Network on Environmental Monitoring (ASNEM) on the *3rd Workshop on Air Quality Monitoring and Analysis with Emphasis on Polycyclic Aromatic Hydrocarbons*, Environmental Research and Training Center (ERTC), Pathumthani, Thailand, 153-176.

Manokaran, N. (1980) The nutrient contents of precipitation, throughfall and stemflow in a lowland tropical rain forest in peninsular Malaysia. *The Malaysian Forester*, Vol. 43, 266-289.

Matsushita, H. (1996) Atmospheric pollution by PAHs in Japan, in Proceedings of the ASEAN Network on Environmental Monitoring (ASNEM) on the *3rd Workshop on Air Quality Monitoring and Analysis with Emphasis on Polycyclic Aromatic Hydrocarbons*, Environmental Research and Training Center (ERTC), Pathumthani, Thailand, 84-104.

Miyake, Y. (1939) The chemistry of rain water. *J. Met. Soc. Japan,II,* Vol. 17, 20-37.

Mo, T., Gu, Q, and Zhao, K. (1988) Criterion of acid rain formation, *Huanjing Kexue Xuebao*, Vol. 8, 32-39.

Murano, K. (1997a) Some part of draft technical manual for monitoring wet deposition. Acid Deposition Network in East Asia, paper presented at the *4th ASEAN Workshop on Air Quality Monitoring and Analysis with Emphasis on Acid Deposition*, Technopolis, Pathumthani, Thailand, 25 February - 4 March 1997.

Murano, K. (1997b) Activity of JEA for East Asian Acid Precipitation Monitoring Network, paper presented at the *4th ASEAN Workshop on Air Quality Monitoring and*

Analysis with Emphasis on Acid Deposition, Technopolis, Pathumthani, Thailand, 25 February - 4 March 1997.

Narita, Y., Chai, H., and Tanaka, S. (1997) A study on acid rain in China on the basis of the literature investigation, paper B9P09, presented at the 7th Asian Chemical Congress, 16-20 May 1997, Hiroshima, Japan.

Nishimura, H. and Sadakata, M. (1989) Emission control technology, in *How to Conquer Air Pollution: A Japanese Experience*, Studies in Environmental Science 38, edited by H. Nishimura, Elsevier, Amsterdam, pp. 115-156.

Park, S.U. (1997) Development of Technology for Monitoring and prediction of Acid Rain, Korean Ministry of Science and Technology, *Research and Development on Basic Technology for Atmospheric Environment in Global Scale. Project G-7*. Seoul National University.

Peart, M. R. (1995) The occurrence of acid rain in Hong Kong, in Proceedings of the *International Symposium on Climate and Life in the Asia Pacific*, University of Brunei Darussalam, 10-13 April 1995, edited by Sirinanda, K.U., pp.10-19.

Rachmawati, E. (1996) NO_2 concentration in local cities in Indonesia. Result of the nationwide monitoring by the TEA plate method, EMC Air Quality Laboratory, Indonesia.

Radojevic, M. (1995) Taking a rain check, *Analysis Europa*, August 1995, 35-38.

Radojevic, M (1996) Sea changes, *Chemistry in Britain*, Vol.32, No. 11, 47-49.

Radojevic, M (1998) Opportunity NO_x, *Chemistry in Britain*, Vol. 34, No. 3, 30-33.

Radojevic, M. and Harrison, R.M. (1992) Atmospheric Acidity: Sources, Consequences and Abatement. Elsevier Applied Science, London.

Radojevic, M. and Lim, L.H. (1995a) A rain acidity study in Brunei Darussalam. *Water, Air and Soil Pollution*, Vol. 85, 2369-2374.

Radojevic, M. and Lim, L.H. (1995b) Short-term variation in the concentration of selected ions within individual tropical rainstorms. *Water, Air and Soil Pollution*, Vol. 85, 2363-2368.

Radojevic, M., Lim, L.B.B., Ling, V.O. and Lim, L.H. (1997) A report on the studies of rain acidity and the concentration of selected ions in Negara Brunei Darussalam, paper presented at the *4th ASEAN Workshop on Air Quality Monitoring and Analysis with Emphasis on Acid Deposition*, Technopolis, Pathumthani, Thailand, 25 February - 4 March 1997.

Radojevic, M., Tan, K.S., Makarimi, A. and Medan, R. (1998). Unpublished data.

Rahman, A. and Chin, T.G. (1984). A preliminary study on the acidity of rainfall in Singapore, in Proceedings of *3rd Symposium on Our Environment*, 27-29 March 1984, edited by Lin, K.L. and Sin, H.C., Faculty of Science, National University of Singapore, pp.344-361.

Rinda, N. (1996) SO_2 concentration in local cities in Indonesia. Result of the nationwide monitoring by the PbO_2 method, EMC Air Quality Laboratory, Indonesia.

Rodhe, H., Galloway, J. and Dianwu, Z. (1992) Acidification in Southeast Asia-Prospects for the coming decades, *AMBIO*, Vol. 21, 148-149.

Ryaboshapko, A.G., Sukhenko, V.V. and Paramonov, S.G. (1994) Assessment of wet sulfur deposition over the former USSR, *Tellus*, Vol. 46B, 205-219.

Saefudin, A. (1997) Analysis of rain water at EMC station Serpong, Indonesia, paper presented at the *4th ASEAN Workshop on Air Quality Monitoring and Analysis with Emphasis on Acid Deposition*, Technopolis, Pathumthani, Thailand, 25 February - 4 March 1997.

Sequeira, R., Peart, M.R. and Lai, K.H. (1996) A comparison between filtered and unfiltered atmospheric depositions from a rural area in Hong Kong. *Atmospheric Environment*, Vol. 30, 3221-3224.

Shim, S.-G., Kim, Y.-P. and Kang, C.-H. (1997) Acid rain research activities in Korea, paper no. B8011, presented at the 7th Asian Chemical Congress, 16-20 May 1997, Hiroshima, Japan.

Shin, E.B., Lee, S.K., and Song, D.W. (1989) Acidity of rainwater in Seoul area, in Proceedings of the 8th World Clean Air Congress, Vol.2, edited by Brasser, L.J. and Moulder, W.C., Elsevier, Amsterdam.

Siador, C.S. Jr. (1996) Philippines country report, in Proceedings of the ASEAN Network on Environmental Monitoring (ASNEM) on the *3rd Workshop on Air Quality Monitoring and Analysis with Emphasis on Polycyclic Aromatic Hydrocarbons*, Environmental Research and Training Center (ERTC), Pathumthani, Thailand, pp. 189-205.

Siador, C.S. Jr. and Calderon, I.G. (1997) Environmental monitoring in the Philippines with emphasis on acid deposition, paper presented at the *4th ASEAN Workshop on Air Quality Monitoring and Analysis with Emphasis on Acid Deposition*, Technopolis, Pathumthani, Thailand, 25 February - 4 March 1997.

Siriswasdi, J. (1997) Acid rain deposition monitoring in Thailand. Monitoring by Pollution Control Department, paper presented at the *4th ASEAN Workshop on Air Quality Monitoring and Analysis with Emphasis on Acid Deposition*, Technopolis, Pathumthani, Thailand, 25 February - 4 March 1997.

Sophy, M. (1997) Kingdom of Cambodia report, paper presented at the *4th ASEAN Workshop on Air Quality Monitoring and Analysis with Emphasis on Acid Deposition*, Technopolis, Pathumthani, Thailand, 25 February - 4 March 1997.

Thuc, T. and Yen, N. H. (1997) Acid rain study in Vietnam, paper presented at the *4th ASEAN Workshop on Air Quality Monitoring and Analysis with Emphasis on Acid Deposition*, Technopolis, Pathumthani, Thailand, 25 February - 4 March 1997.

Vialard-Goudou, A. and Richard, C. (1956) Etude pluviometrique, physiochimique et economique des eaux de pluie a Saigon (1950-1954). *l'Agronomie Tropicale,* Vol. 11, 74-92.

Vongphosy, T. (1997) Lao PDR coutry report, paper presented at the *4th ASEAN Workshop on Air Quality Monitoring and Analysis with Emphasis on Acid Deposition*, Technopolis, Pathumthani, Thailand, 25 February - 4 March 1997.

Wang, W. and Wang, T. (1996) On acid rain formation in China. *Atmospheric Environment*, Vol. 30, 4091-4093.

Yong, L.M. and Eng N.B. (1997) Singapore country report, paper presented at the *4th ASEAN Workshop on Air Quality Monitoring and Analysis with Emphasis on Acid Deposition*, Technopolis, Pathumthani, Thailand, 25 February - 4 March 1997.

Yanagisawa, Y. (1989) Monitoring and simulation, in *How to Conquer Air Pollution: A Japanese Experience*, Studies in Environmental Science 38, edited by H. Nishimura, Elsevier, Amsterdam, pp. 157-196.

Zhang, J., Liu, S.M., Lu, X, and Huang, W.W. (1993) Characterising Asian wind-dust transport to the Northwest Pacific Ocean. Direct measurements of the dust flux for two years. *Tellus*, Vol. 45B, 335-345.

Zhao, D. and Sun, B. (1986) Air pollution and acid rain in China, *Ambio*,Vol. 15, 2-5.

Zhao, D. and Xiong, J. (1988) Acidification in Southwestern China, in *Acidification in Tropical Countries*, ed. Rodhe, H. and Herrera, R., John Wiley & Sons Ltd., 1988, pp. 317-346.

Zhao, D., Xiong, J., Xu, Y., and Chan, W. H. (1988) Acid rain in Southwestern China, *Atmospheric Environment* Vol. 22, 349-358.

Part 3
EXPERIMENTAL STUDIES OF ACIDITY INFLUENCE ON TERRESTRIAL ECOSYSTEMS IN EAST ASIA

Chapter 4

CHANGES IN DISTRIBUTION OF ALUMINUM SPECIES IN SOIL SOLUTION DUE TO ACIDIFICATION

KAZUO SATO, TAKASHI WAKAMATSU AND AKIRA TAKAHASHI
Komae Research Laboratory, Central Research Institute of Electric Power Industry (CRIEPI), 2-11-1, Iwato-kita, Komae, Tokyo 201-8511, Japan

CONTENT

Abstract
1. Introduction
2. Experimental Section
    2.1 Experimental forest
    2.2 Collection of soil solution samples
    2.3 Reagents
    2.4 Determination of cations and anions
    2.5 Aluminum speciation
3. Results and Discussion
    3.1 pH and chemical composition of soil solution samples
    3.2 Speciation of Al in standard solutions
    3.3 Speciation of Al in soil solution samples
    3.4 BC/Al ratios of soil solution samples
References

ABSTRACT

It is well known in Japan that soils close to the stems of Japanese cedar (Cryptomeria japonica) trees are strongly acidic. This is partly due to the leaching of hydrogen ions from the stems. Since till now soil acidification due to acidic deposition has not been observed in Japan, investigation of soils and soil solutions close to stems may be worthwhile to predict the situation that may occur after acidification due to acidic deposition. In a Japanese cedar forest in Gunma Prefecture, Japan, soil solutions have been collected by ceramic porous cups at a depth of 10 cm. Soil solutions far from stems (• 100 cm) are slightly acidic (pH ~

5.8) and contain 1.0 µM of total Al in average. The speciation of Al, using cation exchange chromatography with fluorescence detection of the Al-lumogallion complex, shows that nearly 100 % of the Al consist of species with a charge less than or equal to +1 (possibly, organically chelated Al). The molar ratios of BC (= Ca^{2+} + Mg^{2+} + K^+) to total Al are extremely high, ranging from 66 to 1050. In contrast, soil solutions close to a stem (10 cm) are markedly acidic (pH ~4.5) and contain 47 µM of total Al in average. Furthermore, more than 55 % of the Al are in the form of Al^{3+}. The BC/T-Al ratios in winter decline to as low as 2,

1. INTRODUCTION

Aluminum is toxic to roots of plants (e.g., Godbold et al., 1988). This toxicity can be mitigated by base cations such as Ca^{2+} (Cronan and Grigal, 1995). Thus, it may be reasonable to use the molar ratio of BC (= Ca^{2+} + Mg^{2+} + K^+) to Al in soil solution as an indicator of the potential for forest decline to estimate critical loads of acidity for forest soils (Posch et al., 1995). However, there are still uncertainties regarding the phytotoxic forms of Al. Aluminum in solution occurs as free Al^{3+}, and as complexes with hydroxyl, fluoride, sulfate and organic ligands. Among these, Al^{3+} and its hydrolysis species such as $AlOH^{2+}$ may be much more toxic (Wolt, 1994). Polynuclear hydroxy-Al species such as $AlO_4Al_{12}(OH)_{24}(H_2O)_{12}^{7+}$ (referred to as Al_{13} polymer) are also reported to be phytotoxic (Paker et al., 1989).

In Japan, there were a few studies associated with the speciation of Al in soil solution. Kato et al. (1995) conducted equilibrium calculations for ionic species of Al in soil extracts and estimated that more than 90 % of total Al was $Al_6(OH)_{15}^{3+}$ in the pH range of 4 to 7. Below pH 4, Al^{3+} was the dominant species. Tsunoda et al. (1997) studied Al species in soil extracts by cation exchange chromatography with fluorescence detection using 5-sulfo-8-quinolinol and found that most of the water-soluble Al was complexed with organic substances, whereas KCl-extractable Al was mostly in the form of free Al^{3+}. These studies were conducted in the laboratory using soil extracts and forms of Al in the field are still virtually unknown. The same is true for the BC/Al ratios. In addition, soil acidification due to acidic deposition has not been observed in Japan.

Under these circumstances, this study focuses on an acidification phenomenon in the field - severe soil acidification occurring around the stems of Japanese cedar (Cryptomeria japonica) trees. It has been observed that stemflow samples of Japanese cedar trees are strongly acidic (pH ~3). This is partly due to the leaching of hydrogen ions from the stem (Sato and Takahashi, 1996). Since the zone of soil acidification is limited to within roughly 20 cm of the stem, we have collected soil solution samples at a site close to a stem (10 cm) and a site far from stems (• 100 cm) using porous cups to elucidate the effects of acidification on chemical forms of Al as well as on values of the BC/Al ratios.

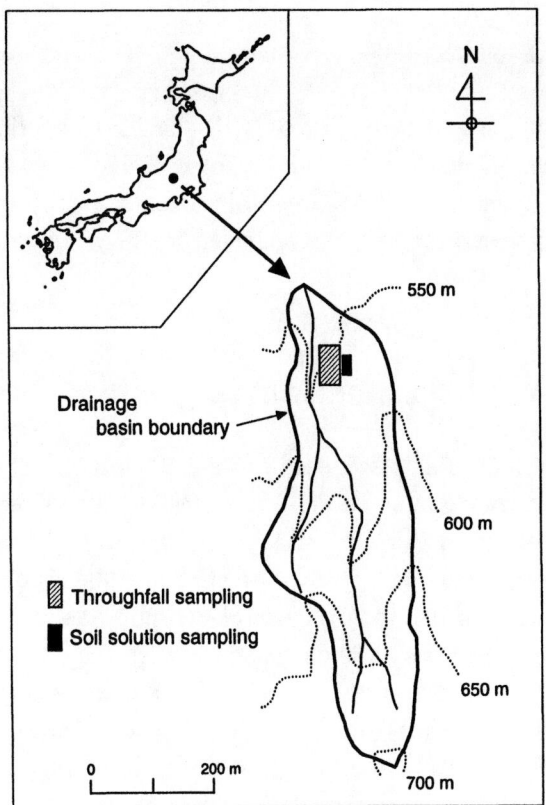

Figure 1. Location of the experimental forest

Table 1 Description of the experimental forest

watershed area	11.4 ha
elevation	520 - 710 m
rainfall	1,112 mm[a]
average air temperature	11.9 ℃[a]
bed rock	Andesite
soil type	Ando soil
vegetation	Japanese cedar[b]

[a] from January to December 1997
[b] 38 years old

2. EXPERIMENTAL SECTION

2.1. EXPERIMENTAL FOREST

Soil solution samples were collected in the CRIEPI experimental forest located at Myogi Town in Gunma Prefecture, Japan (Figure 1). Table 1 shows selected characteristics of the forest. The forest belongs to Gunma Prefecture, is covered by Ando soil whose parent material is volcanic ash, and is dominated by 38 years old planted Japanese cedar trees.

2.2 COLLECTION OF SOIL SOLUTION SAMPLES

To collect soil solution samples close to a stem, a typical cedar tree with DBH of 22 cm was selected, and two ceramic porous cups (Daiki Rika Kogyo DIK-3900-12) were installed to a depth of 10 cm (Ah horizon) within 10 cm from the stem. Two capillary tubes from the cups were connected to a 30 mL polyethylene tube. The tube was then connected to a 500 mL glass flask so that the overflow of the tube was collected. Thus, we sampled soil solution in a tube and a flask. For chemical analyses, the tube sample was used in order to avoid bias due to an increase in pH caused by CO_2 degassing (Suarez, 1986, 1987). To collect soil solution samples far from stems, three porous cups were installed to a depth of 10 cm at a site more than 100 cm distant from the surrounding trees. Connection of three capillary tubes from the cups were made in the same way as stated above. The suction pressure applied to the flask was maintained at 50 cmHg. Sampling was conducted twice a month from May 1997 to January 1998. The duration of a sampling was set to 3-4 days so that more than roughly 200 mL of solution overflowed from a tube to a flask to obtain soil solution, not affected by CO_2 degassing, in the tube.

2.3 REAGENTS

Nitric acid was of special grade for heavy metal analyses from Wako, Japan. Aluminum stock standards were prepared from an Al atomic-absorption standard (1000 ppm) from Wako. Lumogallion was from Tokyo Kasei Kogyo, Japan. The other chemicals were of analytical grade from Wako. Water was distilled and deionized using Yamato Kagaku WA53.

2.4 DETERMINATION OF CATIONS AND ANIONS

Soil solution pH was measured using a glass-electrode without filtration. Cations (Ca^{2+}, Mg^{2+}, K^+, Na^+ and NH_4^+) and anions (SO_4^{2-}, NO_3^-, Cl^- and F^-) were determined by ion chromatography (Dionex DX-500) after rapid filtration of the soil solution using a 0.22 μm Millipore cellulose acetate filter.

Eluent for cations was 20 mM CH_3SO_3H (pH 1.8), and its flow was 1.0 mL/min. Eluent for anions except for fluoride was 2.7 mM Na_2CO_3 - 0.3 mM $NaHCO_3$ (pH 10.6), and its flow was 1.3 mL/min. For fluoride, 1.3 mM $NaHCO_3$ (pH 8.8) was used for eluent to prolong the retention time to separate the peak for fluoride from waterdip. It was confirmed by the analyses of aluminum fluoride standards that the method determined the sum of free F and F complexed with Al^{3+}.

2.5 ALUMINUM SPECIATION

Soluble Al species in soil solution were separated and quantified using cation exchange chromatography with fluorescence detection of the Al-lumogallion complex. This method was originally developed by Sutheimer and Cabaniss (1995b). The principle of this method is to separate Al species in solution into three groups of Al species according to their valence, $AlL_x^{\cdot 1+}$, AlL_x^{2+} and Al^{3+}, where L denotes a ligand. Then, each group of the Al species is quantified fluorometrically. Our method was based primarily on their method except a gradient elution system was not used.

The HPLC system used in this study is diagrammed in Figure 2, and its operating conditions are listed in Table 2. The system, prepared by Shimazu Co., includes two LC-10Ai reagent pumps, a SIL-10Ai autoinjector, a RF-10A_{XL} fluorescence monitor, and a LC-10 chromatography data processor. The cation exchange column (Dionex IonPac CS-2) is maintained at 25 • C in a column oven (Shimazu CTO-10AC$_{VP}$). The mixing coil has a length of 10 m (2 mL), the first half maintained at 60 • C in a reaction oven (Shimazu CRB-6A), while the second half remained at room temperature. All tubing is 0.5 mm i.d. PTFE.

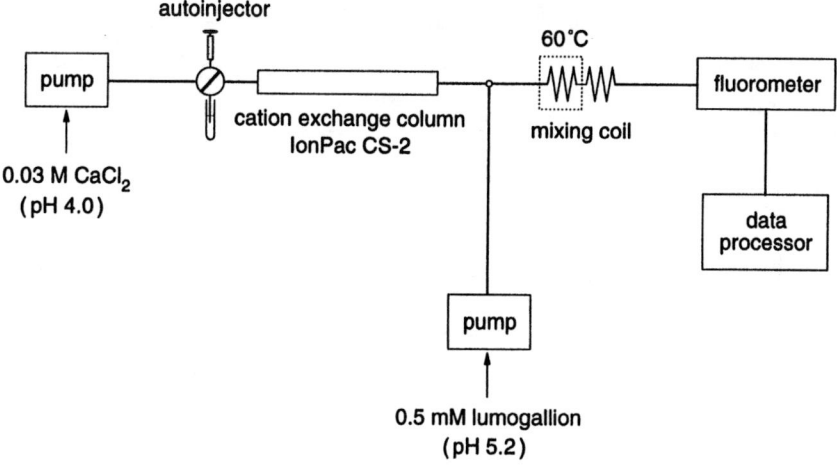

Figure 2. Schematic diagram of the HPLC system

Table 2 Operating conditions for the HPLC system

eluent composition	0.03 M $CaCl_2$ (pH 4.0 with HNO_3)
eluent flow	1.2 mL/min
sample injection volume	30 μL
column temperature	25 °C
postcolumn reagent composition	0.5 mM lumogallion (in 0.2N, pH 5.2 sodium acetate buffer)
postcolumn reagent flow	0.5 mL/min
mixing coil	5 m at 60 °C + 5 m at room temperature
fluorometer	excitation : 500 nm
	emission : 595 nm
	sensitivity : medium
	gain : 16

The original method of Sutheimer and Cabaniss (1995b) uses a gradient elution system that linearly changes the eluent from 100 % water to pH 4, 0.03 N $CaCl_2$ during the first minute of elution time. The advantages of this system and the reason why we did not use it will be described in section 3.2

Eluent (pH 4, 0.03N $CaCl_2$), postcolumn reagent (0.5 mM lumogallion in 0.2 N, pH 5.2 acetate buffer), aluminum standards containing 0, 3.63, 7.18 and 10.7 μM Al in 100 μM citrate, aluminum fluoride standards, and aluminum citrate standards were prepared according to Sutheimer and Cabaniss (1995b).

3. RESULTS AND DISCUSSION

3.1 PH AND CHEMICAL COMPOSITION OF SOIL SOLUTION SAMPLES

Table 3 shows the average pH and chemical composition of 17 soil solution samples close to a stem or far from stems. Hereafter, the former samples are referred to as S samples (meaning to receive stemflow) and the latter as T samples (meaning to receive throughfall).

As expected, pH values for S samples are consistently lower than those for T samples. The average difference between the two is 1.3 pH units which corresponds to a 20 times difference in terms of H^+ concentrations. Stemflow and throughfall samples collected during the same period had pH values of 3.62 and 4.80 in average, respectively. Thus, it is clear that acidification of soils close to a stem is caused primarily by strongly acidic stemflow. This results in extreme increases in total Al concentrations in soil solutions (approximately 30 times).

Table 3 Average chemical composition of soil solutions (n = 17)[a]

		S samples[b]	T samples[c]
pH		4.52	5.82
Ca^{2+}	(μM)	91.7	286
Mg^{2+}	(μM)	22.8	45.6
K^+	(μM)	89.0	119
Na^+	(μM)	41.2	37.0
NH_4^+	(μM)	5.6	3.6
T-Al[d]	(μM)	56.0	1.9
SO_4^{2-}	(μM)	45.1	38.6
NO_3^-	(μM)	265	569
Cl^-	(μM)	147	173
T-F[e]	(μM)	4.1	1.0

[a] from May 1997 to January 1998
[b] samples close to a stem (~ 10 cm)
[c] samples far from stems (≥ 100 cm)
[d] determined by ICP-AES
[e] free F^- + F^- complexed with Al^{3+}

Before forms of Al species in S and T samples are discussed, chromatograms and equilibrium calculations for Al species in standard solutions are described.

3.2 SPECIATION OF AL IN STANDARD SOLUTIONS

Chromatograms

Figure 3 illustrates typical chromatograms of solutions of (a) aluminum fluoride and (b) aluminum citrate. For aluminum fluoride, three different peaks are shown clearly. For aluminum citrate, however, the second peak is small and placed on the tail of the first peak.

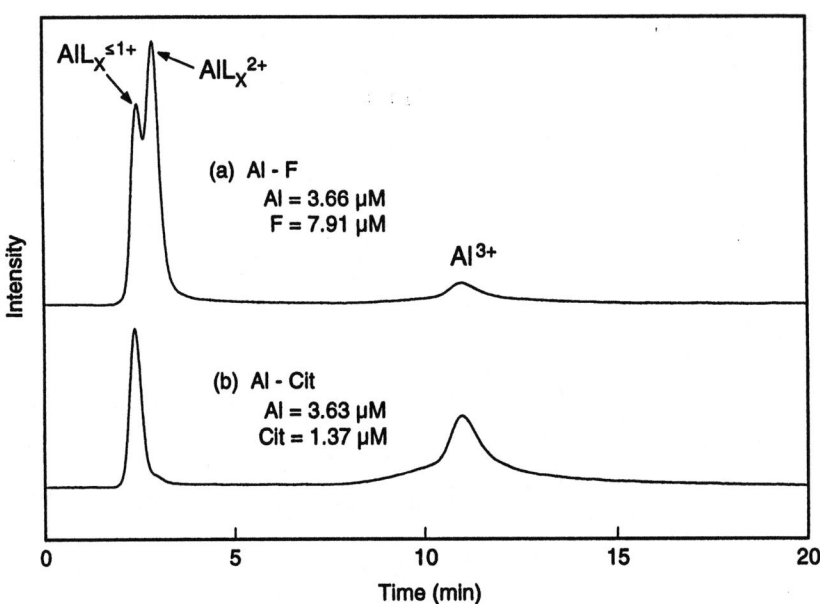

Figure 3. Typical chromatograms of solutions of (a) aluminum fluoride and (b) aluminum citrate.

As stated in section 2.5, the distinctive feature of the method of Sutheimer and Cabaniss (1995b) is to linearly change the eluent from water to pH 4, 0.03 N $CaCl_2$ during the first minute of elution time. With this, the first and the second peaks can be separated completely, because the second peak appears later. To apply this technique, trace aluminum contaminants in the eluent solution, originating from calcium chloride dihydrate, should be removed completely. Otherwise the baseline of the chromatogram will go up during the course of the gradient elution (this was confirmed experimentally by ourselves). Even though the $CaCl_2$ eluent solution is cleaned with Chelex 100 resin, trace aluminum impurities remain in the solution. Thus, Sutheimer and Cabaniss (1995b) place a SynChropak CATPC precolumn (SynChrom Inc.) between the main pump and the injector to remove trace aluminum. But, we could not obtain the precolumn in Japan. Therefore, we gave up applying the gradient elution, and divided the first and the second peaks vertically. This approach was also adopted by Tsunoda et al. (1997) who conducted the speciation of Al using 0.2 M K_2SO_4 as a eluent and 5-sulfo-8-quinolinol as a fluorometric reagent.

Equilibrium calculations

To validate the HPLC system used in this study, aluminum species distributions were estimated by equilibrium calculations and compared with analytical data. Formation constants used in the calculations are listed in Table 4. These constants were corrected for ionic strength using the limited Debye-Huckel equation. Calculations were made using basic programs of our own making.

Table 4   Formation constants used for equilibrium calculations

	log K° (I = 0)	references
$Al^{3+} + H_2O = AlOH^{2+} + H^+$	-5.02	Lindsay (1979)
$Al^{3+} + 2H_2O = Al(OH)_2^+ + 2H^+$	-9.30	Lindsay (1979)
$Al^{3+} + 3H_2O = Al(OH)_3 + 3H^+$	-14.99	Lindsay (1979)
$Al^{3+} + SO_4^{2-} = AlSO_4^+$	3.20	Lindsay (1979)
$H^+ + F^- = HF$	3.00	Lindsay (1979)
$Al^{3+} + F^- = AlF^{2+}$	6.70	Sutheimer and Cabaniss (1995b)
$Al^{3+} + 2F^- = AlF_2^+$	12.09	Sutheimer and Cabaniss (1995b)
$Al^{3+} + 3F^- = AlF_3$	15.60	Sutheimer and Cabaniss (1995b)
$Al^{3+} + Cit^{3-} = AlCit$	9.63	Sutheimer and Cabaniss (1995b)
$Al^{3+} + Cit^{3-} = Al(-H)Cit^- + H^+$	2.93	Sutheimer and Cabaniss (1995b)
$Al^{3+} + Cit^{3-} + H^+ = AlHCit^+$	13.12	Whitten et al. (1992)
$H^+ + Cit^{3-} = HCit^{2-}$	6.44	Whitten et al. (1992)
$2H^+ + Cit^{3-} = H_2Cit^-$	11.32	Whitten et al. (1992)
$3H^+ + Cit^{3-} = H_3Cit$	14.60	Whitten et al. (1992)

The results are shown in Table 5. The Al species distributions analyzed are in good agreement with calculated values. Thus, we believe that the HPLC system provides accurate results even though the first and the second peaks are divided vertically. For aluminum fluoride standards, $AlL_X^{*\,1+}$ and AlL_X^{2+} are dominated by AlF_2^+ and AlF^{2+}, respectively. For aluminum citrate standards, $AlL_X^{*\,1+}$ is dominated by AlCit.

In our calculations, $Al(OH)_2^+$ and $AlOH^{2+}$ are assigned to $AlL_X^{*\,1+}$ and AlL_X^{2+}, respectively. But, Sutheimer and Cabaniss (1995b) include them in the third peak; they assume that the third peak, Al_f, is composed of the sum of Al^{3+} and its hydrolysis species ($AlOH^{2+}$, $Al(OH)_2^+$, $Al(OH)_3$ and $Al(OH)_4^-$). The reason is not shown, but they also report that the second peak may include a fraction of free $AlOH^{2+}$.

3.3 SPECIATION OF AL IN SOIL SOLUTION SAMPLES

Chromatograms

Typical chromatograms of a S sample and a T sample are illustrated in Figure 4. The average concentrations of $AlL_X^{*\,1+}$, AlL_X^{2+} and Al^{3+} are summarized in Table 6. For T samples, the total Al concentrations are 1.0 μM in average, and all of the Al species are determined as $AlL_x^{\cdot 1+}$. In contrast, the total Al concentrations of S samples are significantly high (ave. 47.0 μM), and more than 55 % of the total Al are in the form of Al^{3+}. These results clearly show that severe soil acidification not only causes significant increases in total Al concentrations but also changes the dominant Al species to Al^{3+} that is believed to be highly phytotoxic.

Table 5 Speciation of Al in standard solutions by HPLC analyses and equilibrium calculations

ligand*		Al		AlL_x^{S1+}	AlL_x^{2+}	Al^{3+}
L	μM	μM		%	%	%
fluoride (F⁻)	1.98	3.66	HPLC	n.d.	44	56
			calculation	2	48	50
	3.96	3.66	HPLC	n.d.	66	34
			calculation	10	68	21
	7.91	3.66	HPLC	29	56	15
			calculation	37	59	5
	11.8	3.65	HPLC	50	43	7
			calculation	56	43	2
citrate (Cit³⁻)	0.35	3.65	HPLC	6	3	91
			calculation	8	8	84
	1.37	3.63	HPLC	26	6	68
			calculation	31	6	64
	3.38	3.57	HPLC	61	n.d.	39
			calculation	64	3	33
	5.33	3.52	HPLC	76	n.d.	24
			calculation	81	2	19

*pH of Al - F solutions is 3.88; pH of Al - Cit solutions is 4.00

Table 6 Speciation of Al in soil solutions by HPLC

	S samples		T samples	
	μM	%	μM	%
AlL_x^{S1+}	4.3	9	1.0	100
AlL_x^{2+}	6.9	15	n.d.	-
Al^{3+}	35.8	76	n.d.	-
ΣAl	47.0	100	1.0	100

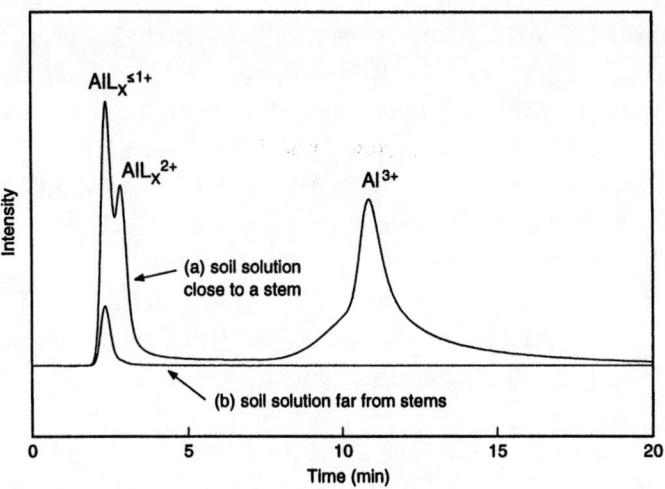

Figure 4 Typical chromatograms of (a) a S sample close to a stem and (b) a T sample far from stems

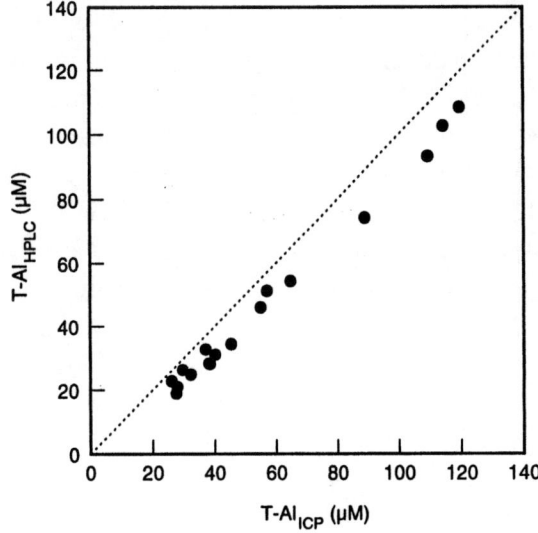

Figure 5 Correlation of the total Al concentrations determined by HPLC and ICP-AES

Total Al concentrations determined by HPLC and ICP-AES

Figure 5 shows the correlation between the total Al concentrations of S samples determined by HPLC and ICP-AES. There is an excellent correlation between the two (r = 0.996), but the HPLC method consistently provides lower Al concentrations by 10-30 %. This phenomenon is believed to be caused by the existence of Al polymers (Sutheimer and Cabaniss, 1995a; Tsunoda *et al*., 1997). According to Kato *et al*. (1995), an Al_6 polymer ($Al_6(OH)_{15}^{3+}$) can exist in soil solutions of forest soils in Japan. Aluminum polymers are thought to be non-reactive with fluorometric reagents such as lumogallion (Sutheimer and Cabaniss, 1995a) and 5-sulfo-8-quinolinol (Tsunoda *et al*., 1997). This may be the cause of lower Al concentrations observed also in this study. In practice, for T samples having higher pH values than S samples (see Table 3), the HPLC method results for total Al concentrations were only 50 % of those determined by ICP-AES.

Equilibrium calculations

Here, the details of $AlL_x^{*\,1+}$ and AlL_x^{2+} species are discussed especially for S samples. Figure 6 compares the concentrations of AlL_x^{2+} determined by HPLC with the concentrations of AlF^{2+} plus $AlOH^{2+}$ estimated by equilibrium calculations using formation constants listed in Table 4. There is a highly positive correlation between the two (r = 0.988), and also both concentrations are almost identical. Thus, AlL_x^{2+} are composed of AlF^{2+} and $AlOH^{2+}$. The average ratio of AlF^{2+} to $AlOH^{2+}$ is 4.1 µM : 2.9 µM. It should be noted that the $AlOH^{2+}$ concentrations were calculated under the condition of pH 4.0, identical to the eluent pH. S samples have higher pH values (~ 4.5), but if these values are used, $AlOH^{2+}$ concentrations become ~ 9.7 µM that make the concentrations of AlF^{2+} plus $AlOH^{2+}$ double. Although the actual separation mechanism of Al species in the cation exchange column used is still unclear, our results strongly suggest that the ratio of Al^{3+} to $AlOH^{2+}$ in the column is governed by the pH of the eluent used. This is warranted by the fact that the equilibrium reaction $Al^{3+} + H_2O = AlOH^{2+} + H^+$ responds very quickly to the changes in pH. Thus, we intend to calculate the concentrations of Al^{3+} and $AlOH^{2+}$ in the sample using their concentrations determined at pH 4.0. For the other Al species such as aluminum fluorides, we postulate that their concentrations in the sample remain unchanged during the course of elution, because their formation constants are relatively high, and so they seem to be sTable 1n the column.

In addition to AlF^{2+} and $AlOH^{2+}$, $AlH_2PO_4^{2+}$ is a likely species with a charge of +2, because its formation constant is relatively large (log K° ~ 3; Bolt and Bruggenwert, 1978). But, phosphate was not detected when S samples were analyzed by ion chromatography. Therefore, the formation of $AlH_2PO_4^{2+}$ is negligible.

It will be worthwhile to mention that more than 99 % of fluoride exist in S samples in the form of AlF^{2+}. This is resulted from the fact that the Al^{3+} concentrations are about 10 times higher than the total fluoride concentrations, and that the formation constant for AlF^{2+} is large (log K° = 6.70). In other words, the AlF^{2+} concentration is regulated by the total fluoride concentration if Al^{3+} concentration is high enough.

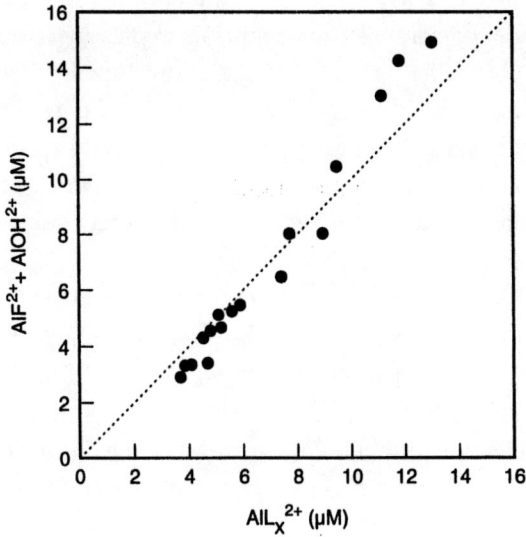

Figure 6   Correlation between AlL_x^{2+} determined by HPLC and $AlF^{2+} + AlOH^{2+}$ estimated by equilibrium calculations

Table 7   Aluminum species in S samples

Al species	μM	%
AlL_x^{S1+a}	3.1	7
$AlSO_4^+$	1.2	3
AlF^{2+}	4.1	9
$AlOH^{2+}$	8.2	17
Al^{3+}	30.4	65

a estimated to be organic Al (see text)

For $AlL_x^{•\ 1+}$ species in S samples, organic and inorganic complexes are likely to occur. Among inorganic complexes, AlF_2^+ and $AlSO_4^+$ have relatively large formation constants ($AlHPO_4^+$ and $AlPO_4^0$ are ignored because of reason mentioned above). The AlF_2^+ concentrations are calculated at ~ 0.02 μM that are negligible levels compared to the analytical concentrations of $AlL_x^{•\ 1+}$ (ave. 4.3 μM). On the other hand, the calculated

$AlSO_4^+$ concentrations are 1.2 µM in average that corresponds to 28 % of $AlL_X^{\bullet\ 1+}$. We think that the remaining 72 % of $AlL_X^{\bullet\ 1+}$ exist in organic forms. To speciate the organically chelated Al, we tried to identify organic ligands in S samples by ion chromatography, but we failed in our efforts probably because of quite low levels of the organic ligands. Anyway, organic Al has been widely detected in soil solutions (e.g., Driscoll et al., 1985).

Table 7 summarizes the equilibrium calculation results for Al species in S samples. The pH values used in the calculations are those of the actual samples. The average concentrations of Al^{3+} and $AlOH^{2+}$ become 0.85-fold and 2.9-fold, respectively compared to those in Table 6 where they are calculated using pH 4.0.

For T samples far from stems, $AlL_X^{\bullet\ 1+}$ was the only Al species determined by HPLC. Strictly speaking, in some cases, the peak for AlL_X^{2+} was found on the tail of the peak for $AlL_X^{\bullet\ 1+}$; the peak for Al^{3+} with a broad shape was found. But, none of peaks were sufficient to determine. The reason why AlF^{2+} and $AlOH^{2+}$ are scarcely produced in T samples is that Al^{3+} levels are negligible. For this reason, the formation of $AlSO_4^+$ should be also negligible. Thus, $AlL_X^{\bullet\ 1+}$ seems to be composed of organic Al species having high formation constants such as aluminum citrates.

Assuming a solution of pH 5.8 (the average pH of T samples) containing 1 µM of total Al and 3 µM of total citrate, then the concentrations of $AlL_X^{\bullet\ 1+}$ (AlCit + Al(-H)Cit$^-$ + AlHCit$^+$), $AlOH^{2+}$ and Al^{3+} are calculated at 0.996, 0.04 and 0.006 µM, respectively. The first value is close to the $AlL_X^{\bullet\ 1+}$ concentrations of T samples (ave. 1.0 µM), and the latter two are too low to determine by the HPLC system used in this study. Now, we are investigating methods to determine low levels of organic ligands to identify organic Al complexes in soil solutions.

Seasonal variations of Al species concentrations

Figure 7 illustrates the seasonal variations of concentrations of the Al species summarized in Table 7. The pH values are also shown. Among the Al species, Al^{3+} concentration varies most significantly. It fluctuates between 8 and 35 µM in warm seasons, but it rises markedly in winter and exceeds 70 µM in January.

If the Al^{3+} concentration is governed by an equilibrium of gibbsite or amorphous $Al(OH)_3$, the considerable increase of the Al^{3+} concentration should be accompanied with an decrease in pH. But, significant pH decline is not observed in winter. Furthermore, the K value defined by the equilibrium equation $p(Al^{3+}) = 3pH - K$ fluctuates between 8.62 and 9.36 $(L/mol)^2$.

On the other hand, values of the ionic strength of S samples increase markedly in winter. In addition, there is a highly positive correlation (r = 0.919) between the total Al concentrations (determined by ICP-AES) and the values of the ionic strength as shown in Figure 8. This high correlation is held even if Al^{3+} is substituted for the total Al. The increases in the ionic strength are caused primarily by the increases in Ca^{2+} concentration. Considering the fact that S samples are collected from an Ah horizon rich in humus, we postulate that Al^{3+} in S samples originates from exchangeable Al retained by organic matter and that its concentration is governed primarily by cation exchange.

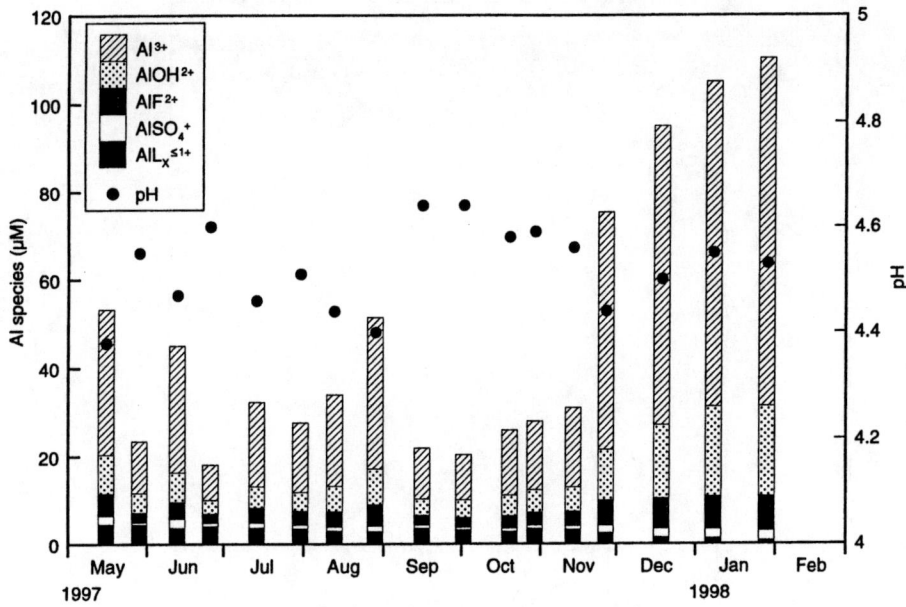

Figure 7 Seasonal variations of Al species concentrations of S samples

Aluminum leaching from porous cups

We have used ceramic porous cups to collect soil solutions. Since ceramic material is made from aluminosilicates, there is concern that Al leaching occurs under acidic conditions and contaminates the soil solution collected. Thus, leaching experiments were conducted using a porous cup.

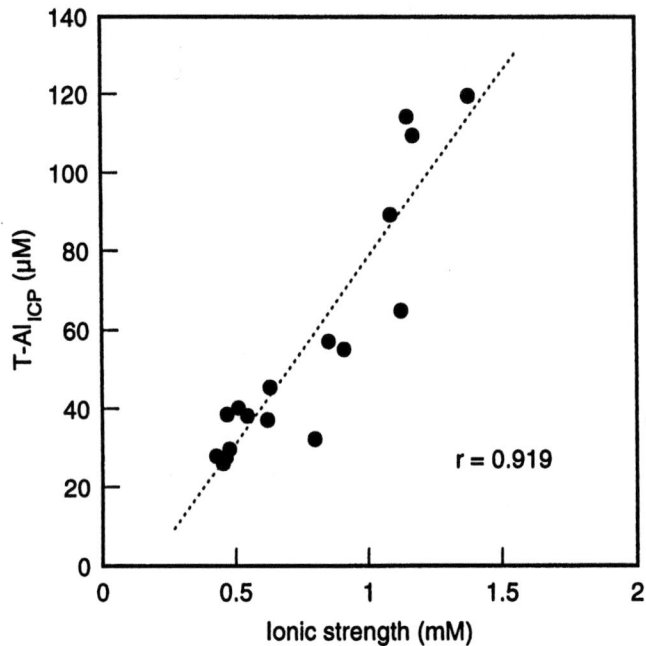

Figure 8. Correlation between the ionic strength and the total Al concentration of S samples.

We had washed porous cups before use with about 200 mL of 0.5 M HCl and then rinsed thoroughly with water according to the recommendation of the manufacturer. The same pretreatment was conducted for a fresh porous cup, and then the cup was immersed in 1 L of a HNO_3 solution of pH 4.47 (close to the average pH of S samples). Then, 25 mL of the solution was collected weekly by suction for a month. As a result, ~ 1 µM of Al was detected. This value corresponds to 2 % of the average Al concentration of S samples. In contrast, when the cup was immersed in a HNO_3 solution of pH 6.05 (close to the average pH of T samples), Al leaching was not detected.

From these results, we presume that Al leaching has not significantly affected the analyzed Al concentrations of S and T samples. For safety, however, it may be better to use materials other than ceramics (e.g., PTFE) for the study of Al.

3.4 BC/AL RATIOS OF SOIL SOLUTION SAMPLES

In Europe, the molar BC (= Ca^{2+} + Mg^{2+} + K^+) to Al ratio of soil solution has been used as an indicator of the potential for forest decline to estimate critical loads of acidity for forest soils (e.g., Posch et al., 1995). The value of 1 is often used for a critical BC/Al ratio for coniferous forests in Europe, below which the risk of forest damage increases. The threshold for Japanese cedar is now under investigation.

Changes in Distribution of Aluminum Species in Soil Solution Due to Acidification

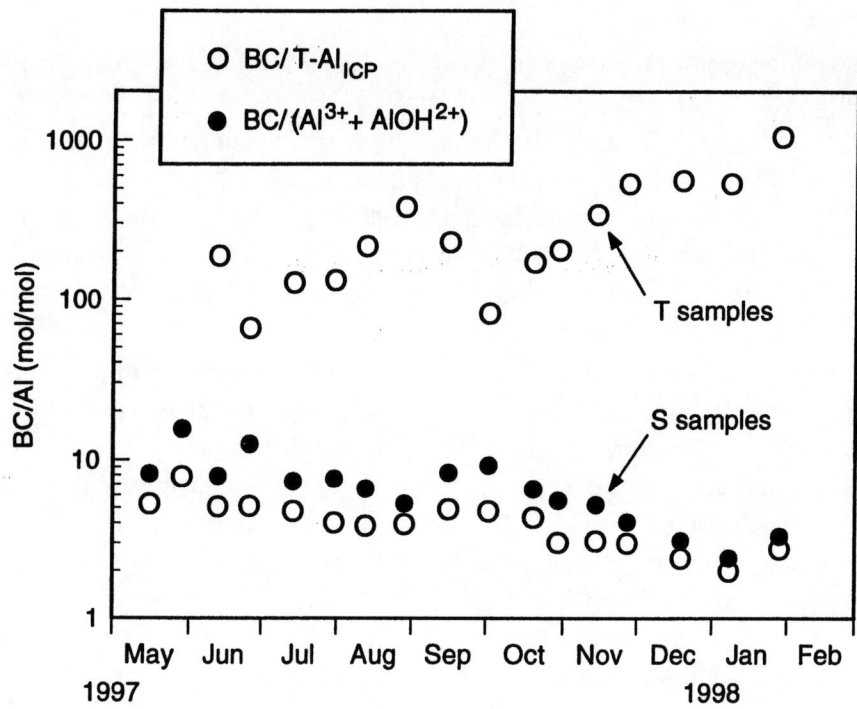

Figure 9. Seasonal variations of the molar BC/Al ratios

Figure 9 shows the seasonal variations of the BC/Al ratios for S and T samples. The Al assigned for T samples is the total Al determined by ICP-AES. For S samples, the total Al as well as Al^{3+} plus $AlOH^{2+}$, which is believed to be highly phytotoxic, are assigned.

For T samples, the BC/T-Al ratios are significantly high, ranging from 66 to 1050 (ave. 320). Furthermore, most of the total Al is thought to be organically chelated as mentioned above. Therefore, Al stress should not be a problem in areas far from the stems of Japanese cedar trees. For S samples, however, the BC/T-Al ratio and the $BC/(Al^{3+} + AlOH^{2+})$ ratio are both below 20 during the period studied and decline in winter to as low as 2, which is close to the threshold of 1 that has been used in Europe. This is due to severe soil acidification caused by strongly acidic stemflow of Japanese cedar tree. The zone of soil acidification is however limited to within roughly 20 cm of the stem and 30 cm deep below the ground surface (Sato and Takahashi, 1996). Thus, most of the roots must be distributed over areas of low Al (and Al^{3+}) levels.

Soil acidification and the resultant decline of the BC/Al ratio observed around the stem imply the situation that may occur when atmospheric acidity loading increases significantly. In that case, the decline of the BC/Al ratio will expand to cover the forest. To prevent such a problem, reduction strategies for emissions of precursors should be vigorously promoted not only in the country but on a East Asian scale.

REFERENCES

Bolt, G. H. and Bruggenwert, M. G. M. (1978) *Soil Chemistry A. Basic Elements*. Elsevier Scientific Publishing Co. (2nd ed.).

Cronan, C. S. and Grigal, D. F. (1995) Use of calcium/aluminum ratios as indicators of stress in forest ecosystems. *J. Environ. Qual.*, **24**, 209-226.

Driscoll, C. T., van Breemen, N. and Mulder, J. (1985) Aluminum chemistry in a forested spodosol. *Soil Sci. Soc. Am. J.*, **49**, 437-444.

Godbold, D. L., Fritz, E. and Huttermann, A. (1988) Aluminum toxicity and forest decline. *Proc. Natl. Acad. Sci. USA*, **85**, 3888-3892.

Kato, H., Shirai, M. and Matsukawa, S. (1995) Ionic species of aluminum in acid soils in terms of chemical equilibrium. *Jpn. J. Soil Sci. Plant Nutr.*, **66**, 39-47 (In Japanese).

Lindsay, W. L. (1979) *Chemical equilibria in soils*. John Wiley & Sons, Inc.

Paker, D. R., Kinraide, T. B. and Zelazny, L. W. (1989) On the phytotoxicity of polynuclear hydroxy-aluminum complexes. *Soil Sci. Soc. Am. J.*, **53**, 789-796.

Posch, M., de Smet, P. A. M., Hettelingh, J.-P. and Downing, R. J. (1995) *Calculation and Mapping of Critical Thresholds in Europe*. RIVM Rep. 259101004, Bilthoven, Netherlands.

Sato, K. and Takahashi, A. (1996) Hydrogen ion leaching from the stem of a Japanese cedar tree and acidification of soils around the stem. *Environ. Sci.*, **9**(2), 221-230 (In Japanese).

Suarez, D. L. (1986) A soil water extractor that minimizes CO_2 degassing and pH errors. *Water Resour. Res.*, **22**, 876-880.

Suarez, D. L. (1987) Prediction of pH errors in soil-water extractors due to degassing. *Soil Sci. Soc. Am. J.*, **51**, 64-67.

Sutheimer, S. H. and Cabaniss, S. E. (1995a) Determination of trace aluminum in natural waters by flow-injection analysis with fluorescent detection of the lumogallion complex. *Anal. Chim. Acta*, **303**, 211 - 221.

Sutheimer, S. H. and Cabaniss, S. E. (1995b) Aqueous Al(III) speciation by high-performance cation exchange chromatography with fluorescence detection of the aluminum-lumogallion complex. *Anal. Chem.*, **67**, 2342 - 2349.

Tsunoda, K., Yagasaki, T., Aizawa, S., Akaiwa, H. and Satake, K. (1997) Determination and speciation of aluminum in soil extracts by high-performance liquid chromatography with fluorescence detection using 5-sulfo-8-quinolinol. *Anal. Sci.*, **13**, 757-764.

Whitten, M. G., Ritchie G. S. P. and Willett I. R. (1992) Forms of soluble aluminium in acidic topsoils estimated by ion chromatography and 8-hydroxyquinoline and their correlation with growth of subterranean clover. *J. Soil Sci.*, **43**, 283-293.

Wolt, J. (1994) *Soil solution chemistry: applications to environmental science and agriculture*. John Wiley & Sons, Inc.

Chapter 5

DIFFERENTIAL SENSITIVITY OF TREES TO SIMULATED ACID RAIN OR OZONE IN COMBINATION WITH SULFUR DIOXIDE

YOSHIHISA KOHNO, HIDEYUKI MATSUMURA, AND TAKUYA KOBAYASHI
Biology Department, Abiko Research Laboratory,
Central Research Institute of Electric Power Industry
1646 Abiko, Abiko City, CHIBA 270-1194, JAPAN

CONTENT

Abstract
1. **Introduction**
   1.1 Background
   1.2 Forest decline in Japan
   1.3 Factors affecting tree growth
   1.4 Air quality and acidic deposition in Japan
2. **Effect of wet deposition on tree growth**
   2.1 Introduction
   2.2 Simulated acid rain generator system
   2.3 Effect of acid mist
   2.4 Visible injuries induced by simulated acid rain
      2.4.1 Coniferous trees
      2.4.2 Broad-leaved trees
   2.5 Growth responses as chronic effects
      2.5.1 Coniferous trees
      2.5.2 Deciduous Broad-leaved trees
   2.6 Effect soil acidification stress due to wet deposition
      2.6.1 pH stress in rhizosphere
      2.6.2 Al toxicity in coniferous trees
      2.6.3 Critical point for Japanese coniferous trees
3. **Effect of ozone and/or sulfur dioxide as dry deposition**
   3.1 Introduction

3.2 Tunnel type open-top chamber (OTC)
3.3 Pollutant exposures
3.4 Visible injuries
3.5 Growth response
    3.5.1 Single effect of ozone or sulfur dioxide
    3.5.2 Combined effect of ozone and sulfur dioxide
4. **Interactive effect of rain acidity and ozone**
5. **Conclusion**
   References

ABSTRACT

Declines of several tree species and forests in the different areas of Japan are noted, as European forests are suffering. However, they may be linked with different causes: forest succession dynamics, diseases and pests, meteorological extreme conditions, air pollutants, acidic deposition, and so on.

Exposure experiments have been conducted to assess cause-effects relationships with acidic deposition and forest decline. Simulated acid rain or ozone in combination with sulfur dioxide was exposed to sixteen potted-tree species for 3 growing seasons. Rain acidity below pH 4.0 could induce deleterious effects on some broad-leaved trees, however, coniferous trees did not show any significant growth reduction. Sulfur dioxide and/or ozone induced complicated differential growth responses than the wet deposition did. Some indicated additive harmful effects of sulfur dioxide. Others showed synergistic adverse effects of ozone and sulfur dioxide. Experimental results suggest that ambient level of ozone may take an important role to reduce tree vitality, since ozone induced chronic changes in carbon allocation will be accelerated by increased nitrogen input.

Assuming that fossil fuel energy consumption will continue at the current rate in the China, sulfur emission will be double at the year of 2010. Without any countermeasures for reduction of sulfur dioxide emission, it may induce possible adverse direct effects on the natural vegetation due to increasing concentration of sulfur dioxide rather than due to increasing wet deposition of sulfate. Exposure experiments suggest that differential sensitivity of plants to primary gaseous pollutants and its critical level will be a more important factor to explain forest decline rather than the soil acidification stress associated with wet acid deposition.

1. INTRODUCTION

1.1 BACKGROUND

Since Scandinavian scientists pointed out that rain acidification was possibly linked with reduction of forest productivity in the early 1970's (Likens & Bormann, 1974; Tamm, 1976), acid rain research activities extensively evolved in the European countries. National Acid Precipitation Assessment Program (NAPAP) in the United States had established in 1980 and had conducted surveys, experiments and review works in different fields. NAPAP narrowed scientific knowledge and raised further research needs to understand forest decline and other fields (Irving, 1990). After the completion of the 10-year NAPAP, many extensive programs in this issue have been re-organized in the North America and European countries. Plant scientists pointed out that photochemical oxidants have still an important and potential toxic factor to reduce plant growth and productivity rather than rain acidity.

In contrast, eastern Asian countries including Japan have a still great concern about effects of acid rain on the environment. Significant increase in SO_2 emission in this area, especially in the China will be a great threat to natural environment as well as agricultural fields (Rodhe *et al.*, 1995). As well as east Asian countries around the China, Japanese Archipelago in the downstream may receive such an increased acidic deposition. Asian vegetation has more diversity than that in Europe and North Eastern America. However, there is little information about long-term chronic effects of current acidic deposition on Asian native species. This article described growth responses in young trees exposed to simulated acid rain or air pollutants: ozone and sulfur dioxide based on the long-term exposure experiments.

1.2 FOREST DECLINE IN JAPAN

In the late 1970's to early 1980's, Western Germany suffered from the new forest decline syndrome in Norway spruce (*Picea abies*) and mass media distributed catastrophic forest devastation and possible linkage of acid rain with such a decline. In Black Triangle Region, where is located along the country borderlines of Czech Republic, former Eastern Germany and Poland, there were many declined forest stands (Cerny & Paces, 1995). However, reforestation in such areas is now recovering forest stands. In the Black Forests and Hartz Mountains in Germany, yellowing or chlorosis of older needles are typical symptoms, but current young needles are not chlorotic (Krahl-Urban *et al.*, 1988). In the northeastern United States fir stands in the top of Great Smoky Mountain and Mt. Mitchell and spruce stands in northern mountain states had been suffered. Also decline of sugar maples and other associated deciduous forest trees in Canada had been reported (UNEP-ECE, 1988, 1991).

Japanese first indication of rain acidification stress was reported as the petal discoloration of morning glory and azaleas in the downtown of the Kanto and Kinki Districts after very small amount of rain events in the early 1970's (Yoshida, 1971; Fujii, 1971). After Tokyo Metropolitan Research Institute reported it as "the wet acid air pollution" (TMRI, 1975), scientists of the local governments volunteered to start the monitoring network of precipitation chemistry. In contrast to European countries where research activities were developed in the both fields of air quality monitoring and its ecological effects, ecological study part was inactive. After Sekiguchi *et al.* (1986) reported decline of Japanese cedar (*Cryptomeria japonica*) in Kanto District in 1986, many kinds of acid rain effect assessment programs started (Katoh *et al.*, 1990).

In the late 1960's dieback of *Cryptomeria japonica* has been recognized in urban areas (Yambe, 1978) and it might be linked with high concentration of sulfur dioxide in those days. Decline of spruce or fir stands distributed in the mountain ridge of the European countries, however, dieback of *Cryptomeria japonica* was different. Nationwide field surveys by Nashimoto (1993) revealed that it distributed in 5 of 9 suburban flat areas in addition to the Kanto Plain, but not in the mountains. He suggested that high photochemical oxidants dose and a small amount of precipitation during the growing season was possibly linked with such dieback from field surveys and analysis of data sets (Nashimoto & Kohno, 1989; Nashimoto, 1993).

Recent public interest has expanded into the decline of natural forest stands in subalpine zone, where *Abies veitchii*, *Abies mariesii*, *Picea hondoensis* or *Picea jezoensis* are dominant. This could be due to strong wind damage such as a typhoon in summer or dry strong wind in early spring and would be possibly linked with forest succession process, however, some scientists suggested a linkage with acidic deposition. Declines of old beech (*Fagus crenata*) or fir (*Abies firma*) stands in the mountains have been also reported (Kohno, 1996). Other widespread one is pine wilt caused by the infection of pine wilt nematodes(*Bursaphelenchus xylophilus*).

1.3 FACTORS AFFECTING TREE GROWTH

There are many stress factors to induce plant growth reduction in the nature (Fig. 1). One is biotic and the other is abiotic. Biotic factors include pathogens, insects, animals, competition of individual plant species in plant succession processes, and so on. Abiotic factors include meteorological conditions such as extremely low and high temperature, strong wind, drought stress, anthropogenic pollutants, and so on. It is very important to understand which factor will be a main or primary cause for plant growth inhibition or reduction, which stress factor will act as a co-factor or secondary one, or which factors will be coupling to induce the growth reduction.

One of possible causes for forest decline in North America and Europe is an effect of acidified rain and/or fog by incorporation of sulfur and/or nitrogen oxides from fossil fuel combustion. As forest in central Europe has been severely damaged and soil has been acidified, large scale clear-cut areas have extended for the last 2 decades (Cerny & Paces, 1995). However, dead tree stands, clear-cut areas, apparent normal forest stands, and reforested areas are co-existing in such areas. This is telling us hardly to explain the forest decline syndrome by only a simple hypothesis such as soil acidification stress. This situation raises discrepancies of acid rain issues: What is acid rain effect? What is soil acidification stress? What is the key factor to induce forest decline? Why can young trees re-grow in such hardly damaged areas? And so on.

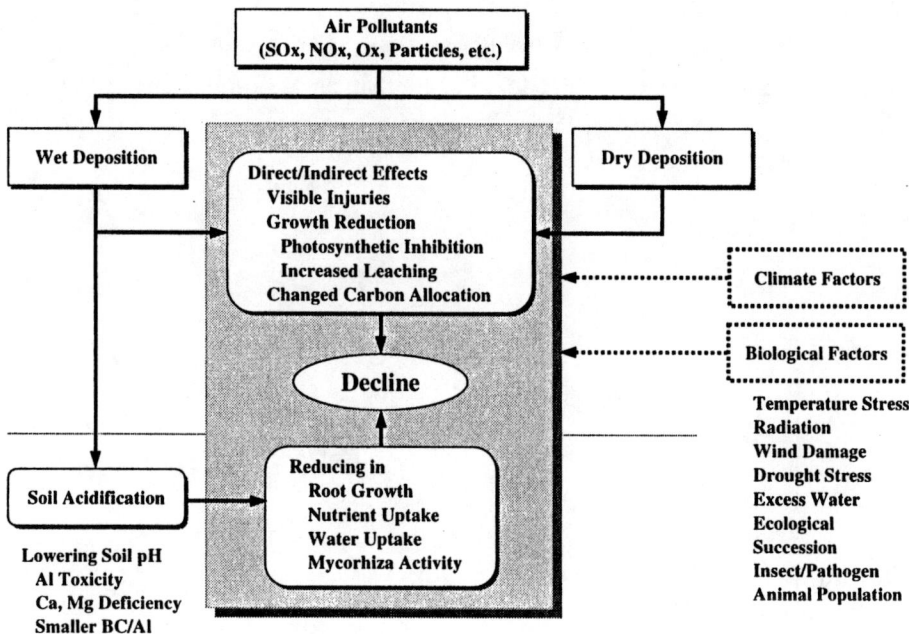

Fig.1. Hypothetical pathways of tree/forest decline emphasized on acidic deposition

1.4 AIR QUALITY AND ACIDIC DEPOSITION IN JAPAN

Acid rain or acid deposition articles reported by mass media have some confusions. Some described acidification of rain, mist, fog, or snow, which should be categorized into wet deposition. Others did air pollution by sulfur dioxide, nitrogen oxides and suspended particulate matters that should be categorized into dry deposition.

"Acid rain" or "rain acidification" is a very part of acid deposition phenomena. Therefore, "Acidic Deposition" has to cover not only acidification of rain and its effects but also other scientific phenomena related with dry and wet deposition. As discussing only wet deposition associated with rain acidification will not contribute to scientific solutions for this issue, effect of air pollution as dry deposition in combination with wet deposition should be discussed in this paper.

Annual mean concentration of sulfur dioxide in the ambient air at the national monitoring stations was extremely high in the middle of 1960's. However, it dramatically decreased by installation of desulfurization units and introducing low sulfur combustion fuels by industries (Fig. 2). Current nationwide annual mean concentration of sulfur dioxide in more than 1000 monitoring stations operated by local governments is about 5 ppb with standard deviation of 1.5 ppb. Mean concentration of sulfur dioxide by weighted land use areas provides 3 ppb (Fujita, 1996).

Concentration of nitrogen oxides is still high in urban areas and that of photochemical oxidants is high enough to induce crop loss in suburban areas. Table 1 suggested that urban areas are high in nitrogen oxides. However, remote area was quite low concentrations of NOx (NO+NO_2) and sulfur dioxide.

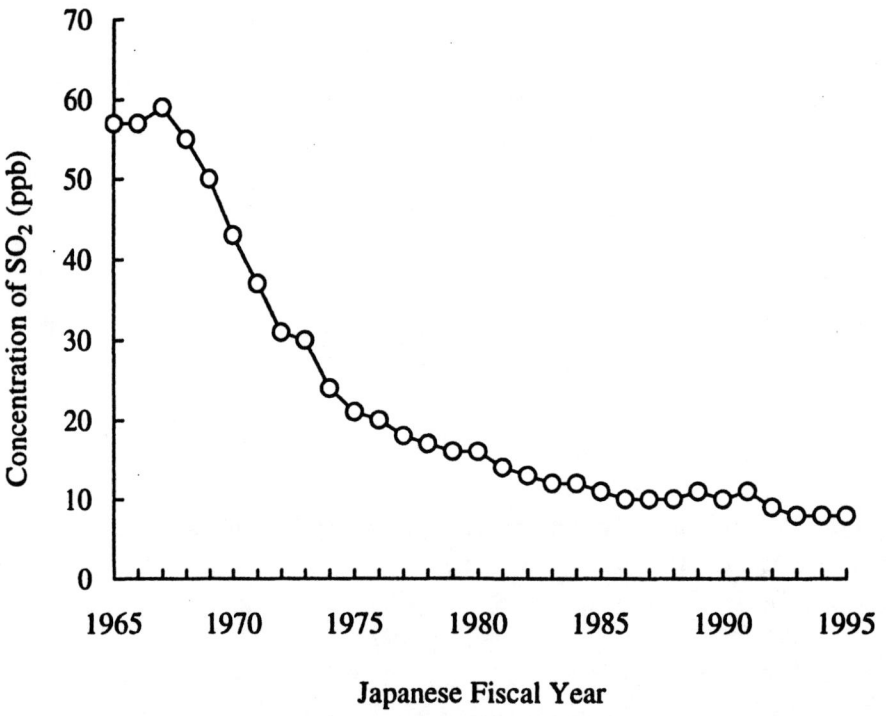

Fig. 2 Trend in annual mean concentration of SO_2 at 15 national monitoring stations (JEA, 1997)

Table 1 Annual mean concentrations of air pollutants in ambient air

Location	SO_2	O_3	NO_x	NO	NO_2
Abiko, Chiba	2.1	20.7	33.6	16.3	16.3
Akagi, Gunma	0.3	33.4	3.5	1.1	2.2

Measured period: Jan. 1991 - Dec. 1996 for SO_2, NO_x and O_3. Jan. 1992 - Dec. 1996 for NO and NO_2.
Values were calculated from means of 24 hours.
Abiko: Abiko Research Laboratory of CRIEPI at 20m above sea level is located in the residential area but along to the national highway. 35 km NE from the central Tokyo.
Akagi: Akagi Testing Center of CRIEPI at 540m above sea level is located in the secondary pine forested area on the south mountainside of Mt. Akagi. 100 km NW from the central Tokyo.

Annual mean acidity (pH) of precipitation at 21 monitoring stations operated by CRIEPI ranged from 4.5 to 5.1 with an average of about 4.8 in Table 2 (CRIEPI, 1992). This was the almost same as the monitoring results from Japan Environmental Agency (JEA, 1992). Concentration of sulfate in precipitation of China was extremely high, however, its acidity was neutralized by high concentrations of basic cations such as calcium and ammonium ions (CRIEPI, 1994). In contrast, Japanese rain contained less sulfate and nitrate, and cations due to low concentration of suspended particulate matters. As an acidity of precipitation represents only an ion balance of cations and anions, this suggests that a composition of ions in precipitation is more important rather than acidity of precipitation.

Fujita (1996) estimated that wet deposition of sulfur in Japan was about 620Gg per year and dry deposition was 430Gg per year, although anthropogenic sulfur emission was 435Gg per year. Ratio of total sulfur deposition to anthropogenic sulfur emission was about 2.4. This suggested that sulfur deposition in Japan was apparently double of the total anthropogenic emission of sulfur in Japan. Sulfate concentration of precipitation in winter season was high in Japan Sea side area where might be affected by seasonal dominant wind from the continent to Japan. Such an apparent budget analysis of sulfate in Japanese Archipelago suggested that total deposition of sulfur would be greater than the total domestic emission and it might have an influence of the import from the outside area.

Table 2 Rain acidity and wet deposition in Japan

Parameter	Mean	Range
Precipitation (mm)	1719	776 - 2722
pH	4.8	4.5 - 5.1
Deposition (kg/ha/yr)		
H^+	0.27	0.06 - 0.47
SO_4	31.9	11.4 - 54.1
nss-SO_4	25.1	10.6 - 40.7
NO_3	12.7	4.3 - 22.1
Cl	57.2	4.3 - 227.6
NH_4	5.8	2.5 - 11.5
Ca	4.4	1.8 - 8.2
Mg	3.8	0.5 - 14.4
K	3.0	0.8 - 9.2
Na	27.0	2.5 - 104.3

Period: Oct. 1987-Sept. 1990 (N=21)
Calculated from CRIEPI (1992)

2. EFFECT OF WET DEPOSITION ON TREE GROWTH

2.1 INTRODUCTION

Various causal factors for the dieback of *Cryptomeria japonica* have been recognized (Nashimoto, 1988). However, its potential causes have not been identified. Annual mean precipitation was 1719 mm with mean pH ranging from 4.5 to 5.1 (CRIEPI, 1992). However, wet deposition monitoring data were not directly correlated with its distribution and severity.

Shriner et al.(1990) and Haines& Carlson (1989) summarized types of visible injury attributed to simulated acid rain (SAR) in many plants and threshold pH value depending on the species, tissue age, and exposure conditions. Tamm & Popovic (1989) and Tveite *et al*. (1990/91) reported the long-term exposure experiments of the artificial rain acidified with sulfuric acid in the field grown *Pinus sylvestris*. However, such field experiments have many limiting factors; less flexibility and uniformity for the experimental design, operation, and result analysis. In contrast, many exposure experiments of SAR to potted tree seedlings were reported by Wood & Bormann (1974, 1977), Percy (1986), Shriner et al.(1990) and so on. Also, there are short-term exposure experiments of sulfuric acid solution to *Cryptomeria japonica* (Izuta *et al.*, 1990; Miwa *et al.*, 1993; Matsumoto *et al.*, 1992). However, there are no comparative studies on the effects of long-term exposure to acidic precipitation in order to assess effect of rain acidity on the growth of Japanese coniferous and deciduous tree species and soil chemistry.

Acid precipitation may induce direct harmful or indirect effects on plants. Hydrogen input to soil will accelerate soil acidification, as soil will consume cations to neutralize the acids. Eventually high concentration of Al toxic to plants will be released and it will inhibit uptake of nutrients, especially Ca and Mg, and root growth (Ulrich, 1989). Thus, it will induce Ca and/or Mg deficiency or will increase susceptibility to drought stress in plants due to reduced uptake of water. This soil acidification stress has been the most powerful and simple hypothesis to explain forest decline in the various hypothetical pathways as described in Fig. 1.

In this section, results of exposure experiments have been summarized to assess whether acidic precipitation currently observed in Japan may take a critical role to induce current decline of *Cryptomeria*, *Abies* and other native species or not. Also, a possible soil acidification stress associated with increased acidic deposition for the explanation of growth inhibition will be discussed.

2.2 SIMULATED ACID RAIN GENERATOR SYSTEM

Semi-automatic rain generator system as described in Fig. 3, generates 2.5mm per hour of simulated precipitation in the form of rain droplets with about 1 mm of diameter for the area of 2m x 5m (Kohno & Kobayashi, 1989a; Kohno *et al.*, 1994, 1995c). Plants will be exposed to a precipitation rate of 10-30 mm (4 to 12 hours) a day, 3 events a week from April to September and at a rate of 10 mm a week from October to March. Exposures were carried out in the evening except for cloudy and rainy days for preventing photo bleaching after the exposures under solar radiation.

Tap water was purified to 0.1 to $0.2\mu S/cm$ by a reverse osmotic pressure membrane system and a mixed ion exchange resin column system. SAR contained sulfate, nitrate and chloride anions in an equivalent concentration ratio of 5:2:3 (Kohno *et al.*, 1989a). It was volumetrically diluted by deionized water to prepare designated pHs for the exposure. Deionized water was used as the control and a diluent.

Mist generating system was also developed by modification of the rain generator system with fogger nozzles instead of needles in the rain generator system. System was intermittently operated for adjusting to designated precipitation rate (Kobayashi *et al.*, 1994). Distribution accuracy of precipitation generated by rain and mist system was below 10% of CV and 36%, respectively (Kohno *et al.*, 1994).

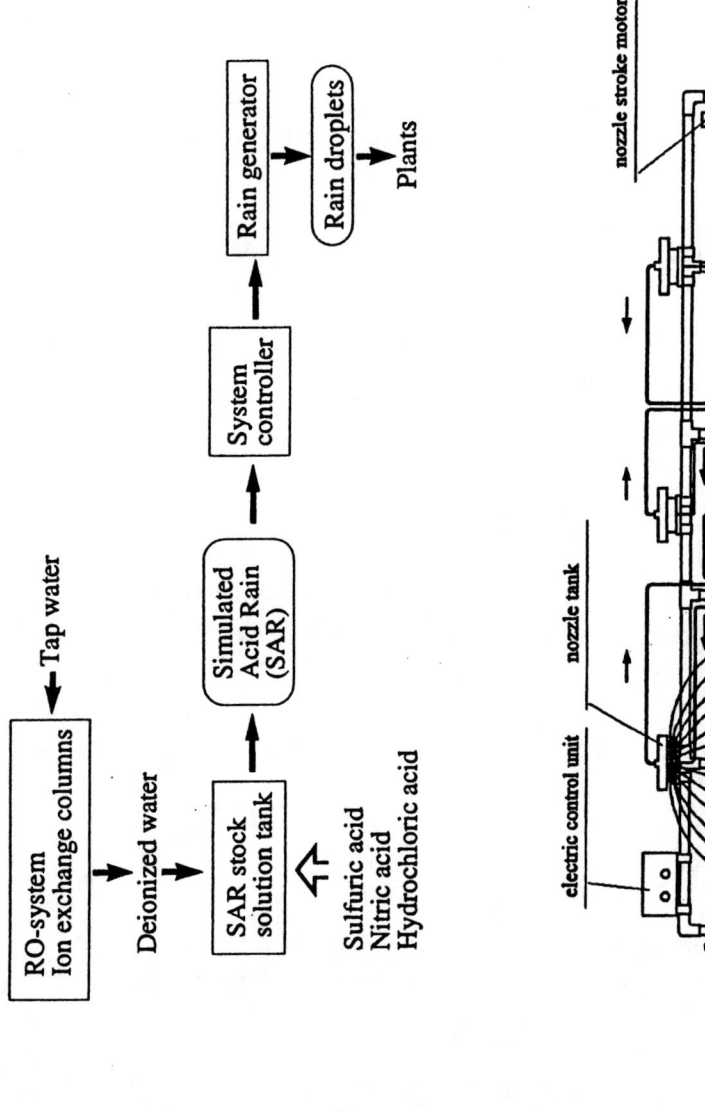

Fig. 3 Schematic diagram of rain generator system for exposure experiments of simulated acid rain. From Kohno et al. (1994)

2.3 EFFECT OF ACID MIST

Wet deposition includes rain, fog or other forms of precipitation. Table 2 showed amount of water remaining on plant surface, when plants were exposed to simulated rain or simulated mist at the same rate of precipitation. Amount of water adhered to leaf surface due to deposition by mist exposure was likely to be greater than that by the rain. Exposure to simulated acid mist (SAM) at pH 2.5 for 9 hours an event at intensity of 2.5mm an hour induced earlier visible injury development than that of simulated acid rain (SAR) did. This may suggest that direct effects of acid mist on plants will be severer than that of acid rain.

Crops including perennial ryegrass (*Lolium perenne*), Kentucky bluegrass (*Poa pratensis*), pea (*Pisum sativum*), perilla (*Perilla frutescens*), turnip (*Brassica rapa*), soybean (*Glycine max*), white clover (*Trifolium repens*) and alfalfa (*Medicago sativa*) developed necrotic symptoms in the leaves by the exposure to either SAR or SAM at pH 2.5. Growth performance showed different patterns in different plant species as shown in Fig. 4. Soybean plants did not show significant growth and yield reduction at pH 3.0 or higher of SAR, but reduced growth and yield reduction significantly at pH 2.0 (Kohno & Kobayashi, 1989a, b). SAM exposure induced visible injuries in plants earlier than SAR exposure did (Kobayashi *et al.*, 1994). Turnip reduced plant dry weight at pH 2.5 of SAR or SAM. White clover, alfalfa, Kentucky bluegrass and perennial ryegrass significantly increased dry weight with increasing acidity. In contrast, pea exposed to SAR at pH 2.5 increased dry weight, but SAM did not show any significant changes. Perilla did not reduce plant growth by the SAR at pH 2.5, but significantly increased by the SAM.

Grasses and crops will respond to nitrogen input through acidic rain or mist containing nitrate as one of major anion components. Also, during acid neutralization process, soil will release cations available for plant nutrition. These may enhance plant growth (Kobayashi *et al.*, 1991). Since rain droplets will strike and injure young seedlings in the early emergence stage, strong rain events with large droplet sizes may have a significant role to cause a growth reduction due to mechanical damage rather than mist or fog with smaller droplet sizes.

Amount of deposition by mist or fog is generally smaller than that of rain. Duration of 9 hours in mist and rain indicating such a extremely high acidity at pH 2.5 is quite unusual. However, this experimental result suggested that droplet size is not an important factor to evaluate effect of wet deposition on plant growth.

In agricultural crop production, soil amendment materials such as calcium carbonate will be generally supplied to adjust soil pH for a proper range before the cultivation. Therefore, soil acidification stress associated with acidic precipitation at the current acidity will not induce a significant effects on crops with relatively shorter rotation rather than trees, if the pH of current precipitation will not always lower below pH 4.0.

Table 3 Amount of water adhered to leaf surface in plants after an exposure to rain or mist

Species	Rain	Mist
Broad-leaved trees		
Betula platyphylla var. *japonica* (g/cm^2, leaf area)	4.74 ± 1.95	6.54 ± 2.32
Quercus mongolica var. *grosseserrata* (g/cm^2, leaf area)	6.58 ± 3.78	12.43 ± 7.10
Coniferous trees		
Cryptomeria japonica (g/g, leaf weight)	0.11 ± 0.05	0.23 ± 0.06
Chamaecyparis obtusa (g/g, leaf weight)	0.15 ± 0.06	0.18 ± 0.09
Pinus sylvestris (g/g, leaf weight)	0.16 ± 0.05	0.25 ± 0.07

Values indicate mean value±standard deviation (n=10).
Exposed period was 2.5 mm/hr x 9 hr., From Kobayashi *et al.* (1994)

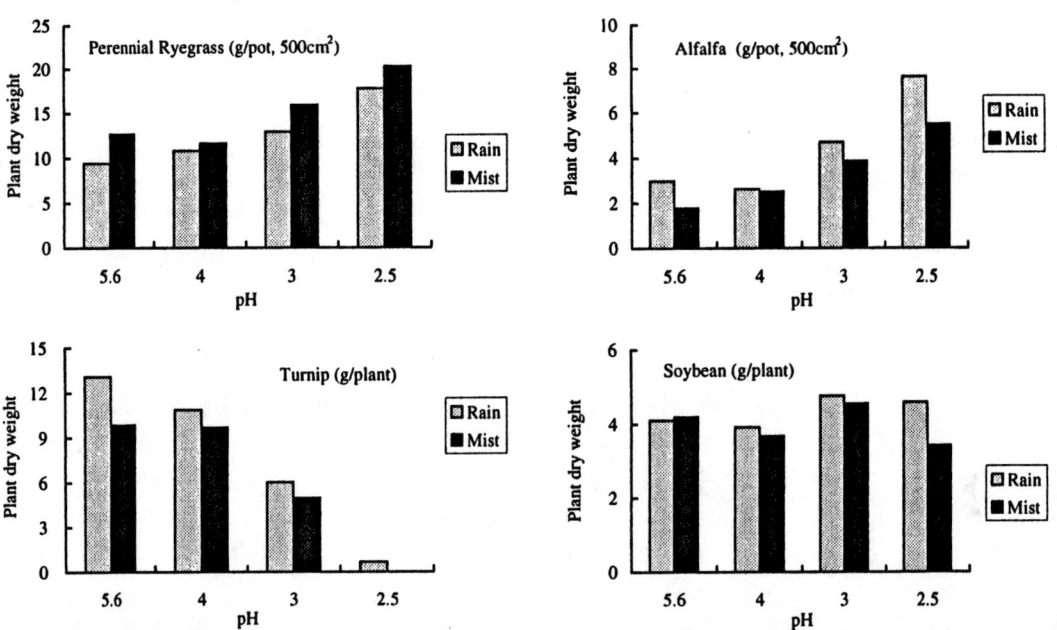

Fig. 4 Effect of acid mist and rain on the growth of crops. Deposition rate was 2.5 mm x 9 hours per event and frequency was 3 events per a week. Exposure period for perennial ryegrass, alfalfa, turnip and soybean was 115, 115, 49 and 65 days, respectively. From Kobayashi *et al.* (1994)

2.4 VISIBLE INJURIES INDUCED BY SIMULATED ACID RAIN

One of acute direct effects of simulated acid rain (SAR) or mist (SAM) are visible injuries. Typical symptoms were necrotic spots, or necrosis, marginal and/or tip necrosis, early senescence, and defoliation (Kohno *et al.*, 1994; Kobayashi *et al.*, 1994, 1996). Exposure to extremely low pH of SAR at pH 2.0 or 2.5 induced necrotic lesion or spots after several events in agricultural crops and broad-leaved trees. Conifer needles also developed tip necrosis as tip burn or blight, but its severity was less than that of crops and broad-leaved trees.

As the exposure to SAR at pH 2.0 or 2.5 caused severe acute leaf necrosis, plants defoliated then regenerated new leaves. However, another continuous exposure to SAR induced another necrosis and leaves did not have enough time to recover from injuries. Thus, leaves became shorter or smaller in size than the normal. This suggested that if the short-term exposure would be terminated, plants could recover from injury due to extremely low pH and they could re-grow as a normal. This could also suggest that it is important to make clarify when plants have been exposed to acidic precipitation, how frequently, how many hours or how many amount of precipitation plants have been exposed.

Differential responses in visible injury development of 46 tree species after the exposure to SAR at a rate of 2.5mm per hour, 8 hr x 3 times per a week for 3 to 4 months during a growing season (Kohno *et al.*, 1994) were summarized in Table 4.

2.4.1 Coniferous Trees

Japanese larch (*Larix kaempferi*), which is the only deciduous conifer species in the 11 conifers listed in Table 4, defoliated at pH 2.0 and developed leaf tip discoloration at pH 2.5. Other species did not show any symptoms at pH 2.5 after the short-term exposure experiment, although the long-term exposure experiments caused light necrosis or defoliation in all species as described in the later section of chronic effect. Thus, larch was classified as the most sensitive coniferous species in the view point of acute visible injury development to extremely low pH of SAR. Conifer including larch did not show any visible symptoms at pH 3.0 or higher for 3-4 months exposure to SAR.

2.4.2 Broad-Leaved Trees

Seven of 14 evergreen broad-leaved trees and 14 of 21 deciduous ones showed foliar injuries at pH 3.0. Among deciduous species exposed to SAR at pH 2.0, 7 species were almost defoliated and three species were dead at the end of the exposure experiments. However, evergreen broad-leaved trees did not show such severe symptoms at pH 2.0.

Exposure to SAR at pH 3.0 caused premature defoliation and shot hole symptoms in *Prunus* species, especially in *P. x yedoensis* and *P. armeniaca*. Any species did not show visible injury symptoms after the exposure to SAR at pH 4.0 for 3-4 months.

These results suggested that coniferous trees were relatively tolerant to exposure to SAR at low pH than broad-leaved trees. Deciduous broad-leaved trees were relatively more sensitive than the evergreen broad-leaved trees.

Above results were from the exposure experiments to SAR at extremely low pH for 3-4 months and this condition was unusual in the nature. Continuous exposure to SAR at pH 4.0 did not cause significant and apparent symptoms. Thus, current acidity of precipitation with few frequency at and below pH 4.0 would not be likely to induce direct injuries in leaves. If trees will be exposed to SAR at pH 2.0 for only a week, they will develop foliar necrosis in broad-leaved trees but not in conifers except a larch. However, injuries do not expand and plants grow normally after the cessation of the exposure to SAR. Therefore, lasting such a condition will be very important factor to develop apparent symptoms.

Table 4 Visible injuries in the leaves of plants exposed to simulated acid rain

Species	Rain pH					Symptoms
	5.6	4.0	3.0	2.5	2.0	
Conifer						
Chamaecyparis obtusa (1)	-	-	-		+	Scale leaf brown necrosis, defoliation
Chamaecyparis obtusa (3)	-	-	-	-	+	Scale leaf brown necrosis, defoliation
Chamaecyparis pisifera (1)	-	-	-		+	Scale leaf brown necrosis, defoliation
Juniperus chinensis var. *kaizuka* (1)		-	-	-	+	Scale leaf brown necrosis
Abies firma (2)	-	-	-		+	Leaf necrosis, defoliation
Abies homolepis (3)	-	-	-	-	+	Juvenile shoot tip necrosis
Picea abies (2)	-	-	-		+	Leaf necrosis, old leaf defoliation
Picea abies (3)	-	-	-	-	+	Old leaf defoliation
Larix kaempferi (2)	-	-	-		Defoliated	Leaf necrosis
Larix kaempferi (3)	-	-	-	+	Defoliated	Leaf tip discoloration
Pinus densiflora (2)	-	-	-		+	Leaf tip brown necrosis
Pinus densiflora (3)	-	-	-	-	+	Leaf tip brown necrosis
Pinus strobus (3)	-	-	-	-		Short leaf at pH 2.5
Pinus thunbergii (1)	-	-	-		+	Leaf tip necrosis
Pinus thunbergii (3)	-	-	-	-	-	
Cryptomeria japonica (1)	-	-	-		+	Leaf necrosis
Cryptomeria japonica (3)	-	-	-	-	+	Leaf necrosis
Evergreen broad-leaved trees						
Eponymous japonicus (1)	-	-	+		+	Necrotic spot, defoliation

Species						Symptoms
Rhododendron indicum (1)	-	-	-	+		Leaf tip necrosis, defoliation
Rhododendron oomurasaki (1)	-	-	-	+		Leaf tip necrosis
Castanopsis cuspidata var. *sieboldii* (1)	-	-	-	+		Necrotic spot
Pasania edulis (1)	-	-	-	+		Marginal necrosis
Quercus phillyraeoides (1)	-	-	-	+		Marginal necrosis
Machilus thunbergii (1)	-	-	-	+		Necrotic spot
Myrica rubra (1)	-	-	-	+		Marginal necrosis
Ligustrum japonicum (1)	-	-	+	+		Necrotic spot
Pittosporum tobira (2)	-	-	+	+		Marginal necrosis
Photinia glabra (1)	-	-	+	+		Necrotic spot
Pyracantha coccinea (1)	-	-	+	+		Whole leaf necrosis, defoliation
Rhaphiolepis umbellata (1)	-	-	+	+		Necrotic lesion, small leaf
Camellia japonica (1)	-	-	+	+		Marginal necrosis

Deciduous broad-leaved trees

Species						Symptoms
Acer buergerianum (3)	-	-	-	+	Defoliated	Marginal necrosis, necrotic spot, defoliation
Alnus firma (1)	-	-	-	+		Small necrotic spot, small leaf, defoliation
Betula platyphylla var. *japonica* (3)	-	-	-	+	Defoliated	Marginal necrosis
Cornus florida (2)	-	-	+		Defoliated	Necrotic spot, marginal necrosis
Enkianthus perulatus (2)	-	-	+		Dead	Brown necrosis, defoliation
Fagus crenata (2)	-	-	+	+		Marginal necrosis, necrotic spot, defoliation
Fagus crenata (3)	-	-	+	+	+	Marginal necrosis, necrotic spot, defoliation
Quercus mongolica var. *grosseserrata* (3)	-	-	+	+		Interveinal necrotic spot, capping
Quercus serrata (2)	-	-	+	+		Interveinal necrotic spot
Quercus serrata (3)	-	-	+	+	+	Interveinal necrotic spot
Cytisus scoparius (1)	-	-	-		Dead	Necrotic lesion
Lespedeza thunbergii (1)	-	-	+		Dead	Necrotic spot
Liriodendron tulipifera (3)	-	-	+	+		Necrotic spot
Fraxinus japonica (3)	-	-	+	+		Necrotic spot
Syringa vulgaris (1)	-	-	+		Defoliated	Necrotic spot
Prunus armeniaca (1)	-	-	+		Defoliated	Necrotic spot, shot hole
Prunus mume (2)	-	-	+	+		Necrotic spot, defoliation
Prunus mume (4)	-	-	+	+		Necrotic spot, defoliation
Prunus jamasakura (4)	-	-	+	+		Necrotic spot, defoliation
Prunus x yedoensis (1)	-	-	Defoliation		Defoliated	Necrotic spot, shot hole
Spiraea cantoniensis (1)	-	-	-	+		Necrotic lesion
Populus maximowiczii (3)	-	-	-	+		Marginal necrosis, vein necrosis
Hydrangea macrophylla var. *otakusa* (1)	-	-	+		Defoliated	Necrotic spot
Zelkova serrata (3)	-	-	-	+		Necrotic spot, small leaf

-: No injury developed, +: Visible injury developed. Precipitation: 20 mm (2.5 mm/h x 8h) / day x 3 times/week

Sulfate:Nitrate:Chloride=5:2:3 (Equivalent ratio). pH5.6 was deionized water.
(1) Exposed from July to October in 1991. (2) Exposed from April to June in 1992. (3) Exposed from April to June in 1993.

2.5 GROWTH RESPONSES AS CHRONIC EFFECTS

Public was concerned about that acid rain might be associated with forest decline in the any places throughout the world, as early forest decline research tried to find out potential linkage with rain acidification. There are many reports about European species, but not for Japanese species. The authors conducted the first 2-year SAR exposure experiment in three important Japanese coniferous trees in wood production for 2 growing seasons during 1990-1992 (Kohno et al., 1995a). The second 3-year exposure experiment has completed to compare sensitivity to SAR in 16 species including domestic and foreign origins for 3 growing seasons during 1993-1995(Kobayashi et al., 1996).

2.5.1 Coniferous Trees

In the first experiment, three Japanese conifers were exposed to SAR for 23 months (Kohno et al., 1995a, c). Fertilized plants showed significant growth reduction at pH 2.0 of SAR, but not at pH 3.0 or higher. Japanese cedar (*Cryptomeria japonica*) with no fertilizer treatment and exposed to SAR at pH 2.0 did not show significant growth reduction, but Jaspanese cypress (*Chamaecyparis obtusa*) and Sawara cypress(*Chamaecyparis pisifera*) did (Table 5). Total dry weights of the latter two species increased significantly at pH 3.0.

Needle dry weight in *Chamaecyparis obtusa* and *Chamaecyparis pisifera* exposed to pH 2.0 decreased significantly due to defoliation of scale leaves associated with the development of necrosis. Root dry weight of the plants without fertilizer did not change significantly. While that of the plants with fertilizer reduced significantly at pH 2.0, reduced root dry weight in plants supplied with fertilizer was almost same level as that of the plants without fertilizer at any pH. This result suggested that soil acidification stress due to acidic precipitation was not sufficient to induce conifer growth reduction.

In the second experiment, coniferous species including domestic and foreign origins as listed in Fig. 5 (top) were exposed to SAR for 30 months including 3 growing seasons. As well as the first 2-year experiment, coniferous species did not show significant growth reduction at pH 3.0 or 4.0. It did fail to induce any significant growth reduction at pH 4.0 that would be the lowest annual mean value of precipitation pH in the heavy polluted areas throughout the world. Even if the pH of SAR was 2.5, maximum percentage of relative growth reduction in the total dry weight to that of plants at pH 5.6 as the control, was about 20% in *Pinus thunbergii*. Others did not show significant growth reduction. This will suggest that rain acidity will not be a direct threat for the coniferous trees, if the acidity in the precipitation will not continue to be always lower than pH 3.0.

2.5.2 Deciduous Broad-Leaved Trees

In contrast to the coniferous trees, deciduous broad-leaved trees showed a wide variation in growth responses at the low pH as shown in Fig. 5 (bottom).

Exposure to SAR at pH 2.5 for 30 months significantly reduced dry weight in *Populus maximowiczii, Quercus mongolica* var. *grosseserrata,* and *Fraxinus japonica.* Although *Betula platyphylla* var. *japonica* and *Liriodendron tulipifera* developed marginal brown necrosis or interveinal necrosis at pH 2.5, they significantly increased dry weight. Except these extremes, about half of broad-leaved trees showed significant growth reduction. This could suggest that broad-leaved trees were relatively sensitive to acid rain than the conifers.

However, broad-leaved trees exposed to SAR at pH 4.0 or 3.0 for 3 growing seasons showed about ±30% of variation in the relative total dry weight to that of plants exposed to pH 5.6. These wide differential responses in broad-leaved trees may be due to different sensitivity to soil acidification stress, due to soil chemical conditions or due to nitrogen uptake efficiency by input of acidic precipitation.

These results suggested that rain with the current acidity level was hardly to induce plant growth reduction in all species at the same time. Especially, *Picea abies* in Europe and *Abies veitchii* and *Cryptomeria japonica* in Japan would be a quite tolerant species to acid input. If their forest/tree decline would be associated with acid precipitation, decline of deciduous tree stands would be apparently more serious and wide spread with a significant extent.

Table 5 Effect of simulated acid rain on the growth of Japanese conifer cuttings grown without fertilizer for 23 months

Species	Total Precipitation (mm)	Rain pH			
		5.6	4.0	3.0	2.0
Cryptomeria japonica	2460	148.7a	168.6ab	185.4 a-c	185.4 a-c
	3960	182.0 a-c	183.6 a-c	203.0 a-c	223.0 bc
	5450	184.2 a-c	179.0 a-c	229.7 c	211.6 bc
Chamaecyparis obtusa	2460	115.5 ab	116.8 ab	152.3 cd	118.3 ab
	3960	111.7 ab	115.4 ab	169.0 d	105.8 ab
	5450	113.2 ab	127.6 bc	175.8 d	87.8 a
Chamaecyparis pisifera	2460	104.7 a-c	115.4 a-c	140.5 cd	84.9 ab
	3960	106.8 a-c	122.6 bc	166.5 d	94.0 ab
	5450	110.9 a-c	118.6 bc	163.0 d	78.7 a

Values indicate means of total dry weight (g/plant). Within a fertilizer treatment in a species, any two means having a same letter in common are not significantly different at the 5% level by the Tukey's HSD multiple range test (n=12).

Initial dry weights were as follows.

Cryptomeria japonica = 93.2g, *Chamaecyparis obtusa* = 51.3g, *Chamaecyparis pisifera* = 42.2g.

From Kohno *et al.* (1995c)

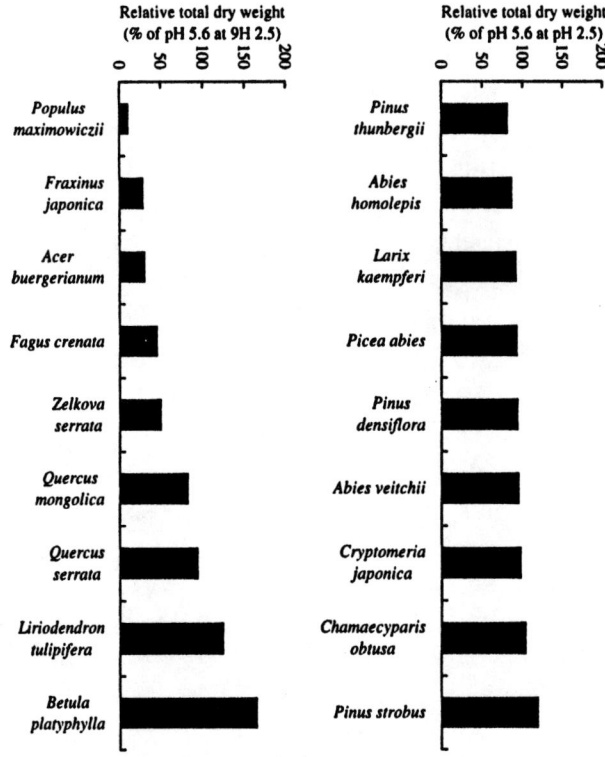

Fig. 5 Growth responses of trees as affected by simulated acid rain exposure for 30 months. *Abies veitchii* and *Zelkova serrata* were exposed to SAR for 18 months. SAR contained sulfate:nitrate:chloride with an equivalent ratio of 5:2:3. Precipitation rate in summer was 20mm x 3 times a week and 10mm a week in winter season. Total precipitation was 4100mm. Vertical scale indicates relative value of total dry weight of plants at pH 2.5 to that of plants at pH 5.6 of deionized water. Soil type was andosol.

2.6 EFFECT OF SOIL ACIDIFICATION STRESS DUE TO WET DEPOSITION

In the above section, direct effect of acid precipitation via surface contact was discussed. Plant growth will be affected by the root environment. If the soil receives large input of acids, soil exchanges cations are able to neutralize hydrogen as acids. In this process, increased soil acidity stress in the rhizosphere and released Al toxicity in root system should be concerned. There are many reports of acid soils and related Al toxicity in crops, grasses and some trees (Andersson, 1988, Foy *et al.*, 1978, Kochian, 1995).

2.6.1 pH Stress in Rhizosphere

Filed survey of dieback in *Cryptomeria japonica* revealed that pHs in soils from stands showing decline were lower than that from healthy (Nashimoto, 1993). Two-year old seedlings of *Cryptomeria japonica* and *Chamaecyparis obtusa* were cultivated in the nutrient culture solution for 15 weeks, to evaluate pH stress in rhizosphere (Kohno *et al.*, 1995b). Acidity of one fifth Hoagland's nutrient culture solution was adjusted to the pH ranging from 3.5 to 6.0 by HCl and NaOH.

Growth performance of both *Cryptomeria japonica* and *Chamaecyparis obtusa* was best in the lowest pH range of 3.5-4.0 (Fig. 6). Contents of K+Ca+Mg in the leaves and roots of both species cultivated at pH 3.5-4.0 were highest. Both species, however, decreased plant dry weights at• higher pH range. Another experiment also confirmed that total dry weight in *Cryptomeria japonica* grown in nutrient culture solution at pH range of 3.5-4.0 was highest in comparison with that at pH 2.5-3.0 or 4.5-5.0 for 15 weeks. In contrast to *Cryptomeria japonica*, *Populus maximowiczii* grown in the nutrient solution at pH range of 2.5-3.0 or 3.5-4.0 showed about 35% growth reduction from that at pH range of 4.5-5.0 for 3 weeks (Fig. 7).

This suggests that these coniferous species have a great adaptability to such a low pH condition. Although *Populus maximowiczii* was only a case study, it suggests that a possible wide significant difference in sensitivity and potential adaptability to soil acidity and its related chemical properties, prior to releasing lot of inorganic Al at low pH range.

2.6.2 Al Toxicity in Coniferous Trees

Large input of acidic deposition will cause soil acidification. Acidified soil will release large amount of soluble Al toxic to plant root growth at low pH zone. Many of nutrient solution culture studies carried out with plants demonstrated that high Al concentration will inhibit uptake of Ca and Mg and physiological activity in the root system (Andersson, 1988; Foy *et al.*, 1978; Hutchinson *et al.*, 1986; Keltjens & Loenen, 1989; Kochian, 1995; McCormick *et al.*, 1978; Rengel, 1992; Rengel & Elliott, 1992; Taylor, 1989)

Young Japanese coniferous tree seedlings were grown in nutrient culture solution containing 0.5 to 20 mM of Al as $AlCl_3$ at the pH range of 3.5 to 4.0 for maintaining designated Al concentration in the solution.

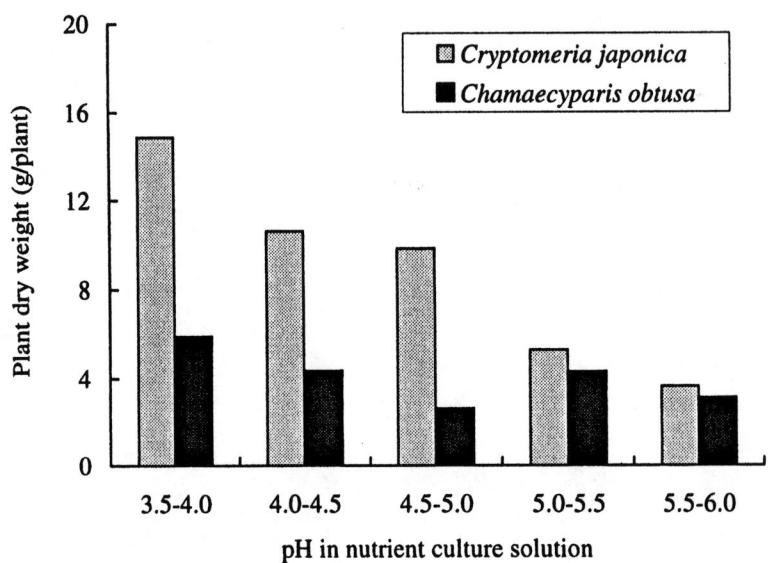

Fig. 6 Effect of solution pH on the growth of *Cryptomeria japonica* and *Chamaecyparis obtusa* grown in nutrient culture solution for 15 weeks. Initial dry weight: *Cryptomeria japonica* = 1.60g, *Chamaecyparis obtusa* = 1.42g. Nutrient solution was 1/5 Hoagland's No.2 culture solution.

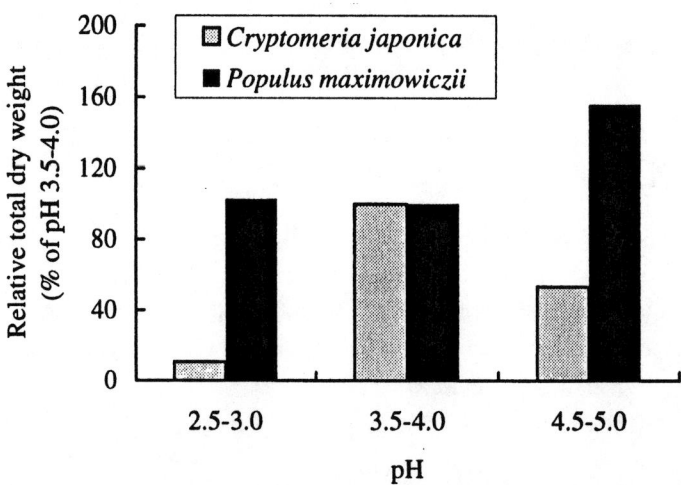

Fig. 7 Effect of solution pH on the growth of *Cryptomeria japonica* and *Populus maximowiczii* grown in nutrient culture solution. Cultivated period was 15 weeks in *Cryptomeria japonica* and 3 weeks in *Populus maximowiczii*. Nutrient solution was 1/5 Hoagland's No.2 culture solution.

Aluminum supplied at or below 1mM had no effect on the mortality of these species. However, the increase of Al concentration higher than 2mM significantly increased their mortality (Kohno et al., 1995b). Fig. 8 showed that Al concentration at 5mM in the root sphere induced significant growth reduction with no significant differences in the response in these Japanese coniferous trees. This figure suggests critical concentration of Al is around 2mM. This concentration was much higher than those reported in the field crops, vegetables and/or herbs (Andersson, 1988; Krizek et al., 1997). Thus, if soil acidification will be induced by acidic deposition in future, herbaceous plants will be influenced earlier than the coniferous trees will suffer. This also means that the significant ground vegetation changes will occur earlier than the conifer decline will be observed due to the soil acidification.

2.6.3 Critical Point for Japanese Coniferous Trees

High amount of cations may alleviate Al toxicity or low cations may aggravate its toxicity (Foy et al., 1978; Kochian, 1995). Previous section described Al toxicity developed coniferous species at and greater concentration of 2mM of Al in the rhizosphere.

The molar ratio of Ca/Al or • K+Ca+Mg• /Al is an important index to calculate critical load (*CL*) (Cronan & Grigal, 1995; Hettelingh et al., 1995a,b; Sverdrup & De Vries• 1994; Sverdrup & Warfvinge, 1993; Posch et al., 1995). Sverdrup & Warfvinge (1993) proposed a critical point of 1.0 for calculate *CL* as expressed as the molar ratio of • K+Ca+Mg• /Al that was mainly from the data set of European species. It has been a main part of RAINS model (Alcamo et al., 1990) and its modified RAINS-ASIA model has been applied to East Asian countries.

In contrast to European vegetation, plant communities in Asia including Japanese Archipelago have much more diversity in their components. Of course, dominant species is different from Europeans and its sensitivity to Al toxicity will be different. Thus, critical point should be discussed prior to calculate *CL* by the model application.

Sigmoidal fitting curve (Fig. 9) was drawn from several Al toxicity experiments shown in Fig. 10 for Japanese conifer species: *Cryptomeria japonica*, *Chamaecyparis obtusa* and *Chamaecyparis pisifera*. It was very similar to that Sverdrup & Warfvinge (1993) had proposed. Its critical point for growth reduction was below 1 for Ca/Al molar ratio in the solution and below 3 for (K+Ca+Mg)/Al. However, it had a wide range variation depending on the experimental conditions, as well as the plotted figures in Sverdrup & Warfinge (1993).

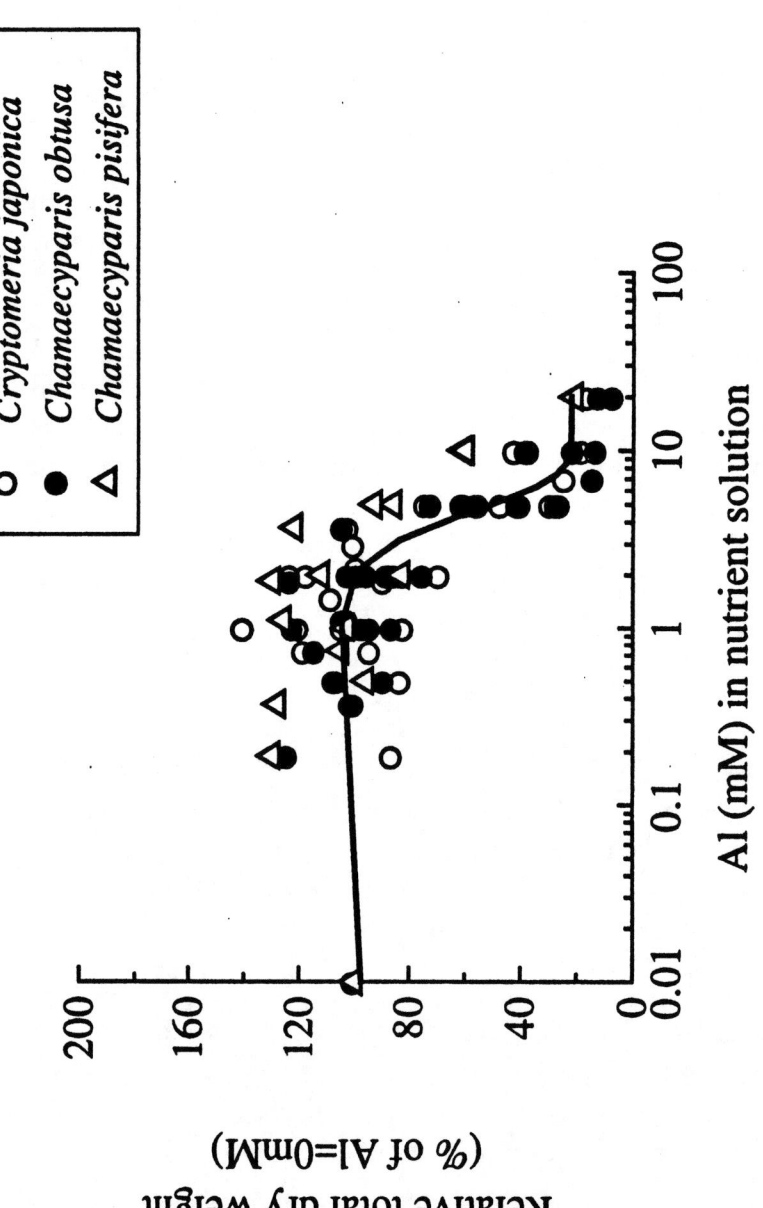

Fig. 8 Effect of Al concentration on the growth of Japanese conifer seedlings grown in nutrient culture solution. Experimental period was 10-14weeks. Nutrient solution pH was maintained in the range of 3.5 to 4.0 by adding HCl and NaOH twice a week. Data plotted are taken from available several different experiments in 1989-1995. Concentration of 0.01mM in the horizontal logarithmic scale was tentatively substituted for 0mM of Al.

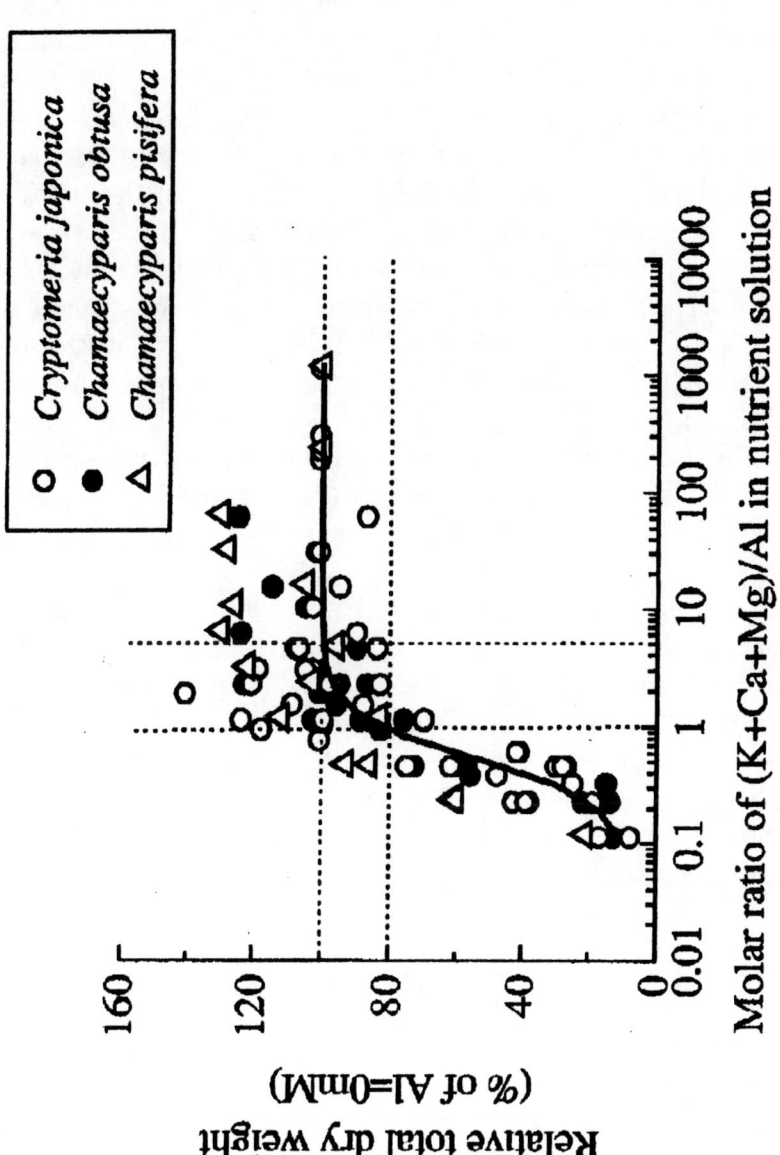

Fig. 9 Effect of (K+Ca+Mg)/Al molar ratio on the growth of Japanese conifers grown in nutrient solution culture. Experimental period was 10-14weeks. Data plotted are taken from available several different experiments in 1989-1995. Concentration of 0.01mM was tentatively substituted to calculate (K+Ca+Mg)/Al at 0mM of Al for the control.

In contrast to the nutrient solution culture shown in Fig. 9, in soil culture, however, they showed different. Plant root responses shown in Fig 10 did not reduce their growth when (K+Ca+Mg)/Al in soil was 0.6 due to receiving SAR at pH 2.0 for 23 months. In this figure, fertilized *Cryptomeria japonica* significantly reduced root growth at pH 2.0, as above ground parts developed necrosis due to direct injury by pH 2.0 of SAR and reduced photosynthetic parts resulted in changed carbon allocation. However, no-fertilized plants did not show significant root growth reduction at pH 2.0. Thus, root growth reduction in fertilized plants was not directly associated with changed soil chemistry.

After the plants received SAR at pH 2.5 for 3 growing seasons, *Populus maximowiczii* showed severe growth reduction in Fig. 5. This may introduce that such a plant growth reduction might be associated with direct and indirect stress by SAR exposure to pH 2.5. When fresh *Populus maximowiczii* was replanted in the same soil but they received only SAR at pH 5.6 for 1 growing season, it also showed significant growth reduction (Fig. 11). This will explain that *Populus maximowiczii* did respond directly to soil chemistry. However, *Picea abies* and *Cryptomeria japonica* did not show such a significant growth reduction but they grew well in the same soil. This suggests that soil condition would be more important rather than the acidification of precipitation in specific trees. Conifers would have a potential adaptability to acid soil condition but *Populus maximowiczii* would be quite sensitive to soil chemical changes. Therefore, it is very important to decide a critical point with consideration of differential sensitivity to soil characteristics without any acidic deposition.

Natural vegetation has adapted and established under the many environmental conditions such as soil conditions and climate, and so on. Assuming that acid deposition will significantly change soil acidity and chemistry due to its large input, plants as *Populus maximowiczii* preferably grown in neutral to calcareous soil conditions will quickly respond to soil chemical change. It will be potential indicator plants for changing soil chemistry due to acid input. In general, conifers may have a potential adaptability to acid soil conditions. Natural vegetation will cause a dramatic change, if they may have a ecological threat due to acid input.

3. EFFECT OF OZONE AND/OR SULFUR DIOXIDE AS DRY DEPOSITION

3.1 INTRODUCTION

Recent analysis of long-term records of tropospheric ozone measurements in the northern hemisphere suggests that it is increasing at a rate of 1 to 2 percent per year (Fishman, 1991). Predicted ozone concentration in the troposphere will double until the next century (Thompson, 1992). In Japan, daily 1-hr maximum concentration of oxidants which main component is ozone, during the summer season has an increasing from 1978 to 1991 in inland area of the Kanto and Kansai Districts (Ohara *et al.*, 1995). As ozone is phytotoxic and occurs at biological significant levels (Matyssek *et al.*, 1995), it has a potential link with forest decline (McLaughlin, 1985; Sandermann *et al.*, 1997) and biomass losses in forestry (Hogsett *et al.*, 1997).

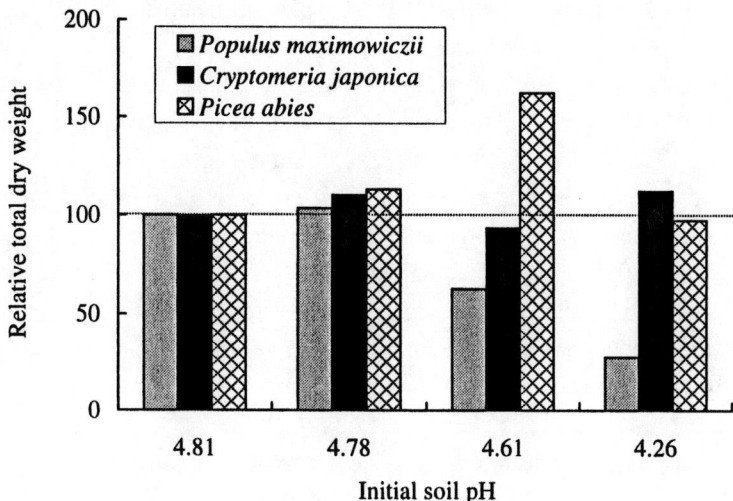

Fig. 10 Root dry weight of *Cryptomeria japonica* exposed to simulated acid rain for 23 months. (K+Ca+Mg)/Al molar ratio in soil exposed to SAR at pH 2.0 was 0.6. Fertilizer(+): N-P_2O_5-K_2O=80-80-80kg/ha/yr, Fertilizer(-): N-P_2O_5-K_2O=0-0-0kg/ha/yr. From Kohno *et al.* (1995c)

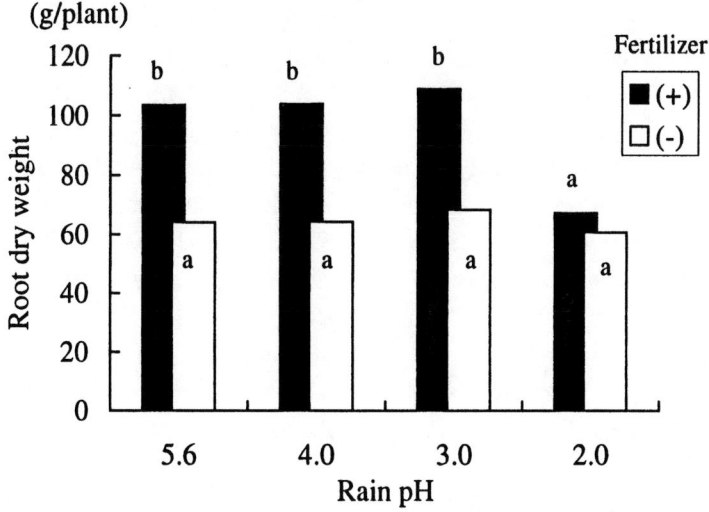

Fig. 11 Growth responses of 3 young trees to acidified soils with different initial soil acidity after receiving simulated acid rain at different pH for 3 years. Plants received deionized water at pH 5.6 as precipitation and were cultivated for 1 growing season.

Industrial activity with an increase of sulfur dioxide emission is rapidly growing in East Asia. In Japan, its concentration in the ambient air has significantly decreased by the strict regulations of sulfur emission for large stationary sources (Fig. 2). However, if the growth rate in the East Asia and the transport of sulfur dioxide to Japan will extend more, the potential risk of adverse effects on terrestrial ecosystems such as forests may increase in the near future.

There are many Japanese reports about visible foliar injuries in tree species and interspecific variations at high concentrations of gaseous air pollutants (e.g., Aoki et al., 1979; Fujiwara et al., 1976; Kadota & Ohta, 1972; Nouchi et al., 1973).

As trees are perennials, tree responses to air pollutants should concern about an alterations in growth by cumulative long-term effects under chronic exposure regimes (Chappelka & Chevone, 1992; Peterson & Mickler, 1994; Pye, 1988; Totsuka, 1989). However, limited information is available concerning about effects of both pollutants on the growth of Japanese tree species. However, they have been conducted for only one growing season in the exposure (Izuta et al., 1996; Matsumura et al., 1996; Miwa et al., 1993; Shimizu et al., 1993; Tsukahara et al., 1985, 1987).

This section described outline results of the exposure experiment of ozone and sulfur dioxide in the potted 16 tree species for three growing season in tunnel-type open-top chambers.

3.2 TUNNEL TYPE OPEN-TOP CHAMBER (OTC)

Modified large tunnel-type OTC originally designed for exposing rice plants in the paddy field by Kobayashi et al. (1994) was developed. The OTC is 15 m in long and has an outer frameworks with a cover for excluding ambient rain (Fig. 12). This type has a large capacity to place many pots enough for statistical analyses in different species at the same time, in contrast to cylindrical OTCs with 2-3 m in a diameter used in North America and Europe.

The OTC had two fans at the east and west sides. Each fanbox was set at offset angle. Air entered from both fan boxes and blew swirly in the center, and it was out through the open top. Sulfur dioxide in intake air was completely removed through charcoal filters because of very low concentration in the ambient air (Table 1). Removal efficiency of ozone by filters was about 70 %. Mean wind velocity at 1m above the ground inside the chamber was 40 cm/sec in the range of 20 to 100 cm/sec (Fig. 13) and air exchange rate per minute in a chamber was 0.5. Air temperature and concentration of gaseous pollutants were not uniformly distributed, however, deviations were about 8 % within a chamber.

During the experiment, monthly averages of daily mean air temperature in the chambers were nearly equal to those in the ambient air. However, those of daily maximum were 1 to 3°C higher than those in the ambient. Monthly averages of solar radiation in the chambers were 40 to 60 % of the full sunlight.

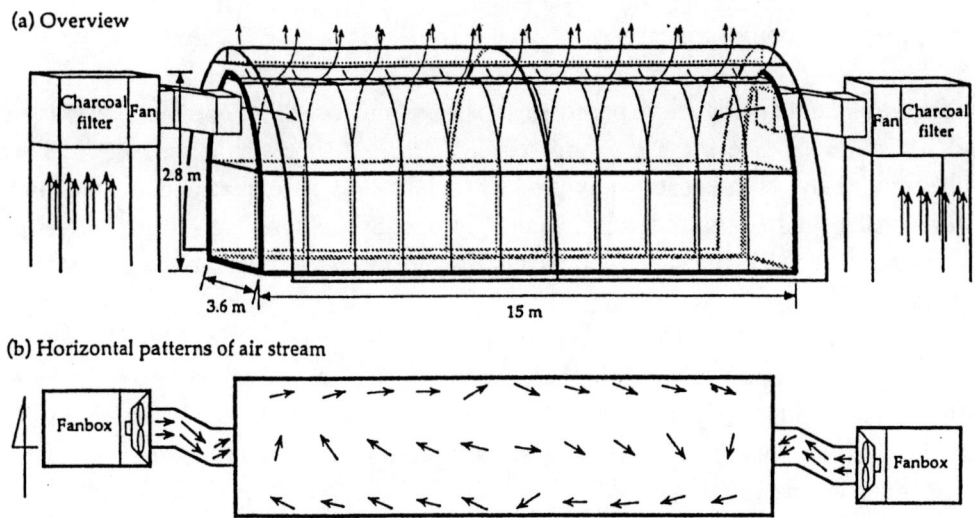

Fig. 12 (a) An overview of a tunnel-type open-top chamber (OTC) with an ambient rain-exclusion top and (b) horizontal patterns of air stream at 1m above the ground. An OTC is made by inner channeled framework, covered with clear polyvinyl chloride film with no cover of top (0.3 m in wide). An ambient rain-exclusion top (1.2 m wide) is made by outer framework, and covered with the same film.

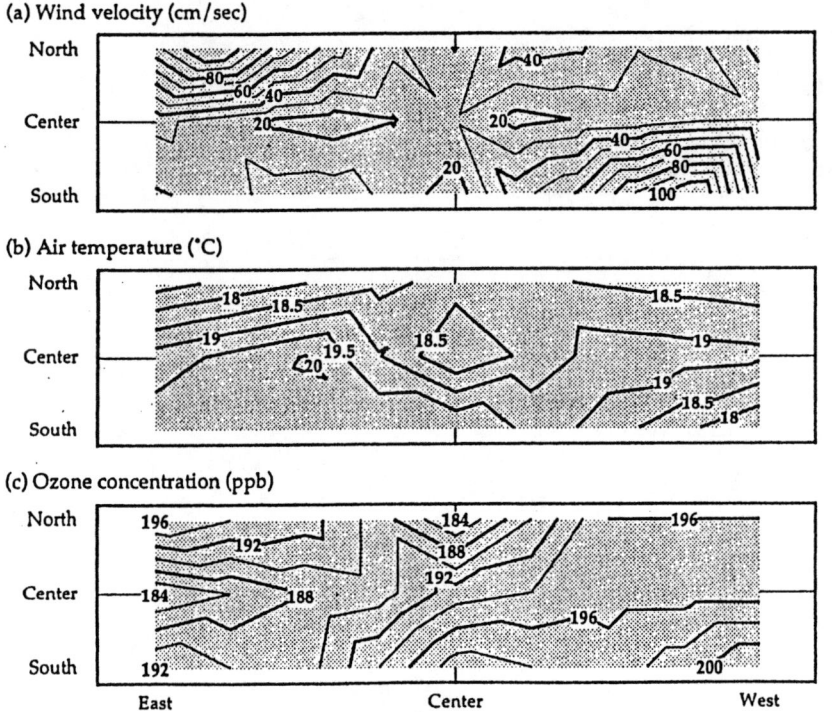

Fig. 13 Horizontal distribution of wind velocity, air temperature and ozone concentration at 1m above the ground in the OTC. Air temperature was measured at 12:00-14:00 in cloudless sunny day. Target concentration of ozone was 200ppb. Values are means of four chambers.

3.3 POLLUTANT EXPOSURES

Young potted trees were exposed to four levels of ozone in combination with four levels of sulfur dioxide for 3 growing seasons. Ozonated air was passed through water scrubber to remove highly reactive by-product such as N_2O_5 before introducing the fan box. Sulfur dioxide source was from gas cylinders of pure SO_2. Supplying sulfur dioxide to the fan box stopped to prevent from its oversupplying, when the mist irrigation system was operating.

Plants were transplanted into 2-, 5-, or 10-liter pots (one seedling per pot, a uniform volume of pot within a tree species) filled with volcanic ash soil (andosol). To reduce the position effects within a chamber, each chamber was separated into ten blocks from east to west. Almost same numbers of pots in a species were arrayed at a random in every block except both end sides.

Every April and July during the three year of the experiment, all pots were fertilized at a rate of 80-80-80 kg/ha of $N-P_2O_5-K_2O$. Plants were irrigated sufficiently both from the top of plants by mist generation system and from the surface of the pot by drip irrigation system with underground water (pH 7.7, 0.14 mS/cm of electric conductivity). They received annual average of 2,200 mm of water.

Table 6 showed mean concentrations and doses of ozone and sulfur dioxide during 3 growing seasons.

Table 6 Concentrations and doses of ozone and sulfur dioxide during the experiment

Treatment		Mean (ppb) 24hr	12hr	1hr max	Total dose (ppm*hr) 24hr	12hr	AOT40 (ppm*hr)	SUM06 (ppm*hr)
O_3	CF	11	11	18	177	94	<1	0
	x 1.0	32	35	53	538	293	60	86
	x 1.5	47	51	78	791	433	172	263
	x 2.0	62	68	104	1054	576	312	482
	AA	34	38	56	579	317	69	101
SO_2	CF	<1	<1	<1	1	1		
	10ppb	9	9	13	146	73		
	20ppb	18	18	24	304	153		
	40ppb	35	35	44	593	294		
	AA	<1	<1	1	6	4		

CF: charcoal-filtered air, AA: ambient air.
12hr: 06:00-17:59.
x 1.0: Equivalent concentration to that of ozone in ambient air. x 2.0: Twice ambient level of ozone.
Total days exposed: 703 days during 3 growing seasons.
AOT40: Accumulated exposure over a threshold of 40ppb for daytime hours (Fuhrer, 1994).
AUM06: Sum of hourly concentrations at or above 60ppb for daytime hours (Lee et al., 1988).

3.4 VISIBLE INJURIES

High levels of ozone or sulfur dioxide induced foliar injuries including early defoliation in *Pinus densiflora*, *P. thunbergii*, *P. strobus*, *Larix kaempferi*, *Picea abies* and eight deciduous broad-leaved trees in the summer seasons (Table 7).

As *Populus maximowiczii* developed black necrosis in the leaves and showed early defoliation at the ambient level (x 1.0) of ozone treatment, it might be the most sensitive species to ozone among the 16 species. On the other hand, three coniferous species: *Abies homolepis*, *Chamaecyparis obtusa* and *Cryptomeria japonica* did not show any visible injury symptoms by ozone or sulfur dioxide. These might be the most tolerant species to ozone or sulfur dioxide. However, combinations of ozone and sulfur dioxide at the high concentration synergistically induced foliar injuries in all species except *Cryptomeria japonica*.

Table 7 Effects of ozone or sulfur dioxide on the development of foliar injuries and growth of young trees

Tree species	Ozone								Sulfur dioxide (ppb)							
	CF	x1.0	x1.5	x2.0	CF	x1.0	x1.5	x2.0	CF	10	20	40	CF	10	20	40
	Visible injury				Total dry weight				Visible injury				Total dry weight			
Coniferous tree																
Pinus densiflora	-	-	+	+	-	-	-	+	-	-	-	+	-	-	-	+
Pinus thunbergii	-	-	-	+	-	-	-	-	-	-	-	+	-	-	-	-
Pinus strobus	-	-	+	+	-	-	+	+	-	-	+	+	-	+	+	+
Larix kaempferi	-	-	-	+	-	-	+	+	-	-	-	+	-	-	-	+
Picea abies	-	-	-	+	-	-	-	-	-	-	-	+	-	-	-	-
Abies homolepis	-	-	-	-	-	-	-	+	-	-	-	-	-	-	+	+
Chamaecyparis obtusa -	-	-	-	-	-	-	-	-	-	-	-	-	-	-	-	-
Cryptomeria japonica	-	-	-	-	-	+	+	-	-	-	-	-	-	-	-	-
Deciduous broad-leaved tree																
Populus maximowiczii	-	+	+	+	-	-	+	+	-	-	+	+	-	-	-	-
Betula platyphylla	-	-	+	+	-	-	-	+	-	-	+	+	-	-	+	+
Quercus serrata	-	-	-	+	-	-	-	-	-	-	+	+	-	-	-	-
Quercus mongolica	-	-	-	+	-	-	-	-	-	-	-	+	-	-	-	-
Fagus crenata	-	-	-	+	-	-	-	+	-	-	-	+	-	-	-	-
Liriodendron tulipifera	-	-	-	+	-	+	+	-	-	-	-	+	-	+	+	+
Acer buergerianum	-	-	-	+	-	+	+	+	-	-	-	+	-	-	-	+
Fraxinus japonica	-	-	-	+	-	+	+	+	-	-	-	+	-	-	-	-

Exposed period for 3 growing seasons: April to December in 1993, May to December in 1994 and April to September in 1995. Actual mean 24hour concentration for 703days for 3 growing seasons (see Table 6). Ozone: CF = 11ppb, x1.0 = 32ppb, x1.5 = 47ppb, x2.0 = 62ppb. Sulfur dioxide: CF = <1ppb, 10 = 9ppb, 20 = 18ppb, 40 = 35ppb. Visible injury: Symptoms included leaf necrosis, leaf discoloration, and early defoliation.
-: no injury, +: apparent symptoms developed.
Growth response: Evaluated by the statistical analyses of total plant dry weight harvested in August to December in 1995 except *Larix* and *Betula* in August,1994. -: no significant difference, +: significant reduction from CF treatment within a pollutant by the Tukey's HSD multiple range test at 0.05 level.

3.5 GROWTH RESPONSE

3.5.1 Single Effect of Ozone or Sulfur Dioxide

Responses of total dry weight (TDW) per plant exposed to either ozone or sulfur dioxide based on the statistical mean separation at the 5% level in pooled data for either ozone or sulfur dioxide treatment were summarized in Table 7.

Twice ambient level of ozone (x2.0) significantly reduced TDW of 5 of 8 coniferous trees and 6 of 8 deciduous broad-leaved trees. *Pinus thunbergii, Picea abies, Chamaecyparis obtusa, Quercus serrata* and *Quercus mongolica* did not show significant growth reduction by such a high concentration ozone. As *Acer buergerianum* and *Fraxinus japonica* significantly reduced TDW even at the ambient level of ozone concentration, these were most sensitive to ozone stress among the 16 species.

Significant reductions in TDW by sulfur dioxide at 40ppb were found in 6 of 16 species: *Pinus densiflora, Pinus strobus, Larix kaempferi, Betula platyphylla, Liriodendron tulipifera* and *Acer buergerianum*. However, others did not show any significant growth reduction at 40 ppb of sulfur dioxide. As *Pinus strobus* and *Liriodendron tulipifera* showed significant growth reduction by the exposure to sulfur dioxide at 10 ppb for 3 growing seasons, these were classified as most sensitive to sulfur dioxide.

These results indicate that tree species have a wide range of variation in growth responses to ozone or sulfur dioxide. There are no correlation between the interspecific variations and the characteristics of needle or broad-leaf type, life span of leaves and growth pattern (determinate or indeterminate).

Relative sensitivity based on the growth response in combination with foliar visible injury development to ozone or sulfur dioxide were presented in Table 8. There were also differences in sensitivity to ozone or sulfur dioxide. *Populus maximowiczii* was most sensitive to ozone by estimating foliar injury development in the 16 species. However, it was intermediate by growth reduction. While *Abies homolepis* and *Cryptomeria japonica* were classified into tolerant by foliar injury development to ozone, they were intermediate by the growth reduction. *Quercus serrata* and *Quercus mongolica* were the other way. *Populus maximowiczii* and *Fagus crenata* being intermediate in the sensitivity to sulfur dioxide by foliar injury, were tolerant by growth reduction. Therefore, sensitivity rankings of species based on the growth reduction and visible foliar injury development is not correlated with each other. It coincided with reports referring ozone and sulfur dioxide effects on trees in North America and Europe (Chappelka & Chevone, 1992; Kozlowski *et al.*, 1991; Pye, 1988; Reich, 1987).

3.5.2 Combined Effect of Ozone and Sulfur Dioxide

Many researchers have already investigated combined effects of ozone and sulfur dioxide on the growth of tree species mainly grown in North America and Europe (Billen *et al.*, 1990; Davis & Skelly, 1992; Holland *et al.*, 1995; Jensen & Dochinger, 1989; Mahoney *et al.*, 1984; Miller & Stolte, 1984; Reich *et al.*, 1984; Taylor *et al.*, 1986). However, no information is available investigating the combined effects of both pollutants on growth of Japanese tree species. Figs. 14 and 15 showed actual growth responses of plants including some European species to ozone in combination with four levels of sulfur dioxide. Horizontal scale indicated modified AOT40 (Ashmore, 1996). While European AOT40 is calculated for 3 months, growing season in Japan is more longer. Therefore, values of AOT40 in this report were calculated for 6 months.

Table 8 Relative sensitivities of 16 tree species to ozone or sulfur dioxide

				Growth response		
			Visible injuries	Sensitive	Intermediate	Tolerant
O$_3$	Coniferous	Sensitive	-	-	-	-
		Intermediate	-		*Pinus densiflora* *Pinus strobus* *Larix kaempferi*	*Pinus thunbergii* *Picea abies*
		Tolerant	-		*Abies homolepis* *Cryptomeria japonica*	*Chamaecyparis obtusa*
	Broad-Leaved	Sensitive	-		*Populus maximowiczii*	-
		Intermediate	*Acer buergerianum* *Fraxinus japonica*		*Betula platyphylla* *Fagus crenata* *Liriodendron tulipifera*	*Quercus serrata* *Quercus mongolica*
		Tolerant	-		-	-
SO$_2$	Coniferous	Sensitive	-		-	-
		Intermediate	*Pinus strobus*		*Pinus densiflora* *Larix kaempferi*	*Pinus thunbergii* *Picea abies*
		Tolerant	-		*Abies homolepis*	*Chamaecyparis obtusa* *Cryptomeria japonica*
	Broad-Leaved	Sensitive	-		-	-
		Intermediate	*Liriodendron tulipifera*		*Betula platyphylla* *Acer buergerianum*	*Populus maximowiczii* *Quercus serrata* *Quercus mongolica* *Fagus crenata* *Fraxinus japonica*
		Tolerant	-		-	-

Classification of sensitivity to ozone or sulfur dioxide was based on Table 3-2.
Sensitive: Visible injury developed or total dry weight significantly reduced at and over ambient level of ozone or at and over 10 ppb of sulfur dioxide.
Intermediate: Visible injury developed or total dry weight significantly reduced at and over 1.5 times higher than ambient ozone level or at and over 20 ppb of sulfur dioxide.
Tolerant: No visible injury developed or no significant reduction in total dry weight at any treatment.

Responses of plants to either ozone or sulfur dioxide were relatively able to classify as in Table 8. However, sensitivity to combination stress of ozone and sulfur dioxide was very complicated as shown in Fig. 14 and 15. *Pinus thunbergii* were tolerant to ozone or sulfur dioxide in Table 8. However, sulfur dioxide caused synergistic growth reduction in combination with ozone. Sulfur dioxide showed additive growth response in *Pinus densiflora* under ozone stress. *Pinus strobus* significantly reduced total dry weight by the ozone exposure, but 40ppb of sulfur dioxide would be a limiting growth factor. While *Cryptomeria japonica* was intermediate to ozone, it was clearly tolerant to sulfur dioxide in combination with ozone. *Picea abies* was also very tolerant to both ozone and sulfur dioxide, if sulfur dioxide was at and below 40ppb.

As well as coniferous species, broad-leaved trees showed different dose-responses individually. As well as European beech (*Fagus sylvatica*) is more sensitive to ozone than *Picea abies* (Skärby & Karlsson, 1996), Japanese beech (*Fagus crenata*) is sensitive to ozone and combination of both ozone and sulfur dioxide (Fig. 15). *Populus mazimowiczii, Quercus serrata, Quercus mongolica, Acer buergerianum,* and *Fraxinus japonica* are sensitive to ozone rather than sulfur dioxide. However, *Liriodendron* and *Betula* were more sensitive to sulfur dioxide than ozone.

The results suggests that combination of sulfur dioxide and ozone for longer time such as 2 to 3 growing seasons has an enough potential to cause plant growth reduction at the relatively low concentration of sulfur dioxide. Most of species did not cause any significant growth reduction at 10 ppb of sulfur dioxide. However, some species may have significant growth deterioration at 20ppb of sulfur dioxide. Current ambient concentration of ozone has a potential threat in domestic species sensitive to ozone. Especially, combination exposure of sulfur dioxide at and higher than 20ppb with current ambient level of ozone may cause significant growth inhibition in many species, although such exposure regime may not be realistic in Japan.

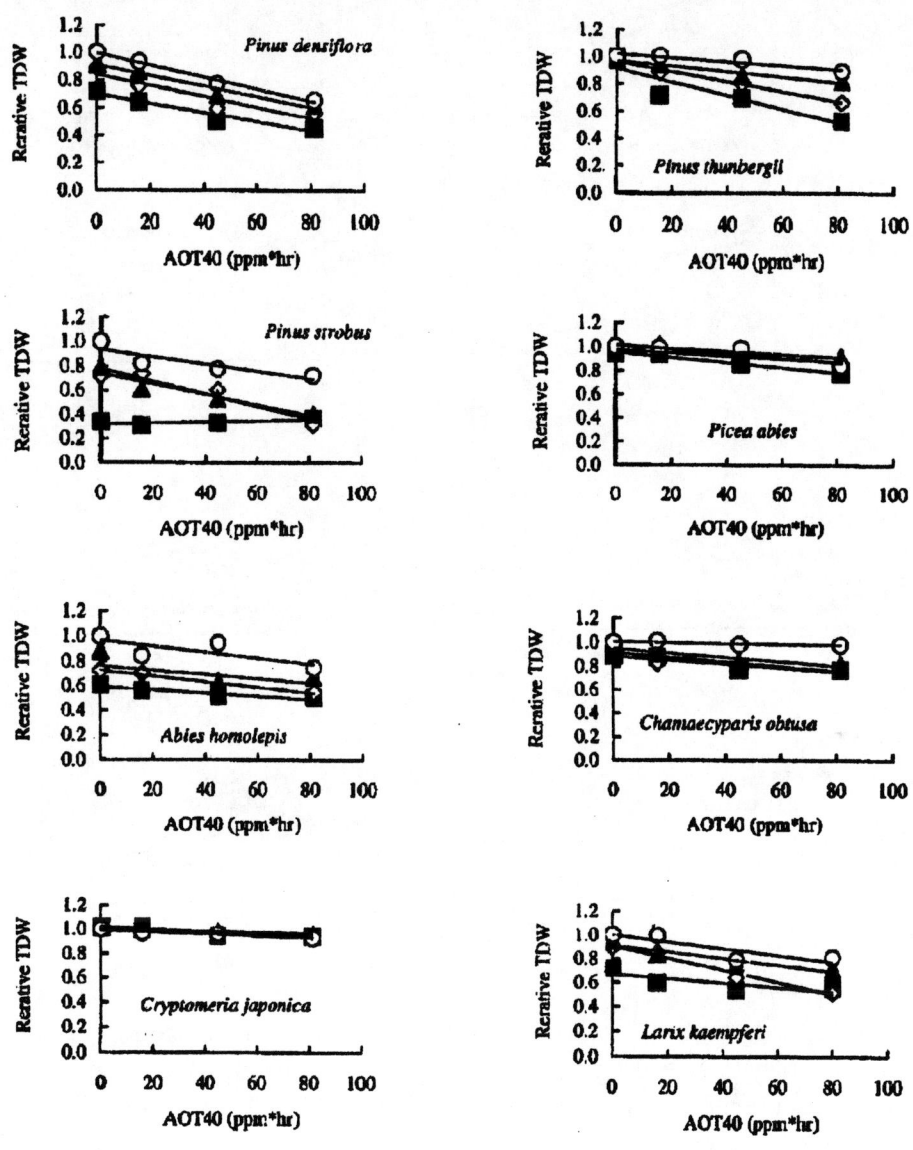

Fig. 14 Growth response to ozone and sulfur dioxide in coniferous trees. Circle: SO_2=<1ppb, Triangle: SO_2=10ppb, Diamond: SO_2=20ppb, Square: SO_2=40ppb. AOT40: Accumulated exposure over a threshold of 40 ppb of ozone in daylight hours were calculated for 6 months: April to September (183 days). Relative TDW: Relative total dry weight of plant to that of plant grown in charcoal-filtered air without sulfur dioxide. Exposed for 3 growing seasons except *Larix kaempferi* for 2 growing seasons.

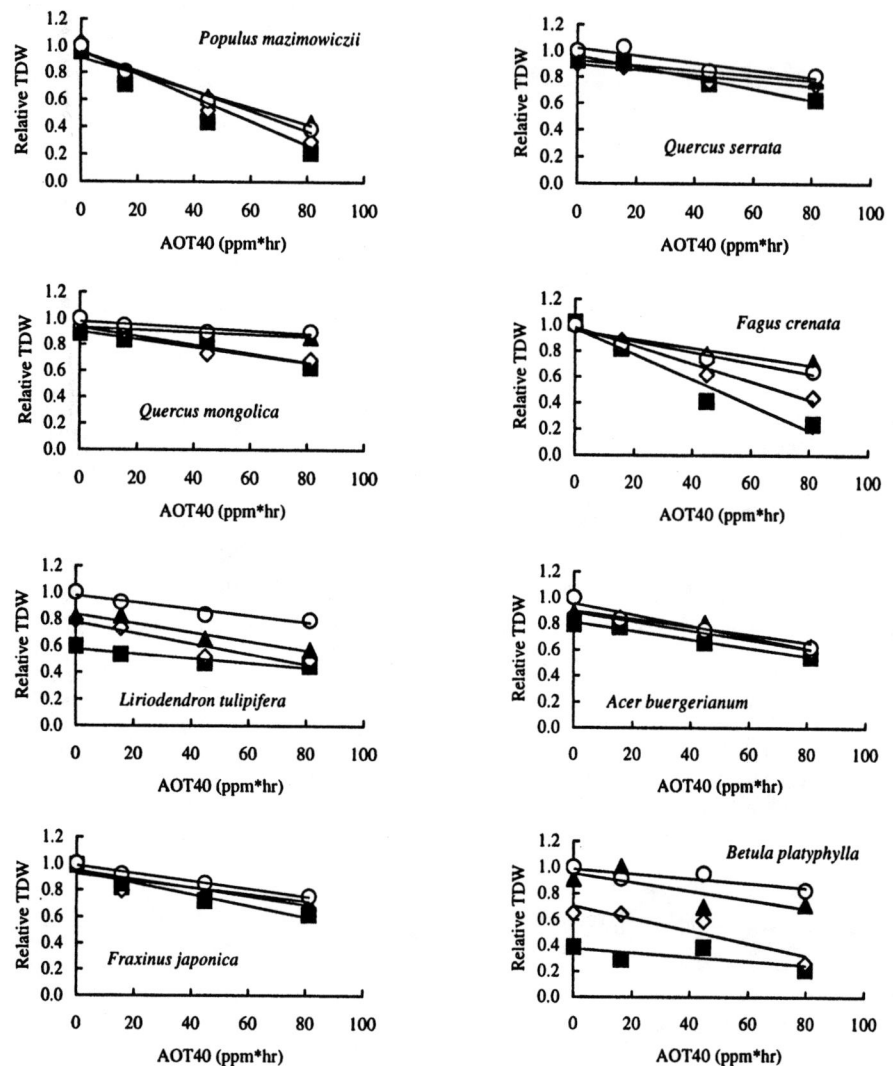

Fig. 15 Growth response to ozone and sulfur dioxide in broad-leaved trees. Circle: SO_2=<1ppb, Triangle: SO_2=10ppb, Diamond: SO_2=20ppb, Square: SO_2=40ppb. AOT40: Accumulated exposure over a threshold of 40 ppb of ozone in daylight hours were calculated for 6 months: April to September (183 days). Relative TDW: Relative total dry weight of plant to that of plant grown in charcoal-filtered air without sulfur dioxide. Exposed for 3 growing seasons except *Betula platyphylla* for 2 growing seasons.

4. INTERACTIVE EFFECT OF RAIN ACIDITY AND OZONE

High concentration of photochemical oxidants and less precipitation during growing season might be linked with a dieback of *Cryptomeria japonica* in the suburban areas of Japan (Nashimoto & Kohno 1989; Nashimoto, 1993). *Cryptomeria japonica* and *Chamaecyparis obtusa* were exposed to simulated acid rain (SAR) and/or ozone for 23 months in the environment controlled chambers under the natural light conditions. The SAR at pH 3.0 consisted of sulfuric, nitric and chloric acids at the equivalent concentration ratio of 5:2:3, respectively. Deionized water at pH 5.6 was as the control and diluent of acids. Plants were received SAR at a rate of 2.5mm/hr x 8hr/day, 3 times/week from April to September, and twice/week from October to March at a rate of 2.5 mm/hr x 8hr per event during the first winter season and 2.5 mm/hr x 4hr per event during the second winter. Total precipitation was 4300mm for 23 months. Dairy ozone exposure has conducted at a constant concentration of 0, 60, 120 or 180 ppb between 09:00 to 15:00 with no exposure between 16:00 to 08:00. However, ozone concentration gradually increased between 08:00 to 09:00 in the morning and decreased between 15:00-16:00 in the evening. Actual mean concentrations of ozone during the experiment were <1, 16, 30 and 46ppb for 24 hours and <1, 47, 90 and 136ppb for 8 hours, respectively.

Chamaecyparis obtusa developed needle chlorosis and early defoliation in the older needles exposed to higher ozone concentrations, however, *Cryptomeria japonica* did not show any visible symptoms for 23 months exposure to ozone. No visible symptoms of exposure to SAR at pH 3.0 were observed.

Dry weight of the plants exposed to SAR at pH 3.0 was greater than that of the plants at pH 5.6. Ozone exposure did not modify the total dry weight of the plants. However, leaf dry weight of plants exposed to higher concentrations of ozone was significantly greater than that of plants exposed to clean air. In contrast, high ozone exposure significantly reduced root dry weights. Thus, top/root dry weight ratio in the plants exposed to higher ozone increased and the exposure to SAR at pH 3.0 enhanced this trend (Fig. 1).

These results suggested that higher ozone concentration interfered with carbon allocation and increased rain acidity probably associated with nitrate, exacerbated this effect. Greater top consumed and required much water via evapotranspiration, however, smaller root biomass could not supply enough water to above ground parts. This imbalance of carbon allocation might accelerated to increase susceptibility to drought stress. As *Cryptomeria japonica* is sensitive to drought stress but *Chamaecyparis obtusa* is tolerant in general, high concentration of ozone and high input of nitrogen in suburban areas may have a possible linkage with dieback of drought sensitive *Cryptomeria japonica*.

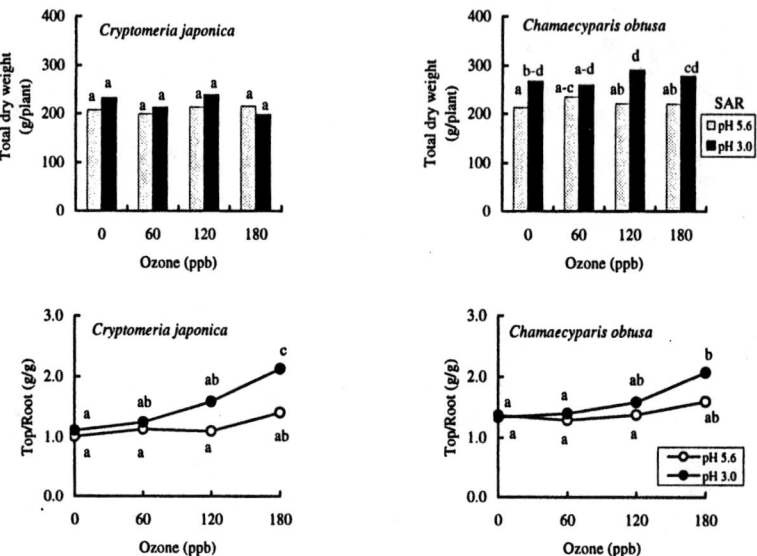

Fig. 16 Effect of simulated acid rain (SAR) and ozone on the growth of *Cryptomeria japonica* and *Chamaecyparis obtusa*. Exposure period was 23 months. SAR at pH 3.0 contains sulfate:nitrate:chloride at equivalent concentration ratio of 5:2:3, but SAR at pH 5.6 was deionized water. Total precipitation of SAR during the experiment was 4300mm.

Daily ozone concentration was constant between 09:00 and 15:00 in the controlled chambers under natural solar radiation. Within a species, any two data having a same letter are not statistically different at the 5 % level by the Tukey's HSD multiple range test (n = 10).

5. CONCLUSION

Air pollution problems mainly due to sulfur dioxides in the western countries including Japan was improved and solved dramatically. In central Europe such as Black Triangle Region, it has started to improve the situation after the civilian revolution (Cerny & Paces, 1995; Galloway, 1995; Hruska et al., 1996; Rodhe et al., 1996). Air quality in south Korea and Taiwan was better than that in China, but still have a big room to improve if compared to that in Japan (Table 9).

Table 10 presented an estimated emission of sulfur dioxide at the year of 1990 in the East Asia. Amount of Japanese emission was less than 1 million tons and was only about 4% in the area. However, China was 21 times higher than that of Japan and was about 83% in the total emission. It is very important that most of them were from stationary sources.

Table 9 Annual mean concentration of sulfur dioxide in East Asia

Country	City	1990	1991	1992	1993
China*	Beijing	38	46	44	
	Taiyuan	151	106	116	
	Qingdao	100	78	79	82
	Changsha	71	66	68	30
	Chongqing	129	134	134	102
	Guiyang	141	130	176	171
	Xiamen	25	3	3	5
	Guangzhou	38	26	23	
Korea*	Seoul	51	43	35	
	Pusan	39	38	33	
	Taegu	41	41	40	
	Kwangju	17	17	17	
	Ulsan	31	38	31	
Taiwan*		26	24	25	25
Japan	15 stations**	10	11	9	8
	Abiko***	3			
	Akagi***	0.3			

* Data from: Cheng & Pai (1996)
** National monitoring stations by Japan Environment Agency (JEA, 1995)
*** Monitored in 1990-1992 by Biology Department, CRIEPI
From Kohno (1996)

Table 10 Emission of sulfur dioxide in East Asia

Country	Total	Stationary	Mobile
Japan*	990	745	245
China	20,951	20,820	131
Taiwan**	583	551	32
S. Korea	1,611	1,422	189
N. Korea	676	671	5
Primorskij***	247	244	3
Mongol	89	89	-
Total	25,147	24,542	605

Unit: kt/Year
*: Fiscal year of 1990, **: Year of 1991, ***: Year of 1989, others Year of 1990
From Kohno (1996)

China is one of the largest emission sources in the world due to large consumption of fossil fuels (Galloway, 1995). If it will continue at a current rate, sulfur emission in China in the year 2010 will be double of the year 1990 (Kohno, 1996). Simulation model analysis of sulfur budget suggested that sulfur deposition in Japan will have a direct influence of the East Asian sulfur emission (Ichikawa & Fujita, 1995; Fujita, 1996).

Mean rain acidity in the Black Triangle Region surrounded with the former eastern Germany, Czech Republic and Poland was around pH 4.2 (Cerny & Paces, 1995). This pH level was not strong enough to induce direct injuries and plant growth reduction in conifers, especially *Picea abies*. However, annual mean concentration or 98 % value of sulfur dioxide in the ambient air of such regions was high enough to induce phytotoxicity in plants (Lux, 1993; Hruska *et al.*, 1996). This could be a very important predisposing factor to induce epidemic forest decline syndrome with other extreme stresses.

As a dry deposition, sulfur dioxide is a primary and important pollutant rather than rain acidification. If stationary sources and mobilization will extend more without any early countermeasures to reduce emission of sulfur dioxide and nitrogen oxides, air quality problems including photochemical oxidants will be more complicated in the East Asia in a quite near future. It also means that a potential risk of an adverse effect on natural vegetation will increase.

Trees had a wide range of variation in growth response to ozone or sulfur dioxide, and there was no correlation between interspecific variations. Sensitivity based on the growth reduction and visible foliar injury development did not have a correlation with each other. However, experimental results suggest that current ambient ozone may accelerate growth reduction in combination with sulfur dioxide at and higher than 20ppb. Thus, at least, sulfur dioxide concentration should be reduced below 20ppb for protecting the Asian vegetation.

An application of critical load calculation by model is a kind of tool to discuss reduction of sulfur emission. Although our experimental results in 16 species were quite limited data set, it showed differential sensitivity and different critical point. Therefore, it should consider differential sensitivity in organisms.

REFERENCES

Alcamo, J., R. Shaw, and L. Hordijk (eds.) (1990)
The RAINS model of acidification: Science and strategies in Europe, Kluwer Academic Publishers, Drodrecht, The Netherlands.

Andersson, M.(1988)
Toxicity and tolerance of aluminium in vascular plants: A literature review. *Water, Air, Soil Pollut.* 39: 439-462.

Aoki, M., T. Ogawa, H. Ishikawa, and Y. Minohara (1979)
Development of leaf injury and accumulation of sulfur in several forest trees exposed to sulfur dioxide at low concentrations. *CRIEPI Report No.477008*, Central Research Institute of Electric Power Industry (in Japanese with English summary).

Ashmore, M. (1996)
    Defining and applying critical levels for ozone and nitrogen dioxide. p.175-183, *In*: Proc. CRIEPI Int'l Seminar on Transport and Effects of Acidic-Substances, Nov. 28-29, 1996, *CRIEPI*, Tokyo, Japan.
Billen, N., H. Schatzle, G. Seufert, and U. Arndt (1990)
    Performance of some variables. *Environ. Pollut.*, 68: 419-434.
Chappelka, A. H., and B. I. Chevone (1992)
    Tree response to ozone. p.271-324, *In*: *Surface Level Ozone Exposures and their Effects on Vegetation*, A.S. Lefohn, ed., Lewis Publ. Inc., Chelsea, MI, U.S.A.
Cheng, W.-L., and J.-L. Pai (1996)
    The state of air pollution and acid precipitation in East Asia. p.304-315, *Proceedings of International Conference on Acid Deposition in East Asia*, Taipei, Taiwan, May 28-30, 1996.
Cerny, J., and T. Paces (Ed.) (1995)
    Acidification in the Black Triangle Region - Acid Reign' 95?, 5th International Conference on Acid Deposition - Science and Policy, Göteborg, Sweden, 26-30 June, 1995.
CRIEPI (1992)
    Acidic deposition in Japan, *CRIEPI Report ET91005*. (in Japanese with English summary)
CRIEPI (1994)
    Sanseiu no eikyou hyouka. *Denchuken Review No.31*. Dec. 1994.
Cronan, C. S., and D. F. Grigal (1995)
    Use of calcium/aluminum ratios as indicators of stress in forest ecosystems. *J. Environ. Qual.* 24: 209-226.
Davis, D. D., and J. M. Skelly (1992)
    Foliar sensitivity of eight eastern hardwood tree species to ozone. *J. Air Waste Manage. Assoc.*, 42: 309-311.
Fishman, J. (1991)
    The global consequences of increasing tropospheric ozone concentrations. *Chemosphere*, 22: 685-695.
Foy, C. D., R. L. Chaney, and M. C. White (1978)
    The physiology of metal toxicity in plants. *Ann. Rev. Pl. Physiol.* 29: 511-566.
Fuhrer, J. (1994)
    The critical level for ozone to protect agricultural crops – an assessment of data from European open-top chamber experiments. p.42-57, *In*: *Critical levels for ozone: A UNECE workshop report (FAC report No.16)*, J. Fuhrer and B. Achermann, eds., Swiss Federal Research Station for Agricultural Chemistry and Environmental Hygiene, Liebefeld–Bern, Switzerland.
Fujii, T. (1971)
    Asagao no kaben no dasshoku, *Air Pollution News Report 62:1*. (in Japanese)
Fujita, S. (1996)

An estimation for atmospheric sulfur budget over the Japanese Archipelago. p.11-25, In: Proc. CRIEPI Int'l Seminar on Transport and Effects of Acidic-Substances, Nov. 28-29, 1996, CRIEPI, Tokyo, Japan.

Fujiwara, T., T. Umezawa, and H. Ishikawa (1976)
Acute effects of combination of sulfur dioxide, nitrogen dioxide and ozone on some plants. *CRIEPI Report No.476001*, Central Research Institute of Electric Power Industry (in Japanese with English abstract)

Galloway, J. N. (1995)
Acid deposition: perspectives in time and space. *Water, Air, Soil Pollution* 85:15-24.

Haines, B.L., and C. L. Carlson (1989)
Effects of acidic precipitation on trees, *In*: Acidic Precipitation 2., Springer-Verlag.

Hettelingh, J.-P., M. Posch., P. A. M. De Smet, and R. J. Downing (1995a)
Deriving critical loads for Asia. *Water, Air, Soil Pollution* 85:2565-2570.

Hettelingh, J.-P., H. Sverdrup, and D. Zhao (1995b)
The use of critical loads in emission reduction agreements in Europe. *Water, Air, Soil Pollution* 85:2381-2388.

Hogsett, W. E., J. E. Weber, D. T. Tingey, A. A. Herstrom, E. H. Lee, and J. A. Laurence (1997) An approach for characterizing tropospheric ozone risk to forests. *Environ. Manage.*, 21: 105-120.

Holland, M. R., P. W. Mueller, A. J. Rutter, and P. J. A. Shaw (1995)
Growth of coniferous trees exposed to SO_2 and O_3 using an open-air fumigation system. *Plant Cell Environ.*, 18: 227-236.

Hruska, J., J. Cerny, and J. Krecek (1996)
Acidification in the Czech part of the Black Triangle Region. p.113-124, *In: Proc. CRIEPI Int'l Seminar on Transport and Effects of Acidic-Substances*, Nov. 28-29, 1996, CRIEPI, Tokyo, Japan.

Hutchinson, T. C., L.Bozic, and G. Munoz-Vega (1986)
Responses of five species of conifer seedlings to aluminum stress. *Water, Air, Soil Pollut.* 31: 283-294.

Ichikawa, Y., and S. Fujita (1995)
An analysis of wet deposition of sulfate using a trajectory model for east Asia. *Water, Air, Soil Pollut.* 85:1927-1932.

Irving, P. M. (Ed.)(1990)
Acidic Deposition: State of Science and Technology. Vol. I-IV, The U. S. National Acid Precipitation Assessment Program, Office of the Director, 722 Jackson Place, NW, Washington, DC 20503, U.S.A.

Izuta, T., M. Miwa, H. Miyake, and T. Totsuka (1990)
Growth response of *Cryptomeria* seedlings to simulated acid rain., *Man & Environ.*, 16:44- 53.

Izuta, T., U. Umemoto, K. Horie, M. Aoki, and T. Totsuka (1996)
Effects of ambient levels of ozone on growth, gas exchange rates and chlorophyll contents of *Fagus crenata* seedlings. *J. Jpn. Soc. Atmos. Environ.*, 31: 95-105.

JEA (1992)
  Kankyo hakusho: Sosetsu. *Japan Environment Agency*. (in Japanese)
JEA (1995)
  Nihon no taiki osen joukyou. *Japan Environmental Agency*. (in Japanese)
JEA (1997)
  Kankyo hakusho: Sosetsu. *Japan Environment Agency*. (in Japanese)
Jensen, K. F., and L. S. Dochinger (1989)
  Response of eastern hardwood species to ozone, sulfur dioxide and acid precipitation. J. *Air Pollut. Control Assoc.*, 39: 852-855.
Kadota, M., and K. Ohta (1972)
  Ozone sensitivity of Japanese plant species in summer, with special reference to a tentative sensitivity grade list for applying to fielded survey on ozone injury. *J. Jpn. Soc. Air Pollut.*, 7: 19-26 (in Japanese with English abstract).
Katoh, T., T. Konno, I. Koyama, H. Tsuruta, and H. Makino (1990)
  Acidic precipitation in Japan. p.41-105, In: Bresser A. H. M., and W. Salomons (eds.), *Acidic precipitation Vol.5: International overview and assessment*. Springer-Verlag, New York.
Keltjens, W. G., and E. V. Loenen (1989)
  Effects of aluminum and mineral nutrition on growth and chemical composition of hydroponically grown seedlings of five different forest tree species. *Plant Soil* 119: 39-50.
Kobayashi, T., Y. Kohno, and H. Matsumura (1996)
  Effect of simulated acid rain on tree species. p.184-189, In: *Proc. CRIEPI Int'l Seminar on Transport and Effects of Acidic-Substances*, Nov. 28-29, 1996, CRIEPI, Tokyo, Japan.
Kobayashi, T., Y. Kohno, and K. Nakayama (1991)
  The effects of simulated acid rain on the uptake of mineral elements in soybean, *J. Agr. Met.*, 48(1): 11-18.
Kobayashi, T., H. Matsumura, and Y. Kohno (1994)
  Effects of simulated acid rain and acid mist on development of visible injury and growth in plants, *CRIEPI Report*: U94025.
Kobayashi, K., M. Okada, and I. Nouchi (1994)
  A chamber system for exposing rice (*Oryzae sativa* L.) to ozone in a paddy field. *New Phytol.* 126:317-325.
Kochian, L. V. (1995)
  Cellular mechanisms of Aluminium toxicity and resistance in plants, *Ann. Rev. Plant Physiol. Plant Mol. Biol.*, 46: 237-260.
Kohno, Y. (1996)
  Summary of CRIEPI acid deposition assessment program: Acid deposition, forest decline and further research needs. p.207-222, In: *Proc. CRIEPI Int'l Seminar on Transport and Effects of Acidic-Substances,* Nov. 28-29, 1996, CRIEPI, Tokyo, Japan.
Kohno, Y., and T. Kobayashi (1989a)
  Effect of simulated acid rain on the growth of soybean, *Water, Air, Soil Pollut.*, 43:

11-19.

Kohno, Y., and T. Kobayashi (1989b)
Effect of simulated acid rain on the yield of soybean, *Water, Air, Soil Pollut.*, 45: 173-181.

Kohno, Y., H. Matsumura, and T. Kobayashi (1994)
Effect of simulated acid rain on the development of leaf injury in tree seedlings, *J. Japan Soc. Air Pollut.* 29(4): 206-219.

Kohno, Y., H. Matsumura, and T. Kobayashi (1995a)
Effect of simulated acid rain on the growth of Japanese cedar, Japanese cypress and Sawara cypress, *J. Jpn. Soc. Atmos. Environ.* 30(3): 191-207.

Kohno, Y., H. Matsumura, and T. Kobayashi (1995b)
Effect of aluminium on the growth of *Cryptomeria japonica* and *Chamaecyparis obtusa* grown in nutrient solution culture, *J. Jpn Soc. Atmos. Environ.*, 30: 316-326.

Kohno, Y., H. Matsumura, and T. Kobayashi (1995c)
Effect of simulated acid rain on the growth of Japanese conifers grown with or without fertilizer, *Water, Air, Soil Pollut.*, 85:1305-1310.

Kozlowski, T. T., P. J. Kramer, and S. G. Palardy (1991)
Air Pollution. p.338-375, *In: The physiological ecology of woody plants*, Academic Press, San Diego, CA, USA.

Krahl-Urban, B., H. E. Papke, K. Peters, and C. Schimansky (1988)
Forest decline: Cause-effect research in the United States of North America and Federal Republic of Germany. Assessment Group for Biology, Ecology and Energy of the Jülich Nuclear Research Center for the U.S. Environmental Protection Agency and German Ministry of Research and Technology.

Krizek, D. T., C. D. Foy, and R. M. Mirecki (1997)
Influence of aluminum stress on shoot and root growth of contrasting genotypes of *Coleus*. J. Pl. Nutr. 20(9):1045-1060.

Lee, E. H., D. T. Tingey and W. E. Hogsett (1988)
Evaluation of ozone exposure indices in exposure–response modeling. *Environ. Pollut.* 53:43-62.

Likens, G. E., and F.H. Bormann (1974)
Acid rain: A serious regional environmental problem. *Science* 184:1176-1179.

Lux, H. (1993)
Ecological monitoring station of Oberbarenburg, Institute of Phytochemistry and Wood Chemistry, Department of Forestry, Technical University of Dresden, Tharandt, Germany.

Mahoney, M. J., J. M. Skelly, B. I. Chevone, and L. D. Moore (1984)
Response of yellow-poplar (*Liriodendron tulipifera* L.) seedlings shoot growth to low concentrations of O_3, SO_2 and NO_2. *Can. J. For. Res.*, 14: 150-153.

Matsumoto, Y., Y. Maruyama, Y. Morikawa, and T. Inoue (1992)
Some negative results of simulative acid mist and ozone treatments to *Cryptomeria japonica* seedlings in explanation of mature *C. japonica* decline in the Kanto plains in Japan, *Jpn. J. For. Environ.* 34: 85-97.

Matsumura, H., H. Aoki, Y. Kohno, T. Izuta, and T. Totsuka (1996)
Effects of ozone on dry weight growth and gas exchange rate of Japanese cedar, Japanese cypress and Japanese zelkova seedlings. *J. Jpn. Soc. Atmos. Environ.*, 31: 247-261 (in Japanese with English abstract).

Matyssek, R., P. B. Reich, R. Oren, and W. E. Winner (1995)
Response mechanisms of conifers to air pollutants. p.255-308, In: *Ecophysiology of coniferous forests*, W.K. Smith and T. M. Hinckley, eds., Academic Press, San Diego, CA, U. S. A.

McCormick, L.H., and K. C. Steiner (1978)
Variation in aluminum tolerance among six genera of trees. Forest Sci. 24(4): 565-568.

McLaughlin, S. B. (1985)
Effects of air pollution on forests: A critical review. *J. Air Pollut. Control Assoc.*, 35: 512-544.

Miwa, M., T. Izuta, and T. Totsuka (1993)
Effects of simulated acid rain and/or ozone on the growth of Japanese cedar seedlings. *J. Jpn. Soc. Air Pollut.* 28: 279-287 (in Japanese with English abstract).

Miller, P. R., and K. W. Stolte (1984)
Response of forest species to O_3, SO_2, and NO_2 mixtures. Proc. Ann. Meet. *Air Pollut. Control Assoc.*, 77: 84.30.5.1-15.

Nashimoto, M. (1988)
Decline of *Cryptomeria japonica* and assessing method by morphological features, *CRIEPI Report* U87091.

Nashimoto, M. (1993)
Decline of *Cryptomeria japonica D. Don* and secondary air pollutants, *CRIEPI Report* U93017.

Nashimoto, M., and Y. Kohno (1989)
Decline of Japanese cedar (*Cryptomeria japonica*) and potential correlation of oxidants and precipitation. *CRIEPI Report* U89017.

Nouchi, I., T. Odaira, T. Sawada, K. Oguchi, and T. Komeiji (1973)
Plant ozone injury symptoms. *J. Jpn. Soc. Air Pollut.*, 8: 113-119 (in Japanese with English abstract).

Ohara, T., S. Wakamatsu, I. Uno, T. Ando, and S. Izumikawa (1995)
An analysis of annual trends of photochemical oxidants in the Kanto and Kansai areas. *J. Jpn. Soc. Air Pollut.*, 30: 137-148 (in Japanese with English abstract).

Percy, K.(1986)
The effects of simulated acid rain on germinative capacity, growth and morphology of forest tree seedlings. *New Phytol.* 104: 473-484.

Peterson, C. E., Jr., and R. A. Mickler (1994)
Considerations for evaluating controlled exposure studies of tree seedlings. *J. Environ. Qual.*, 23: 257-267.

Posch, M., P. A. M. de Smet, J.-P. Hettelingh, R. J. Downing (1995)

Calculation and mapping of critical thresholds in Europe: status 1995. *CCE report*, RIVM, Bilthoven, The Netherland.

Pye, J. M. (1988)
Impact of ozone on the growth and yield of trees: A review. *J. Environ. Qual.*, 17: 347-360.

Reich, P. B. (1987)
Quantifying plant response to ozone: a unifying theory. *Tree Physiol.*, 3: 63-91.

Reich, P. B., J. P. Lassoie, and R. G. Amundson (1984)
Reduction in growth of hybrid poplar following field exposures to low levels of O_3 and (or) SO_2. *Can. J. Bot.*, 62: 2835-2841.

Rengel, Z. (1992)
Role of calcium in aluminium toxicity. *New Phytol.* 121: 499-513.

Rengel, Z., and D. C. Elliott (1992)
Mechanism of aluminum inhibition of net $^{45}Ca^{2+}$ uptake by Amaranthus protoplasts. *Plant Physiol.* 98, 632-638.

Rodhe, H., P. Grennfelt, J. Wisnieski, C. Ågre, G. Bengtsson, K. Johansson, P. Kauppi, V. Kucera, L. Rasmussen, B. Rosseland, L. Schotte, and G. Sellden (1995)
Acid Reign '95? - Conference Summary Statement from the 5th International Conference on Acidic Deposition, Science and Policy", Göteborg, Sweden 26-30 June, 1995. *Water, Air, Soil Pollut.* 85:1-14.

Sandermann, H., A. R. Wellburn and R. L. Heath (eds.) (1997)
Forest decline and ozone (Ecological studies 127), pp.400, Springer-Verlag, Berlin, Heidelberg.

Sekiguchi, K., Y. Hara, and A. Ujiie (1986)
Dieback of *Cryptomeria japonica* and distribution of acid deposition and oxidant in Kanto District of Japan, *Environ. Technol. Lett.*, 7:263-268.

Shimizu, H., Y. Fujinuma, K. Kubota, T. Totsuka, and K. Omasa (1993)
Effects of low concentrations of ozone (O_3) on the growth of several woody plants. *J. Agr. Met.*, 48 (5): 723-726.

Shriner, D. S., W. W. Heck, S. B. McLaughlin, D. W. Johnson, P. M. Irving, J. D. Joslin, and C. E. Peterson (1990)
Response of vegetation to atmospheric deposition and air pollution. NAPAP SOS/T Report 18, *In: Acidic Deposition: State of Science and Technology*, Vol.III, National Acid Precipitation Assessment Program, 722 Jackson Place NW, Washington D.C. 20503, U. S. A.

Skärby, L., and E. Karlsson (1996)
Critical levels for forest trees-best available knowledge from Nordic countries and the rest of Europe. P.72-85, *In*: KärenlampiL., and L. Skärby (eds.) *Critical levels for ozone in Europe: testing and finalising the concepts.* University of Kuopio, Kuopio, Finland.

Sverdrup, H., and W. De Vries(1994)

Calculating critical loads for acidity with the simple mass balance method. *Water, Air, Soil Pollut.* 72:143-162.

Sverdrup, H., and P. Warfvinge (1993)
The effect of soil acidification on the growth of trees, grass and herbs as expressed by the (Ca+Mg+K)/Al ratio. *Report 2:1993*. Department of Chemical Engineering II, Lund University, Sweden.

Tamm, C. O. (1976)
Acid precipitation: Biological effects in soil and on forest vegetation. *AMBIO*, 5(5-6):235-238.

Tamm, C. O., and B. Popovic. (1989)
Acidification experiments in pine forests, National Swedish Environmental Protection Board. *Report 3589*.

Taylor, G. J. (1989)
Aluminum toxicity and tolerance in plants. p.327-361., *In*: D. C. Adriano and A. H. Johnson, eds., *Acidic Precipitation vol.2: Biological and ecological effects*, Springer-Verlag, New York.

Taylor, O. C., P. R. Miller, A. L. Page, and L. J. Lund (1986)
Effects of ozone and sulfur dioxide mixtures on forest vegetation of the southern Sierra Nevada. PB87-106241, U. S. Department of Commerce, NTIS, Springfield, VA, U. S. A.

Thompson, A. M. (1992)
The oxidizing capacity of the earth's atmosphere: Probable past and future changes. *Science*, 256: 1157-1165.

TMRI (1975)
Iwayuru sannseiu" ni kannsuru chousa kenkyu houkoku No. 1-0-28. The Tokyo Metropolitan Research Institute for Environmental Protection. (in Japanese)

Totsuka, T. (1989)
Towards the environmental standards for protecting vegetation from air pollution. *J. Jpn. Soc. Air Pollut.*, 24: 392-396 (in Japanese).

Tsukahara, H., T. T. Kozlowski, and J. Shaklin (1985)
Tolerance of *Pinus densiflora, Pinus thunbergii, and Larix leptolepis* seedlings to SO_2. *Plant and Soil*, 88: 385-397.

Tsukahara, H., T. T. Kozlowski, and J. Shaklin (1987)
Responses of *Betula platyphylla* var. *japonica* seedlings to SO_2. *J. Yamagata Agr. For. Soc.*, 44: 5-12.

Tveite, B., G. Abrahamsen, and A. O. Stuanes, (1990/91)
Liming and wet acid deposition effects on tree growth and nutrition: Experimental results, *Water, Air, Soil Pollut.* 54: 409-422.

Ulrich, B. (1989)
Effects of acidic precipitation on forest ecosystems in Europe, p.189-272., *In:* D. C. Adriano & A. H. Johnson (Eds.): *Acidic Precipitation Vol.2, Biological and ecological effects.*, Springer-Verlag, New York.

UN-ECE (1988)
Forest damage and air pollution. Report of the 1987 forest damage survey in Europe. Convention on long-range transboundary air pollution: International co-operative programme on assessment and monitoring of air pollution effect on forests. Geneva, Switzerland.

UNEP-ECE (1991)
Forest damage and air pollution. Report of the 1990 forest damage survey in Europe. Geneva, Switzerland

Wood, T., and F. H. Bormann (1974)
The effects of an artificial acid mist upon the growth of *Betula alleghaniensis* Britt., *Environ. Pollut.*, 7, 259-268.

Wood, T., and F. H. Bormann (1977)
Short-term effect of simulated acid rain upon the growth and nutrient relations of *Pinus strobus* L., *Water, Air, Soil Pollut.*, 7: 479-488.

Yambe, Y. (1978) Declining of trees and microbial florae as the index of pollution in some urban areas, *Bull. For. & For. Prod. Res. Inst.*, 301: 119-129.

Yoshida, K. (1971)
Sanseiu to asagao, *Air Pollution News Report* 66:1. (in Japanese)

Chapter 6
ACID DEPOSITION EFFECTS ON THE DYNAMIC OF HEAVY METALS IN SOILS AND THEIR BIOLOGICAL ACCUMULATION IN THE CROPS AND VEGETABLES IN TAIWAN

ZUENG-SANG CHEN[1*], JEN-CHYI LIU[1,2], AND CHING-YI CHENG[1]

CONTENT

Abstract
1. Introduction
2. Distribution and Composition of Acid Deposition in Taiwan
    2.1 The mean concentration of ions in rain water
    2.2 The total quantity of acid deposition of ions in rain water
3. Effects of acid deposition on Soil Acidity and Release of Heavy Metals in Taiwan soils
    3.1 Effects of Acid Deposition on Soil Acidity in Taiwan
    3.2 Effects of Acid Deposition on Release of Heavy Metals in Taiwan Soils
4. Acid deposition Influences on the biological accumulation of trace elements in crops by pot experiments
    4.1 Design of Pot Experiments
    4.2 Laboratory Analysis of Trace Elements in Soils and Crops
    4.3 Statistical Analysis
    4.4 Acid Deposition Influence on the Concentration and Total Uptake of Heavy Metals in the Brown Rice
    4.5 Acid Deposition Influence on the Concentration and Total Uptake of Heavy Metals in the Cabbage
5. Acid deposition Influences on the biological accumulation of trace elements in crops and vegetables sampled in the field

---

[1] Department of Agricultural Chemistry, National Taiwan University, Taipei 10617, TAIWAN.
* **Corresponding author**, Dr. Zueng-Sang CHEN, Professor of pedology and soil pollution
Tel: +886-2-2369-8349, fax: +886-2-2392-4335,
E-mail: SOILCHEN@CCMS.NTU.EDU.TW
[2] Department of Agricultural Chemistry, Taiwan Agricultural Research Institute, Taichung 43101, TAIWAN.

5.1 Design of Sampling Crops and Vegetables in the Field
5.2 Laboratory Analysis of Trace Elements in Soils and Crops
5.3 Acid Deposition Influence on the Concentration and Total Uptake of Trace Elements in the brown rice and leaves of Vegetables
5.4 Acid Deposition Influence on the ratio of Concentration of Heavy Metals in the Crops in Two Regions
6. Conclusions
Acknowledgments
References

Abstract

The Environmental Protection Administration (EPA) of Taiwan have established twelve monitoring stations of acid deposition in Taiwan since 1990. Based on the database in the monitoring system, the main compositions of rain water was the sulfate, chloride, sodium and ammonium ions, followed by nitrate and calcium ions. The concentration ratio of sulfate to nitrate of acid deposition in Taiwan is about 5. The mean annual total deposition of sulfate ion from rain water in northern Taiwan is higher than 100 kg/ha/yr which was more than two times compared with that in the eastern USA. Monitoring results also indicated that about 70 kg/ha/yr of sulfate was deposited in southern Taiwan. The mean annual total deposition of nitrate ion from rain water in northern and western Taiwan is 40 to 60 kg/ha/yr, and < 30 kg/ha/yr in eastern and southern Taiwan. The mean annual total deposition of hydrogen ion from rain water is highest in northern Taiwan, about 1 kg/ha/yr, and < 0.4 kg/ha/yr in other regions of Taiwan. The mean annual total deposition of ammonium ion from rain water is highest in southwestern Taiwan, ranged from 20 to 30 kg/ha/yr, and < 10 kg/ha/yr in eastern and southern Taiwan.

Representative two rural soils in Taiwan were selected to compare the extractable concentrations of heavy metals (Cd, Cu, Pb, and Zn) in soils and total concentration of heavy metals in the brown rice and vegetables. These two soils were treated with different grades of concentration of heavy metals and then simultaneously treated with artificial acid rain or without acid rain (treated with top water) by pot experiments in 1996. The Lung-tang area (affected by acid rains) in northern Taiwan and Lung-luan-tang area (unaffected by acid rains) in southern Taiwan were selected for the field monitoring of acid deposition influence on the biological accumulation of heavy metals (Cd, Cu, Pb, and Zn) in the brown rice and in the leaves of 19 vegetables sampled from 1996 to 1997. The results from pot experiments indicated that the treatment of artificial acid rain in two selected soils significantly increase the biological accumulation of trace metals (Cd, Cu, Pb, and Zn) in the brown rice and leaves of pickled cabbage ($p < 0.05$). The results from the field samples indicated that the ratios of relative concentration of Cd, Zn, and Cu, except for Pb, in nineteen vegetable species sampled from the acid rain affected region are almost higher than 1, or higher than 3, compared to the crops grown in acid rain non-affected area. These results suggest that the biological accumulation of Cd, Cu, and Zn, expect for Pb, in the leaves of vegetables was affected by the acid rain, and the rating of effectiveness on the phyto-availability of heavy metals caused by acid deposition followed the trend: CdZn Cu >> Pb. There is no effect on the biological accumulation of trace elements in brown rice sampled from acid rain affected region.

Key words: acid deposition, crops, brown rice, vegetable, heavy metals, biological accumulation.

1. INTRODUCTION

The pH value less than 5.6 in the rain are recognized as acid rain or acid atmospheric wet deposition. The main compositions in the acid rain or atmospheric deposition are mainly connected with sulfate and nitrate ions and some heavy metals such as Cd, Pb, Cu, Zn, and Mn. The controlling strategies are promoted and the status of acidic air pollution have been significantly decreased in the last decade. Some researches also indicated that some vegetation species have strongly tolerant characteristics for atmospheric acidic wet deposition. Many studies have concluded that acidic atmospheric deposition will significantly affect on the growth and production of crops and vegetation species. On a basis of the research results from simulated acid rain or from the monitoring carried out in the field (Irving and Miller, 1981; Dubay and Heagle, 1987; Olsen et al., 1987; Musselman and Sterrett, 1988; Takemoto et al., 1988; Smith et al., 1991; Trumble and Walker, 1991; Rinallo et al., 1993), one can conclude that the injury of leaves of crops or vegetation is the most typical response. The injury of leaves includes: (1) direct injury of surface of leaves, an alteration of physiological, and biochemical functions, such as destroy on leaf structure, resistance on the sexual reproduction, unbalance of pH in cell, and changes on enzymes activities, and (2) indirect influence on the reduced growth rates and reduced yield production owing to the acidic soil properties and lower available soil fertility or higher concentration of toxic substances released from the soils (Su and Yang, 1995).

Root uptake plays an important role in the cycling of essential microelements such as Zn and Cu, and some trace elements such as Cd, Cr, Pb, and Ni, which may led to a certain uptake of these elements into vegetation. Cadmium is chemically similar to Zn, which may lead to a certain uptake of this element into the vegetation (Javis et al., 1976; Cataldo et al., 1983; Pahlsson, 1989).

The ability of different plants to absorb trace elements varies greatly, when compared at a large scale. However, the index of their accumulation ability illustrates some general trends. Some elements such as Cd, B, Br, Cs, and Rb are extremely easily uptaken by the plants. Zinc, Mo, Hg, Cu, Pb, and As are medium available to plant, while Cr, Ni, Se, Ba, Ti, Zr, Sc, Bi, Ga, and Se are slightly available to plants (Chen, 1992; Kabata-Pendias and Pendias, 1992). Chaney et al. (1972) indicated that the reduction step is obligatory in root uptake of Fe. The reduction of other metals such as Mn, Cu, Sn, or Hg in the uptake step apparently has not been clearly observed. Some results also indicated that Cd, Zn, and Pb absorbed by the tops of brown grass were not likely to move readily to the roots, whereas Cu was very mobile (Kabata-Pendias, 1979).

The ratio of metal concentration closed to 1 for Cu and Zn, which was studied in the forest area. This ratio gives the evidence of rather stable Zn and Cu levels in different species and plant compartments from 1982 to 1992, which is further emphasized by a lack of significant changes in plant Zn and Cu concentrations during this period of time (Berthelsen et al., 1995). Cadmium shows a much wide range of ratios, in general indicating no distinct decrease in plant Cd concentration from 1982 to 1992 either in southern or central Norway. Cadmium concentration in vegetation were significantly higher in southern Norway than central Norway at both times. Calculated concentration ratios in 1982 to 1992 for Pb clearly indicate distinct decrease in plant Pb concentration from 1982 to 1992 for most vegetation species and plant compartments both in southern and central Norway. This is further underlined by significant decreases in Pb concentrations in vegetations from 1982 to 1992 (Berthelsen et al., 1995).

Direct effects of acid deposition on soils properties may include both the positive effects due to the fertilization of increasing the sulfur and nitrogen content in soils (Swedish Ministry of Agriculture, 1982; National Atmosphere Deposition Program, 1986), and negative effects such as the increase of the leaching of soil nutrients and mobility of heavy metals (Al, Cd, Cu, Pb, Zn) in soils (Ulrich et al., 1980; Foy, 1981; Rutherford et al., 1985). The Watt Committee (1984) considered that most soils could compensate the increased loss of base cations due to acid precipitation, and also considered that soil acidification was unlike significantly related to other factors about the acidification (Johnson et al., 1986). Hallbaecken and Tamm (1986) have shown that evidence between the pH of forest soils in southwest Sweden under spruce vegetation and the ages of vegetation species, and the acidifying effects on the vegetation growth are clearly apparent, but the soil pH of the later samples is displaced downward. They interpret that this difference are partially due to the acid precipitation. Mapping of sensitivity of soils on acid deposition has become a critical task in order to locate in those areas most susceptible to environmental impact (McFee, 1980). There is some concern that acid precipitation cause significantly soil acidification over that caused by land use practices and natural weathering processes, but it has not yet been unequivocally demonstrated (Binns, 1988, Houng et al., 1990). Many soils have been acidified recently, as a result of human activity, and it display the most obvious effects (Kennedy, 1992). The result also indicated that toxicity in acid soils is often attributed to associated factors such as the presence of aluminum and manganese species. These factors become progressively more soluble as the soil pH decreases (Kennedy, 1992).

The acidity of Taiwan soils was thought to be slightly increased in the last two decades due to the increasing emission of sulfur oxides and nitrogen oxides from the combustion of fossil fuels. The acid deposition from air pollution is more serious in the western, northeastern, and southern Taiwan. According to some investigations, furthermore, results indicated that the main reason of acidification in lake and river indeed was acid rain (Carl, et al., 1982; Chen and Hung, 1987; Chen et al., 1994). The earth's atmosphere has been known to transport both natural and anthropogenically mobilized the trace elements. Wang et al. (1995) also reported that the distributions of aluminum and acid-leached metals in the sediments of some lake of Taiwan, which were suggested that surface enrichments of Cd and Pb have been mainly proposed by the anthropogenic inputs. In northern Taiwan, the soil pH values are higher than mean pH value of acid precipitation, it indicated that further acidification due to the acid deposition maybe possible. As the soil pH decreased, many metals in soils would become more mobile and pollute the agricultural soils by heavy metals (Liu et al., 1993). It also means that soil acidification is accompanied by a great risk of biological effects from excessive heavy metal levels in soil solutions (Fernandez, 1985).

The objectives of this paper are (1) to show the distribution and composition of acid deposition in Taiwan since 1990, (2) to show the effects of acid deposition on soil acidity and release of trace elements in Taiwan soils, (3) to discuss the acid deposition influences on the biological accumulation of trace elements in crops by pot experiments, and (4) to show the acid deposition influences on the biological accumulation of trace elements in brown rice and leaves of vegetables sampled from the acid rain affected areas of Taiwan.

2. DISTRIBUTION AND COMPOSITION OF ACID DEPOSITION IN TAIWAN

The Environmental Protection Administration (EPA) of Taiwan has established twelve monitoring stations (E1 to E12) of acid deposition in Taiwan since 1990. National Central University (NCU) also established three monitoring stations since 1993 and Taiwan Forestry Research Institution (TFRI) also established five monitoring stations (M1 to M5) in mountain areas since 1990 (King et al., 1995). The basic data and distribution of these twenty stations for collection of acid deposition in Taiwan are shown in Table 1 and Fig. 1 (Lin *et al.*, 1997).

The distribution of pH of the rainfall in Taiwan are shown in Fig. 2 (Lin et al., 1997). According to past six-year (1990-1996) data, the northern Taiwan has been widely effected by the acid precipitation with mean values of 4.5. In general, more than 50% of the acid deposition in Taiwan was significantly affected by North-Eastern flow and frontal passage in the dominant precipitation systems. Therefore, the principal composition of rain water was affected by the external materials from other countries, especially from Mainland China.

Table 1. The basic information of monitoring stations for acid deposition established in Taiwan since 1990.

Monitoring Station #	Longitude (East)	Latitude (North)	Elevation (meter)	Precipitation (mm/yr)	Sampling year (month/year)
E1 (Taipei)	121° 31'	25° 01'	12	2114	1990~1995
E2 (Kueishan)	121° 19'	25° 00'	7	1687	1990~1995
E3 (Chungli)	121° 11'	24° 58'	134	1640	1990~1995
E4 (Hsiaokang)	120° 20'	22° 33'	10	1576	1990~1995
E5 (Kengting)	120° 44'	22° 00'	2	1561	1990~1995
E6 (Taimalee)	120° 58'	22° 36'	825	2010	1990~1995
E7 (Lotung)	121° 45'	24° 41'	10	2803	1990~1995
E8 (Taichungkang)	120° 30'	24° 09'	2	1051	1990~1995
E9 (Chaiyee)	120° 23'	23° 36'	2	1328	1990~1995
E10 (Pengfu)	119° 31'	23° 31'	15	756	1990~1995
E11 (Tainan)	120° 12'	23° 00'	2	2801	1993~1995
E12 (Alishan)	120° 49'	23° 31'	2413	1684	1994~1995
N1 (Longtang)	121° 14'	24° 50'	16	1639	11/1993~01/1995
N2 (Taoyuan)	121° 23'	25° 00'	10	1687	02/1994~02/1995
N3 (Chungli)	121° 11'	24° 58'	134	1639	10/1993~10/1994
M1 (Pinglin)	121° 43'	24° 54'	550	2900	05/1990~07/1991
M2 (Fushan)	121° 34'	24° 46'	600	2900	01/1991~12/1994
M3 (Peelushe)	121° 19'	24° 14'	2350	2430	05/1991~12/1994
M4 (Lianfachee)	120° 54'	23° 56'	760	2220	01/1991~04/1994
M5 (Shanping)	120° 41'	22° 55'	750	3560	08/1991~12/1994

#: E1 to E12 were established by EPA/Taiwan, N1 to N3 were established by National Central University(NCU), Taiwan, M1 to M5 were established by Taiwan Forestry Research Institute (TFRI). (Data source from Lin et al., 1997, with permission)

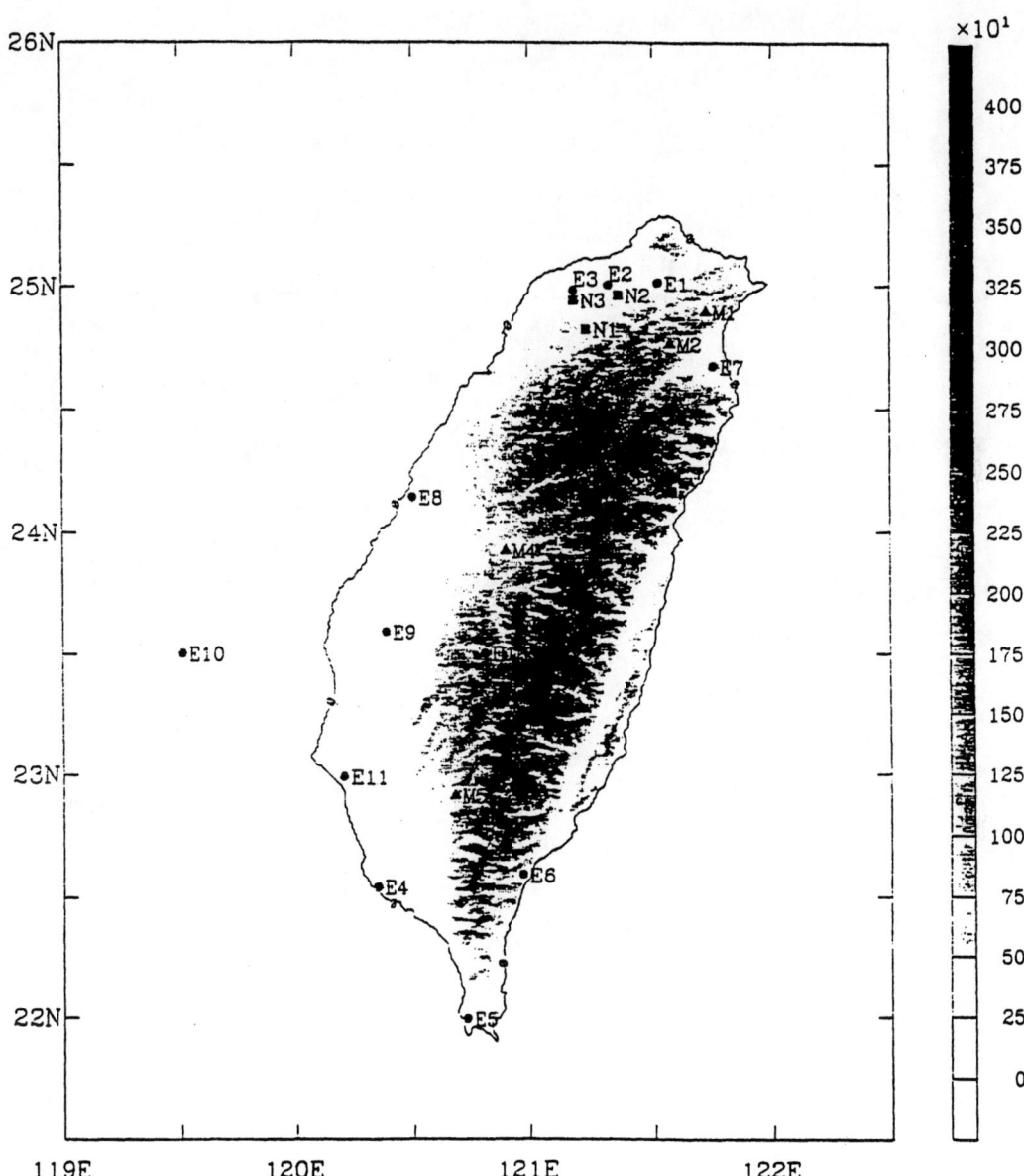

Figure 1. The distribution of monitoring sites of acid depositions established in Taiwan since 1990. (Data from Lin et al., 1997, with permission)

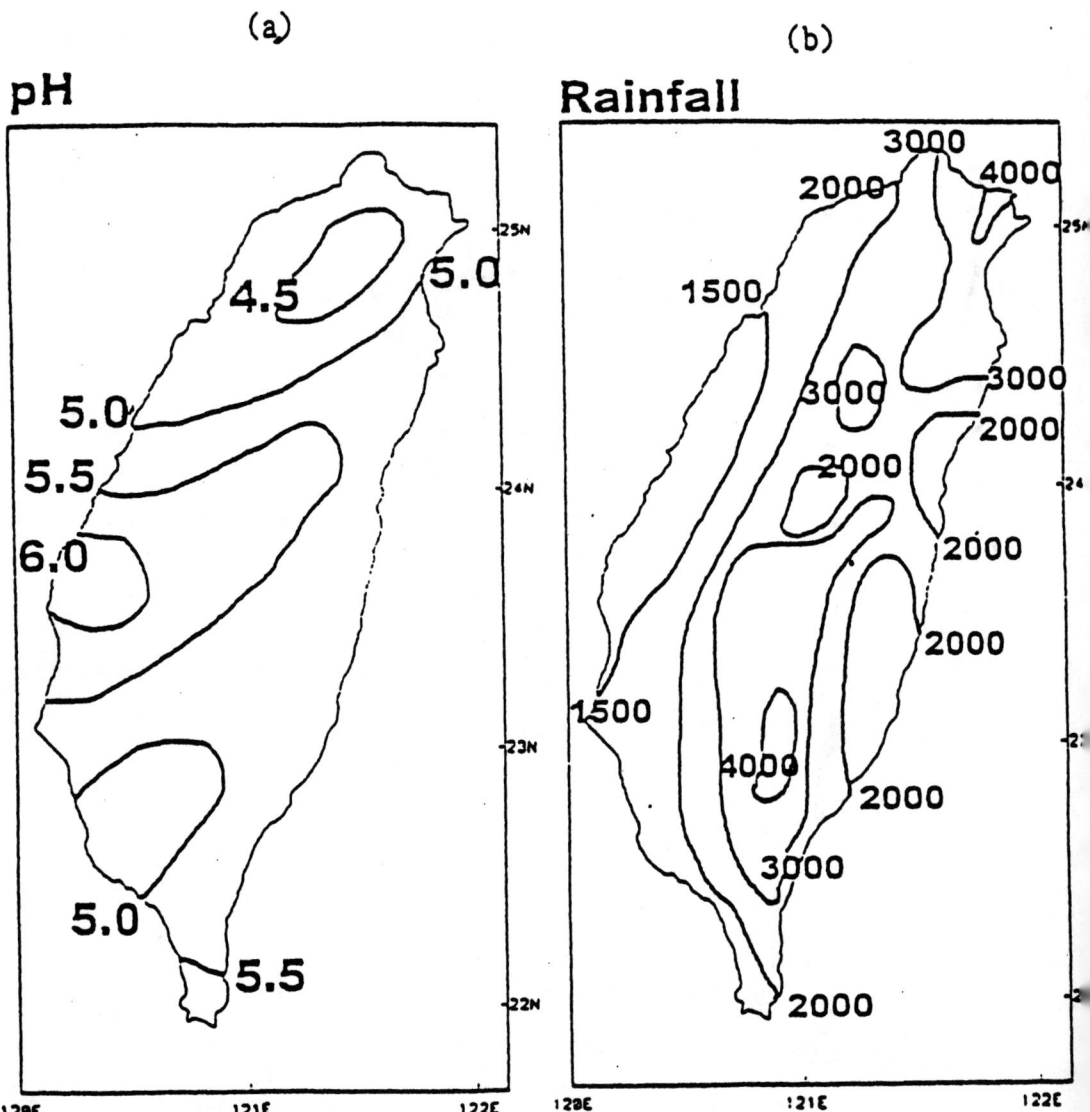

Figure 2. The distribution of annual mean values of (a) pH and (b) total precipitation (mm/yr) of rain in Taiwan. (Data from Lin et al., 1997, with permission)

2.1 THE MEAN CONCENTRATION OF IONS IN RAIN WATER

Based on these monitoring system data, the main composition of rain water was the sulfate, chloride, sodium and ammonium ions, followed by nitrate and calcium ions. The detailed mean concentrations of different ions collected from these 20 stations in last six-year database were shown in Table 2. The distribution of mean concentration of sulfate, nitrate, hydrogen, and ammonium ions in past six-year is shown in Fig. 3. The mean concentration of different ions in the rain water in the twenty monitoring stations can be summarized as follows (Lin et al., 1997):

2.1.1. Sulfate, Nitrate, Hydrogen, and Ammonium Ions

The mean concentration of sulfate in western Taiwan is 100 micro-equivalent/L, and less than 50 micro-equivalent/L in eastern Taiwan. From the analyses of database and model simulation, the content and distribution of sulfate ion are significantly connected with the input from other countries. The mean concentration of nitrate is highest in Taipei, about 50 micro-equivalent/L, and less than 30 micro-equivalent/L in southern and eastern Taiwan. From the databases, the content and distribution of nitrate are significantly related to the input from main capital regions. The mean concentration of hydrogen ion is highest in northeastern Taiwan, around 40-50 micro-equivalent/L, and lowest in southern Taiwan, less than 20 micro-equivalent/L. The distribution of highest concentrations of hydrogen ion is not coincide with those of sulfate and nitrate based on the database. These results indicated that cations in the rain water have a significantly neutral actions with the sulfate and nitrate ions. The mean concentration of ammonium ion is highest in northern Taiwan, around 70-80 micro-equivalent/L, and lowest in southern and eastern Taiwan, less than 30 micro-equivalent/L.

2.1.2. Sodium, potassium, calcium, magnesium, and chloride ions

The mean concentration of sodium ion is highest in Pengfu island, around 300 micro-equivalent/L, and lowest in mountain area, less than 10 micro-equivalent/L. The mean concentration of potassium ion is highest in Taimalee monitoring station, around 60 micro-equivalent/L, and lowest in southern and northern Taiwan, less than 20 micro-equivalent/L. The mean concentration of calcium ion is highest in central Taiwan, around 117 micro-equivalent/L, and lowest in Alishan, southern mountain area, less than 10 micro-equivalent/L. The mean concentration of magnesium ion is highest in Pengfu island, around 80 micro-equivalent/L, and lowest in southern and eastern Taiwan, less than 30 micro-equivalent/L. The mean concentration of chloride ion is highest in Taichung, Pengfu and Lotung monitoring stations, around 200-350 micro-equivalent/L, and lowest in southern and eastern Taiwan, less than 150 micro-equivalent/L.

Table 2. The mean concentration and standard deviation of different ions and total deposition of sulfate and nitrate ions in monitoring stations of acid depositions in Taiwan from 1990 to 1995.

Site	Sample Number (n)	EC mho/cm	pH	Cl$^-$	NO$_3^-$	SO$_4^{2-}$	H$^+$	NH$_4^+$	Na$^+$	K$^+$
						eq/L				
E1	240	54 #	4.41	111	45	112	67	58	84	15
		46 +	0.60	225	46	100	58	47	197	41
E2	136	59	4.51	135	52	119	48	51	111	22
		51	0.50	180	101	85	38	62	168	34
E3	128	70	4.64	235	41	153	46	48	152	54
		69	0.70	379	52	190	48	44	287	88
E4	131	37	4.76	49	25	104	33	52	36	7
		28	0.64	66	33	114	33	50	53	18
E5	75	28	6.01	159	8	35	5	10	115	25
		23	1.05	188	12	35	6	14	141	69
E6	140	37	5.21	160	26	57	18	30	87	64
		39	0.68	178	32	61	25	36	119	104
E7	124	49	5.37	215	29	97	19	48	179	16
		47	0.94	271	33	112	25	50	243	23
E8	127	70	5.07	207	54	224	36	73	254	27
		186	0.96	525	93	670	53	59	1006	70
E9	60	59	6.41	96	38	105	8	79	40	61
		65	1.38	104	51	190	15	57	101	86
E10	50	65	5.62	372	32	90	6	23	299	11
		56	0.73	357	56	85	5	25	283	13
E11	37	25	5.04	50	24	80	20	48	41	5
		16	0.66	62	23	62	25	39	47	5
E12	35	19	5.36	16	13	37	10	34	6	12
		23	0.64	30	9	39	12	39	4	29
N1	36	54	4.37	80	49	130	70	92	59	6
		41	0.58	83	61	123	53	100	66	10
N2	14	34	4.44	50	28	106	43	71	37	4
		11	0.24	44	7	27	25	23	32	1
N3	22	70	4.63	149	62	214	43	90	146	41
		56	0.69	159	60	305	27	72	201	60
M1*		34	4.80	169	27	95	16	26	97	36
M2	77	37	4.69	92	28	81	38	35	74	11
		28	0.61	125	31	61	36	30	111	12
M3	20	18	6.38	44	13	23	1	23	11	22
		16	0.42	24	9	18	1	21	10	17
M4	19	26	5.31	52	37	65	16	46	16	36
		14	0.83	29	29	40	20	26	12	30
M5*		21	4.88	40	20	50	14	52	23	7

#: mean value (First row), +: standard deviation (second row), EC: electric conductivity.

(Table 2 Continued)

Site	Sample Number (n)	Ca^{2+}	Mg^{2+}	$\sum^-$ (sum of anions)	$\sum^+$ (sum of cations)	$\sum^-/\sum^+$ (Ratio)	Total sulfate deposition (calculated)	Total nitrate deposition (calculated)
		---------------- eq/L ----------------					--------- kg/ha/yr ----------	
E1	240	25 #	23	267	271	0.98	110	54
		34 +	46	318	317	0.12	23	9
E2	136	44	35	306	310	1.01	104	51
		67	56	283	303	0.12	20	32
E3	128	75	50	429	425	1.03	119	43
		191	99	551	556	0.13	42	15
E4	131	39	13	179	181	1.00	75	23
		111	20	186	189	0.13	42	15
E5	75	20	28	205	204	1.00	31	9
		31	29	215	202	0.12	8	5
E6	140	19	27	244	243	1.01	58	33
		28	40	242	239	0.13	23	11
E7	124	39	45	341	345	0.98	122	48
		71	57	377	377	0.11	21	10
E8	127	117	56	485	564	1.00	67	24
		613	237	1102	1908	0.13	28	7
E9	60	42	18	239	247	1.00	50	26
		129	51	290	307	0.13	17	19
E10	50	80	82	494	500	0.97	30	15
		112	91	461	447	0.12	21	10
E11	37	29	15	153	158	0.94	46	19
		46	17	129	121	0.12	---	---
E12	35	7	2	65	70	0.94	35	17
		4	2	58	60	0.12	----	---
N1	36	32	20	259	279	0.91	112	54
		34	26	232	225	0.10	48	36
N2	14	44	12	184	211	0.88	82	40
		32	9	56	64	0.13	---	---
N3	22	116	56	425	491	0.88	157	60
		208	78	475	559	0.09	---	----
M1*		41	28	307	244	0.84	132	48
M2	77	39	22	201	214	0.94	113	51
		45	34	252	267	0.95	---	---
M3	20	25	6	79	87	0.91	27	19
		17	9	52	75	0.70	---	---
M4	19	29	10	153	153	1.01	69	50
		19	9	119	117	1.02	---	---
M5*		39	6	124	140	1.27	86	45

#: mean value (first row), +: standard deviation (second row).

*: The data listed for TFRI sites (M1 to M5) are adopted from King et al. (1995)

---: no data.

(Data source from Lin et al., 1997, with permission)

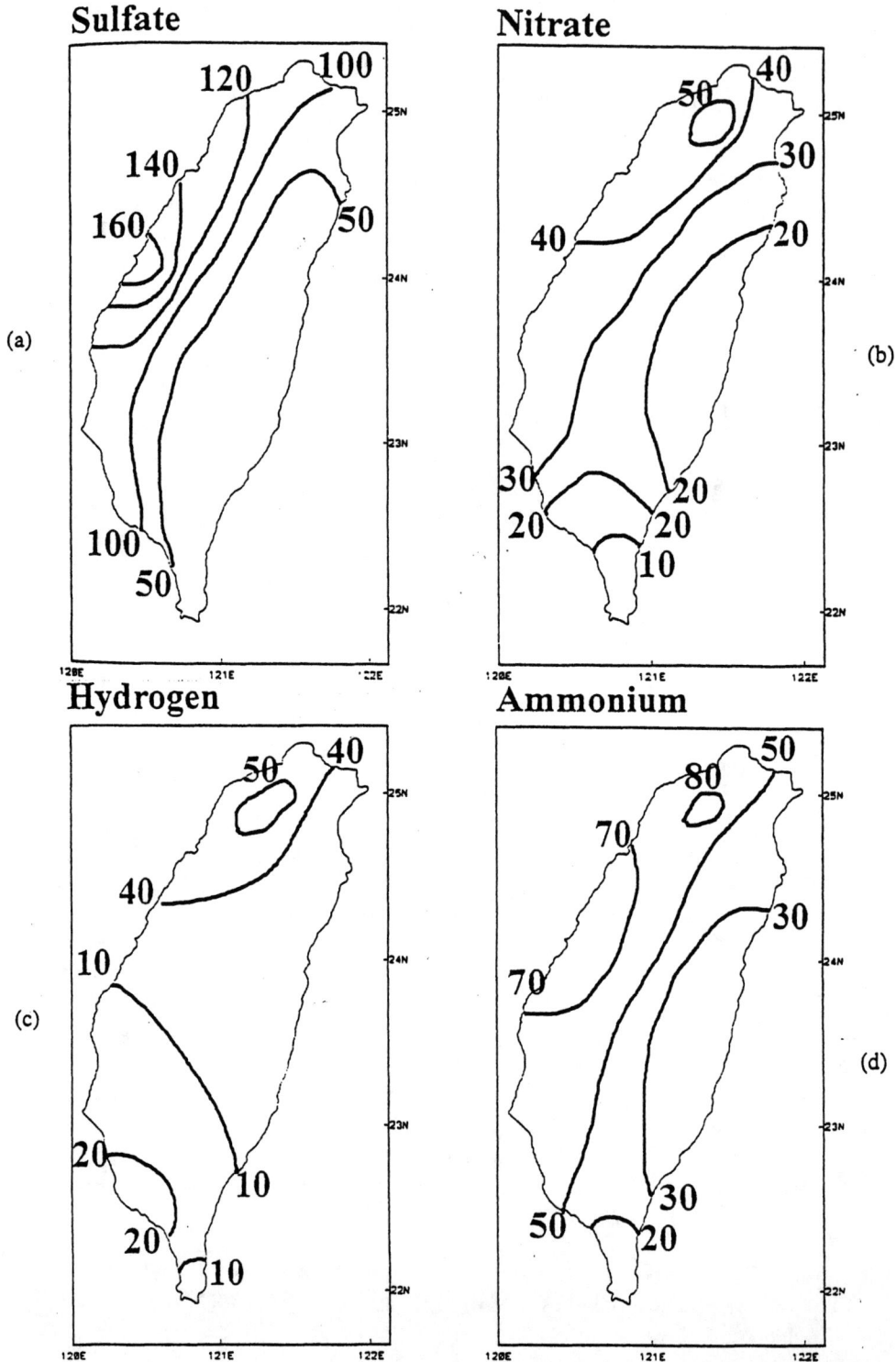

Figure 3. The distribution of annual mean concentration (micro-equivalent/L) of (a) sulfate, (b) nitrate, (c) hydrogen, and (d) ammonium ions of rain in Taiwan. (Data from Lin et al., 1997, with permission)

2.1.3. Ratio of equivalent concentration of sulfate to nitrate

The mean ratio of equivalent concentration of sulfate to nitrate in Taiwan rain is ranged from 2.22 to 8.59 (mean value is about 5) which are almost the same as that in Northeastern United States (2 to 4) and European countries (6.25 to 7.14). This ratio indicates that the acidity of acid deposition in Taiwan mainly contributed by the sulfate ion.

2.2. THE TOTAL QUANTITY OF ACID DEPOSITION OF IONS IN RAIN WATER

We can estimate the total quantity of different ions in rain water from the precipitation and the concentrations of ions calculated from the database. The mean annual total quantity of different ions in the rain water in 20 monitoring stations can be shown in Fig. 4 (Lin et al., 1997):

The mean annual total deposition of sulfate ion from rain water in northern Taiwan is higher than 100 kg/ha/yr which was more than two times compared with that in the eastern USA. Monitoring results also indicated that about 70 kg/ha/yr of sulfate was deposited in Kaohsing area, southern Taiwan. The mean annual total quantity of nitrate ion from rain water in northern and western Taiwan is about 40-60 kg/ha/yr, and less than 30 kg/ha/yr in eastern and southern Taiwan. The mean annual total quantity of hydrogen ion from rain water is highest in northern Taiwan, about 1 kg/ha/yr, and less than 0.4 kg/ha/yr in other regions of Taiwan. The mean annual total quantity of ammonium ion from rain water is highest in southwestern Taiwan, ranged from 20 to 30 kg/ha/yr, and less than 10 kg/ha/yr in eastern and southern Taiwan.

3. EFFECTS OF ACID DEPOSITION ON SOIL ACIDITY AND HEAVY METALS RELEASE IN TAIWAN SOILS

3.1 EFFECTS OF ACID DEPOSITION ON SOIL ACIDITY OF TAIWAN

The effects of acid deposition on the soil acidity are dependent on the buffer capacity of soils (Femandez, 1985). Soil acidity will produce some effects on soil functions including (1) the release of cations or nutrients in soils and produce the deficiency problem of plant nutrition, and (2) the release of toxic materials in soil solution and make some toxic phenomenon in environmental ecology and plant production. Houng et al. (1990) have indicated that the soil acidity of Taiwan soils affecting by the acid deposition are controlled by the soil pH, cation exchange capacity (CEC), base saturation %, organic matter content, content of calcium carbonate, and soil texture. Houng et al. (1990) also indicated that the buffer capacity of acidity in Taiwan soils ranged from 1.88 to 82.92 cmol H/kg soil/pH. They also concluded that the mechanism of the acid buffer capacity of Taiwan rural soils

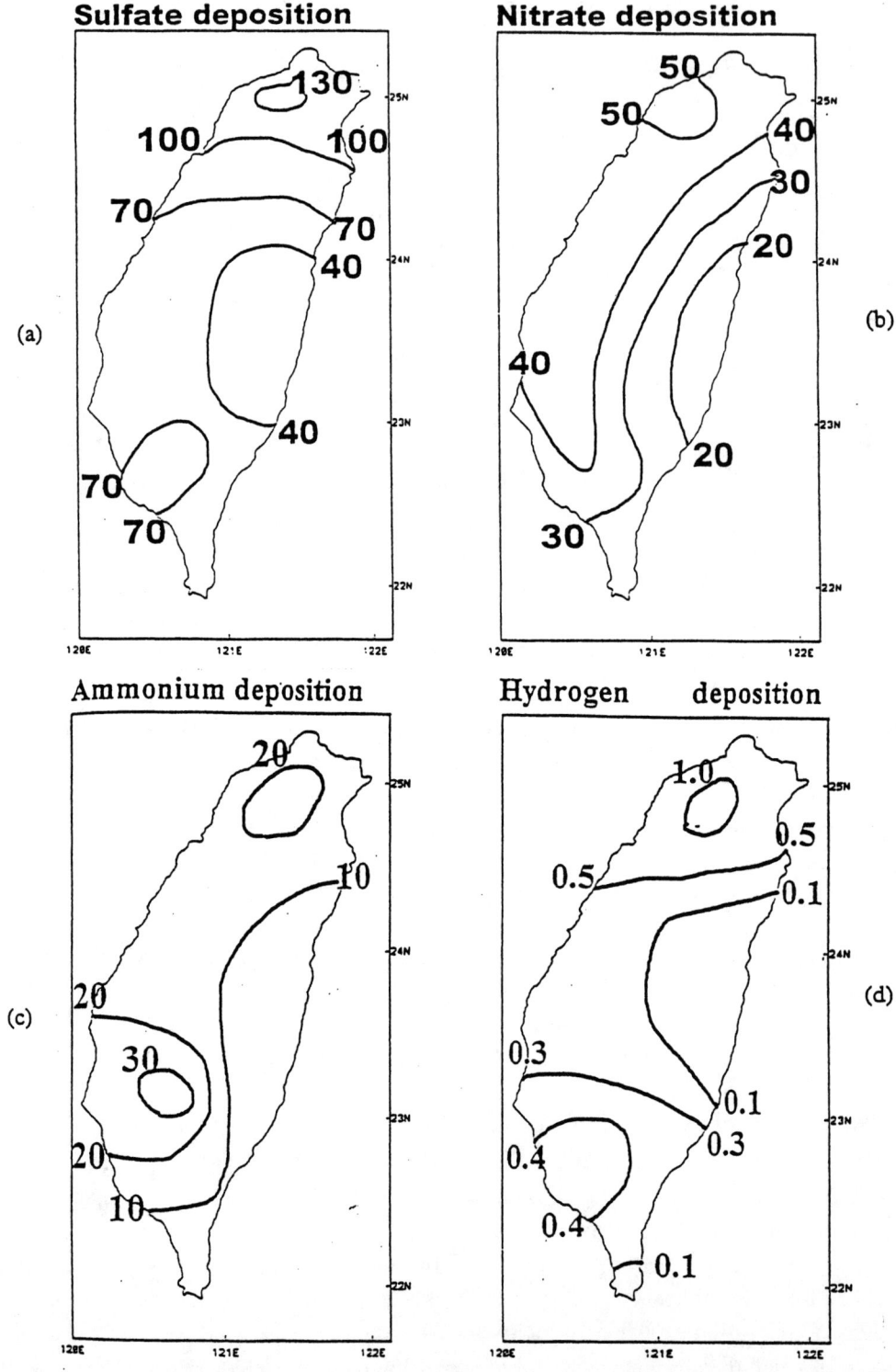

Figure 4. The distribution of annual calculated total depositions (kg/ha/yr) of (a) sulfate, (b) nitrate, (c) ammonium, and (d) hydrogen of rain in Taiwan from 1990 to 1994. (Data from Lin et al., 1997, with permission)

can be classified into four gropes, including (1) carbonate buffer and calcium carbonate solubility buffer system, (2) combined system of carbonate and some ions weathering from silicate minerals, (3) combined system of cation exchange and some ions weathering from silicate minerals, and (4) buffer reaction of release of aluminum in soil solution (Houng et al., 1990)

The other main effect of the acid deposition on the soil was the increase of concentration of aluminum in the soil solution (Cronan and Schofield, 1979; Cronan, 1980; Ulrich, 1980; Johnson et al., 1982). The solubility of aluminum in soils are mostly controlled by the dissoluble organic carbon (DOC), soluble silica, and soil pH value (Litaor, 1987).

3.2 Effects of Acid Deposition on Release of Heavy Metals in Taiwan Soils

Acid deposition contains more than 20 kinds of trace elements and the concentration of heavy metals collected from Taipei capital region are mostly higher than those of surround area of capital or agricultural area (Sposito et al., 1980; McColl et al., 1982, Friedland et al., 1984). The solubility or bioavailability of trace elements in soils are increased with decreasing of soil pH, except for Mo (Tabatabai, 1985). The retention time of trace elements in forest soils was decreased with decreasing of pH (4.2 to 2.8) of artificial acid rain, except of V and Cr (Tyler, 1978).

The first five-year leaching experiments show that the release of heavy metals in Taiwan soils was primarily controlled by soil acidity, acid buffering capacity of soils, and soil availability of heavy metals (Houng, 1997). From these experiments, we can get the following conclusions:

1. The cumulative metals leached from the soils were generally less than 10% of soil metal contents, implying greater affinity of metals in soils.
2. The metal excess leaching (microgram metals/kg soils) in Taiwan representative soils was much greater from pH 3 rain than that from pH 4 rain, and greater from polluted soils than from non-polluted Taiwan soils.

4. Acid Deposition Influences On Biologivcal Accumulation Of Trace Elements In Crops By Pot Experiments

4.1 Design of Pot Experiments

In order to understand the acid rain influences on the biological accumulation of heavy metals (Cd, Cu, Pb, and Zn) in the brown rice and vegetable species, two representative rural soils were selected and treated with different concentrations of heavy metals and simultaneously treated with or without artificial acid rain by pot experiments in 1996 (Chen et al., 1996b, 1997, 1998). Hsulin series was selected from sandstone and shale alluvial soils in northern Taiwan and Lukang series was selected from slate alluvial soils in central Taiwan. The soil was ground to less than 2mm for pot experiments.

The main treatments in the pot experiment are treated with artificial acid rain or without acid rain (top water used) in Taipei city, Taiwan, based on the annual precipitation 1800 mm/yr. The composition of artificial acid rain was prepared with the ratio of sulfate ion to nitrate ion in 5:1 and controlled in pH 4.0. The ionic strength of artificial acid rain treatment is 0.05 mole/L. The sub-treatments of each main treatment are also treated with different concentration of trace elements (Cd, Cu, Pb, and Zn) for two soils. The concentrations of Cd are treated with 0 (check), 0.2, and 2 mg/kg soils. The concentrations of Cu are treated with 0 (check), 20, and 100 mg/kg soils. The concentrations of Pb are treated with 0 (check), 10, and 100 mg/kg soils. The concentrations of Zn are treated with 0 (check), 20, and 100 mg/kg soils. Chemical fertilizers are applied as the recommendation rate in these two soils. Each treatment in these pot experiments has three duplicates in 1/5,000 Wagner's pot and arranged in random completely block design (RCBD) in Agricultural Experimental Station of National Taiwan University located in Taipei, Taiwan. The rice (*Oryza sativa* Linn.) and pickled cabbage were grown in these pot experiments from February to June1996.

4.2 LABORATORY ANALYSIS OF TRACE ELEMENTS IN SOILS AND CROPS

The phyto-availability concentrations of Cd, Cu, Pb, and Zn in Shulin soil extracted by 0.1M HCl (EPA/ROC, 1994) were 0.18, 12.07, 16.28, and 5.04 mg/kg, respectively, and those in Lukang soil are 0.31, 18.15, 13.86, and 14.93 mg/kg, respectively. The pH values of Shulin and Lukang soils were 6.0 and 7.0. The harvested brown rice and pickled cabbage were oven dried at 60 $^\circ$C for 2 days and then ground to less than 0.4 mm. The concentration of trace elements in brown rice and pickled cabbage were digested by $HNO_3/HClO_4$ solutions (Jones and Case, 1990). The determinations of trace elements in extracted soil solution and in brown rice or leaves of pickled cabbage were conducted by flame atomic adsorption spectrometry (Hitachi 180-30 type).

4.3 STATISTICAL ANALYSIS

Analysis of variance using a randomized complete block design was performed for pot experiments. Statistical analyses of data were performed by ANOVA procedure in SAS package (SAS Institute, 1982). Multiple comparisons were also performed using Fisher's protected least significant difference test at level of $p = 0.05$.

4.4 ACID DEPOSITION INFLUENCE ON THE CONCENTRATION AND TOTAL UPTAKE OF HEAVY METALS IN THE BROWN RICE

The mean concentration, mean dry weight, and total uptake of four heavy metals in brown rice grown in Shulin soil treated with or without acid rain by pot experiments were shown in Table 3. The results indicated that concentrations of Cd, Cu, and Pb in the brown rice grown in natural rain condition (irrigated with top water) were significantly increased with increasing the concentrations of heavy metals treated to Shulin soil ($p<0.05$) (SAS Institute, 1982). For the Shulin soil treated with artificial acid rain, the concentrations of Cd, Cu, and Zn in the brown rice were also significantly increased with increasing concentration of heavy metals treated to Shulin soil ($p<0.05$). Table 3 also indicated that total uptake content of four heavy metals in brown rice were all significantly increased with increasing heavy metal concentrations treated with natural or artificial acid rain condition for Shulin soil. These results reveal that the concentration and total uptake of Cu and Zn in brown rice grown in acid rain treatments were significantly higher than those of brown rice grown in natural rain treatments, but the concentration and total uptake in Cd and Pb were almost no significant difference between acid rain or top water treatments.

Table 4 lists the mean concentration, mean dry weight, and total uptake of four heavy metals of the brown rice in slate alluvial soil, Lukang soil, treated with or without acid rain. The Cd concentrations of the brown rice was significantly increased with increasing Cd concentrations under irrigated with top water condition ($p<0.05$), which ranged from 0.10 to 1.27 mg/kg. But under acid rain treated condition, the concentrations of Cd and Cu in brown rice were significantly increased with increasing the concentration of Cd and Cu treated to Lukang soil ($p<0.05$). Table 4 also indicates that total uptake content of Cd, Pb, and Zn of brown rice grown in natural rain condition or those of Cd Cu, Pb, and Zn of brown rice grown in acid rain condition were increased with increasing the concentrations of trace elements in Lukang soil. These results further shows that the concentration and total uptake content of four heavy metals in the brown rice grown in acid rain or top water were significant different by pot treatments. We also found that the dry weight of highest Cu concentration (100 mg/kg) treated in natural rain condition for Lukang soil was only 0.96 g/pot, which was caused by the damage of high concentration of Cu on the roots of rice (Chen and Liu, 1996; Liu and Chen, 1996).

Table 3. Mean concentration, mean dry weight, and total uptake of heavy metals in the brown rice growing in sandstone and shale alluvial soil (Shulin soil) treated with or without acid rain.

Treatments	Mean concentration # --- mg/kg ---	Mean dry weight --- g/pot ---	Total uptake content --- g/pot ---
Natural rain			
Check	0.06±0.00 a	1.50	0.09
Cd 0.2 mg/kg	1.27±0.57 b	1.89	2.40
Cd 2 mg/kg	4.49±1.17 c	2.82	12.7
Check	7.44±0.42 a	1.50	11.2
Cu 20 mg/kg	9.23±0.53 b	0.61	5.64
Cu 100 mg/kg	11.3±0.81 c	1.70	19.2
Check	4.77±0.00 a	1.50	7.14
Pb 10 mg/kg	7.01±3.20 b	1.96	13.7
Pb 100 mg/kg	10.4±5.76 c	1.83	19.1
Check	56.6±0.06 b	1.50	84.8
Zn 20 mg/kg	51.2±4.52 a	1.78	91.1
Zn 100 mg/kg	56.5±0.28 b	2.52	142
Acid rain			
Check	0.06±0.00 a	0.48	0.03
Cd 0.2 mg/kg	1.73±0.04 b	1.14	1.97
Cd 2 mg/kg	3.77±0.48 c	2.20	8.30
Check	8.17±0.00 a	0.48	3.91
Cu 20 mg/kg	8.72±0.02 b	2.64	23.0
Cu 100 mg/kg	13.3±1.62 c	2.97	39.5
Check	8.48±0.00 c	0.48	4.06
Pb 10 mg/kg	5.46±0.69 a	0.78	4.26
Pb 100 mg/kg	7.01±3.20 b	2.31	16.2
Check	58.7±1.00 b	0.48	28.1
Zn 20 mg/kg	51.2±11.5 a	1.72	88.3
Zn 100 mg/kg	69.3±14.5 c	4.03	279

#: Data are expressed as mean value±standard deviation of triplicates and with the same letter within a column are not significantly different ($P < 0.05$).

Table 4. Mean concentration, mean dry weight, and total uptake of heavy metals in the brown rice growing in slate alluvial soils (Lukang soil) treated with or without acid rain.

Treatments	Mean concentration # --- mg/kg ---	Mean dry weight --- g/pot ---	Total uptake content --- g/pot ---
	Natural rain		
Check	0.10±0.04 a	0.91	0.09
Cd 0.2 mg/kg	1.27±0.40 c	3.31	4.22
Cd 2 mg/kg	1.00±0.69 b	3.62	3.63
Check	12.9±7.70 b	0.91	11.8
Cu 20 mg/kg	8.42±0.42 a	2.47	20.8
Cu 100 mg/kg	8.74±0.04 a	0.96	8.36
Check	12.0±7.30 b	0.91	10.9
Pb 10 mg/kg	7.03±3.20 a	3.08	21.6
Pb 100 mg/kg	7.01±3.20 a	3.73	26.2
Check	62.3±6.22 c	0.91	56.6
Zn 20 mg/kg	49.1±5.09 a	4.38	215
Zn 100 mg/kg	52.9±5.18 b	3.29	174
	Acid rain		
Check	0.52±0.34 a	0.74	0.39
Cd 0.2 mg/kg	1.14±0.38 c	2.52	2.86
Cd 2 mg/kg	0.63±0.38 b	3.69	2.27
Check	7.54±0.56 a	0.74	5.61
Cu 20 mg/kg	9.00±0.42 c	1.92	17.3
Cu 100 mg/kg	8.50±0.21 b	3.71	31.5
Check	9.49±4.74 c	0.74	7.06
Pb 10 mg/kg	8.73±3.51 b	1.54	13.5
Pb 100 mg/kg	8.15±2.77 a	2.38	19.4
Check	60.1±3.78 c	0.74	44.7
Zn 20 mg/kg	59.0±1.88 b	2.59	153
Zn 100 mg/kg	52.7±2.54 a	4.34	229

#: Data are expressed as mean value±standard deviation of triplicates and with the same letter within a column are not significantly different ($P < 0.05$).

4.5 ACID DEPOSITION INFLUENCE ON THE CONCENTRATION AND TOTAL UPTAKE OF HEAVY METALS IN THE CABBAGE

Mean concentration, mean dry weight, and total uptake of four trace elements of the pickled cabbage in sandstone and shale alluvial soil, Shulin soil, treated with or without acid rain by pot experiments in greenhouse were listed in Table 5. It indicated that the concentrations of Cd, Cu, Pb, and Zn in pickled cabbage were all significantly increased with increasing the concentrations of trace elements treated with natural rain (top water) ($p<0.05$). A similar tendency was also observed in treatments of acid rain for Shulin soil. Table 5 also reveals that total uptake content of four heavy metals in the cabbage were all significantly increased with increasing the concentrations of trace elements treated with natural rain or acid rain by pot experiments. These results also indicated that the concentration and total uptake of Cd and Zn in cabbage grown in acid rain treatments were higher than those of natural rain treatments, but the concentration and total uptake of Cu and Pb in cabbage grown in Shulin soils were no significant differences between two rain treatments by pot treatments.

Table 6 summarizes the mean concentration, mean dry weight, and total uptake of Cd, Cu, Pb, and Zn in the pickled cabbage for slate alluvial soil, Lukang soil, treated with or without acid rain by pot experiments. The results indicated that the concentrations of four selected heavy metals in pickled cabbage grown in natural rain treatments were all significantly increased with increasing the concentration of heavy metal treated in Lukang soil ($p<0.05$). An analogous tendency was also found in acid rain treatment for Lukang soil. These results also showed that total uptake content of four heavy metals in the pickled cabbage were all significantly increased with increasing the concentrations of trace elements under natural rain or acid rain condition by pot experiments. It indicated that the concentration and total uptake of Pb and Zn in pickled cabbage grown in acid rain treatment were significantly higher than those of natural rain condition by pot treatments, but the concentration and total uptake of Cd and Cu were no significant differences between two kinds of rain treatments by pot treatments.

Table 5. Mean concentration, mean dry weight, and total uptake of heavy metals in the pickled cabbage growing in sandstone and shale alluvial soil (Shulin soil) treated with or without acid rain.

Treatments	Mean concentration # --- mg/kg ---	Mean dry weight --- g/pot ---	Total uptake content --- g/pot ---
	Natural rain		
Check	0.09±0.04 a	0.38	0.03
Cd 0.2 mg/kg	2.24±0.18 b	0.53	1.18
Cd 2 mg/kg	20.9 ±2.74 c	0.47	9.93
Check	2.82±0.61 a	0.38	1.06
Cu 20 mg/kg	16.8 ±1.37 b	0.48	8.09
Cu 100 mg/kg	78.0 ±21.0 c	0.30	23.3
Check	6.78±0.88 a	0.38	2.55
Pb 10 mg/kg	12.0±1.58 b	0.35	4.21
Pb 100 mg/kg	25.2±13.2 c	0.41	10.4
Check	65.3±6.35 a	0.38	24.6
Zn 20 mg/kg	292 ±6.15 b	0.53	168
Zn 100 mg/kg	1102 ±266 c	0.15	171
	Acid rain		
Check	0.26±0.08 a	0.45	0.12
Cd 0.2 mg/kg	2.83±0.58 b	0.37	1.04
Cd 2 mg/kg	28.0±3.00 c	0.47	13.0
Check	6.54±0.58 a	0.45	2.97
Cu 20 mg/kg	13.5±0.79 b	0.40	5.43
Cu 100 mg/kg	69.9±20.9 c	0.24	16.9
Check	5.21±0.31 a	0.45	2.37
Pb 10 mg/kg	5.47±3.90 b	0.34	1.84
Pb 100 mg/kg	12.7±1.49 c	0.36	4.53
Check	103 ±22.2 a	0.45	46.9
Zn 20 mg/kg	317 ±59.6 b	0.36	113
Zn 100 mg/kg	1286 ±321 c	0.26	340

#: Data are expressed as mean value±standard deviation of triplicates and with the same letter within a column are not significantly different ($P < 0.05$).

Table 6. Mean concentration, mean dry weight, and total uptake of heavy metals in the pickled cabbage growing in slate alluvial soil (Lukang soil) treated with or without acid rain.

Treatments	Mean concentration # --- mg/kg ---	Mean dry weight --- g/pot ---	Total uptake content --- g/pot ---
Natural rain			
Check	0.01±0.00 a	0.19	0.19
Cd 0.2 mg/kg	0.94±0.58 b	0.36	0.33
Cd 2 mg/kg	15.3±4.61 c	0.44	6.78
Check	5.38±0.06 a	0.19	1.03
Cu 20 mg/kg	11.7±1.55 b	0.79	9.23
Cu 100 mg/kg	29.1±3.37 c	0.44	12.9
Check	1.25±0.23 a	0.19	0.24
Pb 10 mg/kg	1.51±0.21 b	0.22	0.32
Pb 100 mg/kg	2.91±0.79 c	0.33	0.95
Check	39.6±3.03 a	0.19	7.56
Zn 20 mg/kg	88.5±2.70 b	0.43	37.7
Zn 100 mg/kg	182 ±28.6 c	0.37	67.2
Acid rain			
Check	0.55±0.09 a	0.56	0.31
Cd 0.2 mg/kg	1.40±0.31 b	0.43	0.60
Cd 2 mg/kg	11.9±1.30 c	0.38	4.54
Check	9.49±1.37 a	0.56	5.28
Cu 20 mg/kg	12.9±1.54 b	0.34	4.43
Cu 100 mg/kg	34.1±5.64 c	0.34	11.7
Check	2.55±0.07 a	0.56	1.42
Pb 10 mg/kg	3.73±0.25 b	0.49	1.82
Pb 100 mg/kg	15.0±4.84 c	0.44	6.56
Check	62.8±14.8 a	0.56	34.9
Zn 20 mg/kg	100 ±15.4 b	0.43	43.5
Zn 100 mg/kg	243 ±31.8 c	0.37	90.2

#: Data are expressed as mean value±standard deviation of triplicates and with the same letter within a column are not significantly different ($P < 0.05$).

5. ACID DEPOSITION INFLUENCES ON BIOLOGICAL ACCUMULATION OF TRACE ELEMENTS IN CROPS VARIETIES SAMPLED IN THE FIELD

5.1 DESIGN OF SAMPLING CROPS AND VEGETABLES IN THE FIELD

Two research regions including an acid rain affected region (Lung-tang in northern Taiwan) and an acid rain non-affected region (Lung-luan-tang in southern Taiwan) were selected as the areas for sampling brown rice and different vegetables (Fig. 5). Based on the distribution of Figs. 2 and 4, the mean annual total deposition of sulfate and nitrate from rain water in Lung-tang area is about 120 kg/ha/yr and 50 kg/ha/yr. The mean annual total deposition of sulfate and nitrate from rain water in Lung-luan-tang area is about 50 kg/ha/yr and 20 kg/ha/yr. The pH of rain in Lung-tang area is about 4.5 and the pH of rain in Lung-luan-tang area is higher than 5.5. We can found the significant differences on the pH and composition of rain in these two study regions.

The objectives of this sampling studies in the field are to investigate the acid deposition influences on the biological accumulation of heavy metals in crops including rice and 19 vegetable species sampled from August 1996 to October 1997 in Taiwan (Chen et al., 1997, 1998). Nineteen vegetable species include sweet potato, Welsh onion, pickled cabbage, Chinese chives, mustard, lettuce, chickweed, garlic, kohlrabi, cabbage, tassel flower, celery, spinach, coriander, basil, radish, pepper, kidney bean, and water convolvulus. The results of some investigations have indicated that acid deposition significantly increase the concentration of heavy metals in several agricultural plant species grown in acid rain affected region (Reddy, 1989; Su and Yang, 1995).

Brown rice and 19 vegetables species were sampled in an acid rain affected region (Lung-tang in northern Taiwan) and an acid rain non-affected region (Lung-luan-tang in southern Taiwan) by every season from August 1996 to October 1997. The samples were randomly sampled by triplicates for each crop species.

5.2 LABORATORY ANALYSIS OF TRACE ELEMENTS IN SOILS AND CROPS

Four and seven surface soils (0-20 cm depth) were sampled from acid rain affected region (Lung-tang) and from acid rain non-affected region (Lung-luan-tang) in 1996. The phyto-availability concentrations of Cd, Cu, Pb, and Zn in soils of two regions were extracted by distilled water (Mench et al., 1994), 0.005M DTPA (pH 5.3) (Lindsay and Norvell, 1978), 0.05M EDTA (Mench et al., 1994) and 0.1M HCl (EPA/ROC, 1994) are shown in Table 7. The concentrations of Cd in two regions are varied from 0.06 to 0.11 mg/kg. The concentrations of Cu in two regions are varied from 2.22 to 5.35 mg/kg. The concentrations of Pb in two regions are varied from 2.51 to 7.20 mg/kg. The concentrations of Zn in two regions are varied from 3.68 to 8.28 mg/kg. There are no significant differences for the phyto-availability concentrations of Cd, Cu, Pb, and Zn in the two region soils. The upper limit of background concentration of the phyto-availability of Cd, Cu, Pb, and Zn in Taiwan soils extracted with 0.1M HCl (EPA/ROC, 1994) are 0.43, 26, 18, and 25 mg/kg soil, respectively (Chen et al., 1996a). The results of phyto-availability concentrations of the trace elements in two studied regions are almost lower than the background concentration of trace elements in Taiwan soils, and there are no effect on the biological accumulation of trace elements in the crops and vegetables in Taiwan (Chen et al., 1996).

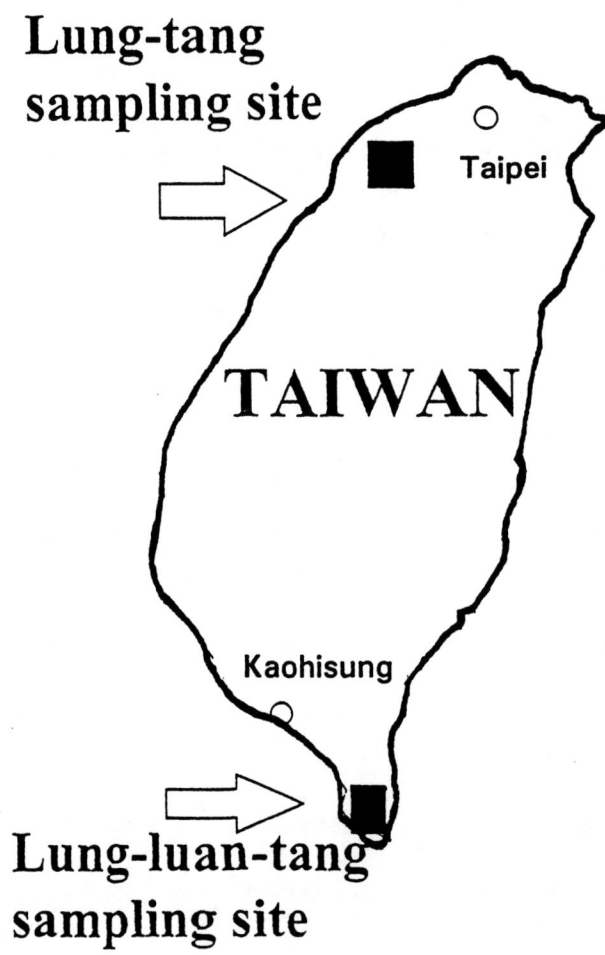

Figure 5. Two research regions including an acid rain affected region (Lung-tang in northern Taiwan) and an acid rain unaffected region (Lung-luan-tang in southern Taiwan) were selected as the field sampling areas for different crops.

Table 7. The concentration of phyto-availability of trace elements in the soils located in acid deposition affecting area or non-affecting area in Taiwan.

Sampling regions #	Sample No. (n)	Phyto-availability * in the soils extracted by			
		water	DTPA	EDTA	HCl
		----------------------- mg/kg soils --------------------			
		Cd			
Lung-tang	4	0.05±0.01	0.11±0.01	0.13±0.06	0.10±0.01
Lung-luan-tang	7	0.07±0.03	0.11±0.01	0.14±0.09	0.11±0.09
		Cu			
Lung-tang	4	2.27±0.27	2.30±0.38	3.76±0.61	4.51±0.55
Lung-luan-tang	7	2.23±2.05	2.89±2.54	5.35±3.83	4.47±4.09
		Pb			
Lung-tang	4	4.46±0.55	4.24±1.00	9.34±1.01	7.19±0.86
Lung-luan-tang	7	2.51±1.03	3.56±1.69	5.90±3.05	4.79±2.53
		Zn			
Lung-tang	4	3.68±1.42	4.34±1.58	5.95±1.84	7.96±2.22
Lung-luan-tang	7	4.33±3.26	3.39±2.37	6.72±5.22	8.28±6.17

#: Long-tang (acid deposition affected region located in northern Taiwan), Long-lun-tang (acid deposition non-affected region located in southern Taiwan)

*: Phyto-availability of trace elements in soils extracted by distilled water (Wench et al., 1994), 0.005M DTPA (pH 5.3) (Lindsay and Norvell, 1978), 0.05M EDTA (Mench et al., 1994), and 0.1M HCl (EPA/ROC, 1994).

Data are expressed as mean value±standard deviation.

The sampled brown rice and leaves of 19 vegetable species were oven dried at 60 ^{o}C for 2 days and then ground to less than 0.4 mm. The brown rice and leaves of vegetables were digested with $HNO_3/HClO_4$ solutions (Jones and Case, 1990). The determination of trace elements in the extracted soil solution and in the digestion solutions of brown rice or vegetables species were determined by flame atomic adsorption spectrometry (Hitachi 180-30 type).

Table 8. Mean concentration of Cd in brown rice and the leaves of vegetable species growing in Lung-tang area (affected by acidic rains) and Lung-luan-tang area (non-affected by acidic rains) from 1996 to 1997.

Rice and vegetable species	Affected by acid rain		Non-affected by acid rain	
	Samples No. (n)	Cd conc. --- mg/kg ---	Samples No. (n)	Cd conc. --- mg/kg --
Rice				
rice (*Oryza sativa* Linn.)	24	0.18±0.20	15	0.14±0.01
Vegetables				
sweet potato *(Ipomoea bataus.)*	14	0.59±0.78	19	0.59±0.75
welsh onion *(Allium fistulosum* Linn.)	10	0.39±0.40	12	0.43±0.49
pickled cabbage *(Brassica chineniss* Linn.)	3	1.59±0.31	10	0.32±0.49
Chinese chives *(Allium tuberosum* Rottlerex Sprengel.)	7	1.14±1.26	5	0.23±0.28
mustard *(Brassica juncea* Coss.)	2	0.94±0.00	4	ND
lettuce *(Lactuce sativa* Linn.)	6	1.43±0.31	8	0.38±0.51
chickweed *(Alsine media* Linn.)	3	1.67±0.36	1	ND
garlic *(Allium sativum* Linn.)	6	1.18±0.58	7	1.25±1.13
kohlrabi *(Brassica campestris* Linn.)	1	1.41	1	0.70
cabbage *(Brassica oleracea* Linn.)	2	1.42±0.00	1	ND
tassel flower *(Amaranthus caudatus* Linn.)	6	1.04±0.99	2	1.08±0.52
celery *(Apium graveolens* Linn.)	2	2.28±1.99	1	ND
spinach *(Spinacia oleracea* Linn.)	2	1.43±0.01	1	ND
coriander *(Coriandrum stivum* Linn.)	1	2.84	4	0.35±0.41
basil *(Ocimum basilicum* Linn.)	1	0.15	3	ND
radish *(Raphanus sativus* Linn.)	4	0.69±0.46	2	ND
pepper *(Capsicum frutescens* Linn.)	3	0.10±0.08	4	0.05±0.06
kidney bean *(Phaseolus vulgaris* Linn.)	3	2.58±3.66	10	1.25±1.97
water convolvulus *(Ipomoea aquatica* Forsk.)	6	0.74±0.78	3	2.66±3.34

Data are expressed as mean value±standard deviation.
ND : not detectable (lower than method detection limit, MDL).

5.3 ACID DEPOSITION INFLUENCE ON THE CONCENTRATION AND TOTAL UPTAKE OF TRACE ELEMENTS IN THE BROWN RICE AND THE LEAVES OF VEGETABLES

Table 8 shows the mean concentration of Cd in brown rice and leaves of 19 vegetable species growing in Lung-tang area (affected by acidic rains) and Lung-luan-tang area (unaffected by acidic rains) and sampled in field. The results indicated that the mean Cd concentration of brown rice in two regions are 0.14 to 0.18 mg/kg, and the mean Cd concentration in leaves of 19 vegetable species growing in Lung-tang area ranged from 0.10 to 2.58 mg/kg, which have high variability of concentration and depend on vegetation species. Higher Cd concentrations were found in celery and kidney bean, which the concentration of Cd are 2.58 and 2.28 mg Cd/kg, respectively. Higher standard deviations (STD) are found probably due to small sampling numbers. This table also shows that mean concentration of Cd in the leaves of 19 vegetable species growing in Lung-luan-tang area are less than that of vegetables growing in the Lung-tang area and more variable (Table 8). Water convolvulus have higher concentration of Cd than that of other vegetable species we sampled.

The mean concentration of Cu in brown rice and leaves of 19 vegetable species growing in Lung-tang area (affected by acidic rains) and Lung-luan-tang area (unaffected by acidic rains) by sampling in field was listed in Table 9. The mean Cu concentrations of brown rice in two regions are 3.56 to 3.75 mg/kg, and the mean Cu concentration in leaves of 19 vegetable species growing in Lung-tang area were ranged from 2.42 to 16.3 mg Cu/kg. Higher mean concentrations of Cu were found in the leaves of some vegetable species including coriander, basil, tassel flower and water convolvulus with 16.3, 11.1, 10.9 and 10.7 mg Cu/kg, respectively. Table 9 also reveals that the mean Cu concentrations in leaves of 19 vegetable species growing in Lung-luan-tang area were around 3 to 5 mg/kg, ranged from 0.81 to 8.10 mg/kg. Chinese chives have higher Cu concentration than that of other vegetable species.

Table 10 summarizes the mean concentration of Pb in brown rice and leaves of 19 vegetable species growing in Lung-tang area (affected by acidic rains) and Lung-luan-tang area (unaffected by acidic rains) by sampling in the field. It indicated that the mean Pb concentrations of brown rice in two regions are 5.06 to 5.51 mg/kg, and the mean Pb concentrations of leaves in 19 vegetable species growing in Lung-tang area ranged from ND to 11.9 mg/kg. Higher concentrations of Pb were found in Welsh onion, lettuce, spinach, basil, and pepper, ranged from 6.28 to 11.9 mg Pb/kg, respectively. Higher standard deviation (STD) is also probably due to small sampling number. Table 10 shows also that mean Pb concentrations in the leaves of 19 vegetable species growing in Lung-luan-tang area ranged from ND to 5.74 mg/kg, except for spinach, pickled cabbage, and Chinese chives which have higher Pb concentration in the leaves, ranged from 8 to 11 mg Pb/kg, respectively.

Table 9. Mean concentration of Cu in the brown rice and the leaves of vegetable species growing in Lung-tang area (affected by acidic rains) and Lung-luan-tang area (non-affected by acidic rains) from 1996 to 1997.

Rice and vegetable species	Affected by acid rain		non-affected by acid rain	
	Samples No. (n)	Cu conc. --- mg/kg ---	Samples No. (n)	Cu conc. --- mg/kg ---
Rice				
rice (*Oryza sativa* Linn.)	24	3.75±1.66	15	3.56±2.42
Vegetables				
sweet potato (*Ipomoea bataus.*)	14	6.81±3.34	19	4.69±3.87
welsh onion (*Allium fistulosum* Linn.)	10	6.25±3.97	12	4.22±3.55
pickled cabbage (*Brassica chineniss* Linn.)	3	8.13±5.87	10	6.62±2.08
Chinese chives (*Allium tuberosum* Rottlerex Sprengel.)	7	5.71±3.51	5	8.10±3.29
mustard (*Brassica juncea* Coss.)	2	6.48±0.05	4	4.08±0.94
lettuce (*Lactuce sativa* Linn.)	6	7.82±5.81	8	4.49±1.47
chickweed (*Alsine media* Linn.)	3	7.80±3.79	1	3.25
garlic (*Allium sativum* Linn.)	6	4.86±2.70	7	2.31±1.58
kohlrabi *Brassica campestris* Linn.	1	6.44	1	3.22
cabbage (*Brassica oleracea* Linn.)	2	6.48±0.00	1	3.25
tassel flower (*Amaranthus caudatus* Linn.)	6	10.9±3.24	2	4.89±0.05
celery (*Apium graveolens* Linn.)	2	6.46±0.02	1	ND
spinach (*Spinacia oleracea* Linn.)	2	4.89±2.29	1	6.51
coriander (*Coriandrum stivum* Linn.)	1	16.3	4	3.25±0.04
basil (*Ocimum basilicum* Linn.)	1	11.1	3	1.370.21
radish (*Raphanus sativus* Linn.)	4	2.42±1.65	2	0.81±0.03
pepper (*Capsicum frutescens* Linn.)	3	8.74±2.28	4	4.29±2.17
kidney bean (*Phaseolus vulgaris* Linn.)	3	8.22±3.63	10	4.62±4.16
water convolvulus (*Ipomoea aquatica* Forsk.)	6	10.7 ±3.94	3	5.41±4.49

Data are expressed as mean value±standard deviation.
ND : not detectable (lower than method detection limit, MDL).

Table 10. Mean concentration of Pb in the brown rice and the leaves of vegetable species growing in Lung-tang area (affected by acidic rains) and Lung-luan-tang area (non-affected by acidic rains) from 1996 to 1997.

Rice and vegetable species	Affected by acid rain		non-affected by acid rain	
	Samples No. (n)	Pb conc. --- mg/kg ---	Samples No. (n)	Pb conc. --- mg/kg ---
Rice				
rice (*Oryza sativa* Linn.)	24	5.51±2.00	15	5.06±2.03
Vegetables				
sweet potato (*Ipomoea bataus.*)	14	3.72±5.41	19	3.49±5.21
welsh onion (*Allium fistulosum* Linn.)	10	6.60±7.21	12	2.14±3.69
pickled cabbage (*Brassica chineniss* Linn.)	3	ND	10	9.16±8.58
Chinese chives (*Allium tuberosum* Rottlerex Sprengel.)	7	0.63±1.49	5	8.27±5.28
mustard (*Brassica juncea* Coss.)	2	ND	4	3.86±4.68
lettuce (*Lactuce sativa* Linn.)	6	6.28±9.72	8	5.74±8.33
chickweed (*Alsine media* Linn.)	3	ND	1	ND
garlic (*Allium sativum* Linn.)	6	3.13±7.76	7	ND
kohlrabi *Brassica campestris* Linn.	1	ND	1	ND
cabbage (*Brassica oleracea* Linn.)	2	ND	1	ND
tassel flower (*Amaranthus caudatus* Linn.)	6	3.96±6.05	2	ND
celery (*Apium graveolens* Linn.)	2	ND	1	ND
spinach (*Spinacia oleracea* Linn.)	2	9.39±13.3	1	11.7
coriander (*Coriandrum stivum* Linn.)	1	ND	4	ND
basil (*Ocimum basilicum* Linn.)	1	11.9	3	ND
radish (*Raphanus sativus* Linn.)	4	ND	2	ND
pepper (*Capsicum frutescens* Linn.)	3	7.85±6.80	4	2.00±2.31
kidney bean (*Phaseolus vulgaris* Linn.)	3	5.69±6.31	10	5.24±4.89
water convolvulus (*Ipomoea aquatica* Forsk.)	6	4.71±5.69	3	1.34±2.33

Data are expressed as mean value±standard deviation.
ND : not detectable (lower than method detection limit, MDL).

Table 11. Mean concentration of Zn in the brown rice and the leaves of vegetable species growing in Lung-tang area (affected by acidic rains) and Lung-luan-tang area (non-affected by acidic rains) from 1996 to 1997.

Rice and vegetable species	Affected by acid rain		non-affected by acid rain	
	Samples No. (n)	Zn conc. --- mg/kg ---	Samples No. (n)	Zn conc. --- mg/kg ---
Rice				
rice (*Oryza sativa* Linn.)	24	23.93.58	15	23.26.43
Vegetables				
sweet potato (*Ipomoea bataus.*)	14	21.6±12.3	19	19.4±19.4
welsh onion (*Allium fistulosum* Linn.)	10	32.9±17.2	12	16.2±7.21
pickled cabbage (*Brassica chineniss* Linn.)	3	32.832.2	10	24.6±9.17
Chinese chives (*Allium tuberosum* Rottlerex Sprengel.)	7	37.620.7	5	24.1±6.07
mustard (*Brassica juncea* Coss.)	2	44.5±46.7	4	20.3±6.78
lettuce (*Lactuce sativa* Linn.)	6	53.8±29.8	8	25.9±10.3
chickweed (*Alsine media* Linn.)	3	32.1±30.2	1	88.4
garlic (*Allium sativum* Linn.)	6	52.3±30.8	7	18.1±8.64
kohlrabi (*Brassica campestris* Linn.)	1	110	1	20.1
cabbage (*Brassica oleracea* Linn.)	2	62.0±24.7	1	20.3
tassel flower (*Amaranthus caudatus* Linn.)	6	44.7±28.4	2	30.5±2.38
celery (*Apium graveolens* Linn.)	2	35.8±37.1	1	23.1
spinach (*Spinacia oleracea* Linn.)	2	15.8±6.21	1	37.7
coriander (*Coriandrum stivum* Linn.)	1	46.1	4	25.6±9.89
basil (*Ocimum basilicum* Linn.)	1	12.1	3	33.8±6.25
radish (*Raphanus sativus* Linn.)	4	41.9±13.9	2	40.3±17.0
pepper (*Capsicum frutescens* Linn.)	3	31.4±31.3	4	35.5±37.1
kidney bean (*Phaseolus vulgaris* Linn.)	3	30.2±3.92	10	21.0±6.09
water convolvulus (*Ipomoea aquatica* Forsk.)	6	20.9±4.23	3	31.6±34.0

Data are expressed as mean value±standard deviation.

The mean concentrations of Zn in brown rice and leaves of 19 vegetable species growing in Lung-tang area (affected by acidic rains) and Lung-luan-tang area (unaffected by acidic rains) by sampling in the field were shown in Table 11. These results also displayed that the mean Zn concentration in the brown rice of two regions are 23.2 to 23.9 mg/kg, and the mean Zn concentrations in leaves of 19 vegetable species growing in Lung-tang area are highly variable, ranged from 12 to 110 mg Zn/kg. Both kohlrabi and cabbage have higher Zn concentration in the leaves, 110 and 62 mg Zn/kg, than that of other species.

Table 11 also indicated that the mean Zn concentrations in leaves of 19 vegetable species growing in Lung-luan-tang area are also highly variable depend on the vegetable species, ranged from 16 to 40.3 mg/kg, except for chickweed which have the highest Zn concentration (88.4 mg/kg) in the leaves.

5.4 ACID DEPOSITION INFLUENCE ON THE RATIO OF CONCENTRATION OF HEAVY METALS IN THE CROPS OF TWO REGIONS

In order to research the acid rain effects on the biological accumulation of heavy metals (Cd, Cu, Pb, and Zn) in the crops, we compared the ratios of relative concentration of four heavy metals in the brown rice and leaves of vegetables sampled from acid rain affected area and non-affected area. The data indicated that the ratios of relative concentration of Cd, Cu, Zn in brown rice and 19 vegetable species growing in acid rain area (Lung-tang) and growing in acid rain non-affected area (Lung-luan-tang) sampled from 1996 to 1997 are almost higher than 1, or higher than 3, except for Pb (Table 12). There is no significant differences on the phyto-availability concentration of Cd, Cu, Pb, and Zn in the soils sampled from two regions (Table 7). The results from Table 12 suggested that there is no effect on the biological accumulation of heavy metals in the brown rice, but there are probably significantly affected on the biological accumulation of Cd, Cu, and Zn in the leaves of vegetables species owing to long term acid deposition in northern Taiwan. Therefore, there is potentially dangerous for those vegetables produced in acid rain affected area.

Table 12 also reveals that the mean concentration of Pb in the brown rice and leaves of 19 vegetable species between acid rain affected area and non-affected area are almost same. On the other word, the ratio is closed to 1. This result indicated that the acid rain can not influence on the biological accumulation of Pb in the brown rice and leaves of vegetables species sampled in Taiwan. Some studies have indicated that concentration of Pb in the crops was only affected when the concentration of Pb in the soils is higher than 500 mg/kg (Kabata-Pendias and Pendias, 1992). Sloan et al. (1997) also indicated that the relative bioavailability of biosolids-applied heavy metals in agricultural soils was Cd $\gg$ Zn > Ni Cu $\gg$ Cr > Pb, for the soils 15 years after biosolids application. It is quite consistent with the results getting from this research. The results from the ratio of heavy metals in two regions indicated that the rating of effectiveness on the phyto-availability of heavy metals caused by acid deposition followed the trend: CdZn Cu $\gg$ Pb.

6. CONCLUSIONS

The mean annual total deposition of sulfate ion from rain water is higher than 100 kg/ha/yr in northern Taiwan and 70 kg/ha/yr in southern Taiwan. The mean annual total quantity of nitrate ion from rain water in northern and western Taiwan is 40 to 60 kg/ha/yr, and < 30 kg/ha/yr in eastern and southern Taiwan. The mean annual total quantity of hydrogen ion from rain water is highest in northern Taiwan, about 1 kg/ha/yr, and < 0.4 kg/ha/yr in other regions of Taiwan. The mean annual total quantity of ammonium ion from rain water is highest in southwestern Taiwan, ranged from 20 to 30 kg/ha/yr, and < 10 kg/ha/yr in eastern and southern Taiwan. The ratio of sulfate to nitrate of acid rain in Taiwan is about 5.

Table 12. The ratios of relative concentration of heavy metals in the brown rice and the leaves of vegetable species growing in Lung-tang area (affected by acidic rains) and Lung-luan-tang area (non-affected by acidic rains) from 1996 to 1997.

Rice and vegetable species	acid rain/non-acid rain affecting area (sampling number)	Ratio in acid rain/non-acid rain area			
		Cd	Cu	Pb	Zn
Rice					
rice (*Oryza sativa* Linn.)	24/15	1.25	1.05	1.09	1.03
Vegetables					
sweet potato (*Ipomoea bataus*)	14/9	1.00	1.45	1.07	1.11
welsh onion (*Allium fistulosum* Linn.)	10/12	0.89	1.48	3.08	2.03
pickled cabbage (*Brassica chineniss* Linn.)	3/10	5.03	1.23	---- #	1.33
Chinese chives (*Allium tuberosum* Rottlerex Sprengel.)	7/5	4.97	0.70	0.08	1.56
mustard (*Brassica juncea* Coss.)	2/4	-----	1.59	-----	2.19
lettuce (*Lactuce sativa* Linn.)	6/8	3.73	1.87	1.00	1.97
chickweed (*Alsine media* Linn.)	3/1	-----	2.40	-----	0.36
garlic (*Allium sativum* Linn.)	6/7	0.85	2.44	-----	4.64
kohlrabi (*Brassica campestris* Linn.)	1/1	2.00	2.00	-----	5.50
cabbage (*Brassica oleracea* Linn.)	2/1	-----	1.99	-----	3.06
tassel flower (*Amaranthus caudatus* Linn.)	6/2	0.97	2.23	-----	1.47
celery (*Apium graveolens* Linn.)	2/1	-----	------	------	1.55
spinach (*Spinacia oleracea* Linn.)	2/1	-----	0.75	0.80	0.42
coriander (*Coriandrum stivum* Linn.)	1/4	8.02	5.01	-----	1.80
basil (*Ocimum basilicum* Linn.)	1/3	-----	8.05	-----	0.36
radish (*Raphanus sativus* Linn.)	4/2	-----	2.76	-----	1.08
pepper (*Capsicum frutescens* Linn.)	¾	1.97	2.04	3.92	0.88
kidney bean (*Phaseolus vulgaris* Linn.)	3/10	2.07	1.78	1.09	1.44
water convolvulus (*Ipomoea aquatica* Forsk.)	6/3	0.28	1.97	3.50	0.66

: The ratios of relative concentration can not calculated because the heavy metal contents of rice or vegetables growing in acidic rain affecting area or in non-acidic rain affecting area is lower than that of method detection limit (MDL) of heavy metals.

The results obtained from pot experiments indicated that the concentrations of Cd, Cu, and Zn of the brown rice grown in sandstone and shale alluvial soils or the concentrations of Cd and Cu of the brown rice grown in slate alluvial soils treated with artificial acid rain were significantly increased with increasing the concentration of heavy metals ($p<0.05$), respectively. The concentrations of Cd, Cu, Pb, and Zn in the pickled cabbage grown in the two tested soils treated with artificial acid rain were also significantly increased with increasing the concentration of heavy metals ($p<0.05$), respectively. It also suggests that low concentrations of four heavy metals in the selected two soils treated with acid rain will still affect the concentration of heavy metals in the pickled cabbage.

The results obtained from field sampling suggested that the ratios of relative concentration of Cd, Cu, Pb, and Zn in the brown rice and leaves of 19 vegetable species collected from acid rain affected and non-affected areas are almost higher than 1, or higher than 3. It also further demonstrated that the biological accumulation of heavy metals in the leaves of vegetables was significantly affected by the acid deposition in Taiwan. The rating of effectiveness of heavy metals followed the trend: CdZn Cu >> Pb.

ACKNOWLEDGMENTS

Funding for this three years studies provided by the Environmental Protection Administration of Republic of China (through contracts of grants no. EPA-85-1404-09-11 and EPA-86-FA44-09-47) and by the National Science Council of Republic of China (through contract of grant no. NSC-87-EPA-P-002-011) is gratefully acknowledged. The authors also thank Mrs. Ing-Yih Leu, and Jse-Ming Lee, and Miss Yeng-Hsuei Liu for their help on the pot experiments, field sampling, and laboratory analysis.

REFERENCES

Berthelsen, B. O., E. Steinnes, W. Solberg, and L. Jingsen, 1995. Heavy metal concentrations in relation to atmospheric heavy metal deposition.. *J. Environ. Qual.* 24: 1018-1026.

Binns, W. O. 1988. Vegetation and soils. pp. 13-21. *In:* K. Mellanby (ed.) *Air pollution, acid rain, and the environment*. The Watt Committee in Energy by Elsevier Applied Science.

Carl, A. L., A. L. Page, A. A. Elseewi, and L. J. Lund. 1982. Input of acidity to agricultural soils from rainfall and other sources. *In: Proceedings of 16th Annual Conference on Trace Substance in Environmental Health*, University of Missouri, Columbia, USA. June 1-3, 1982.

Cataldo, D. A., T. R. Garland, and R. E. Wilding. 1983. Cadmium uptake kinetics in intact soybean plants. *Plant Physiol.* 73: 844-848.

Chaney, R. L., J. C. Brown, and L. O. Tiffin. 1972. Obligatory reduction of ferric chelates in iron uptake by soybeans. *Plant Physiol. 50:* 208-220.

Chen, Z. S. 1992. Metal contamination of flooded soils, rice plants, and surface waters in Asia. pp. 85-107. *In:* Domy C. Adriano (ed.). *Biogeochemistry of Trace Metals.* Lewis Publishers Inc., Florida. USA.

Chen, C. T., and J. J. Hung. 1987. Acid rain and lake acidification in Taiwan. *In: Proceedings, National Science Council Part A: Physical Science and Engineering, 11:* 436-442.

Chen, Z. S., D. Y. Lee, C. F. Lin, S. L. Lo, and Y. P.Wang. 1996a. Contamination of rural and urban soils in Taiwan. pp. 691-709. *In:* R. Naidu, R. S. Kookuna, D. P. Oliver, S. Rogers, M. J. McLaughlin (editors). *Contaminants and the soil environment in the Australasia-Pacific Region.* Proceedings of the First Australasia-Pacific Conference on Contaminants and Soil Environment in the Australasia-Pacific Region. Adelaide, Australia, Feb. 18-23, 1996. Kluwer Academic Publishers, Boston, London.

Chen, Z. S. and J. C. Liu. 1996. Effect of atmospheric acid rain on the heavy metals concentration in vegetation and crops - A review. pp. 501-515. *In: Proceedings of International Conference on Acid Deposition in East Asia.* May 28-30, 1996, Taipei, Taiwan, ROC.

Chen, Z. S., J. C. Liu, and C. Y. Cheng. 1997. The effects of acid rain on the biological accumulation of heavy metals in the crops and vegetables. Final report of second year project of team research project supported by Environmental Protection Administration of Republic of China (EPA/ROC) (grant no. EPA-86-FA44-09-47), Taipei, Taiwan. 97p. (In Chinese, with English abstract and tables)

Chen, Z. S., J. C. Liu, and C. Y. Cheng. 1998. The effects of acid rain on the biological accumulation of heavy metals in the crops and vegetables. Final report of third year project of team research project supported by National Science Council of Republic of China (NSC/ROC) (grant no. NSC-87-EPA-P-002-011), Taipei, Taiwan. 93p. (In Chinese, with English abstract and tables)

Chen, Z. S., J. C. Liu, and Y. S. Liu. 1996b. The effects of acid rain on the biological accumulation of heavy metals in the crops and vegetables. Final report of first year project of team research project supported by Environmental Protection Administration of Republic of China (EPA/ROC) (grant no. EPA-85-1404-09-11, Taipei, Taiwan. 61p. (In Chinese, with English abstract and tables)

Chen, C. T., B. J. Wang, H. C. Hsu, and J. J. Hung. 1994. Rain and lake waters in Taiwan: composition and acidity. *Terres. Atmosph. and Ocean. Sci.* 5: 573-584.

Cronan, C.S. 1980. Controls on leaching from forest floor microcosms. *Plant Soil 56:*301-322.

Cronan, C. S. and C. L. Schofield. 1979. Aluminum leaching response to acid precipitation: Effects on high-elevation watersheds in the Northeast. *Sci. 204:*304-306.

Dubay, D. T. and A. S. Heagle. 1987. The effects of simulated acid rain with and without ambient rain on the growth and yield of field-growth soybeans. *Environ. Exp. Bot.* 27: 395-401..

EPA/ROC. 1994. The standard methods for determination of heavy metals in soils and plants. National Institute of Environmental Analysis (NIEA) of Environmental Protection Administration (EPA) of Republic of China. Taipei, Taiwan, ROC. (In Chinese, with English abstract)

Fernandez, J. J., 1985. Acid deposition and forest soils: Potential impacts and sensitivity. pp. 223-239. *In:* Donale. D., A. Walter. and P. Page (eds.). *Acid Deposition.* Plenum Press, New York.

Foy, C. D. 1981. Effect of nutrient deficiencies and toxicities in plants: Acid soil toxicity. *Manuscript for handbook of nutrition and food,* CRC Press, Boca Raton, FL, USA.

Friedland, A. J., A. H. Johnson, T. G. Siccama, and D. L. Mader. 1984. Trace metal profiles in the forest floor of New England. *Soil Sci. Soc. Am. J. 48:* 442-425

Hallbaecken, L., and C. O. Tamm. 1986. Changes in soil acidity from 1927 to 1982 in a forest area of southwest Sweden. *Scandinavion J. Forest Res. 1:* 219-232.

Houng, C. C., W. C. Liu, S. W. Lee, H. C. Lin, C. H. Tzen, and C. H. Chang. 1990. The effect of acid deposition on the soil acidity in Taiwan. Final report of team research project supported by Environmental Protection Administration of Taiwan (EPA/Taiwan) (Grant No.:EPA-79-022-33-139), Taipei, Taiwan. 97p. (In Chinese, with English abstract)

Houng, C. C. 1997. The effects of acid depositions on the release and percolation of heavy metals in soils and ground water of Taiwan. Sub-project reports of team project of Environmental Protection Administration on *"The management of harm and control of acid deposition influences on the natural environments and human health in Taiwan"* (grant no. EPA-86-FA44-09-47). 95 p. (In Chinese, with English abstract , Figures and Tables).

Irving, P. M., and J. E. Miller. 1981. Productivity of field-grown soybeans exposed to acid rain and sulfur dioxide alone and in combination. *J. Environ. Qual. 10:* 473-478.

Javis, S. C., L. H. P. Tjell, and H. Mosbak. 1976. Cadmium uptake from solution by plants and its transport from roots to shoots. *Plant Soil 44:* 179-191 (1976).

Johnson, D. W., D. W. Cole, H. Van Miegroet, and F. W. Horng. 1986. Factors affecting anion movement and retention in four forest soils. *Soil Sci. Soc. Am. J.* 50: 776-783.

Johnson, D. W., J. Tuner and J. M. Kelly. 1982. The effects of acid rain on forest nutrient status. *Water Resources Res. 18:* 449-461.

Jones, J. B. Jr., and V. W. Case. 1990. Sampling, handling, and analyzing plant tissue samples. p.389-427. *In:* R. L. Westerman (ed.). Soil Testing and Plant Analysis. 3rd edition, Soil Science Society of America, Book series No. 3.

Kabata-Pendias, A. 1979. Effects of inorganic air pollutants on the chemical balance of agricultural ecosystem. Pp. 134-156. *In: Proceeding of United Nations-ECE symposium on effect of airborne pollution on vegetation.* Warsaw, August 20, 1979.

Kabata-Pendias, A. and H. Pendias. 1992. Trace elements in soil and plants. 2nd edition. CRC Press. Boca Raton, Florida, USA. 365 p.

Kennedy, I. R. 1992. *Acid soil and acid rain.* 2nd ed., Research Studies Press Ltd., Taunton, Somerset, England, p254.

King, H. B., Y. J. Hsai, and C. B. Liu. 1995. Chemistry precipitation, throughfall, stem flow and stream water of six forest sites in Taiwan. pp. 335-362. *In:* C.I. Ping and C.H

Chou (editors). *Biodiversity and Terrestial Ecosystem.* Institute of Botany, Academia Sinica. Taipei, Taiwan. Monograph series No. 14.

Lin, N. F., C. H. Chen, C. M. Peng, and M. T. Lin. 1997. The relationship between source and receptor of acid deposition: The analysis of atmospheric flow and rain deposition system. Sub-project report of team project of Environmental Protection Administration of Taiwan on *"The protection, survey and assessment of acid deposition in Taiwan"* (contract no. EPA-86-FA44-09-47). 244 p. (in Chinese, with English abstract, figures and tables).

Lindsay, W. L., and W. A. Norvell. 1978. Development of a DTPA soil test for zinc, iron, manganese and copper. *Soil Sci. Soc. Am. J. 42*: 421-428.

Litaor, M. I. 1987. Aluminum Chemistry: fractionation, speciation, and mineral equilibria of soil interstitial waters of an alpine watershed, Front Range, Colorado. *Geochim. Acta, 51*:1285-1295.

Liu, J. C. and Z. S. Chen. 1996. Effects of atmospheric acid rain on soil properties of Taiwan - A review. pp. 537-549. *In: Proceedinds of International Conference on Acid Deposition in East Asia.* May 28-30, 1996, Taipei, Taiwan, ROC.

Liu, W. C., J. J. Hung, and C. H. Chang. 1993. Effects of acid rain on Cd-contaminated soil from Ruchu, Taoyuan County, Taiwan. pp192-212. *In: The Proceedings of the 4th Symposium of Soil Pollution and Remediation.* Published by EPA of ROC, May, 1993.

McColl, J. G., L. K. Monette, and D. S. Bush. 1982. Chemical characteristics of wet and dry atmospheric fallout in northern California. *J. Environ. Qual. 4:* 585-597.

McFee, W. W. 1980. Sensitivity of soil regions to acid precipitation. EPA-600/3-80-013. US-EPA. Corvallis, OR, USA.

Mench, M. J., V. L. Didier, M. Loffler, A. Gomez, and P. Masson. 1994. A mimicked In-situ remediation study of metal-contaminated soils with emphasis on cadmium and lead. *J. Environ. Qual. 23:* 58-63.

Musselman, R. C., and J. L. Sterrett. 1988. Sensitivity of plants to acidic fog. *J. Environ. Qual. 17:* 329-333.

National Atmospheric Deposition Program. 1986. *Precipitation chemistry in the United States.* 1984. NADP/NTN Coordinator's Office. Natural Resource Ecology Lab., Colorado State Univ., Fort Collins, CO, USA.

Olsen, Jr. R. L., W. E. Winner, and L. D. Moore. 1987. Effect of Pristine and industrial simulated acid precipitation greenhouse-grown radishes. *Environ. Exp. Bot.* 27: 239-244.

Pahlsson, A. M. B. 1989. Toxicity of heavy metals (Zn, Cu, Cd, and Pb) to vascular plants- A literature review. *Water Air & Soil Poll. 47:* 287-319.

Rutherford, G. K., G. W. Vanloon, S. F. Mortensen, and J. A. Hern. 1985. Chemical and pedogenetic effects of simulated acid precipitation on two eastern Canadian soils. II. Metals. *Can. J. For. Res. 15:* 848-854.

Reddy, M. R. 1989. Acid precipitation effects on growth and yield responses of twenty soybean and twelve snap bean cultivars. *J. Environ. Qual. 18:* 145-148.

Rinallo, C., G. Modi, A. Ena, and R. Calamassi. 1993. Effects of simulated rain acidity on the chemical composition of apple fruit. *J. Horti. Sci. 68:* 275-280.

SAS Institute. 1982. SAS user's guide. Statistics. SAS Inst., Cray, NC.

Sloan, J. J., R. H. Dowdy, M. S. Dolan, and D. R. Linden. 1997. Long-term effects of biosolids applications on heavy metals bioavailability in agricultural soils. *J. Environ. Qual. 26:* 966-974.

Smith, C. R., B. L. Vasilas, W. L. Banwart, and W. M. Walker. 1991. Physiological response of two soybean cultivars to simulated acid rain. *New Phytol.* 119: 53-60.

Sposito, G., A. L. Page and M. E. Frink. 1980. Effects of acid precipitation on soil leached quality: Computer calculations. EPA-600/3-80-015.

Su, M. R. and C. M. Yang. 1995. Effects of acid precipitation on crops - A mini-review. *Chinese J. Agrometry. 2:* 47-52.

Swedish Ministry of Agriculture. 1982. *Acidification today and tomorrow.* Environment 1982 Committee, Stockholm, Swedon.

Tabatabai, M. A. 1985. Effect of acid rain on soils. p65-110. Critical review in environmental control. Vol. 15. Issrel. CRC Press. Inc.

Takemoto, B. K., A. Bytnerowicz and D.M. Olszyk. 1988. Depression of photosynthesis, growth, and yield in the field-grown green pepper (*Capsicum annuum* L.) exposed to acidic fog and ambient ozone. *Plant Physiol. 88:* 477-482.

Trumble, J. T., and G. P. Walker. 1991. Acute effects of acidic fog on photosynthetic activity and morphology of *Phaseolus lunatus. Hort. Sci. 26:* 1531-1534..

Tyler, G. 1978. Leaching rates of heavy metal ions in forest soil. *Water, Air and Soil Poll.* 9:137-148.

Ulrich, B. 1980. The production and consumption of hydrogen ions in the ecosphere. pp. 255-282. *In:* Hutchinson. T. C., and M. Havas. (eds.) *Effects of acid precipitation on terrestrial ecosystems.* Plenum Press., New York.

Ulrich, B., R. Mayer, and P. H. Khanna. 1980. Chemical changes due to acid precipitation in a loss-derived soil in central Europe. *Soil Sci. 150:* 193-199.

Wang, J. K., C. T. Chen, Z. M. Lin, and S. J. Jiang. 1995. Sources and sinks of lake and other trace metals enriched in the surface sediments of remote subalpine lakes in Taiwan. *TAO 6:* 379-392.

Watt Committee. 1984. Acid rain. Report No. 14. The Watt Committee on Energy, London.

Part 4
EAST ASIAN ECOSYSTEM SENSITIVITY TO ACID DEPOSITION

Chapter 7

ACID DEPOSITION AND ECOSYSTEM SENSITIVITY IN EAST ASIA

VLADIMIR N. BASHKIN

Institute of Fundamental Biological Problems, Russian Academy of Sciences, Pushchino, Moscow region 142292 Russia, email Bashkin@issp.serpukhov.su; Department of Atmospheric Sciences, College of Natural Sciences, Seoul National University, Seoul 151-742 Korea, e-mail Bashkin@snupbl.snu.ac.kr (*Temporary, till February 28, 1999)

CONTENT

Abstract
1. Introduction
2. Methodology
    2.1. Critical load approach
        2.1.1. General approaches for calculating critical loads
    2.2. Environmental risk assessment under critical load calculations
        2.2.1. Suggested ERA frameworks for development of acidification oriented projects
        2.2.2. Comparative analysis of CL and ERA calculations of acidification loading at ecosystems
    2.3. Biogeochemical approaches to assessment of ecosystem sustainability to acid deposition
        2.3.1. Conceptual ideas
3. Characterization of soil-biogeochemical conditions in East Asia
4. Critical load values of acid forming compounds on ecosystems of East Asia
    4.1. North-Eastern Asia
        4.1.1. Algorithm for critical loads calculations
        4.1.2. Critical loads calculation and mapping
        4.1.3. ERA analysis of critical loads of acid forming compounds at terrestrial ecosystems of the North- Eastern Asia
    4.2. South Korea

4.2.1. Mapping critical levels of acid forming pollutants and their precursors for various ecosystems
4.3. South-Eastern Asia
4.3.1. Example of North Thailand
5. Conclusions
Acknowledgements
References

ABSTRACT

The critical load (CL) and Environmental Risk Assessment (ERA) approaches were used for the evaluation of ecosystem sustainability to acid deposition in the East Asia. Calculation of critical loads for an assessment of the sensitivity of an ecosystem to acidic deposition has been made using biogeochemical approaches including the intensity of biogeochemical cycling and period of active temperature duration. On the basis of these coefficients the soil-biogeochemical regionalization was carried out for the area of East Asia and the values of critical loads (CL) for acid-forming compounds were calculated using modified steady-state mass balance (SSMB) equations. In the north-eastern ecosystems of the Asian part of Russia these values of critical loads for N, CL(N), and S, CL(S), compounds are shown to be less than in Europe due to many peculiarities of climate regime and biogeochemical cycling of elements. The minimum values of both CL(N) and CL(S) are <50 eq/ha/yr. and the maximum ones are >300 eq/ha/yr. being at least a few times less than corresponding European ecosystems. The ERA estimates showed the maximum significance of such endpoints as N content in plant issues and surface water for many North-Eastern Asian ecosystems. For the south-eastern ecosystems of the northern part of Thailand the minimum values were < 200 eq/ha/yr. and maximum values - >700 eq/ha/yr. and the minimum rank is related to more than 75% of the studied area. The exceedances of critical levels for various atmospheric acid forming pollutants and their precursors (O_3, SO_2, NO_x) are shown also for different natural and agricultural ecosystems of South Korea.

Key words: acid deposition, ecosystem sensitivity, critical loads, environmental risk assessment, biogeochemical cycling, East Asia

1. INTRODUCTION

The large population of East Asia and significant growth of both industrial and agricultural production ensure that large increases in SO_2 and NO_x emissions are occurring with the time. The consequences of these growing emissions are closely connected with enhanced acidification loading on ecosystems and the number of such ecosystems at actual and potential risk has sharply increased during last years (Ayers *et al* 1996; Park 1996).

Furthermore, there is agreement both nationally and internationally that long-range transboundary air pollution is not limited to the geographical limits of Europe and pollutants are transferred from Europe to North America and Asia as well as in the opposite directions (Posch *et al* 1996). Consequently, the calculation and mapping of critical loads as the indicators of ecosystem sensitivity to acid deposition outside of Europe are of great scientific and political interest at present and some preliminary attempts have been made to calculate the acidification loading for Asia (Dianwu *et al* 1994; Acid Deposition Survey 1995; Shindo *et al* 1995; World Bank 1994; Kuylenstierna *at al* 1995; Bashkin *et al* 1995, 1996a; 1996b; 1996c; Bashkin 1997b; Kozlov *et al* 1997).

It has been shown (Bashkin *et al* 1995) that the best approach to calculation and mapping of critical loads on ecosystems in East Asia (for the purposes of this paper East Asia includes both East Asian and South-East Asian regions) should be conducted with the use of various combinations of expert approaches and geoinformation systems including different modern methods of expert modeling and environmental risk assessment. These systems can operate using databases and knowledge bases relative to the areas with great spatial data uncertainty. As a rule, the given systems include an analysis of the cycles of various elements in the key plots, a choice of algorithms describing these cycles and corresponding interpretation of the data. This approach requires numerous cartographic materials, for example, maps of soil cover, geochemical and biogeochemical structure, buffering capacity of soil, water, atmosphere, *etc*. It is the most applicable approach for Russia as well as other Asian countries such as China, India, Thailand where, at present, an adequate information on the great spatial variability of natural and anthropogenic factors is either limited or absent (Bashkin *et al* 1996a; 1996b).

This chapter is focused on assessing the applicability of critical load and environmental risk assessment methodology for ecosystem sensitivity in East Asia to acidic deposition on the basis of various biogeochemical approaches.

2. METHODOLOGY

2.1. CRITICAL LOAD APPROACH

It is well known that the biogeochemical structure is the universal feature of the biosphere, which supports its sustainability to different anthropogenic loads, including acid forming compounds. The main basic problem is therefore connected with the quantitative parametrization of the sustainability of biogeochemical cycles of various elements and the determination of reversible deviations of these cycles when the sustainability will be conserved.

Using these biogeochemical principles the concept of critical loads (CL) has been carried out in order to derive the deposition levels at which effects of acidifying air pollutants start to occur. A UN/ECE working Group on Sulfur and Nitrogen Oxides has defined the critical load on an ecosystem as: "A quantitative estimate of an exposure to one or more pollutants below which significant harmful effects on specified sensitive elements of the environment do not occur according to present knowledge" (Nilsson & Grennfelt 1988). On a basis of biogeochemical and ecosystem approaches these critical load values may be characterized as "the maximum input of pollutants (sulfur, nitrogen, heavy metals, POPs, *etc*.) which will not introduce the harmful alterations in biogeochemical structure and function of ecosystems in the long-term, i.e. 50-100 years" (Bashkin 1996).

The critical load concept attended to achieve the maximum economic benefit from the reduction of pollutant emission since it takes into account the estimates of differentiate sensitivity of various ecosystems to acid deposition. So, this concept is considered to be as an alternative to more expensive BAT (Best Available Technologies) concept. The critical load calculations and mapping allow the creation of ecological-economic optimization models with corresponding assessment of minimum financial investments for achieving maximum environmental protection.

Correspondingly, this CL concept has been found as a basis of local, regional and international emission abatement strategy in Europe under Convention on Long-Range Transboundary Air pollution (Posch et al 1996).

In spite of almost global attraction of critical load concept, the quantitative assessment of critical load values is connected till now with many uncertainties. The phrase "significant harmful effects" in the definition of critical load is of course susceptible to interpretation, depending on the kind of effects considered and the amount of harm accepted (De Vries & Bakker, 1996). Regarding the effects considered in terrestrial ecosystems, a distinction can be made in effects on:

- soil microorganisms and soil fauna responsible for biogeochemical cycling in soil (e.g., decreased biodiversity);
- vascular plants including crops in agricultural soils and trees in forest soils (e.g., bioproductivity losses);
- terrestrial fauna such as animals and birds (e.g., reproduction decrease);
- human beings as a final consuments in biogeochemical food chains (e.g., increasing migration of heavy metals due to soil acidification with exceeding acceptable human daily intake *etc*.)
- In aquatic ecosystems, it is necessary to consider the whole biogeochemical structure of these communities and a distinction can be made accounting the whole diversity of food chains:
- aquatic and bentic organisms (decreased productivity and biodiversity);
- aquatic plants (e.g., decreased biodiversity, eutrophication);
- human beings that consume fish or drinking water (surface water) contaminated with mobile forms of heavy metals due to acidification processes (e.g., poisoning and depth).

2.1.1. General approaches for calculating critical loads

The possible impact of a certain load on soil and surface water quality can be estimated by determining:

- the difference between actual load and critical load;
- the difference between the steady state-state concentration (that will occur, when the actual load is allowed to continue Maximum Permissible Concentration, MPC) and increasing levels of pollutant concentration in soil or surface water under permanent pollutant input.

In the first, critical load, approach, the single quality objective is used to calculate a critical load. The second, steady-state, allows comparison with various quality objectives. Both approaches, which are the reverse applications of the same model (Fig.1), have their advantages and disadvantages.

One can see that both algorithms are similar, but steady-state approaches based on MPC values do not practically take into account either ecosystem characteristics or their geographic situation. Furthermore, there are many known drawbacks of traditional approaches applying MPC (Bashkin *et al* 1993; Van de Plassche *et al* 1997). Since the steps in the steady-state approach are similar but in reverse order, they will not be further elaborated and only the various steps of the critical load approach are summarized below.

Select a receptor	Select a receptor
⇓	⇓
Select the environment quality objectives for corresponding ecosystems	Determine the actual load for the corresponding ecosystem
⇓	⇓
Select a computation method (model)	Select a computation method (model)
⇓	⇓
Collect input data	Collect input data
⇓	⇓
Calculate the critical loads	Calculate the steady-state concentrations
⇓	⇓
Compare with the actual load and calculate the exceedances	Compare with environmental quality objectives (MPC)

Figure 1. Flowchart for calculating critical loads (left) or steady-state concentrations (right) of acid forming and eutrophication compounds

1. Select a receptor

A receptor is defined as an ecosystem of interest that is potentially polluted by a certain load of acid forming or eutrophication compounds of sulfur and nitrogen. A receptor is thus characterized as a specific combination of land use (e.g., forest type, agricultural crops), climate, biogeochemical regionalization and soil type or as an aquatic ecosystem, such as a lake, a river or a sea, accounting their trophical status and hydrochemistry. Regarding the terrestrial ecosystems, one should consider an information (environmental quality criteria, methods and data) for both agricultural soils (grassland, arable land) and non-agricultural (forest, bush) soils, where the atmospheric deposition is the only input to the system. The similar information has to be collected for aquatic ecosystems.

2. Select the environmental quality objectives

Quality objectives should be based on insight in the relation between the chemical status of the soil or the surface water and the response of a biological indicator (an organism or population). According to the definition, the critical load equals to the load not causing the irreversible changes in biogeochemical cycling of elements in ecosystems, thus preventing "significant harmful effects on specific sensitive elements of the environment". Consequently, the selection of quality objectives is a step of major importance in deriving a critical load.

3. Select a computation method (model)

In this context, it is important to make a clear distinction between steady-state and dynamic models. Steady-state models are particularly useful to derive critical loads. These models predict the long-range changes in biogeochemical structure of both terrestrial and aquatic ecosystems under the influence of acid deposition such as the weathering rates, base cation depletion, nutrient leaching etc. either in soils or surface waters. Dynamic models are particularly useful to predict time period before these changes will occur. These models are necessary to determine an optimal emission scenario, based on temporal change of pollutants status. The present calculations of critical loads in East Asia are based on steady-state models (Dianwu et al 1994; Shindo et al 1995; World Bank 1994; Kuylenstierna at al 1995; Bashkin et al 1995, 1996a; 1996b; 1996c; Bashkin 1997b; Kozlov et al 1997, see also this book).

4. Collect input data

This includes soil, vegetation, water (surface and ground), geology, land use etc data, influencing acidification and eutrophication processes in the considered ecosystem. For application on a regional scale it also includes the distribution and area of receptor properties (using available digitized information in geographic information systems, GIS).

5. Calculate the critical load

This step includes the calculation of critical loads of sulfur, nitrogen and the total acidity in a steady-state situation for the receptors of choice or for all receptors in all cells of EMEP or LoLa grid (150 x 150 km; 50 x 50 km; 25 x 25 km; 1 x 1°, 10 x 10' etc) of a region using a GIS (to produce critical load maps).

6. *Compare with actual load*

The amount by which critical loads are exceeded and the area in which they are exceeded (using a GIS) can be also included in the calculation when the actual loads (for example, atmospheric deposition data in case of forest) are known. Furthermore, these exceedance values are used for ecological-economic optimization model run the scenario of emission reduction.

2.2. ENVIRONMENTAL RISK ASSESSMENT UNDER CRITICAL LOAD CALCULATIONS

At present there are agreed both nationally and internationally that the process of quantitative predicting the probability of an adverse response in ecosystem health due to exposure to one or more pollutants is collectively known as Environmental Risk Assessment, ERA (US EPA, 1992).

In accordance with this definition, environmental risk assessment process is used especially in case when the probability component presents during the calculation of various parameters due to many reasons: uncertainty of input information; uncertainties in applied algorithm due to lack of knowledge, insufficient knowledge and/or simplification of input information; uncertainties in the defined geographic boundaries of pollutant influence; uncertainties in both computer calculations and management operations based on these calculations.

The principal scheme of ERA (Smith et al 1988) is shown in the Fig.2. The analysis of this principal scheme of ERA is confined that the quantitative risk assessment is possible to be done only based on whole operational flowchart.

2.2.1. *Suggested ERA frameworks for development of acidification oriented projects*

ERA in general is a process, as is EIA (Environmental Impact Assessment), and not the occasional report or document that is published at various steps. The framework for the orderly process which has been developed for various environmental-sound project can be applied also for acidification oriented projects and especially for an evaluation of ecosystem sensitivity to acid deposition and critical load calculations. The close link to management is an essential feature.

Hazard identification is akin to the qualitative prediction of impacts in EIA and is largely accomplished when the EIA is performed independently of, or prior to, an ERA. The potentially significant risks are often identified because of experience elsewhere with similar materials, processes, ecosystems, and conditions. This step is immediately useful to management and helps to sharpen the question posed in the further stages, for example, in the stage Term Of Reference, TOR, for the ERA (ADB 1991).

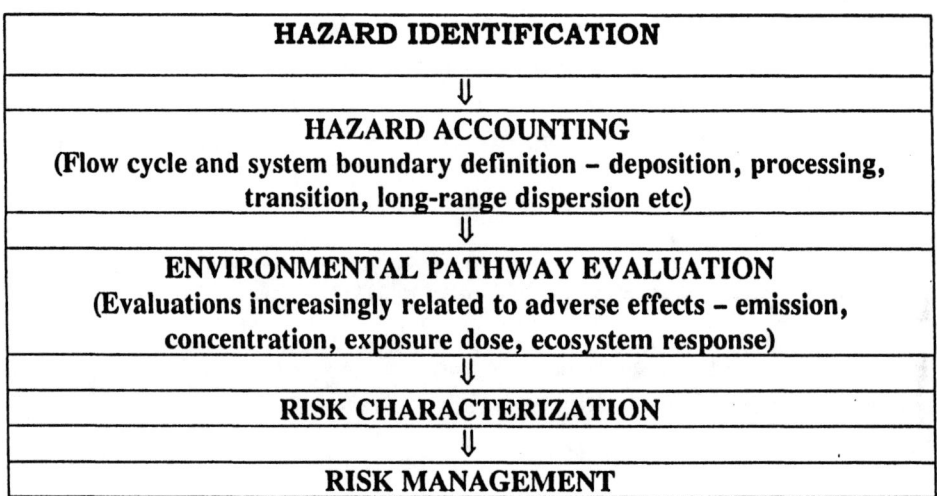

Figure 2. Recommended environmental risk assessment framework (Smith et al 1988)

Hazard accounting considers the total system of which the acidification influence is a part and sets practical boundaries for the assessment. For example, Fig.3 shows the acidification and eutrophication processes identification in both terrestrial and aquatic ecosystems of concern in any of severe points. These two steps are the basis for writing TOR for the ERA to be performed.

The environmental pathway evaluation considers various routes by which ecosystems could be exposed to acid deposition (see Fig.3). Associated with this step is a determination of the degree to which measurements of the hazardous acidification and/or eutrophication effects can be directly related to ecosystem/human health.

Risk characterization estimates the frequency and severity of adverse events and presents the results in a form useful to management, for example, in the form of various scenarios for emission abatement strategy in local or regional scale.

Risk management is the selection and implementation of risk-reduction actions. If the recommendations and findings of the ERA leave important questions unanswered, an iteration to hazard accounting can change the boundaries of analysis and refine the assessment. Although risk management is the use of assessment, it must be integrated to guide the process efficiently.

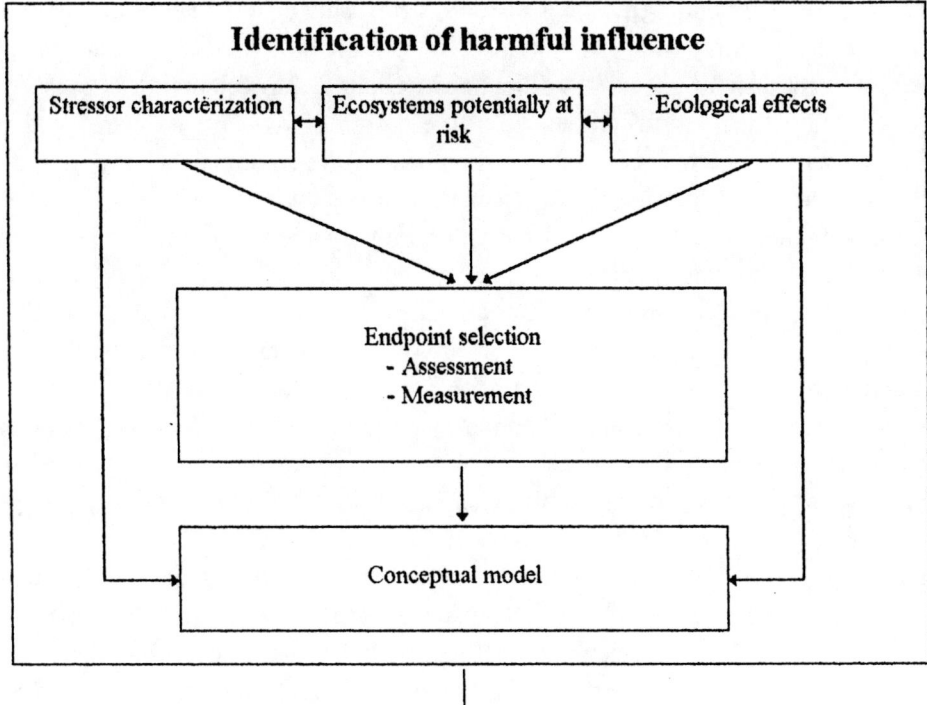

Figure 3. Comparative application of CL and ERA analysis of acidification loading at ecosystems

The purpose of providing risk information to investment responsible managers is to improve decision-making about development of acidification reduction strategy. These decisions are not just a go or no-go statement about a project proposal because the financing agency is usually fully involved in the design of most projects. EIA and ERA advise of unwanted consequences to the environment. If the lender is not comfortable with predictions, there are opportunities to change the project plans such as different sites, alternative technologies, risk reduction measures, and emergency responses. The changes may be made a condition for loan approval or a covenant in the loan.

So, regarding the ecosystem acidification effect assessment, risk management is the evaluation of alternative emission reduction measures and implementation those that appear cost-effective. Management concerns that arise because of substantial uncertainties about major environmental consequences determine the scope of detailed risk assessment. Projects are undertaken for obvious and direct benefits of economic growth, employment, and exploitation of natural resources in various countries of East Asia but the ecosystem sensitivity is of the main environmental concern of many projects. Achievement of these benefits always entails risk, but the risk must be acceptable to the funding agency and the country. Reduction of risk costs money but so does incurring the unwanted impacts (Fig.4). Avoiding one risk may create a new risk; net risk is always a consideration. Thus, the risk assessment analyses trade-off in risk, compares risk levels, and evaluates cost-effectiveness of risk-reduction alternatives.

The inquiry into the presence of hazards is also part of the preliminary assessment for the EIA. It is by the explicit identification of significant uncertainties that the need to extend an EIA to include the ERA is determined. Of course, if uncertainties can be resolved by readily acquiring more information, then the assessor should proceed to do so.

The ecosystem acidification and critical load calculation processes are only partly scientific exercises being connected closely with economic development of all countries of East Asia. So, in different projects the hazards of concern include ecosystem damage due to acidification and eutrophication processes (e.g., decreased productivity and biodiversity, soil erosion, drinking water quality, reproduction losses etc), firstly, in local scale and, secondary, in regional scale that may lead to transboundary pollution. For more details see Fig.3.

Under critical loads calculations the uncertainties arise from:

- lack of understanding of important cause-effect relationships, lack of scientific theory (e.g., biogeochemical cycling of elements; bioaccumulation of toxic chemicals in food chain, reaction of trees and crops to air pollutants);
- models that do not correspond to reality because they must be simplified and because of lack of understanding (see above);
- weakness of available date due to sampling and/or measurement problems, insufficient time-series of data, lack of replication;
- data gaps such as no measurements on baseline environmental conditions at a study site;

- phytotoxicological data that are extrapolated from high dose experiments to relatively low exposure;
- natural variations in environmental parameters due to weather, climate, stochastic events.
-

Consequently, risk assessment process is the obligated continuation of the process of quantitative calculation and mapping of critical loads of sulfur, nitrogen and acidity at various natural and agricultural ecosystems. This is connected with numerous uncertainties *a priori* included in the computer algorithm for CL calculations:

- at *the receptor selection* step the uncertainty is related to the determination of the most sensitive receptor which protection will definitely protect other, less sensitive, ecosystems;
- at *the select environmental quality criteria* step the uncertainty is connected with an assessment of biogeochemical structure of ecosystems and quantitative characterization of biogeochemical cycles of individual elements;
- at *the select computer method (model)* step the uncertainty is related to the applicability of steady-state models to dynamic systems requiring the definite simplification of these systems;
- at *the calculate critical loads* step the uncertainty is usually minimum and related mainly to the possibilities of modern computer pools;
- at *the compare with actual load* step the uncertainty is connected with an assessment of modern deposition and their spatial and temporal conjugation with definite ecosystems at the selected resolution scale.

2.2.2. Comparative analysis of CL and ERA calculations of acidification loading at ecosystems

The existing uncertainty at the all steps of algorithm for critical load calculation and mapping influences the probabilistic character of these values and requires the necessary to joint both approaches. This is illustrated in Fig.3. In the maximum degree the given conjugation is required at *the risk management* step in the ERA flowchart. The probabilistic approach to the critical loads of acid forming compounds allows us to run the set of emission reduction scenarios to minimizing the financial investments for ecosystem protection (Fig.4).

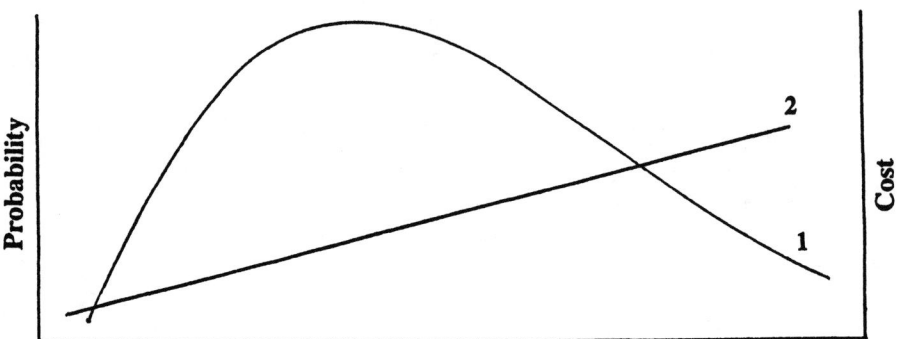

Figure 4. Scheme of comparative analysis of probability distribution function of critical load values of acidity at ecosystems and cost of emission reduction
1- probability distribution function of critical load values;
2- cost of emission reduction.

2.3. BIOGEOCHEMICAL APPROACHES TO ASSESSMENT OF ECOSYSTEM SUSTAINABILITY TO ACID DEPOSITION

2.3.1. Conceptual ideas

Conceptionally, soil and land-use databases characterizing biogeochemical cycling of S and N compounds can serve as a basis for critical load calculation and mapping (Bashkin *et al* 1995). The question arises, however, as to how we can combine steady-state and dynamic approaches during these assessments and calculations to avoid both exaggeration and underestimation of ecosystem sensitivity especially in predominant north Asian areas such as arctic and subarctic permafrost zones and in subtropical and tropical zones where the greatest number of ecosystems occur.

Consequently the idea is connected with the application of two dimensionless coefficients which should be used during ranking of various parameters for CL and ERA calculations. First of all, the coefficient of biogeochemical cycling, C_b, a ratio between litterfall input and decomposition, should be applied to take into account the speed of circulation of various elements including pollutants, as well as the degree of their accumulation in various links of biogeochemical trophical chains. The required C_b values have been calculated for the various ecosystems of East Asia based on literature devoted to an assessment of biological turnover and geochemical mapping (Bashkin *et al* 1995; Bashkin, 1996). Consequently, in accordance with existing data and preliminary estimations, C_b varies from >35-50 for arctic swamps with very depressed (slow) biogeochemical turnover to < =0.1-0.2 for tropical rain forests with very intensive (fast) turnover. Secondly, for northern areas the real duration of any processes (biochemical, microbiological, geochemical, biogeochemical *etc*.) must be taken into account because they are depressed annually for 6-10 months and the influence of acid forming compounds, as well as any other pollutants, occurs during summer. Consideration of duration using the active temperature coefficient, Ct, as a relative part of active temperatures > 5^0C from the total sum, has been applied for correction of Cb values.

The biogeochemical cycling in ecosystems of different scales is to a large extent determined by biota, especially by the primary production of plants and by microbial decomposition. The interest in acidic deposition has resulted in the development of intensive biogeochemical investigations of a large number of ecosystems in North America, Europe, Asia and South America (Moldan & Cherny 1994). The biogeochemical cycling picture is designed to summarize the circulation features in various components of ecosystems such as soil, surface and ground water, bottom sediments, biota and atmosphere. Ecosystem and soil regionalization can serve as a basis for biogeochemical mapping and combining this mapping with the quantitative assessments of biological, geochemical and hydrochemical turnover, as well as climate data, gives an opportunity to calculate the values of biogeochemical cycling, Cb, and active temperature, Ct, coefficients for different ecosystems using soil-geographic regionalization of the East Asia. As a cartographic basis, the soil-geochemical mapping of East Asia shown in Fig. 5 was used (Glazovskaya, 1990).

Table 1 shows the combinations of soil-biogeochemical and temperature conditions in various geographical regions of East Asia. The given combination of factors is represented by ecosystem belts, FAO main soil types, biogeochemical, Cb, and active temperature, Ct, coefficients.

Table 1
The values of biogeochemical cycling (Cb) and active temperature (Ct) coefficients in various soil-ecosystem geographical regions of the East Asia

Ecosystem	Main FAO soil type	Geographic region	Index, Fig1	Cb	Ct
Arctic Deserts	Lithosols, Regosols	Eurasian	1_2	10.0	0.06
Tundra	Cryic Gleysols, Histosols	Eurasian	2_2	18.0	0.15
Boreal Taiga Forest	Podzols, Podsoluvisols, Spodi-Distric Cambisols, Albi-Gleyic Luvisols, Gelic and Distric Histosols, Rendzinas and Gelic Rendzinas, Andosols	North-Siberian	3_6	9.5	0.25
		Central-Siberian	3_7	9.3	0.30
		East-Siberian	3_8	7.5	0.20
		Kamchatian-Aleutian	3_9	5.0	0.25
Taiga Meadow-Steppe	Gleysols, Planosols	Central-Yakutian	4_a1	10.0	0.35

Subboreal Forest	Podzols, Dystric and Eutric Cambisols, Umbric Leptosols, Podsoluvisols	East-Asian	5_a1	2.6	0.67
		East-Chinese	5_a2	1.5	0.81
Forest Meadow-Steppe	Luvic Fhaeozems, Chernozems	South Siberian	6_a4	2.0	0.42
		Amur-Manchurian	6_a5	1.5	0.65
Steppe	Chernozems, Kashtanozems, Solonetzes	Mongolian-Chinese	8_a2	0.8	0.61
Desert-Steppe and Desert	Xerosols, Regosols, Arenosols, Yermosols, Solonetzes, Solonchaks	Pamir-Tibetian	11_a2	0.6	0.62
		Hindukush-Alayean	11-a5	0.4	0.86
		Tan-Shanean	11-a6	0.6	0.60
Savanna, Tropical Forest	Luvi-Plinthic Ferrasols, Luvisols, Vertisols, Subtropical Rendzinas, Ferralitic Cambisols, Nitosols, Arenosols, Ferralitic Arenosols	Sauth-Asian	12_a1	0.3	1.00
Subtropical and Tropical Wet Forest	Subtropical Solonchaks, Ferrasols, Eutric Subtropical Histosols, Gleyic Subtropical Podsols, Plintic Gleysols, Nitosols	South-East-Asian	13_a1	0.2	1.00
		Himalayan	13_a2	0.4	0.80
		Malaysian	13_a3	0.1	1.00

The values of Cb for each geographical region were ranged in order to determine the type of biogeochemical cycling and these ranks are shown in Table 2. Five types of biogeochemical cycling were divided: very intensive, intensive, moderate, depressive and very depressive.

Table 2
The ranges attached to biogeochemical cycling data to assess the migration capacity of soil-ecosystem types

Ranks	Biogeochemical cycling	Biogeochemical cycling coefficient, Cb
1	Very intensive	< 0.5
2	Intensive	0.5 - 1.4
3	Moderate	1.5 - 3.0
4	Depressive	3.1 - 10.0
5	Very depressive	> 10.0

The corresponding values of active temperature coefficients ranged in accordance with the main climatic belts are shown in Table 3.

Table 3
The ranges attached to temperature regime data to assess the duration of active biogeochemical reactions

Ranks	Temperature regime	Active temperature coefficient, Ct
1	Arctic	< 0.25
2	Boreal	0.26 - 0.50
3	Subboreal	0.51 - 0.80
4	"Mediterranean"	0.81 - 0.99
5	Subtropical and tropical	1.00

3. CHARACTERIZATION OF SOIL-BIOGEOCHEMICAL CONDITIONS IN EAST ASIA

ARCTIC DESERTS AND PRIMITIVE TUNDRA ECOSYSTEMS

This zone occurs in the most northern part of the Asian Arctic and includes the northern island of North Earth itself and the islands surrounding North Earth. The biogeochemical conditions of these ecosystems are characterized by arctic hydrothermal regime, related to very severe temperatures, low precipitation (50-150 mm annually) and primitive bush-like, algae and lichen vegetation. The biogeochemical cycling can be termed "very depressed" and the temperature regime - with an annual relative part of active temperatures $>5^0$ C equal to 0.05-0.10. The predominant soil types are lithosols and regosols, in depressions - histosols (Fig.5, Table 1).

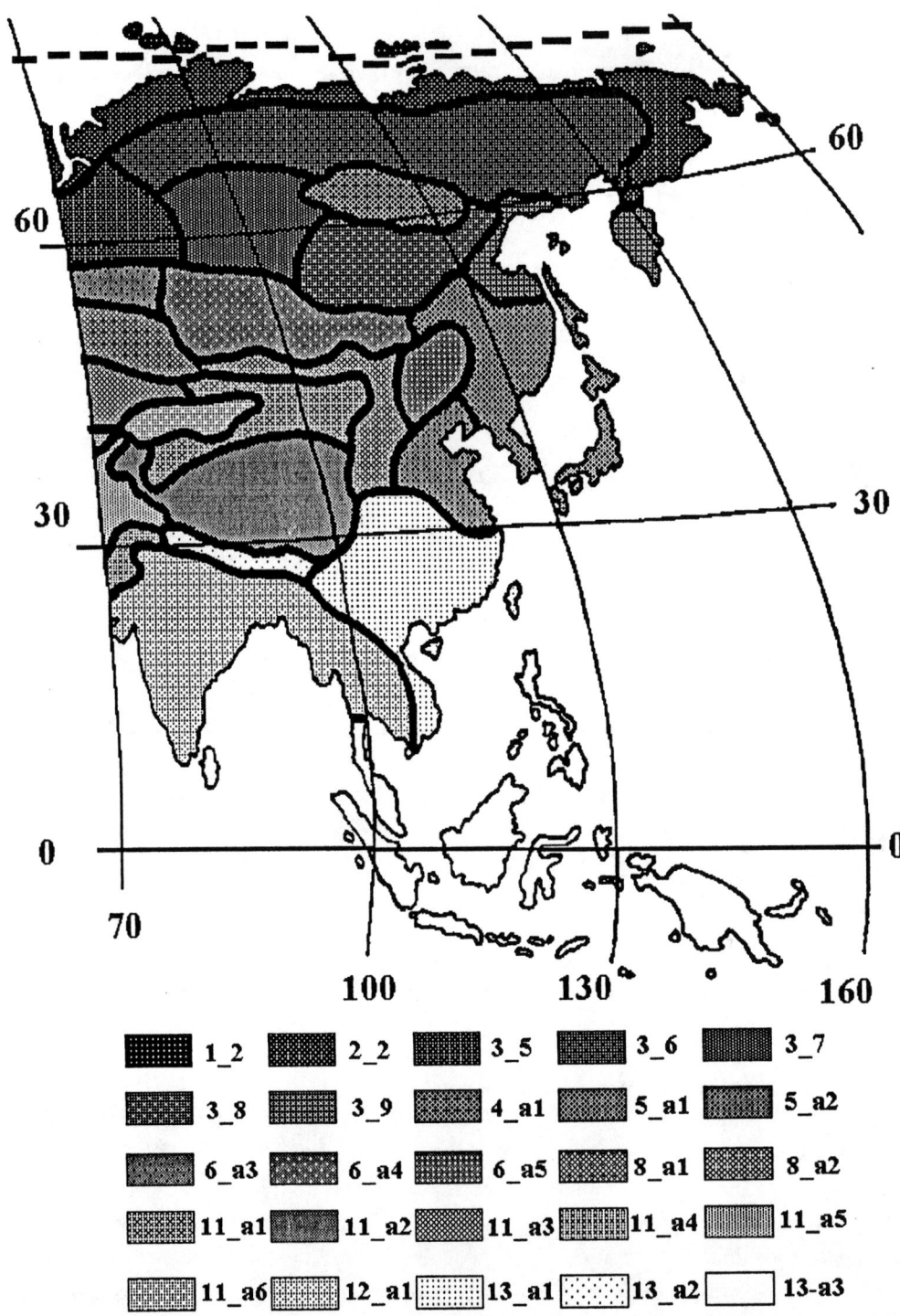

Figure 5. Map of soil-biogeochemical regions in East Asia. Key to legend is presented in Table 1.

TUNDRA ECOSYSTEMS

The tundra ecosystems are represented in Asia by Eurasian geographical region with primitive humid podzols, cryic gleysols, histosols, lithosols and regosols. The biogeochemical processes in these ecosystems are characterized by small heat quantity, a short but very intensive period of active temperatures (the mean values of Ct is equal to 0.15), wide distribution of permafrost, low precipitation, low biological and microbiological activity and low rate of chemical weathering. The mean Cb values are equal to 18 (15-50) which correspond to a very depressed type of biogeochemical cycling. However, the long winter period enhances the accumulation of various pollutants in snow cover with a sharp increases of their rapid influence on different components of ecosystems during the short summer period.

BOREAL TAIGA FOREST ECOSYSTEMS

These are represented in East Asia by North Siberian, Central Siberian, East Siberian and Kamchatian-Aleutian soil-biogeochemical geographical regions. The predominant land use/ecosystem categories are Dense Needle Evergreen Forest, Open Needle Evergreen Forest, Dense Mixed Evergreen Forest and Open Mixed Evergreen Forest. In spite of the differences in species composition, these coniferous forests are characterized by a depressed type of biogeochemical cycling; the Cb values vary from 5.0 (Kamchatian-Aleutian geographical region) to 9.3-9.5 in Central and North Siberian regions. The predominant soils are podzols, podzoluvisols, histosols which have low pH, low base saturation and low cation exchange capacity. The mean values of Ct range between 0.25-0.35. Under these cold climate conditions the additional stress of acidic deposition to the exposed plants may tend to make the vegetation in the taiga forest ecosystems more sensitive to the changes caused by acidification.

TAIGA MEADOW STEPPE ECOSYSTEMS

In East Asia these ecosystems are represented by the Central Yakutian geographical region with planosols. The biogeochemical features of these soils are connected with their localization in the inner part of the Northern Asia where the climate is the most severe and driest and soils are developing under insufficient atmospheric deposition. The ratio of precipitation to potential evapotranspiration (P:PE) is equal to 0.45-1.00 and drops during the summer period up to 0.20-0.45. The permanent permafrost and an abundance of carbonate salinization of parent material lead to the formation of planosol-solonchak-solod-solonetz complexes. Under the long severe winter period and hot, dry short summer, the biogeochemical cycling is depressed; mean Cb is equal to 10.

SUBBOREAL FOREST ECOSYSTEMS

These ecosystems are developing in a monsoon climate with predominant distribution of such land-use categories as Dense Deciduous Forest, Dense Deciduous Broad Leaf Forest, and Open Deciduous Broad Leaf Woodland. Two geographic regions are presented: East-Asian and East-Chinese.

The East Asian geographical region is situated in the continental part of the Far East (Russia, China, Korea) and island parts (Russia, Japan) characterized by different subtypes of cambisols (spodi-distric, spodi-distric cryic, humid, orti-distric, distric) and podzols, especially on Hokkaido, in Manchurian and Sikhote-Alin mountains under Dark Needle Forest land use. In the plains cambisols are located in the most drained areas. The type of biogeochemical processes in these ecosystems is moderate with mean values of Cb equal to 2.5 and Ct - 0.67 which is favorable to soil acidification with input of sulfur and nitrogen acid forming compounds. This process can be especially enhanced in ecosystems with predominant vitric andosols where porosity favors the fast chemical and biogeochemical weathering with allophane-kaolinite formation processes. The abundance of free iron and aluminum oxides under acid soil reactions, reinforced by acidic deposition leads to a release of Al^{3+} ions and toxic influence on the fine roots of trees (Izuta & Totsuka 1996).

The East Chinese geographical region has almost the same spoil-biogeochemical features as above-mentioned but due to some climate characteristics the mean Cb value is 1.5 and the mean Ct – 0.81

FOREST MEADOW STEPPE ECOSYSTEMS

The East Asian part of these ecosystems is represented by South Siberian and Amur-Manchurean geographical regions having predominant luvic phaeozems and luvic chernozems with Open Deciduous Forest natural land use categories. In the wide space of the low plains of Amur-Sunguri drainage basin of the limno-alluvial origin, one can monitor the paleohydromorphic soil features as well as the modern parameters of the given pedogenic process. This determines the biogeochemical cycling of elements in the above mentioned regions as moderate characterized by mean values of Cb - 1.5 and Ct - 0.65. The more continental climate of South Siberian geographical region favors an accumulation of various organomineral compounds in humus biogeochemical barrier, the biogeochemical coefficient Cb is equal to 2. In local depressions salinization processes develop with formation of meadow-steppe solod and solonetz complex and even peaty-swampy solonchaks. The acidification processes are more pronounced in those parts of South Siberian geographical region which are situated in a very complex orographic area of large mountain ranges and vast intermountain depressions. This region occurs in the center of Eurasian continent on the border between boreal taiga forest ecosystems of Siberia and dry steppe and desert ecosystems of Central Asia. It is connected with contrast climatic conditions: the northern and western slopes of mountain ranges have significantly more precipitation than the southern and eastern slopes, intermountain depressions and upwellings. In accordance with climatic conditions the different soil types from podzols to kashtanozems develop but the predominant soils in the slopes of hills and low mountains are luvic phaeozems and calcic chernozems whereas in the highest mountain positions podzols and podzoluvisols are widespread. So, the biogeochemical processes can be characterized as moderate in depressions and as semi-intensive in high mountain forest ecosystems with cambisols, the average Cb is equal 2 and Ct - 0.42. However, the mountain forest ecosystems are very sensitive to acidic deposition.

STEPPE ECOSYSTEMS

The main characteristic features of these ecosystems are connected with continental climate and insufficient precipitation, P:PE ranges between 0.6-0.3. The annual maximum precipitation is during summer, but due to high temperature and evapotranspiration the values of moisture coefficients are in the minimum range during this time. In accordance with the given climatic conditions, the soils of steppe ecosystems (chernozems, kashtanozems, solonetzes) are characterized by the presence of a few biogeochemical barriers such as humus, carbonate, and gypsum that makes them insensitive to actual and potential acidic precipitation. In the East Asian part of the World these steppe ecosystems are represented by Mongol-Chinese geographical region. The biogeochemical cycling is moderate: mean Cb values are between 0.7-0.8 and mean Ct values are between 0.57-0.61.

DESERT-STEPPE AND DESERT ECOSYSTEMS

These ecosystems are widespread in the Asian continent and they are coincided with subboreal and subtropical climatic belts of very strong aridity. There are 3 geographical regions in East Asia: Pamir-Tibetan, Hindukush-Alayean and Tan-Shanean (Table 1, Fig.5). The main soil type are xerosols, arenosols, yermosols, solonetzes, regosols *etc.* and all of them are characterizing by high buttering ability, high pH values, low ratio of P:PE. So, in spite of intensive and even very intensive type of biogeochemical cycling, Cb values are in limits of 0.3-0.6, these soils and corresponding ecosystems are insensitive to actual and potential acidification.

XEROFITIC SAVANNA AND TROPICAL MONSOON FOREST ECOSYSTEMS

These ecosystems are represented in Asia by South-Asian geographical region with predominant ferrosols, vertisols, subtropical rendzinas, ferralitic arenosols *etc.* (Table 1, Fig.5). In spite of very intensive biogeochemical cycling (Cb-0.3; Ct-1.00) the most of soil/ecosystem combinations in Indostan peninsula are insensitive to acidic precipitation. These biogeochemical features are complicated in Sri Lanka island and in plain and low plain areas of Mekong and Menam river basins. These subregions are characterizing by monsoon climate with wet summer (1200-1300 mm) and dry winter periods. The natural land use categories are represented by Dense Drought Deciduous Forest, Open Drought Deciduous Woodland, Tropical Broad Leaf Forest, Tropical Savanna and other subtropical and tropical vegetation types. In the eastern part of the given Mekong-Manam geographical subregion the nitosols and rhodic ferrasols characterizing by very intensive biogeochemical cycling and very high buffering capacity are predominant. These soils are low sensitive to acidic deposition. From other hand, luvi-plinthic and xantic ferrasols, subtropical albi-gleyic luvisols with plinthite are widespread in accumulative low plains of river deltas. The combination of these moisture conditions with very intensive biogeochemical cycling (Cb-0.3) leads to the formation of ecosystems which are very sensitive to actual and potential acidic depositions reinforcing the release of free Al^{3+} in soil-water system.

SUBTROPICAL AND TROPICAL WET FOREST ECOSYSTEMS

These ecosystems are represented by monsoon subtropical and tropical forests (South-East Asian and Himalayan geographical regions) and equatorial wet forests (Malaysian geographical region, Table 1, Fig.5). The main characteristic features of the given regions are the very old type of soil parent materials which are transformed by very intensive geochemical and biogeochemical weathering leading to the destruction of all primary minerals excepted quartz and the accumulation of new formed minerals such as kaolinite, hematite, gibbsite, hydrogillite, *etc*. The predominant soils are ferrasols characterizing by very low buffering capacity, abundance of free Al^{3+} and Fe^{3+}, acid reaction of soil depth, lack of accumulative biogeochemical barrier, very intensive biogeochemical cycling of all elements and especially such nutrients as N, P, K, S, Ca, Mg, *etc*. The combination of these features with monsoon and equatorial climate leads undoubtedly to a shift in the original equilibrium towards acidification under the increasing input of acid forming sulfur and nitrogen compounds.

SOUTH-EAST ASIAN GEOGRAPHICAL REGION occurs the northern part of this zone and this is characterizing by predominant distribution of acric ferrasols. The biogeochemical cycling is very intensive (mean Cb is equal to 0.2) but there are definite differences in this cycling between hilly plains and low mountains up to 400-500 m a.s.l. and middle elevation mountains (up to 1000 m a.s.l.) where the humus biogeochemical barrier presents in the profiles of podzolized ferrasols.

HIMALAYAN GEOGRAPHICAL REGION is situated in the eastern part of Tibet and Chino-Tibetan mountains that determined very complex biogeochemical cycling in vertical biogeochemical catenas of different soils, such as evergreen broad leaf forests on mountainous acric ferrasols (1400-2000 m a.s.l.), evergreen and deciduous forests on transitive ferrasol-cambisol soils (2000-2700 m a.s.l.), mixed coniferous/deciduous forest on cambisols (2500-2800 m a.s.l.), dark coniferous forest on histosols, gleysols and cambisols (2700-3000 m a.s.l.) and mountainous meadows on mountainous phaeozems (3000-3200 m a.s.l.). The given soil-ecosystem consequences are connected with hydrothermal differences and accompanied by biogeochemical cycling forwarding from very intensive in the lowest parts (Cb<0.2) to moderate one in the highest ecosystems (Cb>0.5). The mean values of biogeochemical cycling coefficient are equal to 0.4 and active temperature coefficient - 0.8. The sensitivity of these ecosystems to acidic deposition is varying but the majority of them, especially in the highest elevations is very sensitive and should be protected to avoid the irreversible destruction of their biogeochemical cycling.

MALAYSIAN GEOGRAPHICAL REGION is situated in the Malaysian peninsula, the Indonesian and New Guinea islands. The predominant ecosystems are Wet Equatorial Tropical Forest with ferrasols. In accordance with very intensive biogeochemical cycling (Cb <0.1) and natural acid features of soils, all compounds of the given ecosystems are very sensitive to actual and potential acidic deposition.

Resuming the analysis of biogeochemical turnover in East Asian ecosystems, one should taken into account that these general assessments are complicated inadequacy and uncertainty of initial information that are required to carry out the correct quantitative parametrization and characterisation of biogeochemical cycling even for such elements as nitrogen and sulfur.

4. CRITICAL LOAD VALUES OF ACID FORMING COMPOUNDS ON ECOSYSTEMS OF EAST ASIA

The applicability of biogeochemical approaches for the assessment of acidification loading on the terrestrial ecosystems in the East Asia has been made on the examples of North-Eastern Asia (Asian part of Russia), South Korea and South-Eastern Asia (Northern part of the Thailand). In spite of the great differences in climate, soil and vegetation conditions, these regions can serve as a good test of suggested methodology.

4.1. NORTH-EASTERN ASIA

The calculations of critical load were carried out using the following data for North-Eastern Asia: the inventory of soil types and subtypes, scale 1:5,000,000; biogeochemical regionalization of terrestrial ecosystems, scale 1:4,000,000; annual biomass uptake, scale 1:8,000,000; geology, scale 1:5,000,000; forest data, scale 1:5,000,000; precipitation data and evapotranspiration data, scale $1°\times1°$ LoLa; sulfur and nitrogen deposition, scale 150x150 km Asian EMEP grid cell.

The calculation of wet and dry sulfur and nitrogen deposition in North-Eastern part of Asia were made on the basis of meteodata and emissions for 1991 (Galperin *et al* 1994).

Using the above mentioned biogeochemical approaches and corresponding regionalization, the critical load values for acid forming N and S compounds were calculated using simplified steady state mass-balance (SSMB) equations in accordance with the following algorithm (Bashkin *et al.*, 1995; 1996a).

4.1.1. Algorithm for Critical Loads Calculations

Critical loads of nitrogen

$$CL(N) = {}^*Nu + {}^*Ni + {}^*Nde + {}^*Nl_{(crit)},$$

Where * means that each of the terms refers to the values at the actual total atmospheric deposition at a side. Nu, Ni, Nde and $Nl_{(crit)}$ are permissible nitrogen uptake, soil immobilization, denitrification and leaching, correspondingly.

Permissible atmospheric nitrogen uptake (*Nu) was given as:

*Nu = Nupt – Nu,
Where
Nupt – annual accumulation of nitrogen in biomass and Nu – annual uptake of N from the soil.

Nupt was calculated accounting for the coefficients of biogeochemical turnover (Table 1). Annual Nu from soil was calculated on a basis of nitrogen mineralizing capacity (NMC) of soils which was determined experimentally or calculated using regression equations (Bashkin et al 1997). So,

Nu = (NMC – Ni – Nde) Ct,
Where
$Ni = 0.15NMC$, if C:N<10 $Ni = 0.25NMC$, if 10<C:N< 14
$Ni = 0.30NMC$, if 14< C:N<20 $Ni = 0.35NMC$, if C:N>20;
$Nde = 0.145NMC + 6.447$, if NMC >60 kg/ha/yr
$Nde = 0.145NMC + 0.900$, if NMC<10 kg/ha/yr
$Nde = 0.145NMC + 2.605$, if 10<NMC>60 kg/ha/yr
Permissible immobilization of atmospheric deposition N (*Ni) was found as:
$Ni = [(0.20NH_4 + 0.10 NO_3)/Cb]Ct$, if C:N<10
$Ni = [(0.30NH_4 + 0.20NO_3)/Cb]Ct$, if 10< C:N<14
$Ni = [(0.35NH_4 + 0.25NO_3)/Cb]Ct$, if 14<C:N<20
$Ni = [(0.40NH_4 + 0.30NO_3)/Cb]Ct$, if C:N>20,
Where
NH_4, NO_3 - NO_x and NH_x wet and dry deposition.

Permissible denitrification from atmospheric deposition N (*Nde) was found as:

*Nde = (Nde/NMC) Ntd Ct,
Where
Nde/NMC – denitrification fraction, which depends on many features of soils and calculated on a basis of experimental data and Ntd – total N deposition.

Finally, permissible critical leaching of atmospheric nitrogen (*$Nl_{(crit)}$) was given as
*$Nl_{(crit)} = Q\ C_{Ncrit}$,
Where
Q – annual surplus of precipitation (runoff) and C_{Ncrit} - permissible nitrogen concentration in surface water.

CRITICAL LOADS OF SULFUR

Since for majority of ecosystems in North-Eastern Asia the ratio of precipitation to potential evapotranspiration (P:PE) is equal to < =1.00 or slightly exceeds the 1.00 (except of ecosystems with some cambisols, histosols and andosols) the values of runoff can be neglected in calculation of critical loads of acidity, CL(Ac), and they were found as

CL(Ac) = (BCw Ct)/Cb
And critical loads of sulfur were calculated as
CL(S) = Sf CL(Ac),

Where
BCw – weathering of base cations and Sf – sulfur fraction in total sum of sulfur and nitrogen deposition.

This approach was applied in level "0" and level "I" calculations of critical loads in Europe (Posch *et al* 1993; Bashkin *et al* 1997) as well as for various ecosystems in developing countries (Kuylenstierna *et al* 1995).

The values of BCw were determined on a basis of FAO soil nomenclature, soil parent material and soil texture according to De Vries *et al* (1993) and values of Cb and Ct were applied to accounting biogeochemical cycling intensity and duration of active temperature period. The root zone was assumed to be equal to 0.5 m.

EXCEEDANCES OF CRITICAL LOADS

The values of exceedances were calculated as follows:
$Ex(N) = Ntd - CL(N)$
$Ex(S) = Std - CL(S)$
Where
Ex(N), EX(S) are the values of exceedental input of nitrogen and sulfur compounds above the calculated critical loads, and Ntd, Std are nitrogen and sulfur deposition.

4.1.2. CRITICAL LOADS CALCULATION AND MAPPING

On the basis of above mentioned modified and simplified SSMB equations, the critical loads for nutrient and acidifying nitrogen as well as for sulfur and acidity have been calculated for various terrestrial ecosystems of North-Eastern Asia. Due to the large dimensions of the area all calculation and mapping procedures have been carried out using geoinformation system with elements of simplified expert-modeling system (Bashkin *et al* 1996a; 1996b). Initial information consisted of geological, vegetation, soil, *etc* regionalization. For every elemental taxon the main links of biogeochemical cycles of N and S as well as base cations weathering (BCw) were characterized quantitatively on a basis of available case studies and literature sources. The grid cells were $1° \times 0.5°$.

Regarding the northern and north-eastern areas of Asia, it was calculated that the minimum values of critical loads of nitrogen, CL(N), (<50 eq/ha/yr.) occur for arctic and subarctic ecosystems (Fig.6). The values of CL(N) in the range of 50-100 eq/ha/yr. are typical for the majority of ecosystems in permafrost area. Therefore, these ecosystems are very sensitive to the excessive input of deposition nitrogen. The maximum values of CL(N), >300 eq/ha/yr. are noted for ecosystems of chernozemic and chernozem-like soils of the southern regions of Siberia and the Far East. Exceedances of CL(N) are shown mainly in Ural mountains, in boundary regions with Kazakhstan, in the lower flow of Enisei river and in the Far East (Bashkin *et al* 1996a). The minimum values of sulfur critical loads, CL(S), as well as acidity, are shown predominantly in the northern part of East Siberia and in the Kamchatka peninsula (Fig.7). In the area between the Enisei and Ob rivers these values are shown to increase up to 50-100 eq/ha/yr. and the maximum values (>300 eq/ha/yr.) are shown for ecosystems having neutral and alkaline soils. The corresponding exceedances are indicated for many regions of the north-eastern Asia with maximum values for Altai mountains, for boundary regions within Kazakhstan, in the lower Enisei river flow, and in the Far East, Sakhalin and south-kurilean islands due to both local and transboundary pollution. One can see however that at modern deposition 88.0 and 72.7 % of ecosystems in North-Eastern Asia have no or small (<50 eq/ha/yr) exceedances of CL(N) and CL(S), correspondingly (Table 4). From other side, critical loads for more than 10% of ecosystems are exceeded by nitrogen deposition in the limits of 50-300 eq/ha/yr. and for more than 20% - by sulfur deposition.

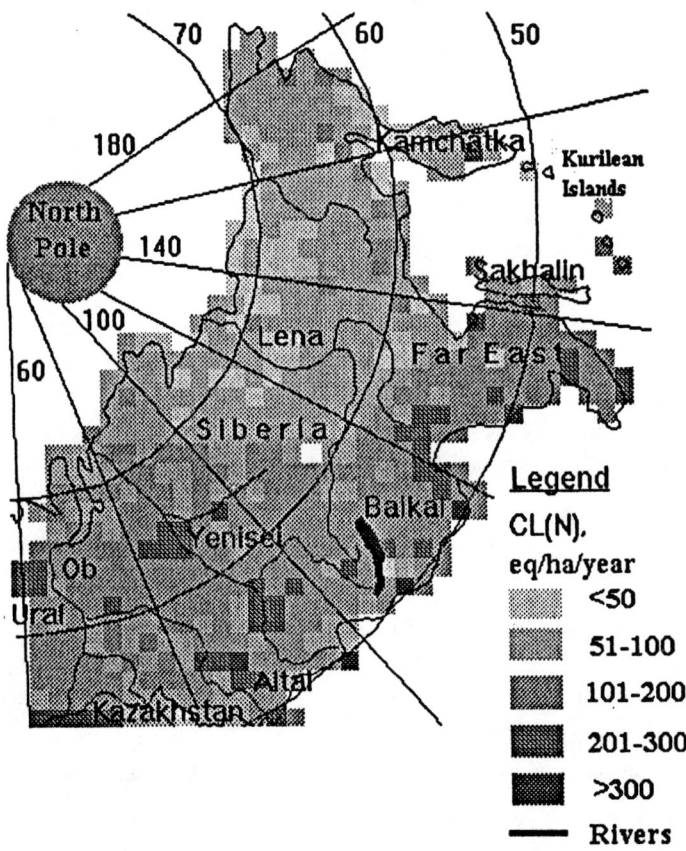

Figure 6. Critical loads of nitrogen on the ecosystems in the North-Eastern Asia (Asian part of Russia).

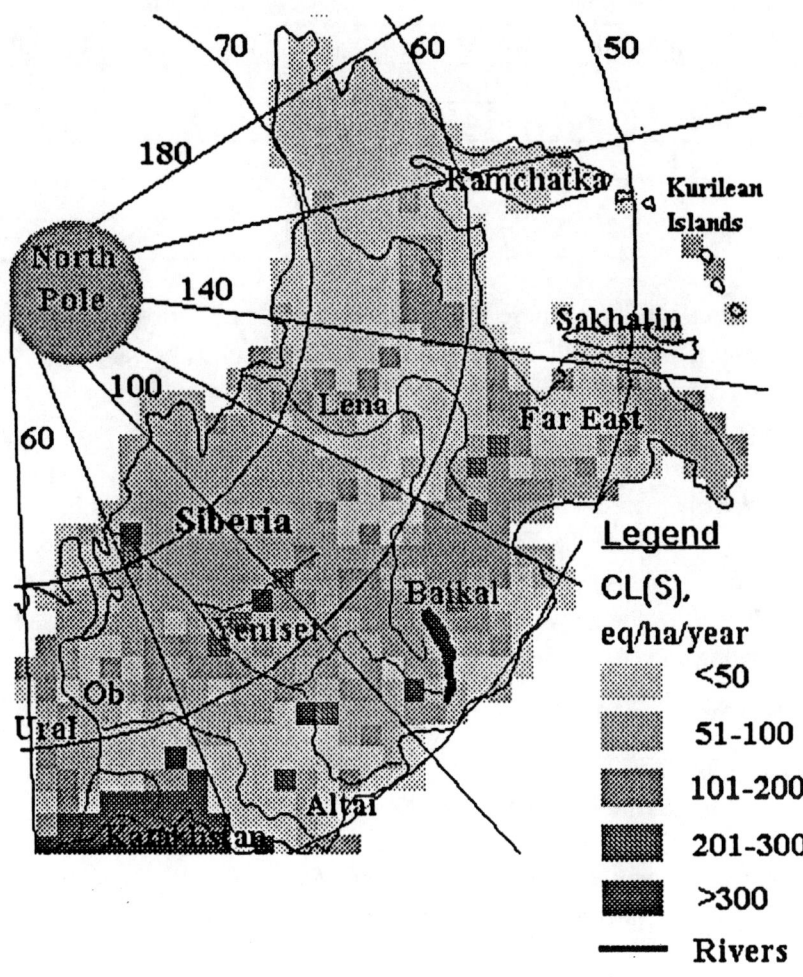

Figure 7. Critical loads of sulfur on the ecosystems in the North-Eastern Asia (Asian part of Russia).

Table 4
Distribution of critical load values of sulfur, nitrogen and their exceedances for the North-Eastern Asian ecosystems

Values range, eq/ha/yr.	Percentage of area under different critical load values		Percentage of area under different exceedance values	
	For nitrogen	For sulfur	For nitrogen	For sulfur
<50	8.3	40.5	88.0	72.7
50-100	40.8	32.4	6.9	14.3
101-200	41.3	18.2	3.9	8.3
201-300	8.0	1.5	1.1	2.6
>300	1.6	7.4	0.1	2.1

The general comparison of these critical load values with those calculated for corresponding ecosystems in Europe (Bashkin, 1997a) reveals the lower values in Asian areas. This can be explained by the more prolonged winter period promoting an accumulation of different pollutants in the snow layer and their enforced influence on various components of ecosystems and biogeochemical cycling of nutrients during short spring and summer period. The values of Ct are within the limits of 0.15-0.57 for the majority of North-Eastern Asian ecosystems whereas in the European part of Russia, for example, the corresponding values are of 0.25-0.87. This points to the shorter but very active periods of biogeochemical turnover in almost all of the above-mentioned soil-biogeochemical regions accelerating the acidification loading on ecosystems.

4.1.3. ERA ANALYSIS OF CRITICAL LOADS OF ACID FORMING COMPOUNDS AT TERRESTRIAL ECOSYSTEMS OF THE NORTH-EASTERN ASIA

Regarding the uncertainty analysis of critical loads of acid forming compound at different terrestrial ecosystems of Asian part of Russia, one can see the following. The strongest influential factor for the values of CL(N) is the parameter Nu. For the biggest part of Siberia and Far East territory this parameter has the first rank and only in the case of Humid Cambisols and Cryic Gleysols ecosystems is its rank second. Among others, Ni and Nde parameters, which are closely intercorrelated, are the weakest ones. The given parameters do not practically have any influence on the values of CL(N), except Dyctric Cambisol ecosystems. The influence of Nl values is decreasing in the following row of ecosystem forming soils: Regosols > Humid Cambisols = Cryic Gleysols = Gelic Podzols > Andosols > Dystric Cambisols > Eutric Cambisols > Luvic Phaeozems = Chernozems > Kashtanozems. These results reflect, in significant degree, the geographical change of ecosystems from north to south and correspondingly an alteration of relationship of temperature and moisture constituents in hydrothermic coefficient. So, the main impact to the assessment of the influence of inputting parameters (Nu, Ni, Nde, Nl) on both uncertainty and sensitivity of outputting values of CL(N) belongs to Nu. It is connected firstly with a deficit of nitrogen as the main nutrient in all studied ecosystems as well as an existing spatial and temporal variability of this parameter that relates to a significance and correctness of experimental and computed values of Nu. In accordance with relatively better knowledge of hydrological picture and relatively homogenous values of critical concentration of nitrogen in surface waters, C_{Ncrit}, included in the calculation of Nl_{crit} values, the input of the given parameter into the uncertainty of CL(N) is expressed in a lesser degree. Furthermore, the runoff processes are practically not significant for ecosystems of Luvic Phaeozems, Chernozems and Kashtanozems due to low P:PE ratio. During the calculations of CL(N) for ecosystems of the North-Eastern Asia, the values of critical immobilisation and denitrification of N depositions both in relative and absolute meanings played a subordinate role that obviously reflects their minor contribution into uncertainty and sensitivity analysis of the computed output values of ecosystem sensitivity to acidic deposition.

So, the ERA estimates shown in Table 5 characterize the significance of such endpoints as nitrogen content in plant issues and surface waters for many ecosystems of North-Eastern Asia. These endpoints have to be taken into account during risk management step of ERA flowchart for emission abatement strategy development.

4.2. SOUTH KOREA

4.2.1. MAPPING CRITICAL LEVELS OF ACID FORMING POLLUTANTS AND THEIR PRECURSORS FOR VARIOUS ECOSYSTEMS

Mapping receptor sensitivity to ozone

The long-term critical level for ozone was expressed as cumulative exposure over a concentration of 40 ppb. This exposure index is referred to as AOT40 (Accumulated exposure Over a Threshold of 40 ppb). The AOT40 was calculated as the sum of the differences between the hourly concentration (in ppb) and 40 ppb for each hour when the ozone concentration exceeds 40 ppb for various land use types/ecosystems (Table 6).

Table 5
Percentage of various endpoints contribution to total environmental risk assessment of ecosystem sensitivity to acid deposition in North-Eastern Asia

Ecosystem forming soils	Endpoint assessment method	Endpoints			
		N content in plant issues	N content in surface waters	Denitrified N	N enrichment of soil pool
Regosols and Lithosols	SRC	50	41	2	7
	RTU	30	27	21	22
Cryic Gleysols and Humic Cambisols	SRC	37	56	2	5
	RTU	22	37	21	20
Gelic Podzols	SRC	47	45	3	5
	RTU	40	39	10	11
Andosols	SRC	40	39	10	11
	RTU	81	17	1	1
Eutric Cambisols	SRC	41	9	25	25
	RTU	63	10	14	13
Distric Cambisols	SRC	96	2	1	1
	RTU	38	32	16	14
Luvic Phaerozems and Chernozems	SRC	96	1	1	2
	RTN	65	6	12	17
Kashtanozems	SRC	95	1	4	5
	RTU	61	5	19	15

Note: SRC – Standard Regression Coefficient
RTU – RooT of the Uncertainty

Table 6
The critical levels of ozone (AOT40 values) for different land use types of South Korea

Land use type	AOT40 ppb h	Period, month
Agriculture	3000	4-10
Shrubs/pasture	4000	3-10
Shrubs/mountain	7000	5-9
Coniferous	10000	1-12
Mixed	10000	1-12
Deciduous	10000	4-10
Evergreen	10000	1-12

The initial information on the ground level ozone concentration in atmosphere above South Korea in 1994 has been calculated by Prof. S.-U.Park (Park 1997). The corresponding distribution of AOT40 exceedances of O_3 for various land use types of South Korea is shown in Fig. 8. One can see that various exceedance values (from <20 to >80% above AOT40) was monitored for almost 50 per cent of Korean ecosytems.

Mapping receptor sensitivity to SO_2

The values of NO_x and SO_2 annual average levels in the atmosphere above South Korea for 1994 (Park 1997) are shown in Fig 9. Comparison of these data with critical levels and ranking of damage for exceedances of critical levels of SO_2 and NO_x are given in Tables 6 and 7.

Table 6
SO_2 average annual levels, ppb

Land use	Critical levels	Safety factor, x2	Degree of damage				Korea, national standard
			Low	Middle	High	Very high	
			Degree of sensitivity*				
			Very sensitive	Sensitive	Intermediate	Resistant	
Agriculture	12	24	28	30	38	>38	30
Shrubs	10	20	26	28	36	>36	30
Shrub/moun	8	16	18	20	24	>24	30
Coniferous	8	16	18	20	22	>22	30
Mixed	8	16	20	22	24	>24	30
Deciduous	8	16	22	24	30	>30	30
Evergreen	6	12	20	22	28	>28	30

Table 7

NO$_x$ average annual levels, ppb

Land use	Critical levels	Safety factor, x2	Degree of damage				Korea, national standard for NO$_2$
			Low	Middle	High	Very high	
			Degree of sensitivity*				
			Very sensitive	Sensitive	Intermediate	Resistant	
Agriculture	20	40	60	80	100	>100	50
Shrubs	15	30	40	52	68	>68	50
Shrub/moun.	12	24	28	30	36	>36	50
Coniferous	12	24	30	36	40	>40	50
Mixed	12	24	30	38	42	>42	50
Deciduous	20	40	42	46	50	>50	50
Evergreen	15	30	36	40	44	>44	50

Note: NO$_x$ alone are not sufficiently toxic to affect different plant species. Thus the definition of critical levels assumed that other pollutants (SO$_2$ and O$_3$) *must present at concentrations close to their critical levels.* For instance, NO$_x$ average annual level >100ppb is very damaged for agricultural crops if SO$_2$ level in the same place is >38ppb and AOT40 for O$_3$ >3000ppbh.

Critical levels for NO$_x$ include the prevention both phytotoxic effects and biodiversity changes.

The degree of damage was expressed (very approximately) as productivity decrease of various species listed below (if other factors like nutrients, temperature and water supply, pests etc are not limiting the productivity) (Table 8):

Low - decrease of very sensitive species bioproductivity in the limits of 5-10%

Middle - decrease of very sensitive species bioproductivity in the limits of 10-15% and sensitive in the limits of 5-10%

High – decrease of very sensitive species bioproductivity in the limits of 15-30%, sensitive in the limits of 10-15%, intermediate resistant in the limits of 5-10%

Very high - decrease of very sensitive species bioproductivity >30%, sensitive in the limits of 15-25%, intermediate resistant in the limits of 10-15% and resistant in the limits up to 10%.

Table 8
Degree of species sensitivity to various air pollutants (SO_2 and $SO_2 + NO_x$)

Land use	Resistant	Intermediate resistant	Sensitive	Very sensitive
Agricultural crops	Barley Potato Corn Spinach Cabbage Rice Lettuce Tomato	Wheat Pasture grasses Celery Carrot Cucumber Beet Turnip	Legumes Peas Rhuward Buckwheat Timophy grass	Soybean Reddish Red clover Alcalfa Wheat grass
Coniferous species	Austrian pine	Larch Red pine	White pine Jaack pine	
Deciduous species	Beech	Willow Large-toothed aspen Oak Balsam poplar Swiss chard	White birch Alder	Quaking aspen
Shrubs (garden and forest)	Raspberry Sugar maple	Witch nazel	Blueberry and other berries	Blackberry

4.3. SOUTH-EASTERN ASIA

4.3.1. EXAMPLE OF NORTH THAILAND

The general principles of the given calculations carried out by J-P. Hettelingh and co-workers (1995) was also applied during the development of corresponding algorithm for calculation of critical loads of acidity. The algorithm description in the whole did not alter from that applied for RAIN-ASIA, phase I, where only acidity loading was calculated without splitting to S and N inputs. In this case study the biogeochemical soil/ecosystem regionalization, more detailed soil and land use types with more tine resolution were applied in order to recalculate the CL(Ac) values. Furthermore, during the subdivision of area under study to elemental taxones soil-biogeochemical characteristics of ecosystems described above was also taken into account. Land use and soil types were applied on a basis of national data with scale 10' x 10' LoLa. For more details see Chapter 5 (Kozlov & Towprayoon, this book). It was shown that the spatial distribution of critical loads corresponds very closely to soil mapping and partly to land use mapping of the area of Northern Thailand. One can see that the calculated CLs across most of the studied area are low and the sensitivity of the majority of ecosystems to acidification loading is very high. To assess the influence of present depositions on the ecosystems of the Northern Thailand the RAIN-ASIA model run values was used (World Bank 1994). The studied area experiences modest exceedances, usually between 375 and 750 eq/ha/yr., although this level of excess deposition is three to six times greater than the calculated CL values. As a result of both the high sensitivity of ecosystems and level of exceedances across the Northern Thailand, more than 75% of the ecosystems across about 50% of this territory is at significant risk from acid deposition.

In spite of a necessary to validate these data due to insufficient experimental data sets, the high sensitivity of tropical forest and mountain ecosystems to acid loading calculated using soil-biogeochemical and land use data is undoubted.

5. Conclusions

1. Critical load and environmental risk assessment methodology were applied for the evaluation of ecosystem sensitivity to acidic deposition in East Asia. The corresponding values of biogeochemical and active temperature coefficients are presented for ecosystems of the area under study in order to take into account the prolonged winter (north-eastern part) and very active summer (south-eastern part) periods in these ecosystems.

2. The values of critical loads for acid forming N and S compounds in the north-eastern part of Asia are shown to be less than in Europe due to many peculiarities of climate regime and biogeochemical cycling of elements. The minimum values both CL(N) and CL(S) are <50 eq/ha/yr. and the maximum values >300 eq/ha/yr., at least a few times less compared to the corresponding European ecosystems.

3. The ERA estimates characterize the significance of such endpoints as nitrogen content in plant issues and surface waters for many ecosystems of North-Eastern Asia. These endpoints have to be taken into account during risk management step of ERA flowchart for emission abatement strategy development.

4. The exceedances of ground level ozone concentration in atmosphere above South Korea in 1994 (from <20 to >80% above AOT40) were monitored for almost 50 per cent of Korean ecosystems.

5. The calculated critical load values across most of the area of the Northern Thailand are low and the sensitivity of the majority of ecosystems to acidification loading is very high. As a result of both the high sensitivity of ecosystems and level of exceedances across the Northern Thailand, more than 75% of the ecosystems across about 50% of this territory is at significant risk from acid deposition.

6. The critical load values for the East Asia territory provide a useful tool for the assessment of modern and future acidification loading on the various ecosystems however more detailed calculation and mapping research will have to be carried out.

ACKNOWLEDGEMENTS

Author wishes to thank Prof. S.-U.Park (Seoul National University, Korea), Dr. M.Kozlov (KMUTT, Bangkok, Thailand) and A.Abramychev (Pushchino State University, Russia) for participating in CL calculations, Stockholm Environmental Institute, Russian Fund for Basic Research (the grant No. 96-05-64368) and MSC-E/EMEP for the financial and scientific support of this project.

REFERENCES

Acid Deposition Survey: Phase 2 (1995) Final Report. Japan Environmental Agency. 87p.

ADB (1991) Environmental Risk Assessment. 182p.

Ayers, G.P., Gillett, R.W., Seleck, P.W., Marshall, J.C., Granek, H., Peng, L., Lim, S.F., Harjanto, P., Mhw, T., Parry, D.(1996) Acid Deposition in South East Asia. In: Proceedings of International Conference on Acid Deposition in East Asia, Taipei, May 28-30,1996, 1-22.

Bashkin, V.N., Evstafjeva, E.V., Snakin V.V. *et al* (1993) Biogeochemical Foundations of Ecological Standardization. Moscow: Nauka Publishing House, 312p.

Bashkin, V.N., Kozlov, M.YA., Priputina, I.V., Abramychev, A.YU. and Dedlova, I.S. (1995) Calculation and Mapping of Critical Loads of S, N and Acidity on Ecosystems of the Northern Asia. Water, Air and Soil Pollution, 85: 2395-2400.

Bashkin, V.N. (1996) Risk Assessment of Computed Critical Loads of Pollutants at Ecosystems. In: Heavy Metals in the Environment, Abstracts of International Symposium, Pushchino, 15-18 October 1996, 172.

Bashkin, V.N., Kozlov, M.YA., and Abramychev, A.YU.(1996a) The Application of EM GIS to Quantitative Assessment and Mapping of Acidification Loading in Ecosystems of the Asian Part of the Russian Federation. Asian-Pacific Remote Sensing and GIS Journal, 8(2): 73-80.

Bashkin, V.N., Kozlov, M.YA., Abramychev, A.YU. and Dedlova, I.S. (1996b) Regional and Global Consequences of Transboundary Acidification in the Northern and Northern-East Asia. In: Proceedings of International Conference on Acid Deposition in East Asia, Taipei, May 28-30,1996, 225-231.

Bashkin, V.N., Kozlov, M.YA., Golinets, O. (1996c) Risk Assessment of Ecosystem Sustainability to Acid Forming Compounds in the North-Eastern Asia - In: Proceedings of International Conference on Acid Deposition in East Asia, Taipei, May 28-30,1996, 347-356.

Bashkin, V.N., Kozlov, M.YA., Priputina, I.V., and Abramychev, A.YU. (1997) Regional Assessment of Ecosystem Sustainability to Atmotechnogenic Deposition of Sulfur and Nitrogen in European Part of Russia. Pt.I. Quantitative Assessment and Mapping of Critical Loads of Sulfur and Nitrogen Compounds at Terrestrial and Freshwater Ecosystems. Regional Ecological Problems, No.1, 57-78.

Bashkin, V.N. (1997a) The Critical Load Concept for Emission Abatement Strategies in Europe: a review. Environmental Conservation, 24: 5-13.

Bashkin, V.N. (1997b) Acid Deposition and Ecosystem Sensitivity in East Asia. In: Proceedings of International Workshop on Monitoring and Prediction of Acid Rain, Seoul, 29.09-1.10.1997, 147-161.

De Vries, W, Posch, M, Reinds, G.J and Kamari J (1993) Critical Loads and their Exceedances on Forest Soils in Europe. The Winand Staring Centre for Integrated Land, Soil and Water Research, Rep. 58, Wageningen, The Netherlands, 116p.

De Vries, W. and Bakker, D.I. (1996) Manual for Calculating Ctitical Loads of Heavy Metals for Soils and Surface Waters. DLO Winand Staring Centre, Wageningen, The Netherlands, Report 114, 173p.

Dianwu, Z., Chuyin, C., Julin, X., Xiaoshan, Z., Zhaohua, D., Jietai, M., Seip, H.M., Vost, R.. (1994) Acid Reign 2010 in China?, 41p.

Galperin, M.V., Erdman, L.K., Subbotin, S.P (1994) Modelling of Pollution of the Arctic by the S and N Compounds and Heavy Metals from the Sources in the North Hemisphere. MSC-E Report, July 1994, 33p.

Glazovskaya, M.A. (1990) Methodological Guidelines for Forecasting the Geochemical Susceptibility of Soils to Technogenic Pollution, ISRIC Technical Report 22, 39p.

Hettelingh, J-P., Sverdrup H. and Zhao Dianwu (1995) Deriving Critical Loads for Asia. Water, Air, and Soil Pollution, 85:2565-2570.

Izuta T. and Totsuka T. (1996) Effect of soil acidification on growth of *Cryptometria japonica* Seedlings. In:Proceedings of the International Symposium on Acid Deposition and Its Impacts, Tsukuba, Japan, 10-12, December, 1996, 157-164.

Kozlov, M.YA., Towprayoon, S. and Sirikarnjanawing S. (1997) Application of Critical Load Methodology for Assessment of the Effects of Acidic Deposition in Northern Thailand. In: Proceedings of International Workshop on Monitoring and Prediction of Acid Rain, Seoul, 29.09-1.10.1997, 141-146.

Kuylenstierna, J.C.I., Cambridge, H.M., Cinderby, S. and ChadwickK, M.J. (1995) Terrestrial Ecosystem Sensitivity to Acidic Deposition in Developing Countries. Water, Air and Soil Pollution, 85: 2319-2324.

Moldan, B. and Cherny, J., Eds. (1994) Biogeochemistry of Small Catchments, Wiley and Sons, 420p.

Nilsson, I. and Grennfelt, P., Eds. (1988) Critical Loads for Sulfur and Nitrogen. Report from a Workshop Held at Stokhoster, Sweden, March 19-24, 1988. Miljo Rapport 1988: 15. Copenhagen, Denmark, Nordic Council of Ministers, 418 p.

Park, S.U. (1996) Estimation of the Anthropogenic Emission of SO_2 and NO_x in South Korea. In: Proceedings of International Conference on Acid Deposition in East Asia, Taipei, May 28-30,1996, 30-44.

Park, S.U. (1997) Development of Technology for Monitoring and Prediction of Acid Rain. SNU, Seoul, 631p.

Posch, M, Hetteling, J-P, Sverdrup, H *et al* (1993) Guidelines for the Computation and Mapping of Critical Loads and Exceedance in Europe. In: Proceedings of 3d CCE meeting, 1993, 15-19 March, Madrid, 1-14.

Posch, M, Hettelingh, I.-P, Alcamo J. and Krol M (1996) Integrated scenario of acidification and climate change in Asia and Europe. Global Environmental Change 6(4):375-394

Shindo, J, Bregt, A.K. and Takamata, T. (1995) Evaluation of Estimation Methods and Base Data Uncertainties for Critical Loads of Acid Deposition in Japan. Water, Air , and Soil Pollution, 85:2571-2576.

Smith, K.R., Carpenter, R.A. and Faulstich, M.S. (1988) Risk Assessment of Hazardous Chemical Systems in Developing Countries. Occasional Report No.5. Honolulu: West-East Environmet and Policy Institute.

Van de Plassche, E., Bashkin, V., Guardans,R., Johansson, K., and Vrubel,J. 1997. An Overview of Critical Limits for Heavy Metals and POPs. Background document. Workshop on Critical Limits and Effect Based Approaches for Heavy Metals and Persistent Organic Pollutants, Bad Harzburg, Germany, 3-7 November 1997. 38p.

US IPA (1992) Framework for Ecological Risk Assessment, EPA/630/R 92/001. Risk Assessment Forum, Washington, DC.

World Bank (1994) RAIN/ASIA. User's Manual, IISAA, Washington, 138p.

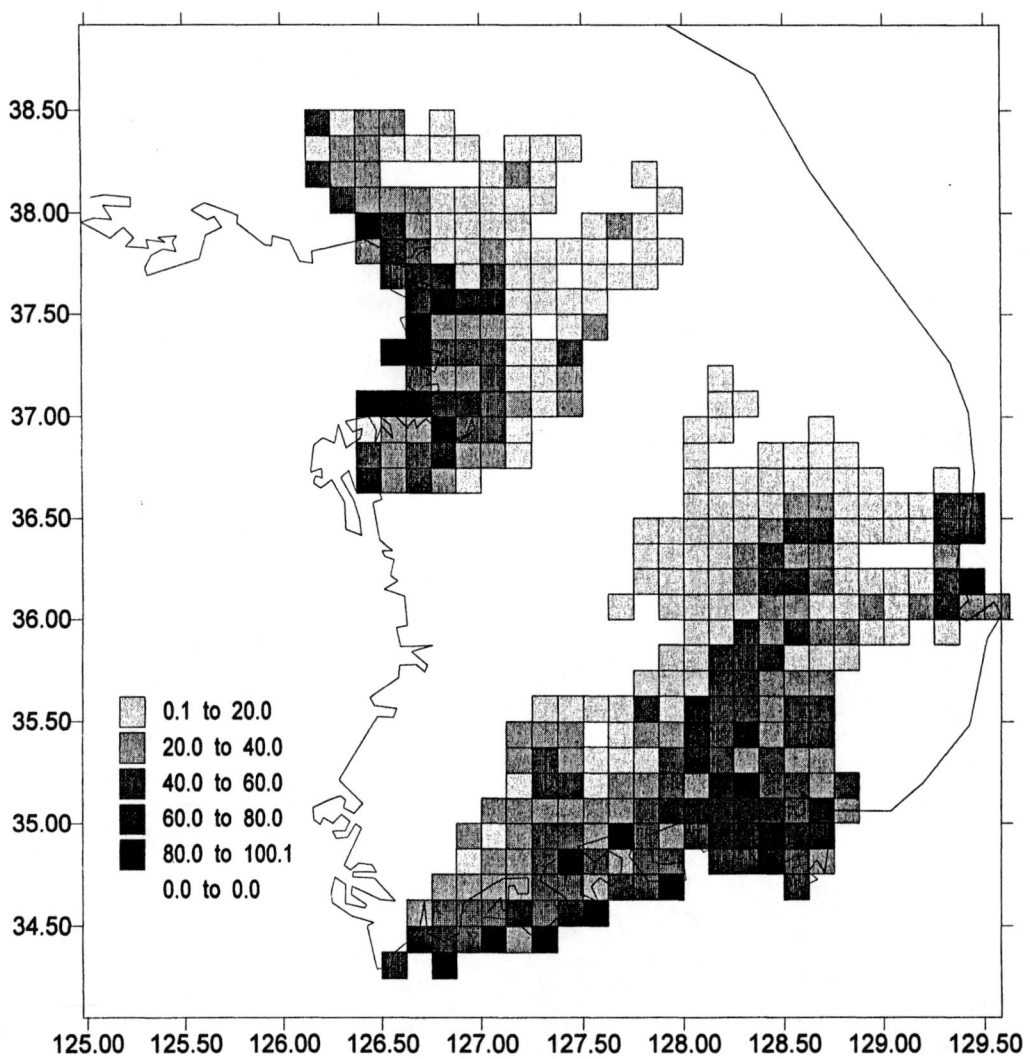

Figure 8. Exceedances of ground-ozone levels for the ecosystems of South Korea (AOT40 ppbh)

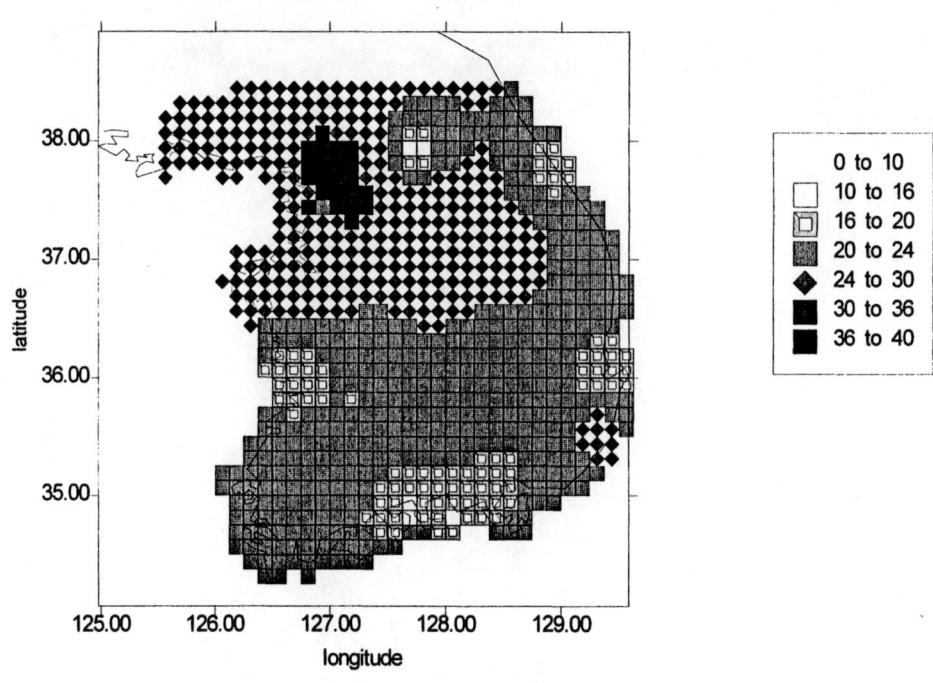

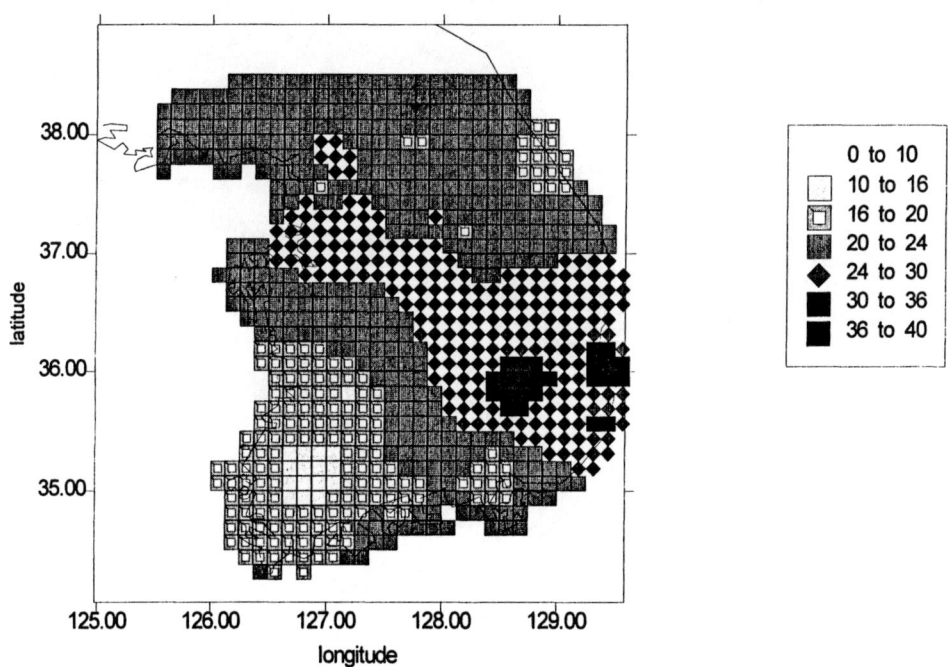

Figure 9. Concentrations of SO_2 and NO_x species in the atmosphere above South Korea in 1994, ppb (Park 1997)

Chapter 8
ACID DEPOSITION AND ECOSYSTEM SENSITIVITY IN CHINA

JIMING HAO, SHAODONG XIE, LEI DUAN, XUEMEI YE

Department of Environmental Sciences and Engineering, Tsinghua University,
Beijing, 100084, P. R. China

CONTENT

Abstract
1. Introduction
2. Current situation of acid rain in China
3. Current situation and characteristics of regional distribution of acid rain in China
    3.1 Geographical distribution of precipitation pH value
    3.2 Distribution of acid rain frequency
    3.3 Current situation and characteristics of the regional pollution caused by acid rain
4. Precipitation chemistry in China
    4.1 Ionic composition of precipitation in China
    4.2 Characteristics of precipitation chemistry in China
5. Essence of the regional distribution of precipitation acidity in China
6. Trend of acid rain in China
7. Effects of acid deposition on ecoenvironment
    7.1 Effects on crops
    7.2 Effects on forest
    7.3 Economic loss of crops and forest
8. Sensitivity of surface water to acid deposition in China
9. Sensitivity of ecosystems and critical loads of acid deposition in China
    9.1 Methods and data
        9.1.1 The modified semi-quantitative method
        9.1.2 Data
    9.2 Results
        9.2.1 Overview of soil distribution in China

9.2.2 Critical load and sensitivity class of each major soil in China
9.2.3 Mapping critical load of acid deposition in China
10. Control strategy of acid deposition in China
10.1 Designating the Acid Rain Control Areas and the Sulfur Dioxide Pollution Control Areas
10.1.1 Fundamentals of designating Control Areas
10.1.2 Principles of designating Control Areas
10.1.3 Designating results
10.2 Control objectives for the Acid Rain Control Areas and SO_2 Pollution Control Areas
10.3 Politics and measures
11. Brief summary
References

Abstract

The areas suffering from acid rain in China have extended northwards from the south of Yangtzee River in 1986 to the whole East China at present. The statistical results from the Acid Rain Survey in 82 cities from 1991 to 1995 indicate that the annual average pH value of the precipitation was lower than pH 5.6 in nearly half of these cities or in 87% of the southern cities which are located in the south to the Qingling Mountain and Huaihe River, and the lowest even reached pH 3.52 in Changsha, Hunan province. In addition, the frequency of acid rain was very high (higher than 60%) in one fourth of these cities. Up to date, acid deposition appears quite severe in wide areas southern to the Changjiang (Yangtze) River and eastern to the Qingzang (Tibet) Plateau, covering 12 provinces/autonomous regions, with the formation of four core zones, namely Chongqing-Guiyang area in the southwest, Changsha-Nanchang area in the south, the southeast coastal area and the area near Qingdao in Shandong province.

The chemical composition of acid rain in China is generally different from that in Europe, with the lower pH value and the higher sulfate, calcium and ammonium concentration. Another difference is that the concentration of calcium relative to sulfate is very high in China, while the nitrate concentration relative to other components is low. In some cases, the fluoride concentration in precipitation appears also high in China, owing possibly to the combustion of coal with high fluoride content. Besides, the alkaline fly ash and soil dust build the capacity for acid neutralizing during washout.

Based on the mineralogy controlling weathering and soil development, sensitivity of ecosystem to acid deposition was assessed with the comprehensive consideration on the effect of temperature, soil texture, land use and precipitation. The results show that the most sensitive area to acid deposition in China is podzolic soil zone in the Northeast, then followed by latosol, dark brown forest soil and black soil zones, the less sensitive area is ferralsol and yellow-brown earth zone in the Southeast, and the least sensitive areas are mainly referred to as xerosol zone in the Northwest, alpine soil zone in the Tibet Plateau, and dark loessial soil and chernozem zone in central China. The reason why the areas in Northeast China are the most sensitive to acid deposition and the ferralsol and yellow-brown earth areas in the south are less can be attributed to the obvious difference in temperature, humidity and soil texture between them.

In order to control acid rain and sulfur dioxide pollution, China will designate areas where acid rain or serious SO_2 pollution occur or may occur as the Acid Rain Control Areas and the Sulfur Dioxide Pollution Control Areas. Thermal power plants and enterprises within these areas must take measures to abate SO_2 emission.

Key words: Acid Deposition, Sensitivity, Forest Soil, Ecosystem, Acid Rain, Acidification.

1. INTRODUCTION

Accompanying with the increasing emission of sulfur dioxide in last decades, acid rain appeared in large areas in China. The continuous extension of acid rain areas has attracted close attention of all aspects. Acid rain monitoring network was first started in the west suburb of Beijing in 1974 and soon afterwards it extended to Shanghai, Nanjing, Chongqing, Guiyang, and other cities in succession. In November 1981, the First National Symposium on Acid Rain was convened in Beijing, which brought about a great advance in research works on acid rain. In 1982, the National Environmental Protection Agency (NEPA) organized and sponsored a National Survey of Acid Rain, in addition to local research projects in the southwest of China. Then comprehensive researches on the formation, effects and control strategies of acid rain were carried out. Based on the findings of these works, the current situation and the characteristics of acid rain in China were preliminarily described in 1985, which was regarded as the foundation of further researches on acid rain in China. From 1986 to 1990, research on acid rain became a national key project, the purpose of which was to make clear the formation, impacts, and trend of acid rain in China, and thereby to formulate befitting control strategies to ensure the healthy and successful development of the national economy. Many research institutes were organized and engaged in this multi-subject and comprehensive project carried out in the southwest and south of China. Later the research on acid rain was placed in the national key project again from 1991 to 1995. As well as to inquire the current state and the future trend of acid rain in China, the main purpose this time was to study the effects of acid deposition on forest, agriculture and soil, and to estimate the economic loss. Moreover, the laws of atmospheric transportation for acidifying pollutants in China was studied, the transprovincial and transboundary transportation and deposition were quantitatively estimated, and the mechanism of atmospheric acidification and acid deposition was expounded. Based on these studies, acid deposition control programming and strategies, which may be suited to the national conditions, were formulated.

2. CURRENT SITUATION OF ACID RAIN IN CHINA

The statistical results from the Acid Rain Survey in 82 cities in 1995 (CNEPA, 1996) indicate that the annual average pH value of precipitation was lower than 5.6 in 39 cities or in 47.6% of the cities, and between 4.5 and 5.0 in 18.3% of them. In addition, the annual average pH value of the precipitation in nearly 86.6% of the southern cities and in 26.5% of the northern ones was lower than pH 5.6. The northern cities with annual average pH value of precipitation lower than 5.6 included Qingdao, Tumen, Taiyuan, and Shijiazhuang, where the frequency of acid rain was 44.3%, 57.8%, 13.8%, and 30% respectively. Among all northern cities, the frequency of acid rain is highest in Qingdao and Tumen. Although the frequency was not high in Taiyuan, there was a great quantity of acidic precipitation at the end of the summer, so the annual average pH value of precipitation may be still low because the acidity was weighted by volume . Moreover, there were five cities with pH value of precipitation lower than 3.2, which were Changsha, Chongqing, Nanchong, Shijiazhuang, and Wuzhou. Table 1 shows the statistical results of acid rain pollution in China in 1995 and 1996.

Table 2 shows some statistical data from the Acid Rain Survey of City from 1991 to 1995. As can be seen from the table, the annual average pH value of precipitation ranged from 3.52 to 7.63 in 1995 in the cities, and the lowest occurred in Changsha. In addition, about 50% (range from 47.56% to 55.56%) of the monitored cities had the pH value of precipitation lower than 5.6.

Table 1 Statistical Results of Acid Rain Pollution in China in 1995 and 1996

Statistical Item	Cities	
	In 1995	In 1996
Annual Average pH Value of Precipitation Lower than 4.5	Changsha, Hengyang, Ganzhou, Xiamen	Changsha, Xiamen, Ganzhou, Yibin
Frequency of Acid Rain Higher than 90%	Changsha, Ganzhou, Yibin	Yibin, hengyang, Ganzhou, Changsha
Frequency of Acid Rain Higher than 80%	Huaihua, Leshan, Nanchang	Wuzhou, Xiamen, Huaihua, Nanchang, Tumen
Frequency of Acid Rain Higher than 70%	Hengyang, Nanchong, Xiamen, Wuzhou	Leshan, Guangzhou

Table 2 Statistical Results from Acid Rain Survey of City in 1991-1995

Year	Number of Cities	Range of Annual Average pH	Cities with pH < 5.6		Range of pH for Single Sample	Highest Frequency of Acid Rain (%)
			Percentage	Number		
1991	54	3.79-7.60	55.56	32	3.00-8.87	97.2
1992	57	3.85-7.43	52.63	30	3.01-11.12	98.9
1993	73	3.94-7.63	49.32	36	3.07-9.52	98.9
1994	77	3.84-7.54	48.05	37	2.50-9.24	94.3
1995	82	3.52-7.73	47.56	39	2.92-10.98	100
Average in Five Years		3.52-7.63	50.62	-	2.50-11.12	94.3-100.0
Middle of 1980's		4.42-7.78	25.7	-	-	-

In the rain water sampled during the five years, pH value ranged from 2.50 to 11.12. Changsha and Xiamen were two cities with precipitation pH value lower than 3.0 and with the highest frequency of acid rain (range from 94.3% to 100%). Among all monitored cities, the average frequency of acid rain in the five years was higher than 90% in five of them such as Changsha, Ganzhou, Yibin, Nanchong, and Xiamen, which are the core zones of two acid rain regions lying in the Central China and Southwest China respectively. Moreover, the frequency of acid rain was higher than 90% in Changsha in all five years, so it was in Ganzhou and Yibin in four years and three years, respectively, and the frequency was higher than 80% in the other years. In addition, the pollution was also severe in a quarter of the monitored cities with frequency of acid rain higher than 60%. However, acid rain was not very serious in half of cities in China because the frequency of acid rain was lower than 20% in about 50% of the cities. The percentage of cities with frequency of acid rain between 20% to 60% was quite stable, with a fluctuation no more than 5%.

Based on the analysis of precipitation pH in each year, conclusion can be drawn that the situation of acid rain pollution in China remains approximately unchanged on a large scale in recent years, with only a slight fluctuation in some areas. Although the pollution is slightly improved in some southern cities, the core zones of acid rain becomes much more obvious in South China. A joined region of acid rain does not occur yet in the north of China because the impacting areas of acid rain are quite small despite that the areas have increased on a local scale.

3. CURRENT SITAUTION AND CHARACTERITICS OF REGIONAL DISTRIBUTION OF ACID RAIN IN CHINA

3.1 GEOGRAPHICAL DISTRIBUTION OF PRECIPITATION PH VALUE

Figure 1 shows the geographical distribution of precipitation pH. It can be seen from the figure that acid rain has become a regional-scale environmental problem in China. The pH contour of 7.0 goes southwards from Heihe in Heilongjiang province to the sea west to the Yalujiang River, then winds its way southwestwards from Beijing to , through Houma, Baoji, Chengdu, Panzhihua, and other cities. The annual average pH value of precipitation is lower than 7.0 in almost 40% of the China's land. Furthermore, the pH value is found lower than 5.6 mainly in the Sichuan Basin and in wide areas south to the Changjiang (Yangtze) River and east to the Qingzang (Tibet) Plateau. In comparison with the geographical distribution of acid rain from 1986 to 1990, the main pattern formed in recent years remains unchanged on a large scale. However, it must be noted that acidic precipitation has appeared in North China near Houma, Beijing, Tianjin, Dandong, Tumen, and so on.

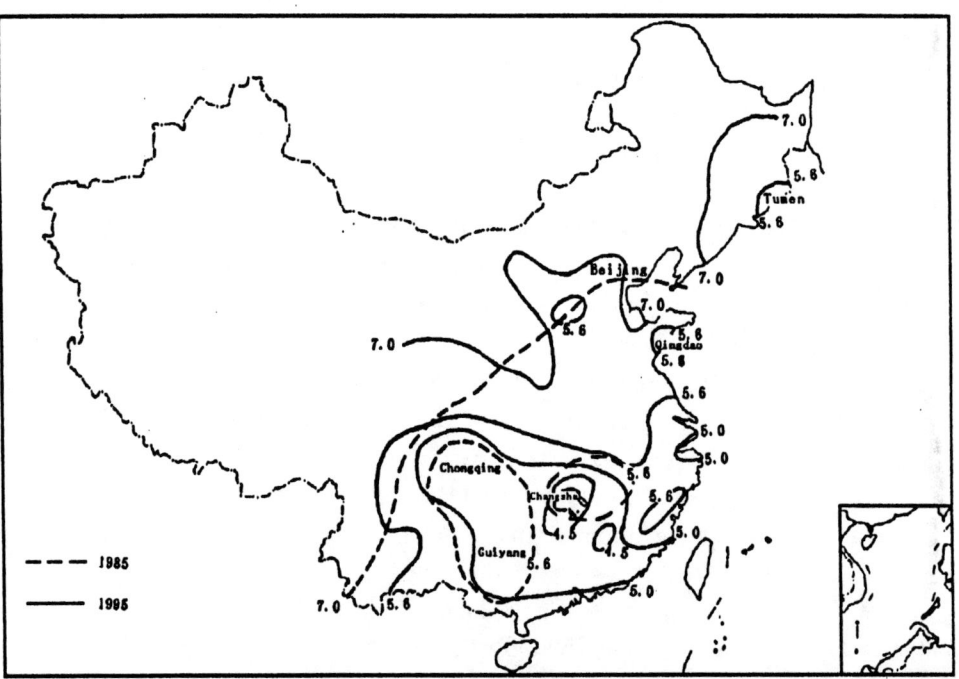

Figure 1. Contours of Precipitation pH in China.

As can be seen from Figure 1, the pH of precipitation is lower than 4.5 in most areas south to the Changjiang (Yangtze) River, which make up heavy polluted regions of acid rain. At present, the precipitation with pH ≤ 4.0 often occurs in Sichuan, Guizhou, Guangxi, Guangdong, Hunan, and Zhejiang province. The areas with the pH of precipitation between 4.0 to 4.5 begin to extend northwards to the north of the Changjiang (Yangtze) River, while the areas with precipitation pH between 4.5 to 5.6 have expanded to wide areas in the east, north, and northeast of China. Up to now, core zones of acid rain with severe pollution have already formed in the center, southwest, south and east of China respectively, and the acidity of precipitation is usually higher in urban areas or near cities than that in other areas. As the seasonal distribution, the acidity of precipitation is higher in the winter than that in the summer. The spatial distribution and the seasonal distribution of ambient SO_2 pollution is similar with that of acid rain.

3.2 DISTRIBUTION OF ACID RAIN FREQUENCY

The frequency of acidic precipitation is an important index to measure the level of pollution caused by acid rain. Figure 2 shows the geographical distribution of the acid rain frequency which is corresponding to the precipitation pH lower than 5.6. It is obvious that the frequency of precipitation pH lower than 5.6 varies in different areas. The frequency of precipitation pH ≤ 5.6 is higher than 50% in areas south to the Yangtze River, while it is between 20% to 50% in areas north to the Yangtze River and south to Baoshan, Huaiyin, Nanyang, Juezhou, Yanan and Chengdu, in Beijing-Tianjin area, and in Northeast China south to Jiamusi, Harbin and Dandong. Among eight provinces including Fujian, Jiangxi, Hunan, Zhejiang, Anhui, Jiangsu and Shandong, the frequency of precipitation pH ≤ 4.5 is lower than 6.5% only in Shandong and Jiangsu Province, and between 15% to 26% in the others. Accordingly, if acid rain defines as the precipitation with pH lower than or equal to 5.6, the acid rain regions have extended northwards, across not only the Yangtze River, but also the Yellow River, and have reached the Northeast China. In most area south to the Yangtze River, the pH value of precipitation is lower than 4.5, and the frequency of acidic precipitation exceeds 50%, with the highest value of 90%.

It can also be seen from Figure 2 that the frequency of acid rain decreases from south to north. Three areas with the highest frequency of acid rain (larger than 80%) in Sichuan-Guizhou, Hunan-Jiangxi, and Zhejiang respectively, belong to heavy acid rain regions. Whereas the areas south to the Yangtze River with the frequency of acid rain higher than 50% constitute the acid rain regions. The light acid rain regions, with the frequency of acid rain of 20%, consist of the Beijing-Tianjin area, the south and east of Hebei province, a large part of Shandong province, and the east of Northeast China including Jilin and Heilongjiang province.

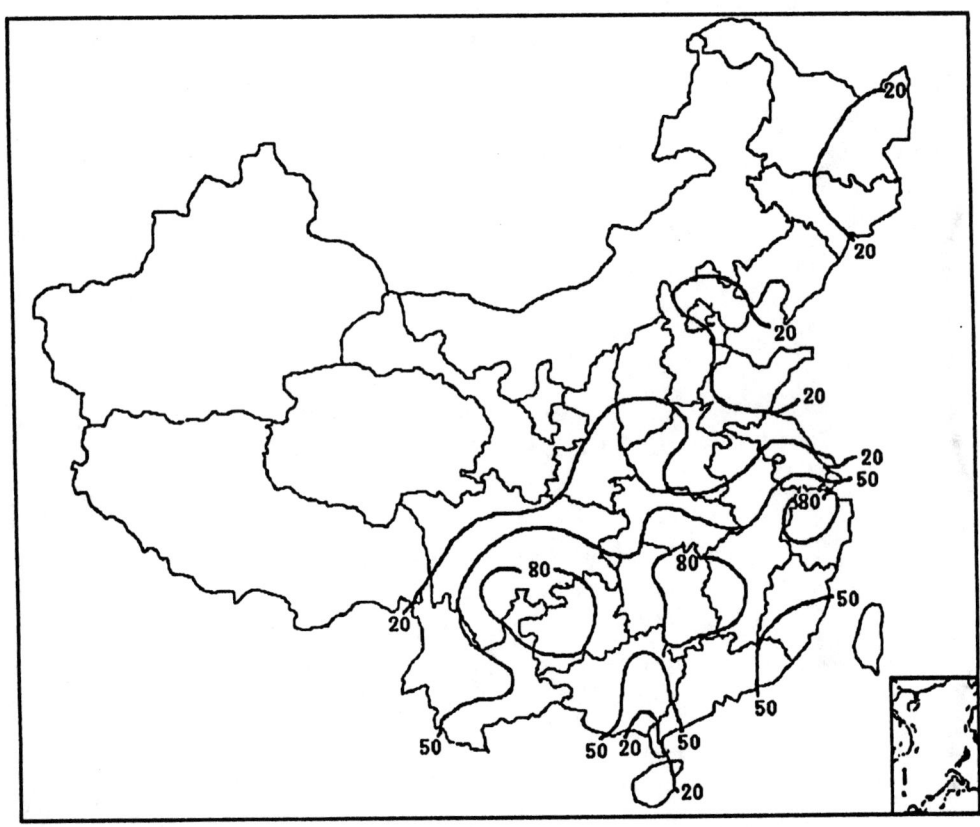

Figure 2. Contours of Acid Rain Frequency Corresponding to pH < 5.6

Table 3 lists the fractional frequency of acid rain in different pH range. As can be seen from the table, the frequency of acid rain with pH ≤ 5.6 is between 60% to 75% in Hubei, Fujian, Anhui, Jiangxi, and Zhejiang province, with the lowest in Shandong and Jiangsu province. The frequency of acid rain with the highest acidity in it, is the highest in Hubei, Anhui and Fujian province.

Table 3 Fractional Frequency of Acid Rain in Some Provinces (%)

pH Ranges	Provinces							
	Fujian	Jiangxi	Hunan	Zhejiang	Hubei	Anhui	Jiangsu	Shandong
Ph≤4.0	7.6	4.9	5.8	4.6	12.8	11.1	1.2	0.6
4.0<pH≤4.5	24.4	16.9	22.6	20.9	25.2	24.6	6.5	1.5
4.5<pH≤5.0	24.5	20.9	16.9	19.8	21.9	18.6	8.3	4.2
5.0<pH≤5.6	15.5	23.6	13.6	18.4	14.5	16.8	8.5	9.2
5.6<pH≤7.0	22.9	32.1	36.3	32.7	23.6	26.1	52.2	63.5
pH>7.0	5.2	1.6	4.9	3.6	2.0	2.8	23.3	21.0
Frequency of Acid Rain	71.9	66.3	58.9	63.7	74.4	71.1	24.5	15.6

3.3 CURRENT SITUATION AND CHARACTERISTICS OF THE REGIONAL POLLUTION CAUSED BY ACID RAIN

As mentioned above, four acid rain regions have formed in the southwest, center, east, and south of China, namely the Southwest Acid Rain Region, the central Acid Rain Region, the East Coastal Acid Rain Region, and the South Acid Rain Region, respectively. Table 4 shows the annual average pH of precipitation and the frequency of acid rain in typical cities in each acid rain region. As can be seen from the table together with the data on yearly changes of each acid rain region, the acid rain region in Central China is steadily deteriorated in pollution, and has been the most seriously polluted region, instead of the acid rain region in Southeast China. Although the pollution in the Southwest Acid Rain Region has been definitely improved, it is very serious anyhow.

The Central Acid Rain Region, with the highest acidity of precipitation and the highest frequency of acid rain, includes such cities as Changsha, Ganzhou, Hengyang, Huaihua, and Nanchang. The pH of acid rain is lower than 4.0 in the center of the region, and the frequency of acid rain is higher than 80%. At the edge of the region, the average pH is lower than 5.0, and the frequency decreases to 60%. According to the statistical results from the National Acid Rain Monitoring Network, the frequency of precipitation with pH ≤ 5.6 (i.e., the frequency of acid rain) exceeds 80% in Hunan and Jiangxi province, while the frequency of precipitation with pH ≤ 5.0 is about 60%.

Table 4 Statistical Results of Annual Average pH of Precipitation and Frequency of Acid Rain in Typical Cities in Each Acid Rain Region

Areas	Annual Average pH of Precipitation	Frequency of Acid Rain %
Qingdao (in North China)	4.5-5.0	40-45
Hangzhou (in East China)	4.8-5.0	65-75
Xiamen (in East China)	4.5	75-90
guangzhou (in South China)	4.3-4.7	70
Guilin (in South China)	4.8	60-70
Wuzhou (in South China)	4.0-4.6	80-95
Chongqing (in Southwest China)	4.5	70-80
Yibin (in Southwest China)	4.2-4.9	85-95
Zhunyi (in Southwest China)	4.3	90
Nanchang (in Central China)	4.6-4.9	70-85
Ganzhou (in Central China)	4.2-4.6	85-99
Changsha (in Central China)	3.5-4.0	90-100

The pollution of acid rain in the Southwest Acid Rain Region is more serious than that in other regions except the Central Acid Rain region. The Southwest Acid Rain Region consists of Nanchong, Yibin, Chongqing, Zhunyi, and other cities. The annual average pH of precipitation is lower than 5.0 and the frequency of acid rain is higher than 80% in the center of the region. The average pH value of rain samples gathered in the National Acid Rain Monitoring Network from 1991 to 1995, is lower than 5.0, and the frequency of precipitation with pH<5.0 is greater than 30%.

The East Coastal Acid Rain Region, including the lower reaches of the Yangtze River and the coastal areas north to Xiamen, is the widest one of the all, but the pollution in it is less serious than that in others. The annual average pH value of precipitation ranges from 4.5 to 5.5 in the key cities, and the frequency of acid rain is between 30% and 90%. Pollution of acid rain is serious in the coastal areas near Fuzhou and Xiamen, with high frequency of acid rain in the south of Fujian province around Xiamen. Typically, the annual average pH value of precipitation is about 4.5 in Xiamen, and the frequency of acid rain ranges from 75% to 90%. According to the monitoring results, the spatial and horaey distribution of the acidity of precipitation and the frequency of acid rain is quite uneven. The acidity of precipitation decreased in many cities of the region in 1995, but the frequency of acid rain became much lower than that from 1986 to 1990.

The South Acid Rain Region is located in the delta of the Zhujiang River and in the east part of Guangxi province. The annual average pH value of precipitation undulates between 4.5 and 5.0 in heavy polluted cities, and the frequency of acid rain ranges from 60% to 90% in the center of the region. Acid rain is also very common in the Guangxi Zhuang Autonomous Region, with the frequency higher than 30% except in the southern coastal areas. The acid rain region extended eastwards in last decades, along the Xianggui Corridor to the delta of the Zhujiang River. According to the statistical results from the National Acid Rain Monitoring Network, the annual average pH value is about 4.7 in Guangdong province and the Guangxi Zhuang Autonomous Region, the frequency of acid rain ranges from 55% to 60%, while the frequency of precipitation with pH ≤ 5.0 is nearly 40%. Now, the distribution of pH and frequency has become relatively stable, without any notable changes in recent years.

Besides the four acid rain regions, the annual average pH value of precipitation is lower than 5.6 in some cities in North China, including Qingdao, Tumen, Taiyuan, and Shijiazhuang. The annual average pH of precipitation in Qingdao kept below 5.6 in recent years, and continued decreasing until 1994. In 1995, the annual average pH was 5.01, a bit higher than that in 1994, but the frequency of acid rain went up to 44.3%. According to the yearly information on the precipitation chemistry in Qingdao, Shandong province, pollution of acid rain has certainly occurred, and tends to be more serious in the future. In 1995, the first report on acid rain monitoring in Taiyuan, Shanxi province show that the annual average pH of precipitation was below 5.0, and the frequency of acid rain was 30%. As can be seen from the relevant data in Shanxi province, the pH value of precipitation decreased in recent years, but the current situation and the future trend of acid rain need to be further studied. In addition, the average pH of precipitation in Tumen, Jilin province is lower than 5.6 in 1993 and in 1994, and it is lower than 5.0 in 1995. Simultaneously, the frequency of acid rain reached 57.8% in 1995. All the evidences indicate that acid rain region on a small scale have formed in the north of China.

As can be seen from Figure 1, the contour of annual average pH equal to 7.0 passes through the Northeast of Heilongjiang province, the center of Jilin province, the east of Liaoning province, the Laizhou Gulf and Zibo area in Shandong province, the north of Heibei province, the center and the south of the Nei Monggol (Inner Mongolia) Autonomous Region, the north of Shanxi province, the southwest of Shanxi province, the center of Shanxi province, the south of Gansu province, the east of Qinghai, and the east of the Xizang (Tibet) Autonomous Rigion. Northwest to the line is the region with annual average pH higher than 7.0. The region with annual average pH higher than 5.6 also includes some southern areas such as the Wuyi Mountain in the west of Fujian province, and some areas in the west of Yunnan and Guizhou province.

4. Precipitation Chemistry in China

The acidity of precipitation depends on the species and the concentrations of ions other than H^+ in the rain water. Therefore, it is very important to study the precipitation chemistry for knowing the sources and the formation mechanism of acidic precipitation. In addition, the knowledge of chemical composition of precipitation is also the foundation of the research works on acid rain such as the regional transportation, the impacts on agriculture, forest and materials, the acidification of soil and water, and the control strategies. The chemical composition of precipitation in China is described in this chapter.

4.1 Ionic Composition of Precipitation in China

The acidity of precipitation caused by acidifying pollutants dissolving in it can be neutralized by the particulate in the atmosphere. In China, the pollution caused by particulate is so serious that the pollution of acid rain is somewhat concealed by it. If the particulate content in the atmosphere decreases, the phenomena of acid rain will soon be more obvious. So the intensity and the potential of the acidification is determined by the chemical composition of precipitation.

Table 5 shows some results of ionic analysis of rain water in 8 provinces. It can be seen from the table that the major ions in the precipitation are in well balance, because the total concentration of the cations is approximately equal to that of the anions, with the deviation less than 20%. The reason, why the concentration of the cations and the anions is always in a ratio higher than 1, may be the omission of some organic acids in precipitation in damp and hot areas in South China and the carbonate in dry areas in North China. In general, the total concentration of ions increases gradually from the south to the north. Among the 8 provinces involved, the concentration is the highest in Shandong province, then followed by Jiangsu. It is the lowest in Fujian, and almost the same in the others.

According to the Table 5, the annual average concentration of H^+ in precipitation ranged from $0.77 \mu eqL^{-1}$ (pH = 6.1) in Shandong province to $33.99 \mu eqL^{-1}$ (pH = 4.77) in Hubei province, and the average value of the eight provinces was $18.1 \mu eqL^{-1}$ (pH = 4.74) in 1992-1993.

In comparison with the chemical composition of precipitation in Europe, North America, and Japan, the concentration of sulfate in precipitation in China is nearly three times as much as that in the other countries. In 1992-1993, the sulfate concentration in precipitation was higher than $160 \mu eqL^{-1}$ in Jiangsu and Shandong province and ranged from $100 \mu eq L^{-1}$ to $130 \mu eqL^{-1}$ in the other provinces, with the average of $125.86 \mu eqL^{-1}$. On the contrary, the nitrate concentration in China is comparative with that in Europe, North America, and Japan. It ranged from $14 \mu eqL^{-1}$ to $23 \mu eqL^{-1}$, with little difference in each province. The average concentration of nitrate in precipitation in the eight provinces was $19.62 \mu eqL^{-1}$.

As can be seen from the table V, the chloride concentration increased as the latitude became high, with the value of $91.38 \mu eqL^{-1}$ in Shandong province and between $14 \mu eqL^{-1}$

and 25µeqL^{-1} in the other provinces. Simultaneously, the concentration of fluoride varied largely from 1.02µeqL^{-1} to 21.45µeqL^{-1}. The average concentration of fluoride in precipitation appeared also high in China, owing probably to the combustion of coal with high fluorine content.

Ammonium and calcium are two major cations in precipitation in China. The average concentration of ammonium and calcium was 67.89µeqL^{-1} an 80.4µeqL^{-1} respectively, which is about twice to four times as much as that in Europe, North America, and Japan.

Table 5. Volume Weighted Annual Average Concentrations (in µeqL^{-1}) of Ions in Precipitation in China (1992-1993) *

Items	Provinces								Average
	Fujian	Jiangxi	Hunan	Zhejiang	Hubei	Anhui	Jiangsu	Shandong	
Precipitation (mm)	1379.3	1554.6	1274.0	1550.1	1108.2	1019.8	1212.1	596.8	1211.9
H$^+$	33.15	32.42	16.61	20.54	33.99	4.30	3.29	0.77	18.13
NH$_3^+$	70.09	51.25	81.74	68.47	100.47	58.54	60.17	54.42	67.89
Ca^{2+}	53.30	64.46	62.98	49.29	64.82	65.17	116.69	167.55	80.41
Mg^{2+}	7.26	11.47	10.12	9.57	9.54	8.58	14.02	52.93	15.44
Na$^+$	14.76	22.54	8.85	24.51	8.86	18.60	29.58	41.41	21.14
K$^+$	5.07	10.42	7.77	8.37	7.30	10.81	8.89	27.98	10.83
SO$_4^{2-}$	104.46	100.33	128.30	109.71	129.46	106.79	166.30	161.51	125.86
NO$_3^-$	14.02	19.62	18.07	17.98	22.40	21.37	20.88	22.64	19.62
Cl$^-$	19.25	14.87	16.33	20.31	16.36	24.43	27.95	91.38	28.86
F$^-$	10.96	16.44	11.67	8.26	3.24	1.02	10.67	21.45	10.46
Σ+	182.63	192.55	188.07	180.75	224.98	166.00	232.65	343.05	213.84
Σ-	148.69	151.26	174.37	156.25	171.46	153.62	225.81	296.97	184.80
Σ+/Σ-	1.23	1.27	1.08	1.16	1.31	1.08	1.03	1.16	1.17

* Data from Wang and Ding (1997)

In summary, the concentration of ions in precipitation is about three times higher than that in Europe and North America, which indicates that air pollution has become a very serious problem in China.

Figure 3 shows the average percentage of each ion in precipitation in China. As can be seen, the percentage of sulfate is the highest of all, then followed by calcium. Chloride and nitrate are another two anions with large fraction, and the percentage of ammonium and sodium is also very high. The percentage of such ions as sulfate, nitrate, ammonium and calcium, which have something to do with the anthropogenic emissions, is 30.9%, 5.0%, 17.2%, and 20.4% respectively. The concentration of sulfate and nitrate in precipitation is in an average ratio of 6.2, which is about twice as much as that in Europe, North America, and Japan. The ratio may be much higher in some southern areas where coal with high sulfur content is also used. However, in some cities with large number of vehicles, the ratio is of course to be a bit lower.

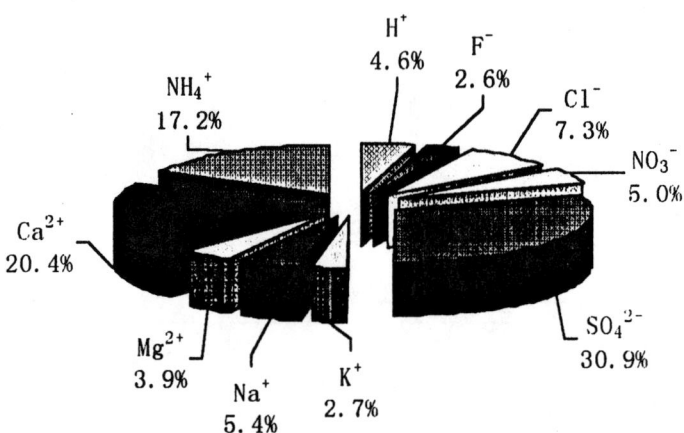

Figure 3. Percentage of Ions in Precipitation in China in 1993.

4.2 CHARACTERISTICS OF PRECIPITATION CHEMISTRY IN CHINA

The ionic composition of precipitation is an indicator of air pollution. Specially, the acidity of precipitation may be lower as the concentration of sulfate and nitrate decreases. In order to learn the trend of acidic precipitation, the ratio of the concentration of the major acidic ions to that of basic ions is studied instead of the absolute concentration of sulfate or nitrate. In general, the major acidic ions are sulfate and nitrate in precipitation and the major basic ions are calcium and ammonium. The former is mainly emitted from anthropogenic sources, but the later may come mostly from natural sources such as soil and the decomposition of organism. Table VI shows the concentration of each ion in precipitation in seventeen cities of China. With further analysis of these data, we can see that the ratio of (sulfate + nitrate)/(ammnium + calcium) in precipitation is well interrelated with the pH value, as shown in figure 4. Therefore, when the precipitation chemistry of typical cities in China was studied, the ratio (sulfate + nitrate)/(ammonium + calcium) was used as an index of precipitation acidification. The difference of precipitation chemistry in 1993 from that in 1986 is shown in table VII (for northern cities) and Table VIII (for southern cities).

As can be seen from Table 7 and Table 8, the ratio sulfate/nitrate in precipitation decreased evidently, wherever in northern cities (from 6.1 to 4.9) or in southern cities (from 7.6 to 5.9), which means that the ratio SO_2/NO_x in the atmosphere also decreased, with the relative increase of NO_x concentration. Accordingly, the type of pollution will probably be changed in China, because the nitrogen compounds, instead of the sulfur compounds, will gradually become the major acidifying pollutant in the atmosphere. In addition, the ratio of ammonium/calcium rose from 0.43 to 0.55 in northern cities and from 1.2 to 1.5 in southern cities, with the increase of ammonium and the decrease of calcium. The reason why the changing trend of ammonium differs from that of calcium may be the difference of the sources. Since calcium is mostly originated from soil dust and particulate emitted by combustion and industry, the effective control of particulate emission may result in the decrease of particulate content in the atmosphere. On the contrary, the emission of ammonium increases every year as a result of the incremental use of fertilizer.

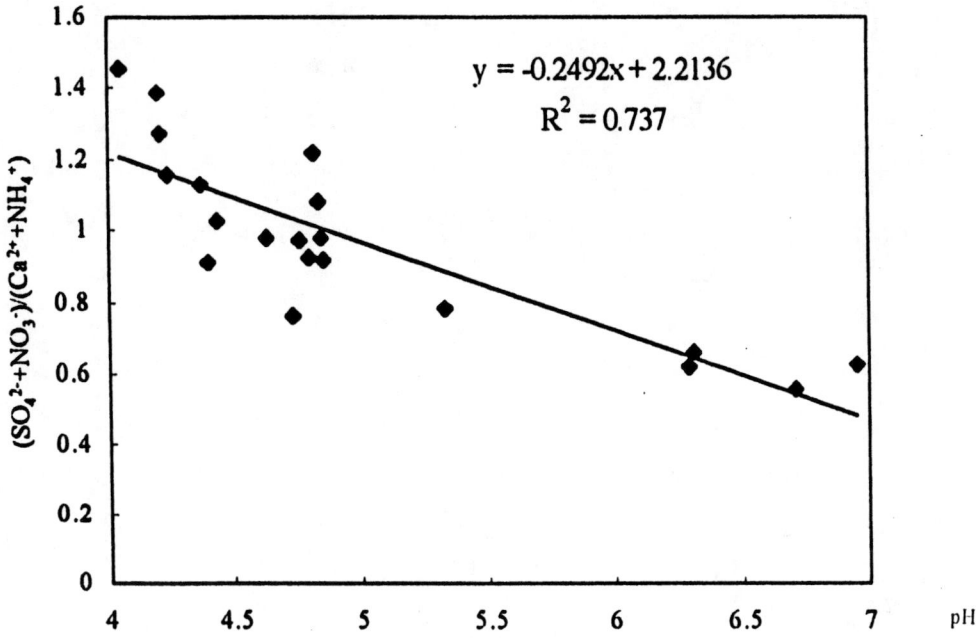

Figure 4. Relationship between pH and Ion Concentration in Precipitation.

According to the tables, the average of (sulfate + nitrate)/(ammonium + calcium) increased from 0.67 in 1986 to 0.75 in 1993 in northern cities, and changed from 0.86 to 0.91 in southern ones. It is clear that precipitation wherever in South China or in North China continues to be acidified. However, the ratio of sulfate/nitrate became lower in 1993 than that in 1986, which indicates that nitrate contributed more and more to the acidity of precipitation in China. The ratio changed just because of the development of desulfurization in coal combustion and the incremental use of oil instead of coal.

Table 6 Concentration of Ions in Precipitation in 17 Cities of China (μeqL^{-1})
(Wang et al, 1993)

Cities	Ca^{2+}	NH_4^+	SO_4^{2-}	NO_3^-	pH	$(SO_4^{2-}+NO_3^-)/(Ca^{2+}+NH_4^+)$
Beijing	151.6	162.8	154.5	39.5	6.29	0.617
Changchun	256.5	61.3	156.5	21.2	6.71	0.56
Jinzhou	340.8	123.8	259.2	49.4	6.31	0.66
Yantai	289.1	39.1	182.5	22.8	6.95	0.626
Pingdingshan	107.7	138.3	152.3	0.4	6.29	0.621
Hefei	110.3	117.3	141.9	31.8	4.73	0.763
Suzhou	125.3	93.6	200.2	14.4	4.63	0.98
Shanghai	104.3	75.3	153.4	12.6	4.85	0.92
Hangzhou	59.9	68.2	112.3	13.5	4.84	0.982
Nanning	26.6	27.7	61.6	4.9	4.82	1.22
Nanning (1988)	131.8	84.9	197.0	14.4	4.76	0.976
Nanning (1989)	150.4	130.9	243.9	17.0	4.80	0.927
Guilin	67.2	50.0	107.2	19.7	4.83	1.08
Chongqing	127.8	151.1	326.6	27.9	4.21	1.27
Guiyang	199.6	174.3	405.2	27.9	4.23	1.16
Maanshan	123.0	73.7	139.2	15.1	5.33	0.784
Guangzhou	175.1	141.1	254.9	33.3	4.39	0.91
Wuzhou (1988)	32.5	46.1	73.7	7.3	4.43	1.03
Wuzhou (1989)	52.6	67.0	141.6	24.9	4.20	1.39
Hengyang (1988)	41.5	55.6	97.3	12.7	4.37	1.13
Hengyang (1989)	46.5	66.1	153.8	10.8	4.05	1.46

Note: other data in 1986

Table 7. Change in Precipitation Chemistry of Typical Cities in North China (µeq L^{-1})

Cities	SO_4^{2-}		NO_3^-		SO_4^{2-}/NO_3^-		NH_4^+	
	1986	1993	1986	1993	1986	1993	1986	1993
Beijing	154.5	151.9	39.5	49.8	3.91	3.05	162.8	113.6
Changchun	156.5	80.9	21.2	38.6	7.38	2.31	61.3	123.6
Shenyang	398.0	626.1	50.3	63.8	7.96	9.81	99.0	83.4
Xian	358.1	377.2	67.3	84.9	5.32	4.44	275.8	176.5
Average	256.8	309.0	44.6	59.3	6.12	4.90	149.7	124.3

Cities	Ca^{2+}		NH_4^+/Ca^{2+}		$(SO_4^{2-}+NO_3^-)/(NH_4^++Ca^{2+})$	
	1986	1993	1986	1993	1986	1993
Beijing	151.6	143.2	1.07	0.79	0.62	0.79
Changchun	256.5	140.4	0.23	0.88	0.56	0.45
Shenyang	305.4	522.6	0.32	0.16	1.11	1.14
Xian	444.8	512.2	0.15	0.35	0.62	0.62
Average	289.5	329.6	0.43	0.55	0.67	0.75

Table 8. Change in Precipitation Chemistry of Typical Cities in South China (µeq L^{-1})

Cities	SO_4^{2-}		NO_3^-		SO_4^{2-}/NO_3^-		NH_4^+	
	1986	1993	1986	1993	1986	1993	1986	1993
Nanjing	144.4	117.7	28.6	30.2	5.00	3.89	104.6	83.2
Hefei	141.9	106.7	31.8	22.4	4.46	4.76	117.3	68.2
Hangzhou	12.3	146.6	13.5	23.1	8.32	6.35	68.2	133.3
Ningbo	169.6	114.6	25.1	16.6	6.76	6.90	74.1	72.8
Wenzhou	48.5	102.5	5.1	17.4	9.50	5.90	85.2	102.8
Chongqing	326.6	239.5	27.9	30.8	11.70	7.78	151.1	150.4
Average	157.2	137.9	22.0	23.4	7.62	5.93	100.0	101.2

Cities	Ca^{2+}		NH_4^+/Ca^{2+}		$(SO_4^{2-}+NO_3^-)/(NH_4^++Ca^{2+})$	
	1986	1993	1986	1993	1986	1993
Nanjing	137.9	102.6	0.76	0.81	0.71	0.80
Hefei	110.3	80.1	1.06	0.83	0.76	0.88
Hangzhou	59.9	94.2	1.14	1.42	0.89	0.76
Ningbo	91.0	42.5	0.81	1.71	1.18	1.14
Wenzhou	36.0	41.0	2.37	2.51	0.60	0.83
Chongqing	199.6	102.5	0.76	1.48	1.01	1.07
Average	105.8	77.2	1.15	1.46	0.86	0.91

5. ESSENCE OF THE REGIONAL DISTRIBUTION OF PRECIPITATION ACIDITY IN CHINA

In China, the major cause of the acidic precipitation may be the emission of sulfur dioxide. Since the sulfur emitted into the atmosphere is mostly from the combustion of coal, the acid rain in China is often called coal-smoke acid rain. According to Wang *et al* (1993), the emission intensity of acidity precursors (SO_2 and NO_x) is the strongest in coastal areas facing the Bo Sea, the Yellow Sea, and the East Sea. Although the emission intensity of SO_2 is more than 7tkm^{-2} in such provinces/municipalities as Jiangsu, Shandong, Liaoning, Hebei, Shanxi, Beijing, and Tianjin, regional distributed acid rain has not yet been observed there. However, the intensity of SO_2 emission is lower than 2.6tkm^{-2} in Fujian, Jiangxi, Guangdong, and Guangxi, where acid rain is quite serious. Even in areas with the most severe pollution in Hunan, Sichuan, and Guizhou province, the emission intensity is no more than 5tkm^{-2}. It is very clear that the formation of acid rain regions depends on a good many natural and anthropogenic factors, and must be the comprehensive effect of them. So acid rain region has not formed in the north of China with high emission intensity of acidity precursors, but appeared in the south.

There are many natural factors, which can affect the acidity of precipitation. The particulate content in the atmosphere and the diffusion capability of the atmosphere are of great importance. Since the atmospheric particulate in the mainland comes mostly from soil (specially 40-60% in North China), the effect of atmospheric particulate on precipitation is actually the effect of soil. Therefore, the acidity of precipitation is directly influenced by the chemical composition and the acidity of soil. Figure 5 shows the geographical distribution of soil pH in China. As can be seen, the pH of soil is low in wide areas south to the Changjiang (Yangtze) River, where acid rain region has formed. The pH value of soil ranges from 5 to 6 in areas north to the Changjiang (Yangtze) River and south to the Qinling Mountain and the Huaihe River, and it is higher than 6 in wide areas in North China and Northwest China. There are also some areas with soil pH around 6 in Northeast China such as the east and north of Heilongjiang province and the east of Liaoning province. It is clear that the geographical distribution of precipitation acidity coincides with that of soil acidity. In addition, according to the distribution of calcium content in soil, the calcium content in soil is approximately 0.5% south to the Changjiang (Yangtze) river, and only 0.1% in South China. As contrast, the calcium content is much higher in soil north to the Changjiang (Yangtze) River, especially it reaches 8% in the northwest of China. However, it must be noted that the calcium content is not very high (no more than 1.0%) in most areas in Northeast China. It is obvious that the acid rain region approximately coincides with the region of low soil pH and the region of low soil basic saturation in China. So the physicochemical properties of soil should be regarded as an important factor that affects the formation of acid rain.

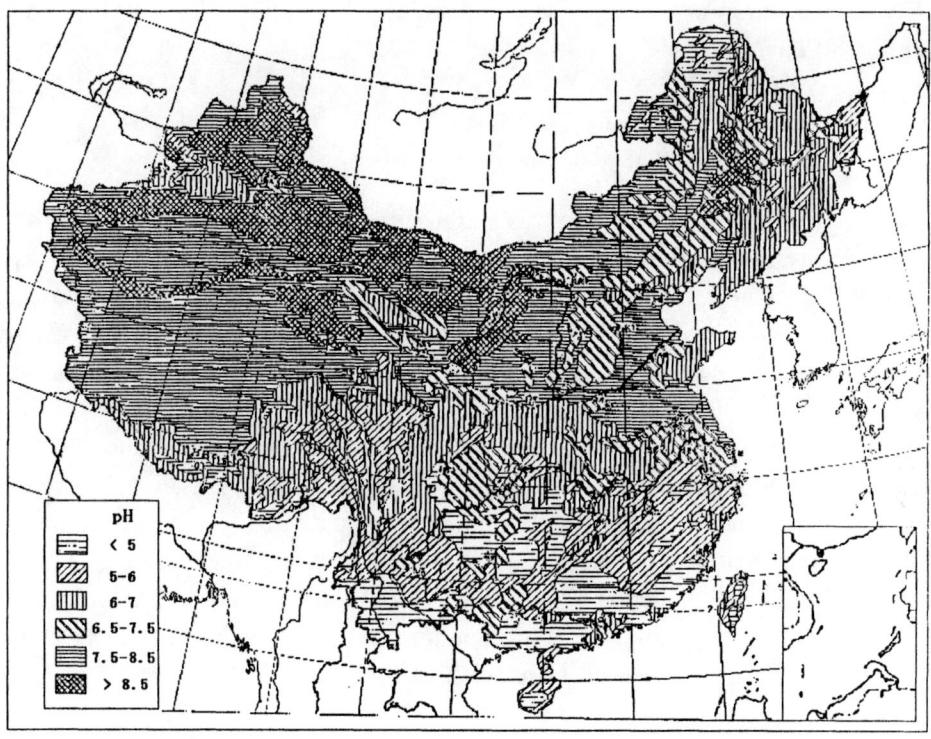

Figure 5. Geographical Distribution of Soil pH in China (Xiong Yi, et al, 1987)

The diffusion capability of the atmosphere is another factor influencing the acidity of precipitation. The accumulation of pollutants in the atmosphere depends directly on the diffusion capability. The strong diffusion region includes the Nei Monggol (Inner Mongolia) Autonomous Region, Hebei province, and Northeast China, while the weak diffusion region consists of Guizhou province, the Guangxi Zhuang Autonomous Region, Hunan province, the east of Sichuan province, the south of Jiangxi province, the west of Hubei province, the north of Guangdong province, and the middle of Yunnan province, where the heavy acid rain region is located (Wang, 1994).

Whether acid rain occurs on a small scale is directly relative to the local conditions. For example, sulfur dioxide is emitted in a great deal in Chongqing and Guiyang, where the sulfur content in combusted coal is very high (higher than 5%). Furthermore, the meteorological and topographical conditions are disadvantageous for the diffusion of pollutants, so sulfur dioxide and sulfate always deposits in the region. Moreover, the soil there is mostly acidic, thus the precipitation is prone to be acidified for the lack of basic particulate to neutralize the acidity in the atmosphere, and the soil can hardly resist acid deposition.

In summary, the regional distribution of precipitation acidity may basically rely on the chemical composition of soil and the geographical distribution of atmospheric diffusion capability in China.

6. TREND OF ACID RAIN IN CHINA

The history of acid rain in China is summarized in Table 9. In the 1980's, acid rain occurred mainly in the southwest of China such as in Chongqing, Guiyang, and Liuzhou, and the area of acid rain region was about 1.7 million km^2. Until 1990's, the acid rain region had extended to the Sichuan Basin and wide areas south to the Changjiang (Yangtze) River and east of the Qingzang (Tibet) Plateau, with total area more than 10 million km^2. Now, the Central Acid Rain Region, typified by Changsha, Ganzhou, Nanchang, and Huaihua, is the heaviest polluted area in China. The annual average pH value of precipitation in the center of this region is lower than 4.0, and the frequency of acid rain is higher than 90%. On other words, whenever it rains, the rain water ought to be acidic. The East Coastal Acid Rain Region is another major acid rain region in China, typified by Nanjing, Shanghai, Hangzhou, Fuzhou, and Xiamen. Moreover, acidic precipitation often appears in the Beijing-Tianjin area in North China, and the Dangdong-Tumen area in Northeast China. It is obvious that the developing rate of acid rain region is very quick in China, and the pollution will be more and more serious in the future.

Table 9. Development of Acid Rain Regions in China

Year	Distribution of Acid Rain
1983	Pollution Occurred in Cities, Most Serious in Chongqing, Nanchang, Guiyang, Nanning, and Hangzhou
1985	Acid Rain Region formed: the Southwest Acid Rain Region including Chongqing, Guiyang, Zigong, Liuzhou, and Nanning; the Central Acid Rain Region consisting of Changsha, Huangshi, and Zhuzhou; the Fujian Coastal Acid Rain Region containing Fuzhou and Xiamen
1990	the Southwest Acid Rain Region, the South Acid Rain Region, and other acid rain regions
1992	the Central Acid Rain Region extending eastwards to Nanchang; the East Coastal Acid Rain Region replacing the Fujian Coastal Acid Rain Region and extending northward to Qingdao; the South Acid Rain Region
1995	the Central Acid Rain Region, the Southwest Acid Rain Region, the East Coastal Acid Rain Region, and the South Acid Rain Region

By comparing the distribution of precipitation acidity in 1995 to that in 1985 (see Figure 1), we can see that the acid rain regions have extended markedly in last decade. The areas with pH lower than 5.6 have expanded from small areas in the southwest and center of China in the past to wide areas south to the Changjiang (Yangtze) River and east of the Qingzang (Tibet) Plateau, covering more than ten provinces/autonomous regions. Moreover, the areas with annual average pH of precipitation lower than 4.5 contained only scattered urban areas such as Chongqing, Changsha, Guiyang, Liuzhou, and Nanchang in 1985, and in 1995 they contained already most areas in Southwest China, and some areas in Central China and South China such as Sichuan province, the east of Yunnan province, Guizhou province, the north of the Guangxi Zhuang Autonomous Region, Hunan province, and the north of Guangdong province.

Following the fast development of China's economy, a acid rain region with the widest area and the strongest intensity of pollution has formed in Central China, where the frequency of acid rain keeps increasing and the annual average pH of precipitation decreases every year. Although the pollution in the Northwest Acid rain Region has somewhat been mitigated in recent years, especially the annual average pH of precipitation in key cities is a bit higher than before and the frequency of acid rain is somewhat lower, acid rain is still a serious problem in the region. At the same time, the South Acid Rain Region keeps almost unchanged, with heavy pollution, too. Moreover, it must be noted that the acidity of precipitation and the frequency of acid rain is very high in some northern cities such as Qingdao and Tumen. There are some cities in the north of China with pH of precipitation close to or lower than 5.6, but the frequency of acid rain there is not more than 30%. The data now available in these areas is not sufficient for determining the level of pollution caused by acid rain.

In China, the acidic precipitation is mainly contributed by the emission of sulfur dioxide. The high concentration of sulfate in precipitation is coincide with the fuel structure of China dominated by coal. Along with the high-speed development of economy, the consumption of coal and oil will increase, and the emission of SO_2 and NO_x will keep growing too. Consequently, the pH of precipitation in some cities will probably continue going down in the near future. If the techniques of pre-combustion desulfurization and fuel gas desulfurization are widely applied in China, the geographical distribution of SO_2 concentration in the atmosphere and the emission intensity of SO_2 in the future may maintain as the state quo. However, the NO_x concentration in the atmosphere will increase for the incremental combustion of oil. Simultaneously, the particulate concentration in the atmosphere will decrease because of the effective measures of particulate control and the increase of plant cover. Under these conditions, the acidity of precipitation in the future is predicted, as shown in figure 6 (Li and Tang,1998). It can be seen that the area of acid rain region and the heavy acid rain region will increase gradually.

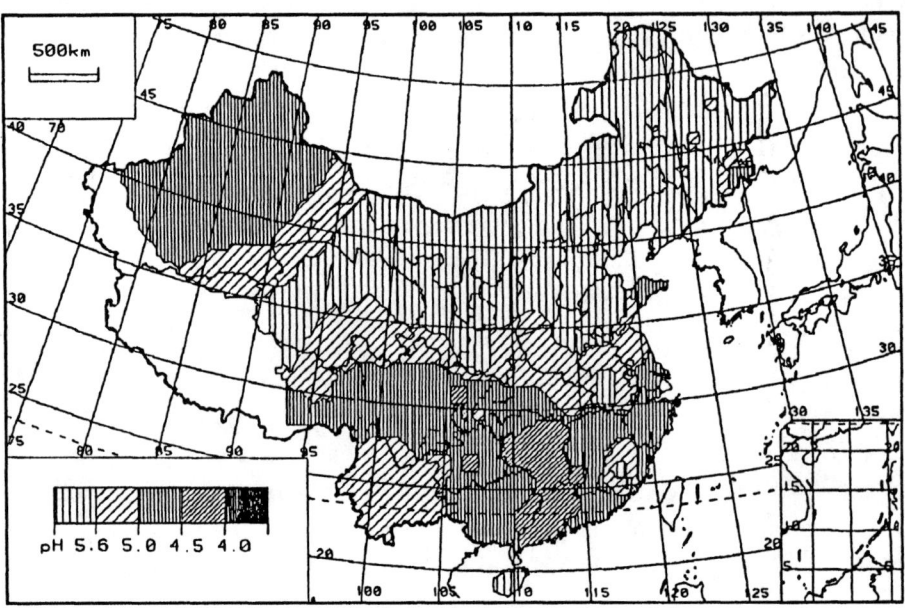

Figure 6. Prediction Result of Precipitation Acidity in China

According to the analysis above and the preliminary prediction of the developing trend of acid rain in the last decade, it is predicted that the area of acid rain region will continue increasing, and the acid rain region will extend northwards and westwards, with continuous increase of precipitation acidity. Especially, heave acid rain with precipitation pH lower than 4.0 will occur more frequently in wider areas south to the Changjiang (Yangtze) River. Consequently, the ecoenvironment and materials will be more seriously damaged by acid rain.

7. EFFECTS OF ACID DEPOSITION ON ECOENVIRONMENT

7.1 EFFECTS ON CROPS

The effect of acid deposition on crops is often shown as the visible damage during a short-term exposure, the disturbance of the physiological and metabolic processes, and the impacts on the yield and the quality. When crops come into contact with precipitation with high acidity, the foliage may be visibly injured in short times (from 24 hours to 72 hours), with the typical symptom of withered, dot-shaped, little, white spots. Generally, the pH value at which 5% of the foliage is visibly damaged, is considered to be the criterion of visible damage. The research results of the criteria for short-term damage are shown in Table 10. As it can be seen, the critical pH is 3.0 for sensitive crops such as wheat, barley and spinach, 2.5 for middle species such as soybean, kidney bean, and cotton, and 2.0 for resistant ones such as rice.

If the pH value at which the yield of crops is lost by 5% is adopted as the criterion of long-term effect, the experimental results for the most common crops in China are shown in Table XI. The critical pH for vegeTable 1s higher than 5.0, that for rice and peanut is lower than 2.8, and that for the others ranges from 4.0 to 5.0.

Table 10. Criteria of Short-term Damage by Simulative Acid Rain for 10 Crops

Crops	Wheat	Corn	Cabbage	Soybean	Rice	Kidney Bean	Rape	Cucumber	Tomato	Barley	Cotton	Spinach
Criteria (pH)	3.0	3.0	2.5	2.5	2.0	2.5	2.5	2.5	2.5	3.0	2.5	3.0

Table 11. Long-term Criteria of Simulative Acid Rain for Crops

Crops	Rape	Kidney Bean	Carrot	Tomato	Soybean	Cotton	Barley	Wheat	Cabbage	Corn	Rice	Peanut
Criteria (pH)	5.01	5.24	5.04	4.36	4.49	4.37	4.44	4.59	5.01	4.1	<2.8	<2.8

Actually, plants often grow where acid rain and ambient SO_2 pollution occur simultaneously. In China, areas with acid rain are always regions with serious SO_2 pollution. Researches indicate that the compound pollution of acid rain and ambient SO_2 concentration may aggravate the reduction of yield. The long-term criteria of the compound pollution are shown in Table 12. The investigated crops have decreasing sensitivity to acid deposition as follows: vegetables > industrial crops > grains, and the ranking of the sensitivity for vegetables is: root > leaf > fruit.

Table 12. Criteria of Multiple pollution for Crops

Crops		Vegetable	Soybean	Cotton	Wheat
Criteria	pH of Precipitation	4.64-4.79	4.63	4.54	4.46
	SO_2 (mg m^{-3})	0.114-0.120	0.176	0.15	0.166

Generally, the acid rain and SO_2 pollution affect the agriculture together. In seven provinces in the east of China, SO_2 is the major factor for the reduction of yield, i.e., the major pollutant that cause the reduction of agriculture. Acid deposition do not affect the yield of rice which is widely cultivated in China, but wheat, soybean, cotton, and vegetables are generally influenced by acid deposition. Especially, vegetables are the most sensitive of all common crops, and the reduction of yield caused by acid deposition is the most of all, too.

According to the statistical results, the yield of almost 19% of the seeded land, i.e. 992 hectares of field is reduced because of the acid deposition. The annual reduction in the seven provinces is 562 tons, being about 4.34% of the total yield. On other words, acid deposition causes the average reduction of 4.34% in China. Among these yield-reduced fields, the area of wheat is the largest, then followed by vegetables. The area is 457ha, 233ha, 159ha, and 143ha for wheat, vegetables, soybean, and cotton, respectively, and the percentage is 46%, 24%, 16%, and 14%. The average reduction rate of vegetables is the largest, then followed wheat. The reduction rate is 7.77% and 5.41% respectively.

7.2 EFFECTS ON FOREST

Since 1980's, forest decline has been observed in large areas of China, involving the areas of firs on the Emei Mountain in Sichuan province, the masson pines (*Pinus massoniana*) on the South Mountain in Chongqing city, the masson pines in the suburb of Liuzhou city in the Guangxi Zhuang Autonomous Region, then Wanxian district of Chongqing, and the plums in Hangzhou, Zhejiang province (Liu, 1988; Chao, 1989; Yu, 1985; Du and Liu et al., 1988). These areas are all heavy polluted by acid deposition, with high concentration of acidic pollutants in the atmosphere, high frequency of acid rain, and high acidity of precipitation. Based on inquiries and researches, it is widely accepted that the death of forest in these areas is mainly caused by acid deposition.

There used to be 270km^2 of masson pines, half of which is afforested, on the South Mountain in Chongqing. Since 1982, half of the masson pines have withered. Moreover, the remnants are still faced with death. According to the research results of Chinese Research Academy of Environmental Sciences, the frequency of acid rain there is 100%, and the pH of rain water rages from 3.6 to 4.8. As a typical result of the soil acidification, the concentration of active Al^{3+} and the ratio of Ca/Al in soil is high. The root of masson pine has been obviously damaged by Al^{3+}. An unanimous viewpoint now is that the death of masson pines may be the comprehensive effects of all adverse factors caused by acid deposition.

There are 6500km^3 of masson pine in Wanxian district in Chongqing, among which 150km^2 has withered and 380km^2 shows the symptoms of decline. Especially, there are 60km^2 in Fengjie county, which were planted in 1958 and grew well at early stage, but the forest has declined by 90%, hitherto. The remnant is weak and faced with death.

A national key project was carried out for monitoring the effects of acid deposition on forest in South China from 1986 to 1990. The effects of ecological factors on the productivity of masson pines such as soil depth, soil nutrient, precipitation, illumination, temperature, humidity, sloping direction, and slope gradient, were studied in some areas with different acid depositions in Guangdong province and the Guangxi Zhuang Autonomous Region. The effect of ambient SO_2 concentration, and that of the precipitation acidity was also studied. According to the results of multi-factor analysis, the major factors affecting the growth of masson pines are SO_2 concentration and precipitation acidity. The areas suffering from acid deposition had reached 37270km^2 until 1990, which is nearly half of the two provinces. The annual economic loss was about 1 billion Renminbi yuan.

According to the findings of the national key project carried out from 1991 to 1995 (Zhang Linbo, et al., 1998), forest suffering from acid deposition in seven provinces (Jiangsu, Zhejiang, Anhui, Fujian, Jiangxi, Hunan, and Hubei) was 12821km^2, or about 4.18% of all forest land and 6.52% of the commercial forest. Since forest declined as acid deposition did harm to the trees, the reduction of yield was 1 million m^3 in the seven provinces, with the economic loss of 0.6 billion Renminbi Yuan. Moreover, the economic loss of ecological benefit was about 0.54 billion Yuan. According to the comprehensive assessment with seven indices such as the area of forest damaged, the percentage of damaged forest by forest land, the percentage by commercial forest, the loss rate of wood, the wood loss, the economic loss of wood, and the economic loss of ecological benefit, the effects of acid deposition on forest is the most severe in Zhejiang province, then followed by Jiangxi, Hunan, Hubei, Jiangsu, and Fujian, and that in Anhui is the lightest.

7.3 ECONOMIC LOSS OF CROPS AND FOREST

The estimation of economic loss is an effective method to quantify the effects of acid deposition and it is very important for implementing the control strategies of acid deposition and evaluating the benefit of them. The economic loss of crops and forest caused by acid deposition has been estimated in the national key projects carried out in last decades. Table 13 shows the wood loss and the economic loss caused by acid deposition in 11 provinces. As can be seen from the table, the total economic loss of wood is nearly 5 billion Renminbi yuan. If the economic loss of ecological benefit is eight times as more as that of wood, the annual loss of ecological benefit is about 45.9 billion yuan. In summary, the total loss is 51 billion yuan per year.

Table 13. Economic Loss of Forest in 11 Provinces

Province	Loss of Wood (10^3 m^{3}a^{-1})	Economic Loss (10^6 yuana^{-1})
Guangdong	24664	1.454
Guangxi	4787	2.82
Sichuan	271	0.16
Guizhou	132	0.078
Jiangsu	50.1	0.029
Zhejiang	384.1	0.210
Anhui	15.9	0.009
Fujian	109	0.071
Jiangxi	215.2	0.134
Hunan	127.9	0.080
Hubei	112.3	0.067

Note: the price of wood was 590 yuanm^3

The effect of acid deposition on crops is directly shown as the reduction of yield. If the price of grains is 0.532 yuan kg^{-1}, and the price of vegetables is 0.562 yuan kg^{-1} while estimating the economic losses, the economic loss is estimated, as shown in Table 14. The total economic loss of crops caused by acid deposition and SO$_2$ pollution is nearly 4.39 billion yuan a^{-1}.

Table 14. Yield Loss and Economic Loss of Crops in 11 Provinces

Provinces	Vegetables		Grains		Others	
	Reduction of Yield (t a^{-1})	Economic Loss (10^6 yuan a^{-1})	Reduction of Yield (t a^{-1})	Economic Loss (10^6 yuan a^{-1})	Reduction of Yield (t a^{-1})	Economic Loss (10^6 yuan a^{-1})
Guangdong	316000	177	620	0.33	-	177.3
Guangxi	218600	122	19600	10.43	-	132.4
Sichuan	240000	134	250000	133	0.60	327
Guizhou	40000	22	25000	13.3	0.200	55.3
Jiangsu	498261	280	314126	221.8	1.753	677.1
Zhejiang	411108	231	36389	25.7	0.283	285
Anhui	338207	190.1	385943	272.5	1.419	604.5
Fujian	563787	316.8	10717	7.6	0.074	331.8
Jiangxi	540533	303.8	3435	2.4	0.542	360.4
Hunan	866736	487	14803	1.05	0.963	593.8
Hubei	729317	410	275452	19.45	2.418	846.3

8. SENSITIVITY OF SURFACE WATER TO ACID DEPOSITION IN CHINA

The sensitivity of surface water to acidification, which means the neutralizing capacity to acid input, often defends on the alkalinity of the water body. J. Galloway *et al* (1978) account that water body with HCO$_3^-$ concentration lower than 0.5meqL^{-1} is sensitive to acid deposition, and that with HCO$_3^-$ concentration higher than 0.5meqL^{-1} is not sensitive. Based on this classification, the sensitive surface waters are mainly in catchments in the southeast of China, including the Beihai area in the south of the Guangxi Zhuang Autonomous Region, the Maoming area, the Yangjiang area, and large areas east to Guangzhou and Shaoguan in Guangdong province. In addition, very few areas in Heilongjiang province, the Xinjiang Uygur Autonomous Region, and Yunnan province, are sensitive too. Moreover, some water bodies, such as the Poyang Lake in Jiangxi province, the Yongjiang River, the Oujiang River, and the Qiangtang River in Zhejiang province, the Hanjiang River and the Yuejiang River in Guangdong province, the Xijiang River and the Qinjiang River in the Guangxi Zhuang Autonomous Region, are more sensitive to acid deposition with the HCO$_3^-$ concentration even lower than 0.3meqL^{-1}. Especially, 52% of the surface water in Guangdong province is sensitive to acid deposition, while only 3.8% in Guangxi municipality is (Jiang et al, 1992). It can be seen that the water bodies in Guangdong province are faced with more severe menace of acidification than that in Guangxi.

The belt-like distribution of surface water's sensitivity to acidification is quite obvious in Guangdong province and the Guangxi Zhuang Autonomous Region, with higher sensitivity in the southeast and lower in the northwest. The distribution of the sensitivity is corresponding to the distribution of bedrock and soil in the region. In Guangdong province, the stratum contains lots of metamorphic rock, granite, and acidic lava. Since the magma is active, there exist more than 500 pieces of granite bedrock in the stratum, and 50% of the stratum is covered by lava with the main composition of acidic granite. Additionally, the round water is almost soft water with low mineralization. Therefore, many water bodies are sensitive to acid deposition. However, the carbonate rock is widely distributed from southwest to northeast in the Guangxi Zhuang Autonomous Region, with very deep sediment. In the karst areas, the concentration of HCO_3^-, Ca^{2+}, and Mg^{2+} is very high in groundwater, and the surface water is supplied by the groundwater all the year round. Consequently, most of water bodies in Guangxi are not sensitive to acid deposition.

The preliminary mapping work of surface water's sensitivity to acidification in the southwest of China (Feng Zhongwei, 1993) indicates that natural water bodies in Sichuan and Guizhou province are mostly sensitive, middle sensitive, and non-sensitive, which share 26.4%, 53.0 and 18.3% of the land in the two provinces, respectively, and 97.7% for the whole. Besides the three classes, the extremely sensitive water bodies and the extremely non-sensitive water bodies are rare, and shores only 1.2% and 1.1% of the total land respectively. In brief, most of water bodies in the southwest of China is not very sensitive to acid deposition. Therefore, acidification of surface water has not been found yet on a large scale, and most of the natural water bodies will not be acidified in the near future. Although the acid deposition has not done obvious harm to the water bodies in the southwest of China, the potential impacts of acid deposition on surface water should not be ignored, because there do exist some sensitive water bodies with the trend of acidification.

9. SENSITIVITY OF ECOSYSTEMS AND CRITICAL LOADS OF ACID DEPOSITION IN CHINA

Critical load is often define as the maximum input of acid deposition to an ecosystem that will not cause long-term damage to ecosystem structure and function. The concept of critical load is widely used in Europe as a guideline to formulate abatement strategies of SO_2 and NO_x. In China, the study on critical load and target load of acid deposition became a National Key Project in the 8th Five-year Plan (from 1991 to 1995). It, however, was confined to a few areas heavily polluted by acid deposition such as Liuzhou and Guiyang, and the mapping of all the country is not carried out. In order to formulate ecosystem-based control strategies, critical loads for different types of soil in each areas in China must be estimated.

Critical loads of acid deposition can be derived through several methods, which may be divided into three levels according to their complexity (Sverdrup et al., 1990): the semi-quantitative methods using existing date to assess the sensitivity of ecosystems to acidic deposition, the steady state mass balance (SSMB) method by mass balance calculation or the application of a steady state model such as PROFILE (Sverdrup and Wafvinge, 1988), and the dynamic modeling method such as MAGIC (Cosby et al., 1985). The dynamic modeling method, which is only applied in estimating the critical load on individual site, is less suitable for mapping critical loads as compared to the steady state mass balance method, because the former requires more input date and some of which are more difficult to obtain. Although the SSMB method requires less data than the dynamic modeling method, data from field investigation that can represent the characteristic of each soil have not been sufficient yet for such a wide area as in China. Therefore, in the study of mapping critical load in China, critical loads were estimated by means of a semi-quantitative method (Nilsson and Grennfelt, 1988), which is based on the parent materials controlling weathering and soil development. According to the actual condition in China, the method was modified by introducing an Arrhenius relationship to quantify the effect of temperature and by using a weighting average approach to evaluate the influence of natural conditions such as soil texture, land use and precipitation.

9.1 METHODS AND DATA

9.1.1 The Modified Semi-quantitative Method

The rate of chemical weathering, which is primarily dominated by the content of weatherable minerals in soil, is the most important factor to consider when estimating critical loads to forest ecosystems. Thus, the critical loads for soils are closely correlated to the dominant weatherable minerals occurring in the soils and therefore can be assessed by soil mineralogy. At a workshop on Critical loads held in Skokloster, Sweden, in 1988, five sensitivity and corresponding critical loads classes of soil on the basis of mineralogy that controls weathering rates were proposed (Nilsson and Grennfelt, 1988), as shown in Table 15. Soils in the most sensitive class (class 1) is derived from highly siliceous parent rocks such as quartzite and K-feldspar-rich granite, and the least sensitive class (class 5) of soil is from parent materials with free carbonates (Marls, limestone and line-rich proluviun, sediment and aeolian deposit etc.). Between these extremes are soils derived from plagioclase-rich granite, gneisses, etc. (class 2), granodiorite, schist, etc. (class 3), and gabbro and basalt, etc. (class 4).

The content of weatherable minerals in soil is the basic factor in determining the weathering rate, while temperature is an unneglible external factor which can influence the weathering rate. So when critical load is estimated by the semi-quantitative method, the effect of temperature must be considered. It is assumed that Arrehnius relationship can be used to correct the effect of temperature to critical load.

Table 15. Mineralogical and Petrological Classification of Soil Material and the Corresponding Critical Loads for Forest Soils (0-50 cm) (Nilsson and Grennfelt, 1988)

Sensitivity class	Minerals controlling weathering	Usual parent rock	Critical loads for total acidity, CL(Ac) (keq H^+ha^{-1} yr^{-1})	Equivalent amount of sulphur, CL(S) (g m^{-2} yr^{-1})
1	Quartz, K-feldspar	Granite, Quartzite	0.2	0.3
2	Muscovite, Plagioclase, Biotite(5%)	Granite, Gneiss	0.2-0.5	0.3-0.8
3	Biotite Amphibole (5%)	Granodiorite, Greywakee, Schist, Gabbro, Shale	0.5-1.0	0.8-1.6
4	Pyroxene, Epidote, Olivine (5%	Gabbro, Basalt	1.0-2.0	1.6-3.2
5	Carbonate	Limestone, Marlstone	2.0	3.2

In addition to soil mineralogy and parent rock-type, the sensitivity of an ecosystem to acid deposition may be influenced by several other factors, such as vegetation type, precipitation, soil depth, soil texture, chemical and hydrological conditions, biological process, landform, land use and chemical characteristics, etc., all of which will affect the critical load in any particular region. If conditions of these factor enhance sensitivity of soil, the lower limit of the rang of critical loads from Table 1 should be used. On the contrary, the upper limit should be applied if the conditions mitigate soil sensitivity. In this study, the influence of these factors to critical load is quantitatively described by weighted averaging approach instead of qualitatively considered as later shown.

In summary, the semi-quantitative method was modified by considering the effect of temperature, soil texture, land use and precipitation. Critical loads for soil were estimated by this modified method through three steps stated as followed:

Step 1: Assign initial value of critical load for each soil according to Table 15

Based on the mineral component of soil in tropical and subtropical regions, and/or by comprehensive considerations of mineral component and parent rock types in other areas, the initial critical load class for each soil was determined according to Table 15. Then, the initial value of critical load for soil was assigned using the lower limit of the corresponding range.

Step 2: Correct critical load by considering the effect of local temperature

To establish critical loads to the upper 50 cm of forest soil which approximately corresponds to the rooting zone of forest vegetation, the critical load values given in Table 15 were estimated according to the ranges of weathering rate of soil minerals. However, since the weathering rate of the soil minerals is affected by the ambient temperature, the values of critical load shown in Table 15 depend on the temperature at which the weathering rate was measured. Obviously, the results of Table 15 need to be corrected by temperature while they are extensively applied to estimate critical load for soils and classifying the sensitivity of ecosystems in other countries, especially in China, where the annual mean temperature is quite different from the south to the north, and the effect of temperature can not be ignored. The values in Table 15 are based on the results from watershed budget studies and laboratory experiment in Sweden, taking the annual mean temperature of Sweden, 8°C as the reference temperature. In China another reference temperature is required.

Therefore, the Arrhenius relationship was assumed to be applicable in describing the effect of temperature on critical loads. Thus, the initial value of critical load assigned from the Step 1 were corrected for the effect of temperature according to:

$$CL_j(T) = CL_i(T_0) \cdot e^{(\frac{A}{T_0} - \frac{A}{T})} \qquad (1)$$

where $CL_i(T_0)$ is the lower limit of the critical load range for soil in class i assigned according to Table 15 at the reference temperature T_0 (281K), $CL_j(T)$ is the critical load of the soil at the local annual-mean temperature T(K), the subscript j means that the critical load of soil is change from class i to class j after the correction, and A is the Arrhenius preexponential factor with the approximate value of 3600K (Sverdrup H. 1990).

Sometimes, the corrected value of critical load may be out of the range assigned from Step 1, which is determined according to the soil mineralogy. The sensitivity class of soil must be re-determined according to the result from equation (1). It means that the soil belongs to the class in which the range of critical load content that result.

Step 3: Correct critical load by local environmental factors

As mentioned before, critical load can be influenced by some environmental factors such as precipitation, vegetation, elevation/slope, soil texture, soil drainage, soil/till depth, soil sulfate adsorption capacity and base cation deposition, etc.. Base on the actual situation of usable data in China, only the effects of soil texture, land use and precipitation were took into account in this study. The effect of precipitation is often expressed as humidity, which is the ratio of precipitation to the maximum vaporization (P:PE). The level at which these factors affect critical load was distinguished by weights, which was determined according to literature on this topic (e.g., Kuylenstierna and Chadwick, 1989). Table XVI shows how the three site factors were divided into categories and also the associated weights used in their combination to arrive at overall assessment of sensitivity to acid deposition and critical loads.

Table 16. Division of Site Factors Influencing Critical Loads of Acid Deposition for Forest Soil into Categories and the Associated Weights for Use in Combination

Factor	Weight	Category	Weighting factor	Weight value for combination
Land use	3	I Coniferous forest, Tropic broad leaf forest, Tropic rain forest	0	0
		II Bush, Grassland	1/3	1
		III Deciduous forest	2/3	2
		IV Arable, Barren	1	3
Humidity	1	I PPE<0.5 (dry)	0	0
		II PPE>0.5 (wet)	1	1
Soil texture	2	I Arenaceous	0	0
		II Doras	1/2	1
		III Argillaceous	1	2

Therefore, by comprehensively considering the effect of soil texture, land use and precipitation, the following equation was established to correct CL_j, the critical load obtained from equation (1) again:

$$CL = CL_j + \frac{\sum_{k=1}^{3} R_k \cdot x_k}{\sum_{k=1}^{3} R_k} \times (CL_{Uj} - CL_{Lj}) \qquad (2)$$

where CL is the final value of critical load determined by this method, R_k is the weight of the kth factor (the 2nd column in Table 16), x_k is the weighting of the kth factor (the 4th column in Table 16), CL_{Lj} is the lower limit of the critical load range of soil class j in Table XV, and CL_{Uj} is the upper.

It is necessary to point out that the correction of critical load above may also cause a change in sensitivity class. Accordingly, the final sensitivity class of soil depends on the resulting critical load from equation (2).

9.1.2 Data

Data applied in estimating critical loads are based on the results of the 2nd National Soil Survey and some literature (e.g., Xiong Yi, et al., 1987). Special maps collected include: (1) 1:18, 000,000 soil map of China and soil maps of each province; (2) 1:18,000,000 geology map of China; (3) the 1:18,500,000 map of soil pH value and base saturation; (4) the 1:18,500,000 map of the distribution of clay minerals; (5) the 1:18,000,000 vegetation maps; (6) the 1:30,000,000 maps of climate (annual mean temperature and precipitation); and (7) the 1:18,500,000 map of soil texture. Among these seven, map (1-4) were used to classify the sensitivity of each soil in China (step 1), and the others were applied in the correction of critical loads (step 2 and step 3).

9.2 RESULT

9.2.1 Overview of Soil Distribution in China

There are 10 classes, 46 species and 128 subspecies of soils in China. Among these 10 classes, Alpine soil occupies the largest area, then followed by Ferralsol, Alfisol, Pedocal soil and Regosol, and Semi-aquitic soil, Semi-alfisol, Aquatic soil and Saline-alkali soil are less. In the east of China, from south to north, there are belts of Latosol, Lateritic red earth, Red earth, Yellow earth, Yellow-brown earth, Brown forest earth, Dark brown forest earth and Podzolic soil. This spatial distribution of soils is called the Humid-oceanic Belt Spectrum. On the contrary, soil belts in the west of China running from north to south form the Arid-inland Belt Spectrum, which include Chestnut soil, Brown soil, Sierozem and Desert soil. Between the two spectrums is the Interim Belt Spectrum, which contains areas of Cinnamon soil, Dark loessial soil, Chermozem Chestnut soil, Gray-cinnamon soil, Gray forest soil and Black soil, stretching northeastward from the Loess Plateau to the west of Da-xing-an-ling Mountain.

9.2.2 Critical Load and Sensitivity Class of Each Major Soil in China

Critical load and corresponding sensitivity class of each major soil in China was estimated by this modified method. The initiatory classification of each soil obtained are shown in Table 15. Data used in this step were adopted from the 2nd National Soil Survey and maps such as geological maps, soil mineral maps and base saturation maps. Then according to equation (1), the initial values were corrected by local temperature, which was found in an annual-mean temperature map of China. Finally, the values are corrected once more by equation (2) for the effect of environmental factors such as land use, soil texture and precipitation, data of which are collected respectively from vegetation map, soil texture map and annual precipitation map. The final classification of each soil by sensitivity is shown as in Table 17.

Table 17. Sensitivity Class of Soil by Initiatory Classification
(without consideration on temperature and other factors)

Sensitivity class	Soil type
1	Latosol, Lateritic red earth, Podzolic
2	Red earth, Yellow earth, Torrid red earth, Yellow-brown soil, Dark brown forest soil, Black soil
3	Brown forest soil, Gray forest soil, Albic bleached soil, Purplish soil, Subalpine meadow soil, Alpine meadow soil
4	Cinnamon soil, Paddy soil, Meadow soil, Bog soil, Alpine frozen soil
5	Yellow cultivated loessial soil, Stratified old manured loessial soil, Dark loessial soil, Fluvo-aquic soil, Oasis soil, Gray-cinnamon soil, Chernozen, Chestnut soil, Brown soil, Sierozem, Gray desert soil, Gray brown desert soil, Brown desert soil, Solonchak, Solonetz, Limestone soil, Aeolian soil, Subalpine steppe soil, Alpine steppe soil

Table 18 Sensitivity Class and Critical Load of Acid Deposition for Each Soil in China

Critical load class	Soil type	CL(Ac) (keq H^+ ha^{-1} yr^{-1})	CL(S) (g m^{-2} yr^{-1})
1	Podzolic	<0.2	<0.3
2	Latosol, Dark brown forest soil, Black soil	0.2-0.5	0.3-0.8
3	Lateritic red earth, Red earth, Yellow earth, Torrid red earth, Yellow-brown soil, Brown forest soil, Gray forest soil, Albic bleached soil, Subalpine meadow soil, Alpine meadow soil	0.5-1.0	0.8-1.6
4	Purplish soil, Cinnamon soil, Paddy soil, Meadow soil, Bog soil, Alpine frozen soil	1.0-2.0	1.6-3.2
5	Yellow cultivated loessial soil, Stratified old manured loessialsoil, Dark loessial soil, Fluvo-aquic soil, Oasis soil, Gray-cinnamon soil, Chernozen, Chestnut soil, Brown soil, Sierozem, Gray desert soil, Gray brown desert soil, Brown desert soil, Solonchak, Solonetz, Limestone soil, Aeolian soil, Subalpine steppe soil, Alpine steppe soil	>2.0	>3.2

9.2.3 Mapping Critical Load of Acid Deposition in China

The critical loads of acid deposition have been mapped for Chinese soils, as shown in Figure 7. It can be seen that the majority of areas sensitive to acid deposition are in the Southeast China, and that insensitive is in the Northwest. The sensitive areas, including the catchment of the Changjiang (Yangtze) River and the wide areas to the south of it, are warm and rain-abundant. The natural vegetation is the tropic rain forest, seasonal rain forest, and subtropic evergreen forest. Ferralsol, the dominant soil in these areas, is acidic, with iron and aluminum obviously accumulated. It can tolerate approximately 0.8-1.6 $gSm^{-2}yr^{-1}$ acid deposition and belongs to the intermediate sensitivity class 3. In Northwest China, it is semiarid or arid. There is mainly open land and arable land, seldom covered with deciduous forest or coniferous-deciduous forest. The soils, dominated by soil class of Xerosol and Alpine soil, are carbonate-rich and saline on low and level land. Consequently, they are resistant to acidification.

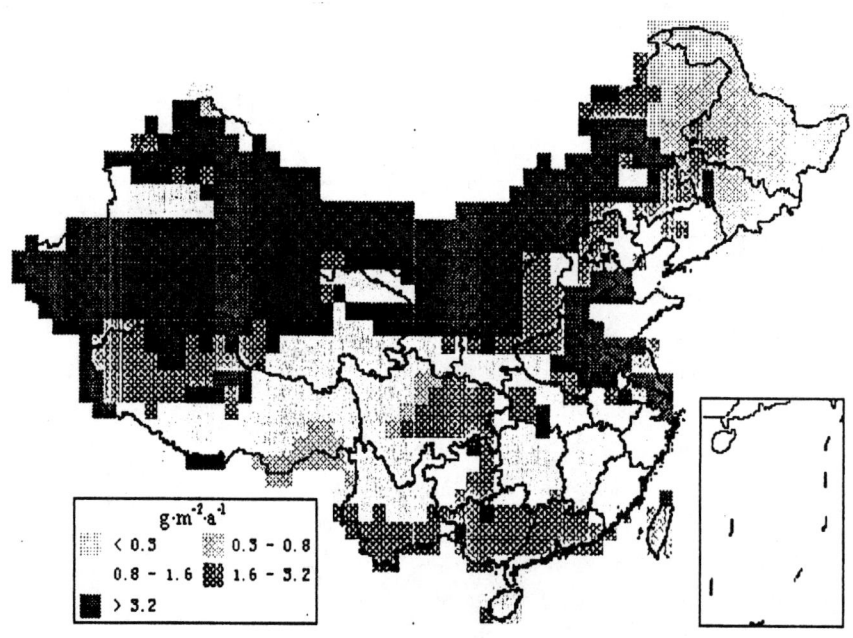

Figure 7. Critical load map of sulfur deposition in China. Sulfur is assumed to be the only acidifying agent.

The critical load class 1, namely the most sensitive class of soil, is chiefly referred to Podzolic soil, which is also called Brown coniferous forest soil. It occurs in small areas in the northeast of China, mostly on the north part of Da-xing-an-ling Mountain, which is covered by coniferous forest, with precipitation of 400-500 mm and annually-mean temperature of -4.9-0°C. Podzolic soil is derived from granite or quartz and its formation is often influenced by leaching and settling of organic acid complex compound. Hence, the soil shows acidic reactions and its base saturation is very low. The clay minerals in the soil are composed of hydrous mica and small amounts of other unweatherable minerals such as montmorillonite, kaolinite, roseite and chlorite. The low temperature and the coarse texture of soil constitute the importance to the low weathering rate of soil minerals there. Therefore, these areas must be paid great attention, even if acid deposition has not yet appeared.

Class 2 is found in Dark brown forest soil and Black soil areas in the Northeast, and Latosol areas in the south of Taiwan province, in the north of Hainan province and near Hekou in Yunnan province. The Dark brown soil area, with annual-mean temperature of -1-5°C, is covered by coniferous forest. The coarse soil is derived from granite and the chief clay minerals include hydrous mica. As in the Dark brown soil area, the temperature in Black soil areas is very low. The Black soil contains clay minerals such as hydrous mica, fulonite, gibbsite and kaolinite. It is obvious that the meteorological conditions and the physical character of the two kind of soil above are not conducive to chemical weathering of minerals. Contrarily, Latosol areas are temperature-high and rain-abundant (the annual-mean precipitation is 1900 mm, the annual-mean temperature is 23.5symbol 176 \f "Symbol" \s 12°}C). The texture of Patosol is fine and the fraction of clay granules in the soil is high, which is advantageous to weathering. However, the clay minerals in the soil are dominated by kaolinite and gibbsite. Anorthite has completely decomposed and K-feldspar is rare. The weathering rate is still low as a result of the lack of weatherable minerals.

Soils of class 3 include Lateritic red earth to the south of the Nanling Mountain, Red earth and Yellow earth between Nanling Mountain and the Changjiang (Yangtzee) River, Yellow-brown earth in the lower reaches of the Changjiang(Yangtzee) River, Subalpine meadow soil and Alpine meadow soil on the Plateau of Tibet.

Class 4 is found in Paddy soil areas sporadically distributed throughout China and in the Purplish soil area in the Sichuan Basin.

Class 5 (the least sensitive) soil include Chestnut soil, Brown soil and Sierozem areas on the Plateau of Inner Mongolia and the Loess Plateau, Desert soil areas in He-xi-zou-lang and the Talimu Basin, Subalpine steppe soil, Alpine steppe soil and Alpine desert soil on the Plateau of Tibet. These kind of soils, belonging to the soil class of Xerosol or Alpine soil, consist of easy weathering minerals such as carbonate. They show alkaline reactions, with weak leaching and sparse vegetation. Those kinds of soils are insensitive to acid deposition.

Figure 8 illustrates the percentage of areas shared by each critical load class. As can be seen, the most sensitive soil (class 1) shares only 2% of the whole area in China, and the sensitive soil (class 2) is no more than 8.7%. The intermediate sensitivity and the least sensitive soils are the most extensive, which account for 35.5% and 42.4%, respectively. Class 4 covers 11.4%.

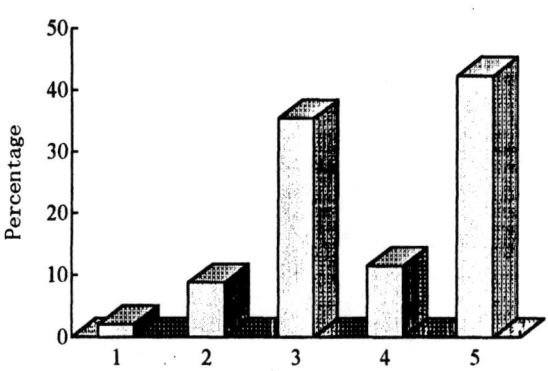

Figure 8. Percentage of areas shared by each critical load class.

In summary, the sensitivity of an area to acid deposition and the critical load values are depended on soil and vegetation types and meteorological conditions. In China, acid deposition often occurs in the Allite areas in the southeast of China, where soils are of intermediate sensitivity except Latosol. The annual mean temperature in these tropic or subtropic areas is 16-25°C, the accumulated temperature of ≥10°C is 5000-9000°C, and precipitation more than 1500 mm. The characteristics of temperature-high, rain-abundant, and wet-warm-in-same-season, promote reactions of soil minerals and cause the rapid weathering of soil minerals and the rapid circulation of biological materials. Furthermore, it is because of the rapid weathering reactions that there are more fine clay granules in soil, which means that the specific surface area of soil increases. This causes more rapid weathering in soil. As a result, the weathering rate is not very low, and the corresponding critical load is high, even if there is only a small amount of weatherable minerals in soil. On the other hand, the areas of Podzolic soil, Dark brown forest soil and Black soil are in frigid temperate zone. The accumulated temperature of ≥10°C is about 2000, the annual mean temperature is lower than 5°C, and the precipitation is low. These kinds of soil consist of a large amount of sand and only a few clay granules. Thus, the weathering rate is quite low despite the fact that there are some weatherable minerals, and the acceptable acid deposition is very low.

It can be concluded that high temperature, high humidity, wet-warm-in-same-season and high concentrations of fine granules, are important reasons why soils and water bodies have not been found acidified in heavy acid rain areas such as Chongqing, Guiyang and provinces of Guangdong and Guangxi. However, it should be noted that the weatherable materials in soils in south China are in scarcity at all, and the potentiality of acidification still exists. Acid deposition must be controlled in Northeast areas, or acidification of soils and decrease of forest production can occur following those in North Europe and North America.

Comparison of the critical loads obtained to the sulfur deposition of 1995 in China (Hao *et al*, 1996) led to the critical load exceedance map of sulfur deposition in China in the year 1995 (see Figure 10). As can be seen from the map, sulfur deposition exceeds critical load in a wide land area that amounts for 25% of Chinese totals, which mainly referred to the southeast of China. Among these areas, the exceedances are especially serious in the lower reaches of Changjiang (Yangtzee) River, in the Sichuan Basin, and in the Delta of Zhujiang River.

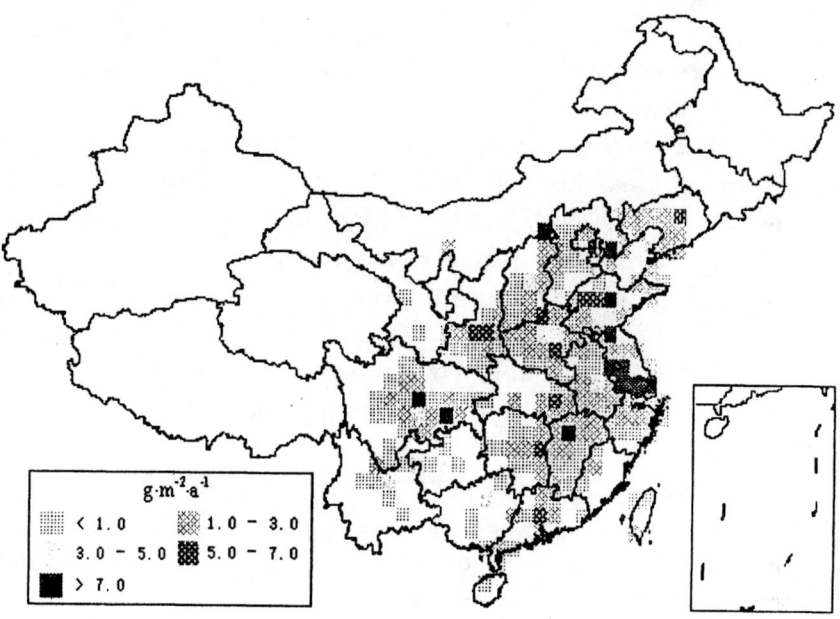

Figure 9. Critical load exceedance map of sulfur deposition in China in 1995. Data of sulfur deposition come from Hao JM. Et al. (1996). Free space cells mean zero exceedance.

10. CONTROL STRATEGY OF ACID DEPOSITION IN CHINA

The government of China attaches great importance to the pollution caused by acid rain and ambient sulfur dioxide. After the implementation of *The Suggestion on Controlling Acid Rain Development* in 1990, pollutant fee for SO_2 emission from industrial coal-combustion was charged in Guizhou and Guangxi province, and nine cities of Liuzhou, Nanning, Guilin, Hangzhou, Qingdao, Chongqing, Changsha, Yichang, and Yibin since 1992 to set the experience of the comprehensive prevention and control of acid rain. These measures took positive effect on the prevention and control of SO_2 pollution too. For further control of acid rain and SO_2 pollution, the Standing Committee of the National People's Congress amended the Law on the Prevention and Control of Atmospheric Pollution in August 1995, the 27th article of which entitles the National Environmental Protection Agency jointly with other ministries, approved by the State Council, to designate the Acid Rain Control Areas and the Sulfur Dioxide Pollution Control Areas (two Control Areas called for short) for those areas which are or could become affected by acid deposition or ambient sulfur dioxide concentrations.

10.1 DESIGNATING THE ACID RAIN CONTROL AREAS AND THE SULFUR DIOXIDE POLLUTION CONTROL AREAS

10.1.1 Fundamental of Designating Control Areas

According to the prescripts on the designation of the Acid Rain Control Areas and Sulfur Dioxide Pollution Control Areas such as the 27th article of *the Law on the Prevention and Control of Atmospheric Pollution*, and the demand of *the State Council's Decision of Some Problems on Environmental Protection, the Outline of the Ninth Five-Year-Plan and the Perspective Objective on National Economy and Social Development*, and *the Ninth Five-Year-Plan and the Perspective Objective on National Environmental Protection*, the designation of Control Areas is to achieve the objective of environmental protection and the programming of gross control.

The designation of the Control Areas must embody the request of the sustainable development. The acid rain regions and the cities with serious pollution of ambient SO_2 concentration, must be emphasized.

The designation of the Control Areas must make full use of the monitoring and research results on acid rain and SO_2 pollution obtained since 1980's. For the convenience of carrying out the environmental management and supervision, the administrative division should also be considered while determining the extent of Control Areas.

10.1.2 Principles of designating Control Areas

Principles of designating the Acid Rain Control Areas

According to the results of the relevant researches (Feng, 1993), precipitation with pH ≤ 4.9 may do notable harm to forest, agriculture, and materials, but in practice, only acidic precipitation with pH ≤ 4.6 should be controlled. In addition, the resistibility of ecosystem to acid deposition, which varies from site to site, can be quantified by critical load. Since the major acidifying precursor is sulfur dioxide in China, sulfur deposition should be preferentially controlled.

Based on the results of the national key projects carried out in last decades (CRAES, et al., 1995), SO_2 emitted from local sources is the major cause of the formation of acid rain in China, whereas SO_2 from long-range transportation can simultaneously affect this process. Therefore, regional gross control of SO_2 emission must be carried out, i.e., as well as areas suffering seriously from acid rain, the adjacent areas with high emission of SO_2 must be controlled together. However, in the acid rain regions, there are some poverty areas (such as the National Poverty Counties), where the SO_2 emission is very little, and the pollution of acid rain is caused by SO_2 emitted from foreign sources. On the other hand, it is not practical for the poverty counties to strictly control the SO_2 emission because large investment is needed for the abatement of SO_2 emission. Since the most urgent affairs for them are to develop the economy, the National Poverty Counties will temporarily not be included in the Control Areas.

(2) Principles of Designating the SO_2 Pollution Control Areas

The Ambient Air Quality Standard includes three classes, among which the second class indicates the basic requirement of protecting human and ecoenvironment from any harms, and the third one is to protect human and ecoenvironment from short-term damages. The ambient SO_2 pollution, which is caused by SO_2 emission from local sources and seldom affected by foreign ones, always occurs in cities. Thus, Local sources of SO_2 should be firstly controlled in cities, and the SO_2 Pollution Control Areas should focus on cities. In brief, the SO_2 Pollution Control Areas should include the cities with the annual average concentration of ambient SO_2 higher than the second class of the national standard and the daily average concentration exceeding the third class, except the National Poverty Counties. The southern cities, where SO_2 pollution and acid rain both serious, should designated in the Acid Rain Control Areas.

10.1.3. DESIGNATING RESULT

The range of the Acid Rain Control Areas and the SO_2 Pollution Control Areas, as shown in figure 10, is about 1.09 million km^2, and shares 11.4% of the whole area in China. The Acid Rain Control Areas involves 14 provinces/autonomous regions/municipalities south to the Changjiang (Yangtze) River, with the area of 0.806 million km^2, while the SO_2 Pollution Control Areas includes 63 cities north to the Changjiang (Yangtze) River, with the total area of 0.29 million km^2. The Acid Rain Control Areas and the SO_2 Pollution Control Areas shares 8.4% and 3% of the Chinese land respectively.

In 1995, the SO_2 emission is about 23.7 million tons in China, while the emission is 14 million tons and nearly 60% of the total in Control Areas. Thus, the acid rain and SO_2 pollution in China will not be deteriorated if the SO_2 emission is well controlled in the Acid Rain Control Areas and the SO_2 Pollution Control Areas.

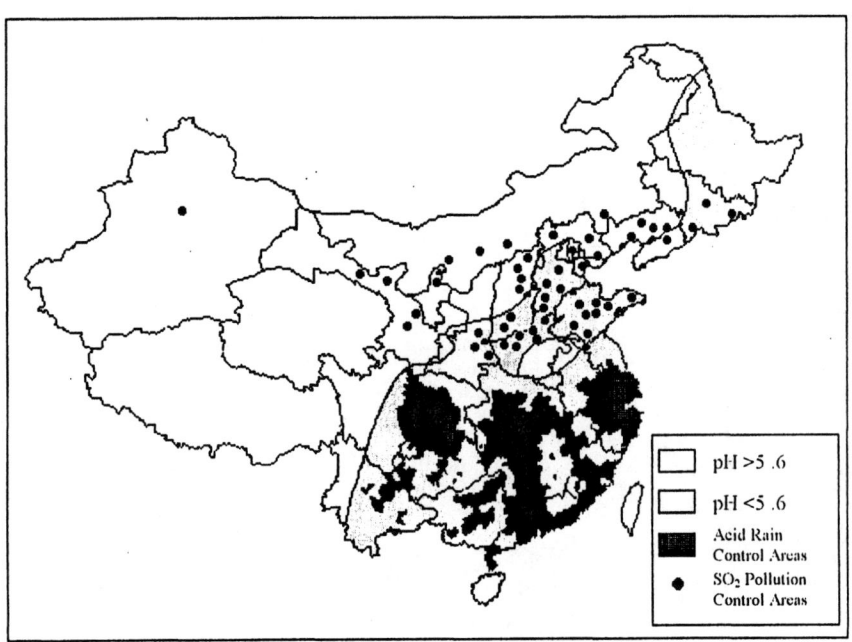

Figure 10. Sketch Map of the Acid Rain Control Areas and the SO$_2$ Pollution Control Areas in China (Liu Bingjiang, et al., 1998)

10.2. CONTROL OBJECTIVES FOR THE ACID RAIN CONTROL AREAS AND THE SO$_2$ POLLUTION CONTROL AREAS

According to *the State Council's Decision of Some Problems on Environmental Protection* and *the Ninth Five-Year-Plan and the Perspective Objective on National Environmental Protection*, the control objectives of the Acid Rain Control Areas and the SO$_2$ Pollution Control Areas are submitted as follows:

The deteriorating trend of acid rain and SO$_2$ pollution should have been controlled until 2000. Thus, the SO$_2$ emission from industrial sources must follow the emission standard in the Acid Rain Control Areas and the SO$_2$ Pollution Control Areas, and the gross of SO$_2$ emission needs to be controlled within a target prescribed by the government. At the end of the century, the ambient SO$_2$ concentration in key cities of environmental protection should be kept below the National Ambient Air Quality Standard, and the pollution of acid rain in the Acid Rain Control Areas should not be deteriorated.

The environmental situation should be evidently improved in 2010. The SO$_2$ emission must be kept below the level of 2000, and the ambient SO$_2$ concentration must follow the National Ambient Air Quality Standard. Consequently, the area with pH ≤ 4.5 in the Acid Rain Control Areas should obviously decrease.

10.3. POLICIES AND MEASURES

Formulate comprehensive control programming of acid rain and SO_2 pollution in the Control Areas

The programming of comprehensive protection and control of acid rain and SO_2 pollution and the stepwise plan of gross control should be formulated by local government and departments concerned, and should be carried out together with the plan of National Economy and Social Development. According to the principle that who pollutes must be responsible for treating the emission, money can be raised and the projects of pollution control may be put into effect.

Limit the mining and the use of coal with high sulfur content

At present, the annual yield of coal with sulfur content higher than 3% is nearly 90 million tons, which is about 7% of the total. Since this kind of coal is almost used in the acid rain regions, the mining, production, transport, and use of the high sulfur content coal must be limited for controlling SO_2 emission. New collieries with sulfur content of the coal-bed larger than 3% should not be approved by local government and departments concerned, while the already constructed ones should gradually reduce or stop production. Simultaneously, corresponding coal washing and dressing facilities must be installed in reconstructed collieries as well as the new ones with sulfur content larger than 1.5%. Coal with low sulfur content and washed power-coal are preferentially supplied to the Control Areas. Row coal without being washed, desulfurized, or shaped should not be used for cooking after 2000.

Control preferentially the SO_2 emission from thermal power plants, abate the gross emission of SO_2

For controlling SO_2 emission from thermal power plants, the layout of power plants should be optimized. Thermal power plants should not be approved by local government to be built in large cities or middle cities (including urban district and close suburbs) in the Control Areas. When the power plants using coal with sulfur content higher than 1% are constructed or reconstructed in the Control Areas, corresponding facilities of flue gas desulfurization (FGD) should simultaneously be installed. In addition, effective measures of SO_2 emission abatement should be taken before 2000 in already constructed power plants using coal with sulfur content higher than 1%, and installations of desulfurization or other installations with the same effect should gradually be constructed before 2010.

Control SO_2 emission during the whole producing process of chemical industry, metallurgy industry, and building material industry, etc.

SO_2 emission from producing processes of chemical industry, metallurgy industry, and building material industry, which is about 20% of the total, is an important cause of acid rain and SO_2 pollution in China. All measures should be aimed at the pollution control during the whole industrial process, and the clean production, which means selecting the raw materials without causing much pollution, using advanced techniques and equipment, increasing the energy efficiency, intensifying the management of each process, and necessarily treating the pollutants at end-of-pipe. At the same time, measures of emission control must be carried out for industrial boilers and kilns with SO_2 emission exceeding the national discharge standard. Enterprises failing an emission reduction within the time limit required may be ordered to suspend its operation or closed down.

Develop and utilize advanced techniques and installations for SO_2 pollution control

The research, development, spread, and use of desulfurizing techniques and installations, such as the limestone-gypsum flue gas desulfurization, the lime spray drier, the wet scrubber, the lime injection FGD, the Cycled Fluidized Bed, and the industrial briquette, etc., must be speeded up. Simultaneously, advanced techniques and equipment from other countries should also be introduced and utilized.

Charge the SO_2 emission fee, and use economic means to control pollution

The works of charging, managing and using of SO_2 emission fee must be seriously accomplished. More than 90% of the bankroll raised should be used for SO_2 control in key factories.

Intensify the environmental supervision and management in the Control Areas

The concentration control, which has been adopted in air pollution control for a long time, now becomes inadequate along with the development of economy, the increase of energy consumption, and the rising requirement of environmental quality. Consequently, the gross control of pollutant emission, as well as the concentration control, must be carried out. Firstly, the objective for gross control or gross abatement must be set, and effective measures should be carried out together with severe supervision and management. The gross of SO_2 emission must be diminished while a project is constructed or reconstructed.

In order to find out the developing trend of acid rain and SO_2 pollution and provide the scientific foundation to the policy-making, the monitoring network of acid rain and SO_2 pollution must be established. Especially, online equipment with continuous monitoring and computation must be installed on the key souses of SO_2 in the Control Areas, and keeps on operating for long time.

11. BRIEF SUMMARY

To sum up, preliminary conclusions on acid deposition and its effects may be drawn as follows:

The annual average pH value of precipitation is lower than 5.6 in nearly 50% of the statistical cities or in 87% of the southern cities which are located south to the Qingling Mountain and the Huaihe River, and the lowest even reaches 3.52 in Changsha. In addition, the frequency of acid rain is higher than 60% in 25% of the monitored cities.

Acid rain regions have formed in the Sichuan Basin and in wide areas southern to the Changjiang (Yangtze) River and eastern to the Tibet plateau, covering 12 provinces/autonomous regions, with four core zones in the center, southwest and south, and east of China.

The Central Acid Rain Region, with the highest acidity of precipitation and the highest frequency of acid rain in China, keeps on deteriorating in recent years, while the pollution state in the South Acid Rain Region, which is also serious, does not change on a large scale. The pollution of acid rain in the Southwest Acid Rain Region is more severe than that in other regions except the Central Acid Rain region, with some fluctuation on a local scale and the frequency of acid rain higher than 80% in some key cities. The spatial and temporal distribution of the precipitation acidity and the frequency of acid rain is quite uneven in the East Acid Rain Region. Since acid rain is widely distributed in China, with different geographical and climatic conditions and various sources of acidifying precursors, the formation of acid rain is very complex.

According to the analysis of precipitation chemistry, the acidity of precipitation is increasing in China. Moreover, the ratio of SO_4^{2-}/NO_3^- is decreasing and the ratio of NH_4^+/Ca^{2+} is increasing, which means that nitrogen compounds contribute more and more to acid rain, and the ammonium emission increases very fast relative to the calcium.

The regional distribution of acid deposition is the comprehensive effect of many natural and anthropogenic factors. The occurrence of heavy acid rain in the areas to the south to the Yangtze River, is due to the low neutralizing capacity of atmospheric particulate, the acidic soil, the high humidity, the high temperature, the strong solar radiation, and a certain emission of acidifying precursor. Although the emission intensity of acidifying precursor is very high in the north of China, the natural conditions are disadvantageous to the formation of acid rain. However, the emission must be controlled, otherwise acid rain will probably occur in the north of China in the summer.

Acid deposition has done great harm to the agriculture and forest in China, and caused economic losses and ecological damages. SO_2 is regarded as the major pollutant being responsible for all the consequences.

The sensitive surface waters are mostly in the coastal areas and the catchment of the Zhujiang River in the southeast of China, including Guangdong, Fujian, Zhejiang, and Jiangxi province, and the Guangxi Zhuang Autonomous Region. In the ares, acidity input is high as acid rain often occurs. However, the buffering capacity is low because of the acidic soil, the granite-dominated bedrock, and the low HCO_3^- concentration in the water bodies. Consequently, acidification of surface water may be a potential environmental problem in the region. Although there are some sensitive water bodies with the potential of acidification, most of the surface waters in the southwest of China are non-sensitive and will not be acidified in the near future.

The critical loads of acid deposition and the corresponding sensitivity classes of each major soil in China were estimated. The most sensitive area to acid deposition is the belt of Podzolic soil, which shares only 2% of Chinese land, then followed by Latosol, Dark brown forest soil and Black soil areas, the total of which is no more than 8.7%. The intermediate sensitive areas, which can tolerate about 0.8-1.6 g S m^{-2} yr^{-1} sulfur deposition and shared 35.5% of Chinese land, are almost Ferralsol areas in South China, where the acid rain is quit serious. The least sensitive areas are mainly referred to Alpine soil areas on the Plateau of Tibet and areas of Xerosol and other types of soil in Northwest China, which account for 42.4% of total land area. The reason why the critical load in the Northeast is lower and that in South China is higher can be attributed to the difference of temperature, humidity and soil texture among these areas. Therefore, the effects of high temperature, high humidity, humid-warm-at-same-season and high concentration of fine granules are important reasons why soils and water bodies have not been found acidified in south China where acid rain is serious.

REFERENCES

Chao Hongfa, Gao Yingxin, and Shu Jianmin: 1989. Study on Relationship Between Fir Decline and Acidic Precipitation on the Emei Mountain. In: *Chinese Environmental Science Lyceum's Collection on Acid Rain*, 381-391. (in Chinese)

Chinese National Environmental Protection Agency: 1996. *The Report on Environmental Quality in 1991-1995*. (in Chinese)

Chinese Research Academy of Environmental Sciences, Tsinghua University, and Chinese Academy of Sciences: 1995. Study on Programming and Strategy for Acid Deposition Control, *research report of National Key Project in the Eighth Five-year plan*.

Du Xiaoming, and Liu Houtian: 1988. Characteristics of Mist Water and Its Effects on Masson Pines in Chongqing. Journal of Environmental Sciences (China), 8(4),467-473.

Feng Zhongwei (ed.): 1993. *The Effects of Acid Rain on Ecosystem: Study on Acid Rain in Southwest China*. Chinese Science and Technology Publishing House, Beijing. (in Chinese)

Hao J. M., Zhou X. L., Fu L. X., and Li Q. L.: 1996. Sulfuric Deposition Modeling Research in the East Part of China. *China Environmental Science*, 16 (5), 345. (in Chinese)

Jiang Jingrong, Zhou Xiuping and Qin Wenjuan: 1992. Sensitivity of the Surface Water to Acid Rain in Guangdong and Guangxi Provinces. *Acta Scientiae Circumstantiae*, 12(1),119-123.

Li Jinhui, and Tang Hongxiao: 1998, Model for Predicting the Acidity of Precipitation in China. *China Environmental Science*, 18(1),8-11. (in Chinese)

Liu Bingjiang, Hao Jiming and He Kebin et al.: 1998. Study on Designation of Acid Rain and SO_2 Pollution Control Areas and Policy Implementation. *China Environmental Science*, 18(1),1-7. (in Chinese)

Liu Houtian: 1988. Relationship Between Masson Pine Decline and Air Pollution in Chongqing. *Journal of Botany (China)*, 30,319-325.

Wang Wenxing: 1994. Study on the Origin of Acid Rain Formation in China. *China Environmental Science*, 14(5),323-329. (in Chinese)

Wang Wenxing and Ding Guoan: 1997. The Geographical Distribution of Ion Concentration in Precipitation Over China. *Research of Environmental Sciences*, 10(1),1-7.

Wang Wenxing, Zhang Wanhua, Shi Quan and Hong Shaoxian. 1993, Study on Factors Related to Acidity of Rain Water in China. *China Environmental Science*, 13(6),401-407. (in Chinese)

Wei Mingsheng, Wang Mingxia, Wang Ruibin: 1989. The Spatial and Horary Distribution of Precipitation Acidity and Chemical Composition. In: *Chinese Environmental Science Lyceum's Collection on Acid Rain*, 203-207. (in Chinese)

Xiong Yi and Li Qingkui, et al., 1987. *China's Soils*, the second edition, Science Book Concern, Beijing. (in Chinese)

Yu Shuwen, Bian Yongmei, and Ma Guangjing: 1990. Discussion on Relationship Between Masson Pin Decline and Air Pollution in Chongqing. *Journal of Environmental Sciences (China)*, 10(3), 378-383.

Zhang Linbo, Cao Hongfa and Shen Yingwa, et al.: 1998. Effect of Acidic Deposition on Agriculture of Jiangsu, Zhejiang, Anhui, Fujian, Hunan, Hubei and Jiangxi Province – Damage to Agricultural Ecosystem. *China Environmental Science*, 18(1):12-15. (in Chinese)

Chapter 9
MODEL APPLICATION FOR ASSESSING THE ECOSYSTEM SENSITIVITY TO ACIDIC DEPOSITION BASED ON SOIL CHEMISTRY CHANGES AND NUTRIENT BUDGETS

JUNKO SHINDO

National Institute of Agro-Environment Sciences
Kannondai 3-1-1, Tsukuba, Ibaraki 305 Japan

CONTENT

Abstract
1. Introduction
2. A dynamic model and soil buffering capacity due to chemical processes estimated with the model
    2.1. Description of the model
    2.2. Application of the model to the soil acidification experiments
    2.3. Results of model estimation
3. Evaluation of ecosystem sensitivity with a steady-state model
    3.1. Long-term buffering capacity to the acidic substances in Japanese ecosystems
        3.1.1. Mineral weathering rate BC_{we}
        3.1.2. Base cation and nitrogen uptake rates (BC_{gu}, N_{gu})
    3.2. Estimation of an impact indicator of acidic deposition
4. Conclusions
References

ABSTRACT

Models were developed to evaluate soil acidification caused by acidic deposition and ecosystem sensitivity to it in different spatio-temporal scales. For the prediction of catchment scale acidification, a dynamic model was made. The model took rapid chemical reactions into consideration as the (quasi-)equilibrium processes and processes such as chemical weathering, nutrient uptake, nitrification etc. as element flux at a constant rate to the soil system. Application of the model to the soil acidification experiments using simulated acid rain showed that changes in soil chemistry could be well expressed by the model and the acid loaded into the soil was neutralized in the top 10 cm horizon and soil acidification occurred there. Against the urgent acidification, main buffering mechanisms were cation exchange and dissolution of Al hydroxides whose relative contributions differed according to soil characteristics and its acidification stage. For larger scale estimation, a steady state mass balance model was employed. Magnitudes of acid neutralizing capacity and acid production due to chemical weathering, base cation and nitrogen uptakes and base cation deposition were estimated for Japanese forest ecosystems based on existent data bases of geology, soil, vegetation etc., some measured data on soil properties and parameters derived from literature. The mineral weathering rate appeared to have the most significant effect on the neutralizing capacity of ecosystems, because the values estimated for individual grid points varied over the wider range than the other processes. An indicator was proposed to evaluate the possible ecosystem impact by acid deposition based on the steady state model under the condition of current or predicted acid deposition rates.

1. INTRODUCTION

Ecosystem sensitivity to acidic deposition depends on acid buffering capacity of soils as well as vegetation sensitivity, meteorological condition and so forth. Acidic substances loaded to the soil is neutralized there, depending on its buffering capacity. At the same time, soil is acidified by losing the corresponding amount of acid neutralizing capacity (ANC) (Van Breemen et al, 1983, 1984). Lake acidification happened in Scandinavian countries in early 1970s was appeared to have a strong relation to the weak buffering ability of surrounding soils. Soil acidification has also been suspected as one of the reasons of forest decline that has gathered a great attention since 1980s.

Hallbacken and Tamm (1986) demonstrated that pH of forest soils decreased with 0.3-0.9 unit by comparing the current pH measurements to the one measured in 1927 for 90 soil profiles. Falkengren-Grerup (1987) also showed the topsoil had become more acidic in 1984 compared to the original pH measured in 1949-1970. In soil systems, various kinds of buffering mechanisms are taking place and changes in chemical property of soils are quite slow. Detection of soil chemistry change requires such long-term monitoring.

Soil measurements have a large spatial variability. Chemical property of soils and its temporal changes are different from each other by parent materials, geographical features, vegetation types etc. Figure 1 shows box plots for pH, exchangeable calcium and exchangeable aluminum measured at 50 sites each within the five watersheds of 1.5 to 10 km^2 in Japan (Shindo and Hakamata, 1991). The values scattered in a quite wide range. The spatial variation was partly explained by geographical feature (ridges or valleys) and soil types. Spatial variability in the smaller scale is also large: a value measured at one site was sometimes quite different from the value at the site several meters apart. The large spatial variability makes the detection of the soil chemistry changes difficult.

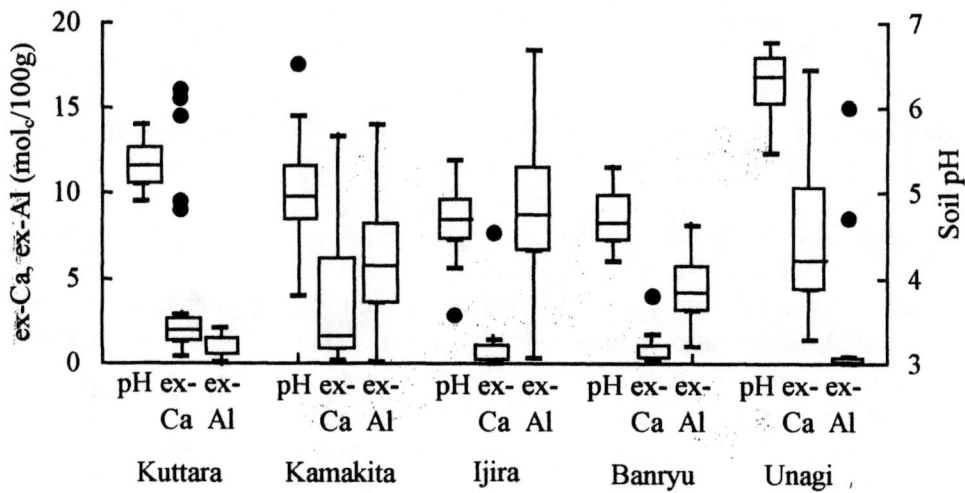

Figure 1 Spatial variability of soil parameters of 50 points each in small catchments

Modeling of soil processes and element cycling within the ecosystem is thought helpful for evaluating the ecosystem sensitivity of each location to the acidic deposition and for predicting future effect of acidic deposition on the ecosystem. Acidic condition of soil is controlled by element cycling within the biogeosphere through the various processes. Some models were already developed taking specific processes into mainly in Europe and North America. They are classified into two groups: dynamic models and steady state models. Dynamic models include rapid chemical reactions as the main processes to estimate the temporal change of soil chemistry. Existing models aim at prediction in various scale in time and space: from the prediction of daily change in soil and water chemistry within a small catchments to the evaluation of yearly trend of ecosystem acidification at the country and regional scale (Goldstein et al., 1985 (ILWAS model); Cosby et al., 1985 (MAGIC model); De Vries et al., 1994 (SMART model); de Vries et al., 1995 (RESAM model), Warfvinge et al.,1995 (SAFE model) etc.). Processes that are taken into consideration in the models and methods to quantify them are different from each other according to the spatio-temporal scale, characteristics of the objective ecosystem and also to the objective of the estimation. Although the modeling approach possibly provides the useful insight on the ecosystem change caused by acidic deposition, it is difficult to obtain the reliable result for the objective area by using the model that has been made for the other area and for the other objectives. In order to predict acidification of Japanese soils, we made a dynamic model, which is described in the following chapter. The steady state model has been developed mainly for the usage of the critical load estimation at the European scale (Hettelingh et al., 1991; Sverdrup and de Vries, 1994) It evaluates the proton budget by calculating balance of acid input, net neutralization in the ecosystem due to mineral weathering, biological processes etc. and acid output at the steady state condition. In this paper, the sensitivity of Japanese ecosystem is evaluated with the steady state model to explore the problems to be solved.

2. A Dynamic Model and Soil Buffering Capacity Due to Chemical Processes Estimated with the Model

2.1. Description of the Model

A dynamic model was made which calculated the element concentration in soil solution and the adsorption phase assuming one or two layer(s) of soil. It was assumed that soil solution was completely mixed within the each layer and was electrically neutral. The mass balance equation during one time step (from t-1 to t) was expressed as equation (1).

$$n([X^{n+}]_t-[X^{n+}]_{t-1})Vw+(A(X^{n+})_t-A(X^{n+})_{t-1})Ws/100+(S(X)_t-S(X)_{t-1})$$
$$=D(X^{n+})_t+P(X^{n+})Ws-n[X^{n+}]_tVo_t \quad (1)$$

where Vw indicates water volume (m^3) in the soil of 1 m^2 of area and d m of depth, Ws : soil weight (kg), Vo_t : leachate volume (m^3), $[X^{n+}]$: concentration of X^{n+} in soil solution (mol/m^3), $ex(X^{n+})$: concentration of X^{n+} in an adsorption layer (cmol(+)kg^{-1}). $D(X^{n+})$ denotes the deposition rate (eq/m^2) when objective soil is a surface layer, and it means input by leaching from the upper layer for sub-layers. In case of aluminum, $D(Al)$ is considered only for the sub-layer. $P(X^{n+})$ denotes consumption or supply of elements within the soil system (eq/kg) due to mineral weathering, nutrient uptake, organic matter mineralization, nitrification etc. In the current model, $P(X^{n+})$ does not vary with time and the constant yearly flux must be specified for each objective soil and ecosystem. $S(X)$ is took into account only for Ca and Al, which indicates the amount of Ca carbonate and Al hydroxide (eq/kg), respectively. Weathering of these minerals is exceptionally treated as a equilibrium process depending on the pH and concentration in the soil solution.

Under these conditions, the model took the several chemical processes into consideration, that were carbonate/bicarbonate equilibrium, Al hydrolysis, ammonium dissociation, carbonate dissolution, Al hydroxide dissolution, cation exchanges, sulfate adsorption and organic acid dissociation. The first three processes were treated as equilibrium processes and the common equilibrium constants derived from literature (Lindsay, 1979) were used. The equations in Table 1 were introduced to the model for the other processes which are treated as quasi-equilibrium processes, and the reaction constants were specified for the individual soil.

Table 1 Process descriptions used in the model

Al hydroxide dissolution: $[Al^{3+}] = K_{Alox}(\gamma_1[H^+])^3 / \gamma_3$ (2)

Cation exchange: $A(H^+)^n / A(X^{n+}) = S_{H/X} \gamma_1^n [H^+]^n / \gamma_n [X^{n+}]$ (3)

   X^{n+}: Ca^{2+}, Mg^{2+}, Na^+, K^+, NH_4^+, Al^{3+}

Sulfate adsorption:

   Langmuir's adsorption isotherm

   $A(SO_4^{2-}) = A(SO_4^{2-})_{max} [SO_4^{2-}] / \{[SO_4^{2-}]_{1/2} + [SO_4^{2-}]\}$ (4)

   $A(SO_4^{2-})_{max}$ maximum adsorption capacity

   $[SO_4^{2-}]_{1/2}$ equilibrium SO_4^{2-} concentration for half saturation

   Freundlich's adsorption isotherm

   $\log(A(SO_4^{2-})) = K_s [SO_4^{2-}]^a [H^+]^b$ (5)

Organic acid dissociation: Oliver's equation (Oliver et al. 1983)

   $[RCOO^-] = K_{org} [RCOO] / (\gamma_1 [H^+] + K_{org}) / \gamma_1$ (6)

   $-\log K_{org} = a_0 + a_1 pH - a_2 (pH)^2$,

   $[RCOO] = [RCOO^-] + [RCOOH]$

$\log \gamma_i = -0.509 z_i^2 (\mu^{1/2} / (1 + \mu^{1/2}) - 0.3\mu)$ (Davies' equation)

μ : ionic strength ($= \Sigma_i c_i z_i^2 / 2$), c_i : concentration (mol/l), z_i : valence

2.2. APPLICATION OF THE MODEL TO THE SOIL ACIDIFICATION EXPERIMENTS

Japan Environment Agency conducted the soil acidification experiments to characterize the sensitivity of typical Japanese soils to the acid deposition with the aid of prefectural research institutes. The model was applied to these experiments in order to evaluate the model applicability. Table 2 shows initial soil properties of twelve soil samples used for model application. Soils of 8 to 10 kg was packed into a pot with 500 cm² of base area (with soil depth of about 20 cm) and was treated with five liters of simulated acid rain, which was the pH adjusted sulfuric acid or distilled water, every week for 10 or 20 weeks a year. Soil pH and concentration of exchangeable cations were measured just before and after the treatment every year. For the soil samples of Gunma Prefecture and Kanagawa Prefecutre that are indicated as group I in Table 2, the soil was sampled after mixing up entire soil in a pot each year. For the other soils (indicated as group II), on the other hand, soil was sampled from top 10 cm. Percolated water was divided into the former half and the latter half every year and ion concentrations were analyzed. After the treatment with simulated acid rain, the pot with the soil was covered and left alone in the laboratory rest of the year.

Parameters required by the model such as K_{Alox}, $S_{H/X}$, parameters of sulfate adsorption etc. shown in Table 1 were calculated based on the soil measurements after the acid rain application and concentration of the latter half of percolated water in each year, and their averages over four years were

Table 2 Soils used for acidification experiments

Prefecture	Stand Group	Soil weight kg	Soil pH	CEC cmol(+) kg^{-1}	Base saturation %	Ex-Al cmol(+)kg^{-1}	Water content %	Soil type
Hokkaido	II Asahikawa	10	5.71	17.0	54.4		32.2	Gray Upland soils
	II Forest station	10	6.01	14.6	41.9		50.8	Andosols
	II Tokachi	10	5.83	12.1	27.6		35.8	Andosols
Gunma	I Haruna	10	4.68	13.8	4.3	1.84	42.0	Andosols
	I Hanamigahara	10	4.76	32.8	3.1	3.89	58.3	Andosols
Kanagawa	I Fudakake	10	5.66	28.2	43.8		41.5	Brown Forest soils
	I Shimoyasiro	10	5.63	21.0	45.1	0.3	33.2	Brown Forest soils
	I Miharashidai	10	5.91	20.9	44.2		46.0	Andosols
	I Yanagisima	10	5.82	10.4	72.20.	88	8.4	Sand-dune Regosols
Fukuoka	II Yahata	8	5.00	31.8	10.7	5.64	28.2	Brown Forest soils
	II Ogohri	8	4.50	17.7	2.8	7.59	26.6	Brown Forest soils
	II Ohmuta	8	4.60	34.1	8.5	14.46	32.4	Brown Forest soils

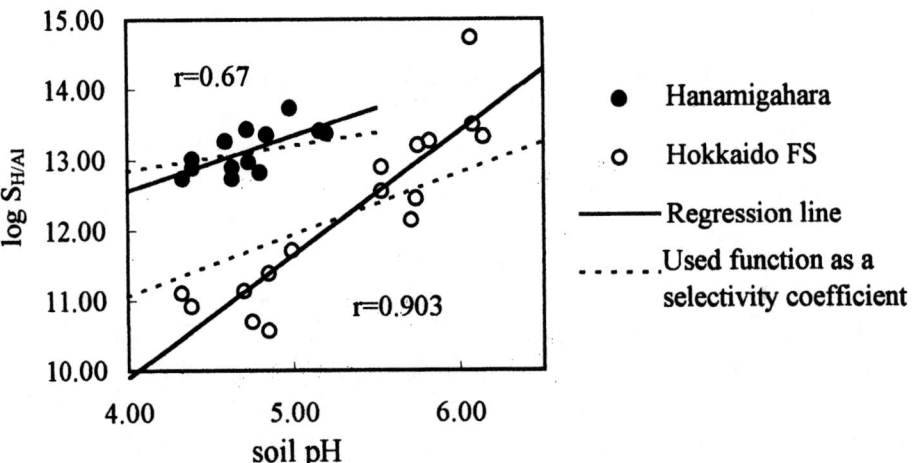

Figure 2 Dependence of the selectivity coefficients on soil pH

used for the model inputs. As for the nitrate ion and base cations, net production in soil, P(X), was taken into account. Net production of nitrate was calculated by assuming that total amount of nitrate in percolated water was produced by nitrification during the experimental period. Net production of base cations due to mineral dissolution and organic matter decomposition was calculated as a difference between total amount in leachate and decrease of exchangeable cations.

Although the minerals supplying solution with aluminum ion were not known for each soil, Al hydroxide was assumed. The estimated values of K_{Alox} ranged from $10^{8.0}$ to $10^{9.5}$ (mol L^{-1})$^{-2}$ which approximately corresponded to the dissolution constant for gibbsite and amorphous, respectively.

In many existing models selectivity coefficients for the cation exchange reaction is assumed constant. As shown in Figure 2, however, the selectivity coefficient of aluminum to proton derived from the results of experiments for each year and for each pH of simulated acid rain were strongly correlated with the soil pH. Such remarkable correlation was shown in other cations and for almost all soils used for the estimation. One reason of the correlation was considered that the activity coefficient (f_X) of adsorbed ion X^{n+} increased with the increase of its molar ratio, whereas the selectivity coefficient was expressed as

$$S_{H/X} = f_x / f_H^n \, K_{H/X} \qquad (7)$$

where $K_{H/X}$ denotes the equilibrium constant. The increased ratio of exchangeable hydrogen ions with accordance to the pH decrease causes the increase of its activity coefficient (f_H) and the decrease of the selectivity coefficient. Another possible reason is that soil contains several kind of colloidal constituents with different selectivity coefficients. Assuming the mixture of imaginary constituents whose $\log S_{H/X}$ for divalent ions are 7.0, 8.0 and 9.0, the apparent $\log S_{H/X}$ showed also positive correlation with soil pH (Figure 3). As the dependence of the selectivity coefficient on soil pH was considered realistic, the equation (3) was extended to the following formula to express the cation exchange process:

$$A(H^+)^n / A(X^{n+}) = S_0 \gamma_H^{n-b} [H^+]^{n-b} / \gamma_X [X^{n+}] \qquad (8)$$

$$S_{H/X} = S_0 \, 10^{b*pH}. \qquad (9)$$

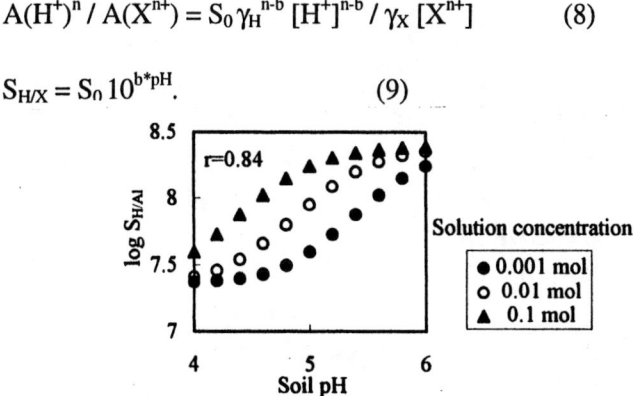

Figure 3 Relation between selectivity coefficients and soil pH for the imaginary soil which contains constituents with selectivity coefficients of $10^7, 10^8,$ and 10^9

2.3. RESULT OF MODEL ESTIMATION

One layer model and two layers model were applied to soils in Group I and Group II, respectively. Predicted soil chemistry changes were compared to the measurements during four years experiment with acid rain of pH 3 and 4 for Hanamigahara and Hokkaido Forest Station soils in Figure 4. "B" and "A" indicate the "'before" and "after" the treatment with simulated acid rain of each year, respectively. In the figure of the Hokkaido Forest Station, solid line indicating the predicted values of top-soil should be compared with the measurements. The dashed lines are prediction for sub-soil. The trend of soil chemistry change could be predicted with the model to some extent especially for the cases using simulated acid rain of pH 3. In these predictions, dependence of the selectivity coefficients on pH was assumed only for aluminum exchange ($S_0 = 10^{11.42}$, b=0.36 for Hanamigahara soil and $S_0 = 10^{7.55}$, b=0.88 for Hokkaido FS soil. Values of b for other cations were set to 0). When $S_{H/Al}$ was assumed constant, the increase of exchangeable aluminum was not realized with the model estimation. Figure 4(b) showed that the changing patterns of soil chemistry of top-soil and sub-soil were estimated quite different each other: rapid acidification occurred in top-soil while acidification in sub-soil was relatively moderate, though the changes in the sub-soil could not be validated based on the measurements. The final soil pH after four years experiments predicted with the model well corresponded to the measurements as shown in Figure 5. The correlation coefficients were 0.62 and 0.77 for the cases with simulated acid rain of pH 3 and pH 4, respectively.

In addition to the acid rain, some processes such as nitrification supplied soil solution with hydrogen ions, which were neutralized by several processes. Figure 6 indicates the contribution of the acid neutralizing processes during the changes of soils shown in Figure 4 for the cases that the pH of acid rain was 3. Input and produced acids were mainly neutralized by dissolution of Al hydroxides and cation exchange processes, and rate of proton leaching was quite low. Input acid was almost neutralized in the top-soil layer and nitrification was the main origin of acid supply in the sub-soil. According to the estimation with the model, 0.37 mol(+)/5 kg(soil) of aluminum ions were leached in the top-soil and were discharged passing through the sub-soil in the case of Hanamigahara soil. In Hokkaido Forest Station soil, amount of Al discharge was little: contribution of the cation exchange was relatively large and aluminum ion leached in the top-soil was adsorbed on the exchangeable sites or precipitated in the sub-soil.

The similar estimation was shown in Figure 7 for the entire soil mass of 12 objective soils. For soils of Group II, sum of the contributions in top-soil and in sub-soil are indicated. The soils of Gunma Prefecture and of Fukuoka Prefecture, of which pH and exchangeable base cations were initially low, showed the large buffering capacity due to Al hydroxide dissolution. To the contrary, base cation exchange reaction was dominant in the soils of Hokkaido and Kanagawa Prefecture. In some soils such as

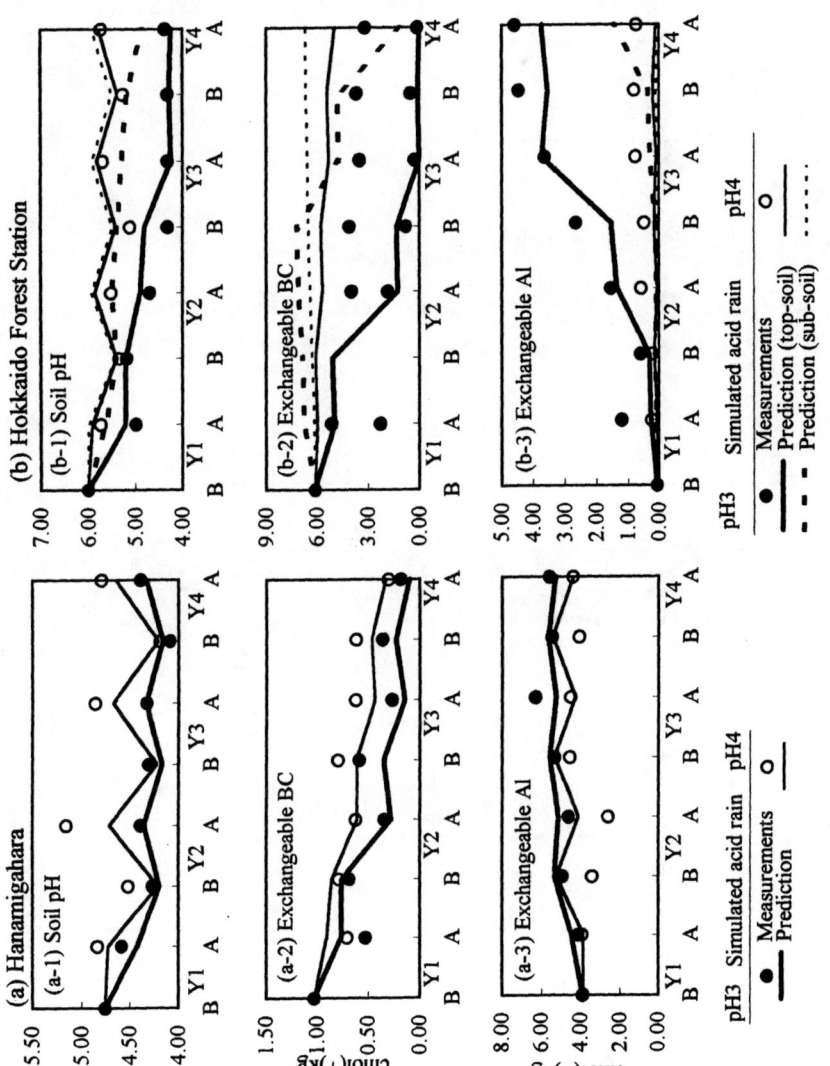

Figure 4. Comparison of the predicted temporal changes of soil chemistry to the measurements

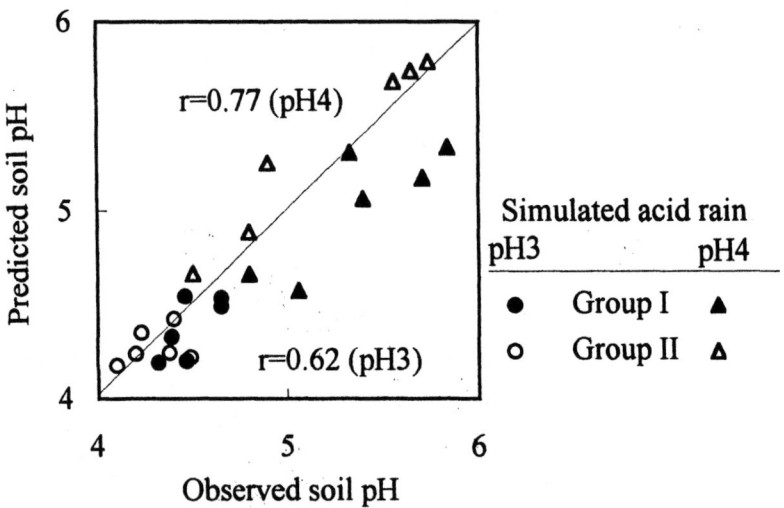

Figure 5 Comparison of the final soil pH predicted with the model to the measurements
In case of Group II soils, predicted values for the top layer are plotted.

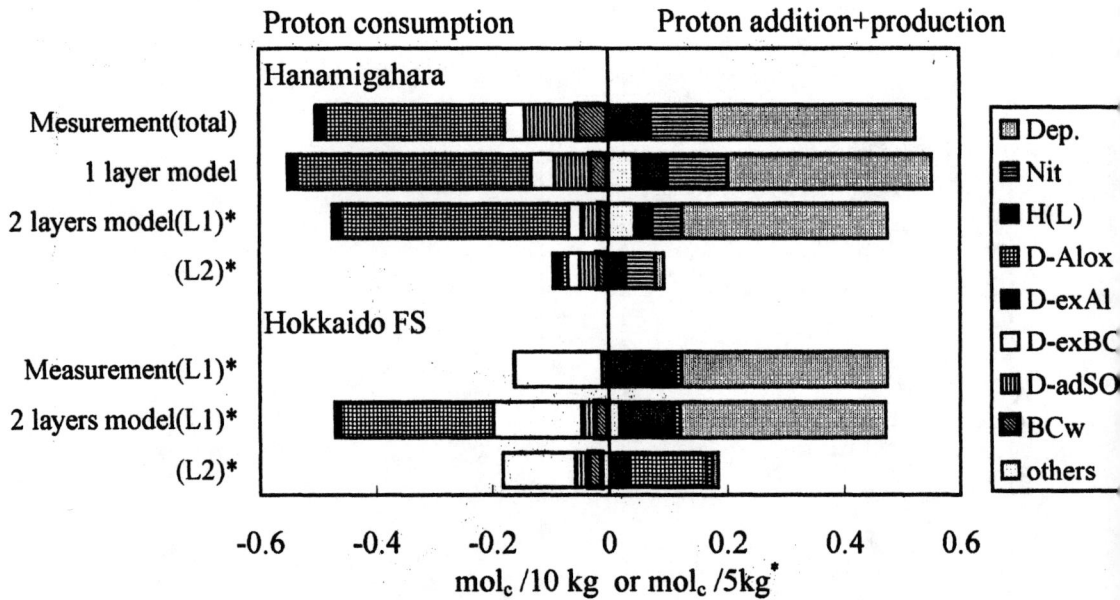

* The values are for the 5 kg soils of top or sub layer with 2 layers model

Figure 6 Proton budget for Hanamigahara soil and Hokkaido FS soil during 4 years experiment with acid rain of pH3

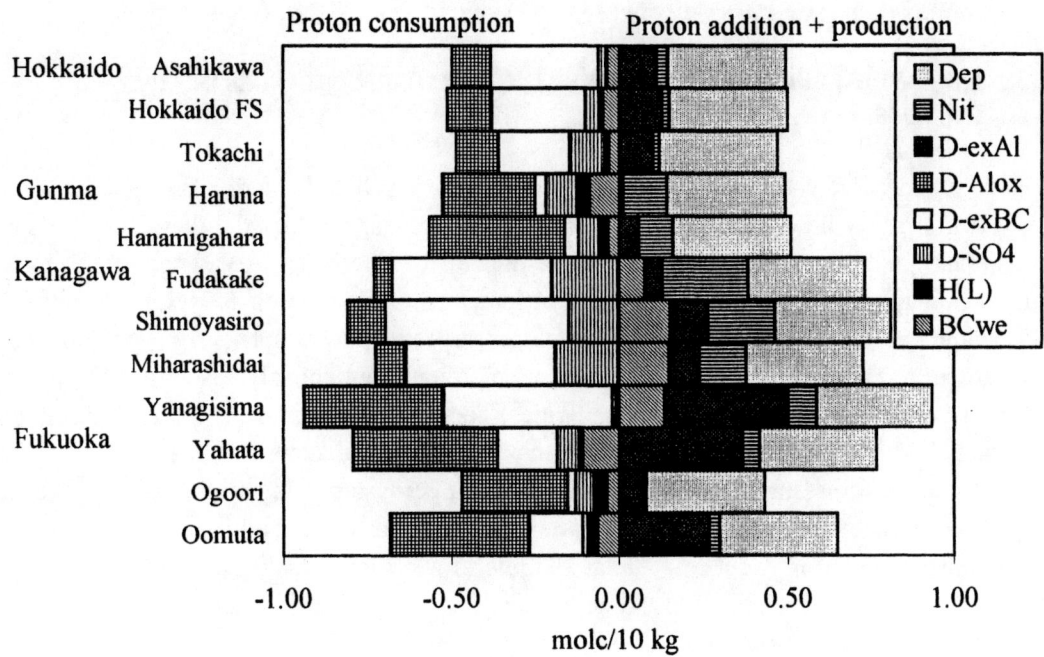

Figure 7 Proton budget for all objective soils during 4 years experiment with acid rain of pH3

Andosols in Tokachi, Haruna, Shimoyashiro etc. and Brown Forest soils in Kanagawa Prefecture, sulfate adsorption had a substantial contribution. In the third and the forth year, however, sulfate adsorption capacity almost saturated and sulfate ions added as acid rain were discharged with little adsorption.

In spite of the rough estimation of parameters, changes of soil chemistry caused by acid rain of pH 3 corresponded to the actual changes. For the experiment with the acid rain of pH 4, however, prediction accuracy was worse than that with the pH 3 rain. The discrepancies between predictions and measurements were considered to be caused by the errors in estimation of nitrification rate, base cation release rate due to mineral dissolution etc. Although constant values were used for these rates, they actually varied from year to year and also with the pH of simulated acid rain. The changing trend was not clear but the larger nitrificaition rate and the smaller mineral dissolution rate tended to be observed when the pH of simulated acid rain was higher. Estimation errors in these rates had a significant influence to the results of prediction of soil acidification by acid rain with pH 4 because the acid input rate of simulated acid rain was smaller comparing to the acid production or consumption rates by these processes. This fact indicates that it would be required to obtain accurate estimates of these rates in the real fields in order to make a reliable prediction for a real ecosystem where rain pH usually ranges from 4 to 5.

3. EVALUATION OF ECOSYSTEM SENSITIVITY WITH A STEADY STATE MODEL

3.1. LONG-TERM BUFFERING CAPACITY TO THE ACIDIC SUBSTANCES IN JAPANESE ECOSYSTEMS

Exchangeable cations that work as one of the immediate buffering processes against the acid input will be depleted unless base cations are supplied by the dissolution of soil minerals or soil organic matters. Therefore mineral weathering is considered the ultimate neutralizing process from the view point of long-term evaluation on ecosystem sensitivity to acidic deposition. Some biological processes such as nutrient uptake by trees, nitrification, denitrification etc., that supply or consume protons continuously, also affect the acidic condition of ecosystems and their sensitivities. The steady state model which evaluates the element budget by these processes, has been proposed to evaluate the critical load of acidic substances as a measure of long-term sensitivity of ecosystems. In the previous chapter, it was shown that the flux estimation for proton and related elements due to these processes were also essential in order to evaluate the rate of short-term acidification. As the steady state model is usually applied to the wide area, it is required to get data for the national or regional basis for instance.

In our application of a steady state model, the following methods were used for estimation of mineral weathering rates and nutrient net uptake rates mainly on the basis of the several national data bases whose spatial resolutions are one square kilometer. Ammonium ion loaded from atmosphere is taken up by the vegetation and it was assumed that remaining ammonium was nitrified completely in a soil layer in our estimation. Other processes like denitrification, nitrogen immobilization and nitrogen fixation were not taken into consideration because their contribution to the entire proton budget were considered relatively small except for the particular limited ecosystems and because of the difficulty to obtain reasonable estimation of the rates. The following is a preliminary estimation using the simplest methods and improvement of the model itself and the method for parameter derivation should be done.

3.1.1. Mineral weathering rate (BC_{we})

There are quite little data on weathering rates of soil minerals in Japan. In European countries and in USA, on the other hand, numerical estimation of the weathering rates have been derived for many soils (and for ecosystems) based on the field measurements and/or model estimations, although there have been a serious discussions on the reliability of the estimation (Bain and Langan, 1995; Langan et al., 1995, Hodson et al, 1997).

As the first step of the estimation, a table produced by De Vries (1991) was used as a transfer function. The table contains the round number of weathering rate at the temperature of 5 degree for each group defined by the acidity classes of parent material (acid, intermediate and basic) and by the texture classes with six levels. In the application of this transfer function to Japanese soils, surface geology data was used to identify acidity classes and clay contents were derived from soil type data. Surface geology and soil type data were provided as the Digital National Land Information data bases (Geographical Survey Institute, 1992). Values derived from the (European) transfer function was corrected with the temperature according to the following Arrhenius' equation to estimate the mineral weathering rate ($BC_{we(E)}$) at each grid point.

$$BC_{we}(T) = BC_{we}(T_0) \exp(A/T_0 - A/T) \qquad (10)$$

$$A = E / R$$

where T and T_0 denote absolute temperature, E is activation energy of mineral dissolution (J mol^{-1}) and R is gas constant (8.315 J K^{-1}mol^{-1}). Activation energy usually ranges from 20 to 32 kJ mol^{-1} and 30 kJ mol^{-1} (A=3600 K) was used after the European application.

In order to evaluate the uncertainty of the weathering rate estimation and to improve the transfer function, a soil survey was carried out at 42 sites in Hiroshima and Shimane Prefectures (Figure 8) . For each site surface soil just below a litter layer and sub-soil (at 5 to 15 cm depth) were collected at three sampling points that were 3 to 5 m apart each other. These sampling sites could be classified into six surface geology groups (granite, rhyolite, andesite, gabbro, volcanic ash and sediment rock), and into six soil type groups (Brown forest soils(dry), Brown forest soils, Andosols, Regosols, Red soils and Yellow soils).

Particle-size distributions were measured with sieving method and by density measurements of the suspension of soil particles. Element contents in both of smaller and larger than 2mm fractions were analyzed with ICP spectrometry. PROFILE model (version 3.2, Sverdrup and Warfvinge, 1993) was applied to these soils by assuming two soil layers of 10 cm and 40 cm to estimate weathering rates ($BC_{we(P)}$). In this estimation, UPPSALA model was used to specify the mineralogy based on element content data and to calculate surface area of soil particles from size distribution data (Sverdrup et al., 1992). Other input data were the same for all sites that were rough averages of the area.

Figure 9 shows average and standard deviation of $BC_{we(P)}$ for each surface geology group. According to the classification method used for weathering rate estimation with the transfer function described above, surface geology of sampling sites were classified into tree acidity groups as indicated in Figure 9. This classification was considered appropriate qualitatively because average rates based on the soil measurements ($BC_{we(P)}$) showed a significant difference between soils on acidic rocks (granite and rhyolite) and other soils, though the difference between intermediate and basic rocks were not clear. Comparison of $BC_{we(E)}$ and $BC_{we(P)}$ in the corresponding grid point showed, however, that estimation has large uncertainty due to spatial variability within the same geology group especially for the intermediate group (Figure 10). In this figure, two times of $BC_{we(P)}$ was regarded as the weathering rate by the PROFILE model of soils with 1 m depth. It had also biases: $BC_{we(P)}$ is generally larger than $BC_{we(E)}$ for soils with intermediate acidity. For acidic soils, on the contrary, $BC_{we(P)}$ was smaller than $BC_{we(E)}$ and had no relation to clay content derived from soil types. On the basis of this Figure, the transfer function was revised and weathering rates were re-calculated which was indicated as $BC_{we(R)}$.

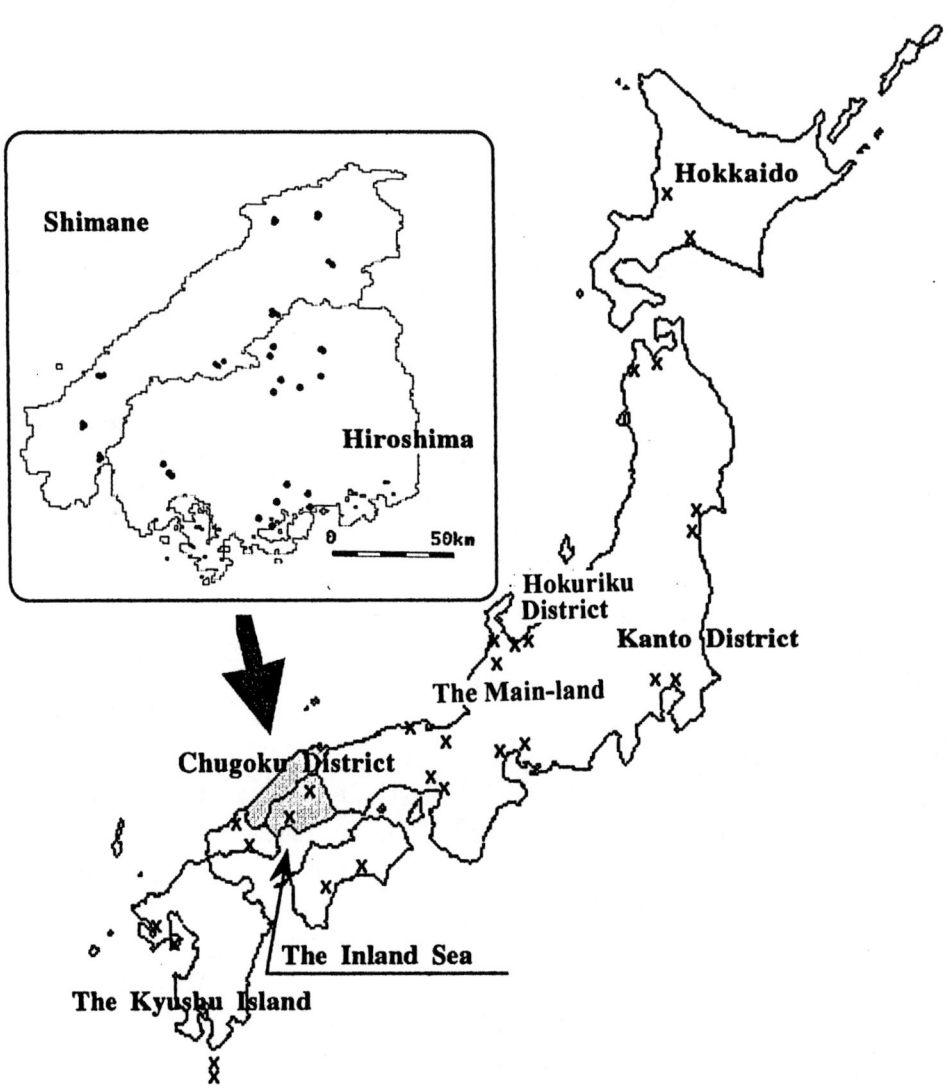

Figure 8. Locations of soil sampling (●) in Hiroshima and Shimane Prefectures and monitoring stations for acidic deposition (x).

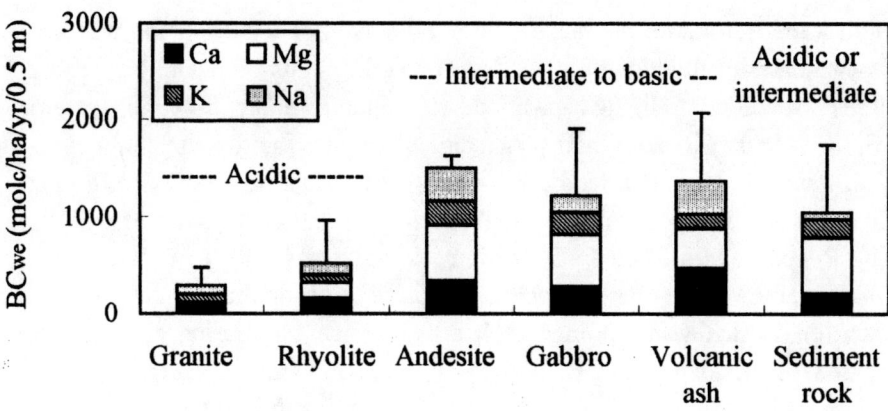

Figure 9 Average and standard deviation of base cations release rate due to mineral weathering estimated with the PROFILE model for each geology group

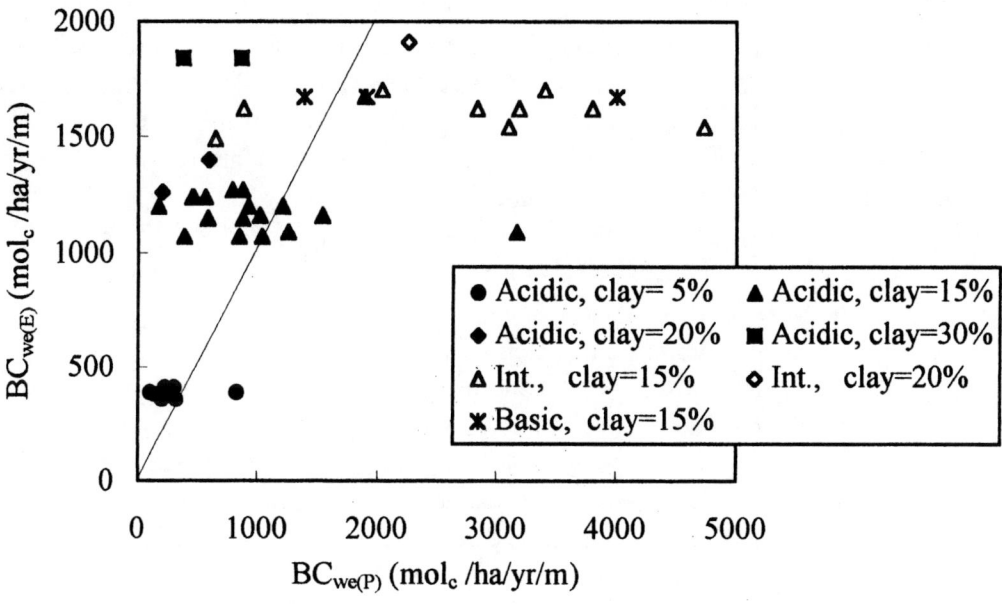

Figure 10 Comparison of weathering rate estimation with the different methods
BCwe(E): based on surface geology and soil types with the European transfer function
BCwe(P): based on soil measurements and the PROFILE model

Table 3 shows the minimum, the maximum and the average values of the weathering rates based on two transfer functions for all terrestrial grid points of Japan. According to the estimation results, soil neutralizing capacity due to mineral weathering varies extremely by location. Soils with small neutralizing capacity distributed at the mountainous area around the central part of the Main-land covered with Lithosols, in the Chugoku district including Hiroshima and Shimane Prefectures (especially in the north coast of the Inland Sea covered by Regosols originated from acidic rocks) and some area in Hokkaido etc. Spatial distribution of $BC_{we(R)}$ quite resembles the distribution of the exchangeable base cation concentration, which was obtained with an extensive monitoring carried out by Forest Agency of Japan (Ishizuka and Taoda, 1997), indicating that the estimation method is adequate qualitatively. It is uncertain, however, whether the estimated results are credible or not quantitatively. Large part of Japanese soils are affected by volcanic ash and contain easily weatherable minerals. The used transfer function and the PROFILE model were based on the data of European soils and could not regard such property of Japanese soils, for example.

3.1.2. Base cation and nitrogen uptake rates (BC_{gu}, N_{gu})

We have rather a lot of domestic data on forest production and nutrient cycling in forest ecosystems that can be used to estimate base cation and nitrogen uptake rates. Growth uptake of element X (X_{gu} (mol$_c$ha^{-1}yr^{-1})), that was accounted as the amount of the element contained in the yearly increment of tree bodies (stems and branches), was calculated according to the following equation;

$$X_{gu} = P_{body} * \{X_{st} + f_{br/st} * X_{br}\} / (1+f_{br/st}) \qquad (11)$$

where P_{body} denotes production of stems and branches (kg ha^{-1}yr^{-1}), $f_{br/st}$ is branch to stem ratio (kg kg^{-1}), and X_{st} and X_{br} are concentration of X in stems and branches (mol$_c$ kg^{-1}), respectively. Based on the data in literature, element content in stems and branches were shown in Figure 11 for several types of the forest in Japan (DB : deciduous broad leaves, DN : deciduous needles, EB : Evergreen broad leaves, and EN : Evergreen needles, T and SA in parentheses indicate the Temperate forest and the Sub-arctic forest, respectively). Values are trimmed means of the measurements of 20 to 30 stands for DB(T) and EN(T) forests and of about 5 stands for other types of forests. In the figure, $f_{br/st}$ calculated based on stem and branch biomass is indicated for each forest type. P_{inc} was estimated as the product of net primary production (NPP: kg ha^{-1}yr^{-1}) and net growth to NPP ratio that was also evaluated on the basis of data in the literature and was shown in the Table. Net primary production varies with temperature, precipitation, soil productivity and so on. The following equations proposed by Lieth (1975) were used and the smaller value of those calculated with the two equations was taken as the NPP values for each grid point.

$$\text{NPP}(\text{t ha}^{-1}\text{yr}^{-1}) = 30 / \{1 + \exp(1.315 - 0.119\ T)\} \qquad (12)$$

$$NPP(t\ ha^{-1}yr^{-1}) = 30\ \{1 - \exp(-0.000664\ P)\} \qquad (13)$$
(T: Yearly average temperature (degrees centigrade), P: annual precipitation (mm))

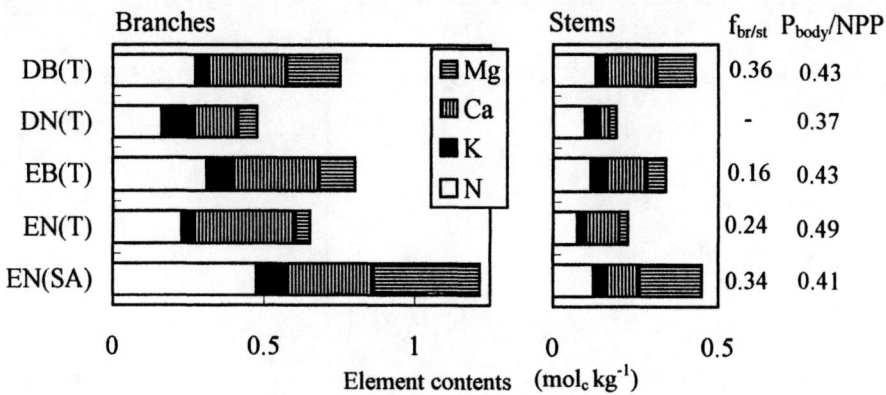

Figure 11 Element contents in stems and branches for the several types forests $f_{br/st}$: Ratio of branch biomass to stem biomass
P_{body}/NPP: Ratio of net growth of stems and branches to net primary production

Figure 12 shows the change of NPP with temperature and with precipitation expressed with the above equations and ranges of measured aboveground NPP for Japanese forest (about 90 stands in total) appearing in the literature. In the natural forests, it was considered that the forests were at a mature stage and growth uptake was set to zero. Information on the forest types in each grid was derived from vegetation data base of the Japan Environmental Agency made from the 1 : 50 000 vegetation map of Japan (Japan Environment Agency, 1983). Number of grids covered by natural forests was about 57,000 that was one forth of the forested area in Japan. Precipitation data (average of 24 years; 1953-1976) and temperature data (average of 30 years; 1953-1982) were produced by the Meteorological Agency of Japan and provided by the National Land Agency of Japan.

Table 3 shows the statistics of the calculated growth uptake rates of base cations and nitrogen and the difference between them, that corresponds to the net acid production by nutrient uptake. Larger uptake rates were observed in the southern part of Japan reflecting higher NPP. According to the Figure 11, element concentration is high for the evergreen needles forest in Sub-arctic zone and the deciduous broad leaves forest in Temperate zone. Excess uptake of base cations over nitrogen had the same tendency. In these types of the forests, soil acidification is relatively accelerated through vegetation growth if the trees are harvested.

The uptake rates calculated according to the equation (11) was regarded as a potential rate and they were replaced with nutrient supply due to deposition and weathering if the calculated uptake rate was larger than the supply in the application of the steady state model of the following section.

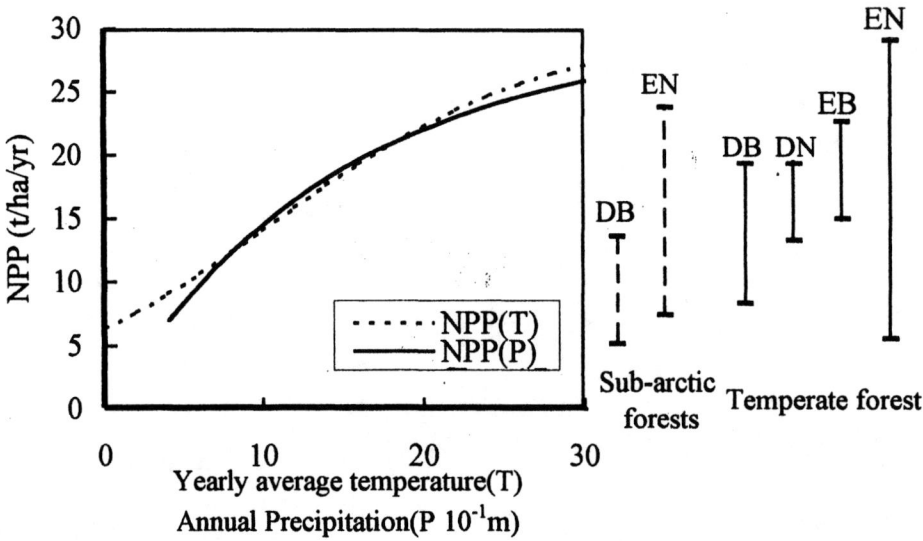

Figure 12 Net primary production estimation with a model and ranges of measured values for Japanese forests derived from literature

Table 3 Statistics of estimated parameters for element budget calculation in Japan (mol$_c$ha^{-1}yr^{-1})

	Minimum	Maximum	Average
Base cation weathering ($BC_{we(E)}$)*	96	3564	1382
Base cation weathering ($BC_{we(R)}$)*	74	5198	1729
Base cation uptake (BC_{gu})**	674	2839	1663
Nitrogen uptake (N_{gu})**	337	1359	819
Acid production due to uptake ($BC_{gu}-N_{gu}$)**	337	1480	858
Deposition rates*** H$^+$	34	1343	335
NH$_4^+$	151	1690	407
BC (Ca^{2+}+Mg^{2+}+K$^+$)	226	2678	908

* Statistics for whole Japan (about 360,000 grid points),
** Statistics for the secondary or planted forests (about 180,000 grid points)
***Statistics for the 29 monitoring stations

3.2. ESTIMATION OF AN IMPACT INDICATOR OF ACIDIC DEPOSITION

Assuming the steady state condition where soil chemistry such as soil pH, concentration of exchangeable cations etc. are stable, the acid leaching rate from the soil layer ($Acid_{le}$) is calculated as a difference between acid atmospheric input rate ($Acid_{de}$) of acidic substances and net acid neutralization rate due to the processes described above.

$$Acid_{le} = Acid_{de} - BC_{we} - N_{gu} + BC_{gu}$$

where all terms are expressed as the annual flux (unit : $mol_c ha^{-1} yr^{-1}$). $Acid_{de}$ was calculated as a sum of proton deposition rate and two times of ammonium deposition rate, because ammonium ions produces 2 times of protons by nitrification in soils. $Acid_{le}$ was assumed to consist of leaching of protons and aluminum ions on the basis of the assumption that dissolution of aluminum hydroxide (or aluminosilicates) was taking place to neutralize acid. Concentrations of protons and aluminum ions ($mol\ m^{-3}$) in the leached water were calculated by solving the following cubic simultaneous equations:

$$3\ [Al^{3+}] + [H^+] = Acid_{le}\ /\ Q$$

$$[Al^{3+}]\ /\ [H^+] = K_{Alox}$$

Q denotes the runoff water flux through soil layers ($m^3 ha^{-1} yr^{-1}$) and K_{Alox} is the dissolution constant of Aluminum compound ($mol^2 m^{-6}$). Amount of protons in the leached water estimated through this method is the excess of acid which can not be neutralized by chemical weathering, net uptake of nutrients and aluminum dissolution. And if the pH of the leached water is much lower than the current soil pH, it would be feasible that other buffering mechanisms such as cation exchange, sulfate adsorption etc. are actually working. This means that the ecosystem is not at the steady state and soil acidification is going on by losing exchangeable cations, closing to the sulfate saturation etc. Therefore we define the difference between soil pH and the calculated pH of leached water on the steady state condition as an impact indicator of the ecosystems caused by acidic deposition.

In order to apply the steady state model described here, the following databases and methods were used to derive the additional parameters. As $Acid_{de}$, two year averages from 1986 to 1987 of deposition rates measured with bulk samplers at 29 locations plotted in Figure 8 were interpolated to the 1 km square grid data. Though the data is rather old, there are no significant difference between them and the current data. A bulk sampler collects wet deposition and a limited part of dry deposition. We do not have nationwide data on dry deposition but some reports showed that the amount of dry deposition is as much as the amount of wet deposition. We doubled the values of the database to make total (wet and dry) deposition rates. Summarized values of the deposition rates that was used for estimation were also shown in Table 3. Areas with high deposition rates were located from Kanto to Hokuriku Districts and the southern part of the Kyushu District. In the former area, ammonium deposition rate was especially high. Runoff Q, was extracted from the Japanese Environmental Data Base (JEDB) compiled by Nogami (Nogami, 1992). In this database, runoff was calculated as the difference between the annual precipitation and the annual evapotranspiration estimated from yearly average temperature by the Thornthwaite method. K_{Alox} was assumed to have a value corresponding to the amorphous dissolution ($K_{Alox} = 10^{9.66}$) for the volcanic ash soil areas and to the gibbsite dissolution ($K_{Alox} = 10^{8.04}$) for the other areas.

The result of the steady state estimation showed that soil neutralizing capacities due to mineral weathering were sufficiently large to eliminate acid input from atmosphere and acid produced by nutrient uptake in about 25 % of the forest area in Japan. Such areas were mainly appeared in Hokkaido, northern part of the Main-land and Kyushu District. In the remaining area, dissolution of Al compounds took place to neutralize the acid under the steady state assumption. Leached water with low pH were observed in some localized areas. Leachate pH resulted in under pH4 at the north coast of the Inland Sea where buffering capacity of soils were small as described before. But it would not be probable that the ecosystem acidification progresses drastically there, because the soils are originally acidic with pH around 4.0. Areas with very low pH estimation of leached water were also identified at some areas in the central mountainous district and around the northern edge of the Kanto District. Original soil pHs of these areas are not very low and it was considered that there was a possibility of soil acidification going on caused by acidic deposition especially by ammonium deposition.

4. CONCLUSIONS

Acidification and neutralization processes taking place in ecosystems are very complicated and diversified. It is important to determine which processes should be extracted and in which level of accuracy these processes should be quantified in the modeling approach. It depends on the estimation objectives, temporal and spatial scale of the estimation.

A dynamic model developed for prediction of soil acidification in Japanese ecosystems and application of a steady state model were described in the paper. Change of chemical properties of soil samples could be estimated with the model as the result of chemical reactions to a certain extent. It is necessary to verify the model by apply it to the real ecosystems. The model will be helpful to understand the acidification process in an objective ecosystem when it is improved adaptively by introducing the characteristics process of that ecosystem.

As for the steady state model, its objective is considered to provide general insight on the effects of acidic deposition to the broad areas. It would be too rough to describe the acidification state of a specific small catchment. The structure of the model is rather simple and reliability of the estimate with the steady state model depends on the accuracy of the estimates of element fluxes by various processes relating to the proton budget in the ecosystems. Weathering rate estimation dose not have the sufficient accuracy in the estimation in this paper. It based on the European soil data that is thought quite different ecosystems. Weathering rate estimation dose not have the sufficient accuracy in the estimation in this paper. It based on the European soil data that is thought quite different from Japanese soils. Improvement of the weathering rate estimation is necessary on the basis of field measurements in Japan. The same is true for the nutrient uptake processes and other biological processes that were not used. The objective of the previous section is proposing the evaluation method of the possible ecosystem impact by means of the steady state model. The speculation on the ecosystem impact made there should be re-examined based on more accurate data.

REFERENCES

Bain, D.C. and S. J. Langan (1995) Weathering rates in catchments calculated by different methods and their relationship to acidic input. *Water, Air and Soil Pollution* 85, 1051-1056

Cosby, B. J., RF. Wright, G. M. Hornberger and J. N. Galloway (1985) Modeling the effects of acid deposition: estimation of long-term water quality responses in a small forested catchment. *Water Resources Research* 21, pp.15911601

De Vries, W. (1991) Methodologies for the assessment and mapping of critical loads and the impact of abatement strategies on forest soils. *Report 40*, DLO the Winand Staring Centre, Wageningen.

De Vries, W., G. J. Reinds, M. Posch and J. Kamara (1994) Simulation of soil response to acidic deposition scenarios in Europe, *Water, Air and Soil Pollution* 78, 215-246

De Vries, W., J. Kros and C. van der Salm (1995) Modelling the impact of acid deposition and nutrient cycling on forest soils, *Ecological Modelling* 79, 231-254

Falkengren-Grerup, U. (1987) Long-term changes in pH of forest soils in southern Sweden, *Environmental Pollution* 43, 79-90

Geographical Survey Institute (1992) *Users guide of digital national land information.* pp.494 (in Japanese)

Goldstein, R. A., Chen, C.W. and Gherini, S.A. (1985) Integrated lake-watershed acidification study: summary, *Water, Air and Soil Pollution* 26, 327-337

Hallbacken, L. and C. O. Tamm (1986) Change in soil acidity from 1927 to 1982-1984 in a forest area of south-west Sweden, *Scand. J. For. Res.* 1, 219-232

Hettelingh, J. P., R. J. Downing, P. A. M. de Smet (1991) Mapping critical loads for Europe, *CCE Technical report No.1*, RIVM, Bilthoven, the Netherlands, 86pp.

Hodson, M. E., S. J. Langan, M. J. Wilson (1997) A critical evaluation of the use of the PROFILE model in calculating mineral weathering rates, *Water, Air and Soil Pollution* 98, 79-104

Ishizuka, K. and H. Taoda (1997) The sate of Japanese forest and soil - Monitoring of forest damage caused by acid rain and air pollution. *Proceedings of the International Symposium on Forest Decline Caused by Air Pollution* - Photo-oxidants, Acid Rain and Fog -, Japan Science and Technology Corporation, Hiroshima, Japan, 6-9.

Japan Environment Agency (1983) *User s manual for the data base of the national state of nature survey*. Environmental Agency, Tokyo, p 377. (in Japanese)

Langan, S. J., M. E. Hodson, D. C. Bain, R.A. Skeffington and M. J. Wilson (1995) A preliminary review of weathering rates in relation to their method of calculation for acid sensitive soil parent materials. *Water, Air and Soil Pollution* 85, 1075-1081.

Lieth, H. (1975) Modeling the primary productivity of the world. In: H.Lieth and R. H. Whittaker (eds.) Primary productivity of the biosphere, *Ecological Studies* 14, Springer-Verlag, New York, pp.339

Lindsay, W.L. (1979) *Chemical equilibria in soils*, John Wiley & Sons Inc., p449, New York

Nogami, M. (1992) Files in Japan environmental data base (JEDB). *News letter GIS for environmental change*, 14.

Oliver, B. G., E. M. Thurman and R. L. Malcolm (1983) The contribution of humic substances to the acidity of colored natural waters. *Geochim*. Cosmochim. Acta 47, 2031-2035

Shindo, J. and T. Hakamata (1991) Survey on soil response to acidic deposition - Spatial variability of soil pH and concentration of exchangeable Ca and Al -, *Journal of Center for Environmental Information Science* 20(2), 67-74 (in Japanese)

Sverdrup, H., P. Warfvinge and C. Jonsson (1992) Critical loads of acidity for forest soils, groundwater and first order streams in Sweden, HMSO *Proceedings from the British Critical Loads conference*, Institute of Terrestrial Ecology, Merlewood Research Station, Cumbria, Great Britain.

Sverdrup, H. and P. Warfvinge (1993) Calculating field weathering rates using a mechanistic geochemical model PROFILE, *Applied Geochemistry* 8, 273-283.

Sverdrup, H and W. de Vries (1994) Calculating critical loads for acidity with the simple mass balance method, *Water, Air and Soil Pollution* 72, 143-162.

Van Breemen, N., J. Mulder, C. T. Driscoll (1983) Acidification and alkalinization of soils, *Plant and Soil* 75, 283-308

Van Breemen, N., C. T. Driscoll, J. Mulder (1984) Acidic deposition and internal proton sources in acidification of soils and waters, *Nature* 307,16 February 1984, 599-604

Warfvinge, P., U. Falkengren-Grerup, H. Sverdrup, B. Andersen (1993) Modelling long-term cation supply in acidified forest stands, *Environmental Pollution* 80, 209-221

Chapter 10
SENSITIVITY OF THAILAND'S ECOSYSTEMS TO ACIDIC DEPOSITION

MICHAEL KOZLOV AND SIRINTORNTHEP TOWPRAYOON
School of Energy and Materials, KMUTT, Bangkok, Thailand, imiczlor@cc.kmitt.ac.th

CONTENT

Abstract
1. Introduction
    1.1. Acidic emission and deposition in Thailand
2. General description of the study area
    2.1. Soil and Climate Database for Thailand
    2.2. Ecological resources
    2.3. Main occurring and potential impacts
3. Methodology
    3.1. Experimental part
    3.2. Calculation and mapping part
        3.2.1. General approaches and uncertainties
        3.2.2. New Methodology of SEI
4. Results
    4.1. Calculation and mapping of ecosystem sensitivity to acidic deposition
    4.2. Experimental results
        4.2.1. Soil properties
        4.2.2. Impact of acidity on tropic soil
        4.2.3. Impact on soil microorganisms
5. Conclusions
6. References

ABSTRACT

The objectives of this study are: i) to provide and express results of sensitivity assessment in term of maps for the studied area; ii) to determine and identify sensitive receptors and locations where abatement strategies would be implemented to reduce environmental impacts of acidification on forest and agriculture. New national data have been picked up for some Thailand ecosystems, and revised methodology for assessing ecosystem sensitivity to acidic loading has been applied as well as the experimental research concerning of the sulfur dioxide impact on the forest soils in the Northern Thailand has been conducted. The assessment of ecosystems sensitivity to acidic deposition has been carried out by using two steps procedure. Firstly, the sensitivity mapping according to revised methodology of Stockholm Environment Institute has been conducted for Thailand conditions. The aim of such sensitivity mapping is to define the distribution of ecosystems with the same relative level of reactions to the given rate of acidic deposition. On the second stage the sensitivity of Thailand ecosystems to acidic deposition has been described by means of Critical Loads (CL) and exceedances, using modified Steady-State Mass Balance model with simplified expert-modeling approach. The maps for CL of sulfur derived according to this methodology have been overlaid with current (or projected future) deposition maps in order to show areas where the CL of sulfur are (or will be) exceeded. Using a geographical information system (GIS) for manipulating with the numerous maps and data, the vast areas of high sensitivity to acid deposition have been determined in the Northern Thailand.

1. INTRODUCTION

At present the assessment of acidic deposition influence on the environment is very critical for Thailand since the anthropogenic emission of acidforming compounds has being increased significantly during last decades (Kozlov and Towprayoon, 1995; 1997). Recent monitoring studies in Thailand have documented the occurrence of acid rains over some parts of the Northern Thailand, even though there are no observable damages caused by acidic deposition on an ecosystem level. In the early 1990's, the issue of acidic depositions in the northern part of the country became a great concern for the government. The regional/local implications of acidic deposition in Thailand touch not only the natural environment, but also have far-reaching implications for important commercial and cultural activities such as forestry, agriculture and tourism.

Therefore, because of the increasing use of coal and heavy fuel oil in the energy sector and industry, the country continues to face severe challenges connected with acid rain problems. In particular, the dramatic growth of these energy systems carries with the continuing rapid growth in levels of sulfur emissions and acidic depositions in many areas of Thailand. The modern energy sources are represented mostly by fossil fuels with high sulfur content, through their combustion responses ultimately for the sulfur dioxide emission. The sulfur content of various fossil fuel in Asian countries are illustrated in Table 1 (Kato and Akimoto, 1991).

Table 1. Sulfur content of fuel used in South East Asia (unit : weight %), Kato and Akimoto, 1991.

Country	Hard coal	Brown coal	Motor Gasoline	Kerosene	Industrial diesel oil	Road diesel oil	Residual oil
Brunei	0.62	-	0.005	0.150	0.50	0.50	2.8
Cambodia	0.62	-	0.120	0.032	0.40	0.16	1.50
China	1.35	-	0.120	0.032	0.40	0.16	1.50
Indonesia	0.60	0.60	0.005	0.160	0.50	0.50	2.80
Laos	0.62	-	0.120	0.200	0.40	0.80	1.50
Malaysia	0.62	-	0.140	0.160	0.96	0.96	3.2
Myanmar	0.86	0.86	0.180	0.200	1.44	0.80	3.2
Philippines	0.65	0.98	0.035	0.020	1.0	1.0	3.2
Singapore	0.62	0.93	0.140	0.020	0.46	0.46	1.6
Thailand	2.8	2.8	0.035	0.020	0.50	0.66	2.92
Viet Nam	0.20	-	0.120	0.032	0.40	1.16	1.50

Hard and brown coal (mostly lignite) and residual oil in Thailand possess higher sulfur content than other fuel. One can see the higher sulfur content of those coal and residual oil in comparison with among Asian countries. However, Thai government has realized this situation and issues the fuel quality improvement policies in particular on these high sulfur coal, fuel oil and diesel oil. The sulfur content of diesel fuel is to be reduced from the level of about 1 % sulfur in 1990 to 0.05 % in 2010 and beyond. Sulfur content in heavy fuel oil will also be reduced from 3 % in 1990 to less than 2 % in 2010 and beyond. The implementation of these two policies is likely to occur sooner than it has been scheduled earlier.

Preliminary studies carried out by one of the authors with the help of RAINS/ASIA model have indicated that a base line scenario in Thailand could lead by year 2020 to acidic depositions considerably above the threshold at which many ecosystems begin to experience damage. The expected value of the sulfur dioxide emissions will be more than 2 million tons per year for the whole country if the base line scenario will be implemented. The great concern about sustainability of Thailand's ecosystems is timely arisen due to conclusion of a joint Thai-Swedish research that some parts of the country are as sensitive to acid deposition as those in Scandinavia (which are among the most sensitive ones in Europe).

Historically there have been two main approaches to controlling atmospheric pollutants from large point sources such as fuel or coal-fired power plants. The first approach simply employs the best available control technology with further monitoring of the environmental consequences arise from the resulting emissions. In the other approach, the emissions control program is designed to meet specified ambient air quality objectives. Generally, the latter approach is effective in protecting human health, especially in the region near the point source. But it is usually ineffective in preventing distant impacts to sensitive ecosystems. During the last decade a third approach - known as critical or target loads - has been developed in Europe and North America to overcome deficiencies in the other two strategies.

The Critical Load (CL) concept is receiving much attention as a measure for developing national emission abatement strategies. Although practical definition of CL depends upon the pollutant of concern and the target ecosystems under observation and can not be agreed easily, the concept is quite simple. It could be generally regarded as the highest load that will not cause harmful effects on the receptors under concern (Nilsson and Greenfelt, 1988). Since several receptors coexist in a given region (e.g., forests, crops, aquatic biota, human beings, etc.), a common strategy is to select a critical load, which protects the most sensitive receptor. The intent of this study, for instance, is to move as far as possible towards the assessment of critical loads for sensitive terrestrial ecosystems.

Research throughout the world has shown that both SO_2 and NO_x can play a significant role in acidic deposition, depending on their particular abundance and the many parameters of the relevant ecosystems. Over the past several years greater attention and priority has been devoted to sulfur, because of its much greater emissions at the current stage of industrial development in Thailand. The research conducted recently by Thai-Canadian research team in the northern Thailand showed that measured wet nitrate deposition rates represent less than 5 percent of the corresponding sulfate deposition rates. As concerning the dry deposition the ratio between sulfate and nitrate has a lot of uncertainties due to lack of valid and representative experimental data.

1.1. ACIDIC EMISSION AND DEPOSITION IN THAILAND

In 1990-1995, annual sulfur emissions in Thailand were around 1000-1300 metric kilo tons of SO_2, of which 45 % was estimated to originate from sources in northern Thailand and 30 % from the Bangkok Metropolitan Region (DEDP, 1995). The nitrogen emissions for the whole country were raising steadily from 508 metric kilo tons of NO_x in 1990 up to 815 metric kilo tons of NO_x in 1995 with more than 50 % originated from the Bangkok Metropolitan Region. The main sources for this emission were transportation (40-45 %) and power (21-23 %) sectors.

The main point source of SO_2 emission in the northern Thailand is Mae Moh power generation complex. This complex is located in the Mae Moh coal basin, 25 km to the east from the Lampang city in the northern region of Thailand. The existing project comprises the large open pit lignite mine and 13 operating lignite fired generating units with a total

Table 2. Anticipated Trends in Primary Energy Consumption for 1996-2010 years in PetaJoules by type of fuel, Khummongkol et al., 1996.

type of fuel	1996	1998	2000	2002	2004	2006	2008	2010	2020
Lignite									
Power Generation	158.8	163.5	163.5	163.5	163.5	163.5	163.5	163.5	72.1
Industry	92.7	109.4	128.7	151.7	178.7	210.1	246.7	289.4	485.95
Total	*251.5*	*272.9*	*292.2*	*315.2*	*342.2*	*373.6*	*410.2*	*452.9*	*558.05*
Imported Coal									
Power Generation	-	-	-	132.7	286.3	505.1	725.0	963.0	950.1
Industry	3.2	3.5	3.7	3.9	4.1	4.3	4.4	4.5	7.3
Total	*3.2*	*3.5*	*3.7*	*136.6*	*290.4*	*509.4*	*729.4*	*967.5*	*957.4*
Fuel oil									
Power Generation	259.0	260.6	120.9	119.7	111.0	65.5	45.7	117.5	613.0
Industry	155.1	181.6	214.3	244.3	281.8	323.8	370.9	430.9	768.7
Agriculture	0.2	0.1	0	0	0	0	0	0	0
Commercial	1.6	1.6	1.7	1.8	1.9	2.0	2.1	2.3	3.6
Total	*415.8*	*443.9*	*336.9*	*365.8*	*394.7*	*391*	*418.7*	*550.7*	*1385.3*
Natural Gas									
Power Generation	313.5	368.2	624.0	631.4	644.5	639.6	619.9	517.9	1679.0
Industry	13.8	16.0	18.3	21.0	24.0	27.3	31.0	35.0	60.39
Total	*327.3*	*384.2*	*642.3*	*652.4*	*668.5*	*666.5*	*650.9*	*552.9*	*1739.4*
Diesel									
Power Generation	27.6	44.1	5.6	7.4	8.0	9.9	11.4	14.7	6.1
Transport	341.8	373.9	403.4	430.2	455.9	480.4	503.7	592.6	656.8
Industry	27.2	31.2	36.9	43.1	50.1	58.2	67.4	77.81	131.6
Agriculture	89.0	95.6	102.2	105.2	108.1	111.1	114.1	117.0	190.9
Commercial	0.1	0.1	0.1	0.1	0.1	0.1	0.1	0.1	0.1
Total	*485.8*	*544.9*	*549.1*	*586.0*	*622.2*	*659.7*	*696.7*	*802.2*	*985.5*
Gasoline									
Transport	180.4	210.8	238.2	254.9	271.2	287.0	302.3	323.8	394.7
Industry	0.2	0.2	0.2	0.3	0.3	0.4	0.4	0.5	0.7
Agriculture	2.5	2.6	2.7	2.8	2.9	2.9	3.0	3.1	5.3
Total	*183.1*	*213.6*	*241.1*	*258*	*274.4*	*290.3*	*305.7*	*327.4*	*400.7*
Coke									
Industry	3.1	3.7	4.5	5.4	6.5	7.9	9.5	11.4	18.9
Total	*3.1*	*3.7*	*4.5*	*5.4*	*6.5*	*7.9*	*9.5*	*11.4*	*18.9*
Kerosene									
Industry	1.0	1.2	1.3	1.5	1.7	2.0	2.3	2.6	3.93
Commercial	1.1	1.1	1.2	1.2	1.3	1.4	1.4	1.5	2.1
Residential	1.0	0.9	0.8	0.7	0.6	0.5	0.4	0.3	0.1
Total	*3.1*	*3.2*	*3.3*	*3.4*	*3.6*	*3.9*	*4.1*	*4.4*	*6.1*

output capacity of 2,625 MW. The sulfur content of local lignite is 3.3 % in average. Total consumption of lignite in Mae Moh power plant during 1993-1995 years was 11.5 million tones per year in average. Taking into account the environmental issues the FGD control technology was applied for the new units during 1995-1996 years. This decreases the SO_2 emission on 95 %.

The anticipated future consumption of some important fossil fuel are shown in Table 2 (Khummongkol et al., 1996). High sulfur content fuel consumption like lignite, although trend to increase but the consumption in power generation will be constant until 2010 while the increasing is found in industrial sector. Moreover coal used in power generation in the future will be imported coal with less sulfur content. Natural gas is highly demanded for electric production in the future as well. Fuel oil, one of the high sulfur content fossil fuels in Thailand, also shows the increasing trend in industrial sector but, on the other hand, the reduction occurs in power generation. In transport sector, the major fuels used are diesel and gasoline. Although table 2 shows high demand of these two fuels in the future, the compulsory to limit sulfur in the fuel have been implemented from 1996.

The National Energy Policy Office (NEPO) of Thailand has utilized the Rains-Asia Model to determine present and projected levels of sulfur dioxide emissions and deposition (NEPO, 1997). The estimated emissions of SO_2 for the baseline scenario from this report are shown in Tables 3 and 4. Under this scenario, emissions are estimated to increase from about 957 kilo ton (kt) in 1990 to 1,998 kt in 2020. This represents nearly a doubling of emissions over the 30-year period with a corresponding average annual growth rate of 2.5%. Emissions decrease from 1990 to 2000, and then increase sharply again from 2000 to 2020. Emissions from 2000 to 2020 will increase at a rate of about 5.6% per year – a rate slightly faster than projected for total energy over this same period (5.0% per year). The largest share of this emissions growth is contributed by heavy fuel oil (HF). Emissions from this fuel increase from 303 kt in 1990 to over 1,100 kt in 2020 – an increase of over three times. Increases in emissions are also prominent in the case of coal (HC) which grows from just 2 kt in 1990 to about 500 kt in 2020. This increase is largely due to increased combustion of coal in the power sector associated with the strong projected growth of Independent Power Producer (IPP). Also noTable 1s the modest increase in emissions from domestic lignite. This increase occurs despite the fact that new lignite-fired power generation will be not be developed in Thailand after the period 2000-2010 and all existing lignite capacity will be controlled by highly efficient wet FGD technology. The reason behind this increase is the modest increase in consumption of domestic lignite assumed for industry which, when combined with the poor quality of this fuel, results in significant emissions of sulfur.

Table 3. Total SO$_2$ Emission by Sector in Thailand [kt SO$_2$/year]

Sector	1990	2000	2010	2020
Conversion	1	9	17	18
Industry	180	341	668	1,170
Domestic	62	27	37	45
Transport	157	59	75	90
Power	557	239	379	674
TOTAL	957	675	1,177	1,998

Table 4. Total SO$_2$ Emissions by Fuel in Thailand [kt SO$_2$/year]

Fuel	1990	2000	2010	2020
Lignite	483	312	482	709
Coal	3	3	193	551
Fuel Oil	303	575	810	1,581
Other	210	100	92	115
TOTAL	956	675	1,177	1,998

The decrease in emissions noted early in the study period is due to the aggressive pollution control policies for the power sector implemented in the 1990s. These policies serve to reduce power sector emissions by 57%. Fuel desulfurization policies for the transport and domestic sectors also contribute to emissions reductions in on the order of 60%. Following these early gains in emission reductions in the 1990s, emissions increase as a consequence of several key factors. First, energy growth over this period remains high, about 5% per year. Second, the fuel mix for industry and power sector includes increasing shares of higher-emission fuels, including heavy fuel oil and coal, which tend to replace the high utilization of natural gas. Third, new pollution control initiatives are not yet assumed for implementation in this period. Thus, with emission policies constant, emissions tend to continue to rise in relation to the growth in energy consumption.

A geographical representation of the deposition calculated by the RAINS/ASIA Model for the baseline emissions scenario for Thailand is shown in Figures 1 and 2 for the year 1990 and 2020 respectively.

Deposition is calculated in mg S/m^2-yr and is shown in the figures as shaded ranges. The darker the shading, the greater the average annual sulfur deposition. As can be expected, sulfur deposition in 1990 is greatest in the area directly surrounding the Mae Moh power plant in the Northern Thailand. The grid cell 18° latitude/100° longitude has the greatest calculated sulfur deposition at 4.7 g S/m^2-yr. Several other grid cells in this immediate area have average annual depositions that exceed 0.7 g S/m^2-yr. Also significant is the deposition calculated by the model in the Bangkok region and the eastern portions of the Central Valley region. Deposition of sulfur in this area ranges from about 2.7 g S/m^2-yr to about 0.3 g S/m^2-yr.

By the year 2020, the distribution and magnitude of sulfur deposition as calculated for the Baseline emissions scenario will have changed significantly. In general, increased levels of sulfur deposition are observed throughout the country with the exception of several areas of the North Highlands region. The previously large deposition values associated with emissions from the Mae Moh plant have diminished greatly due to installation of emission controls on all units remaining in operation and the retirement of the older units. In fact, deposition in the grid cell 18° longitude/100° longitude has decreased by 82% to 0.9 g S/m^2-yr from the 1990 levels.

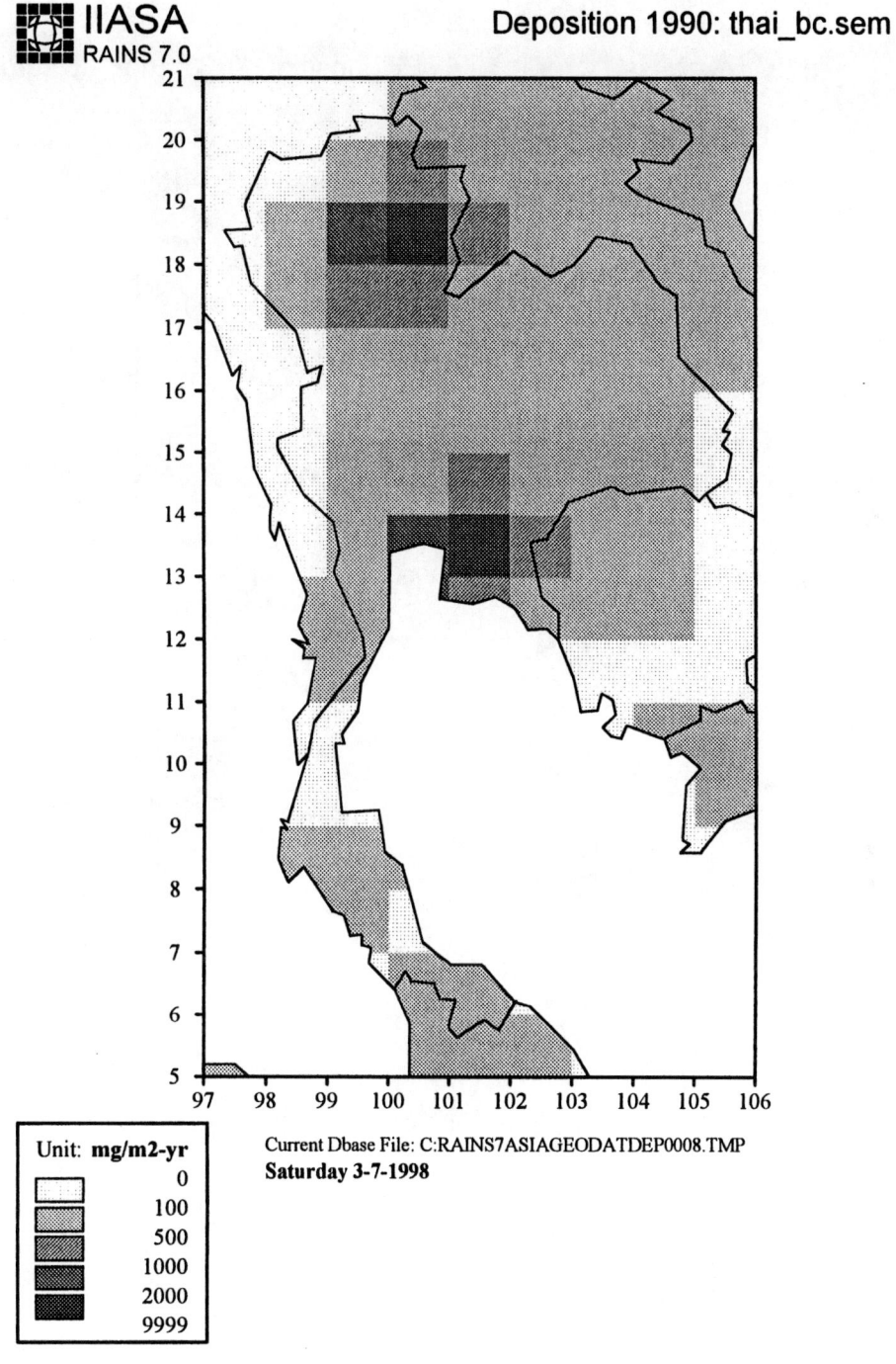

Figure 1. Sulfur Deposition for Thailand in 1990 Calculated by Rains/Asia Model.

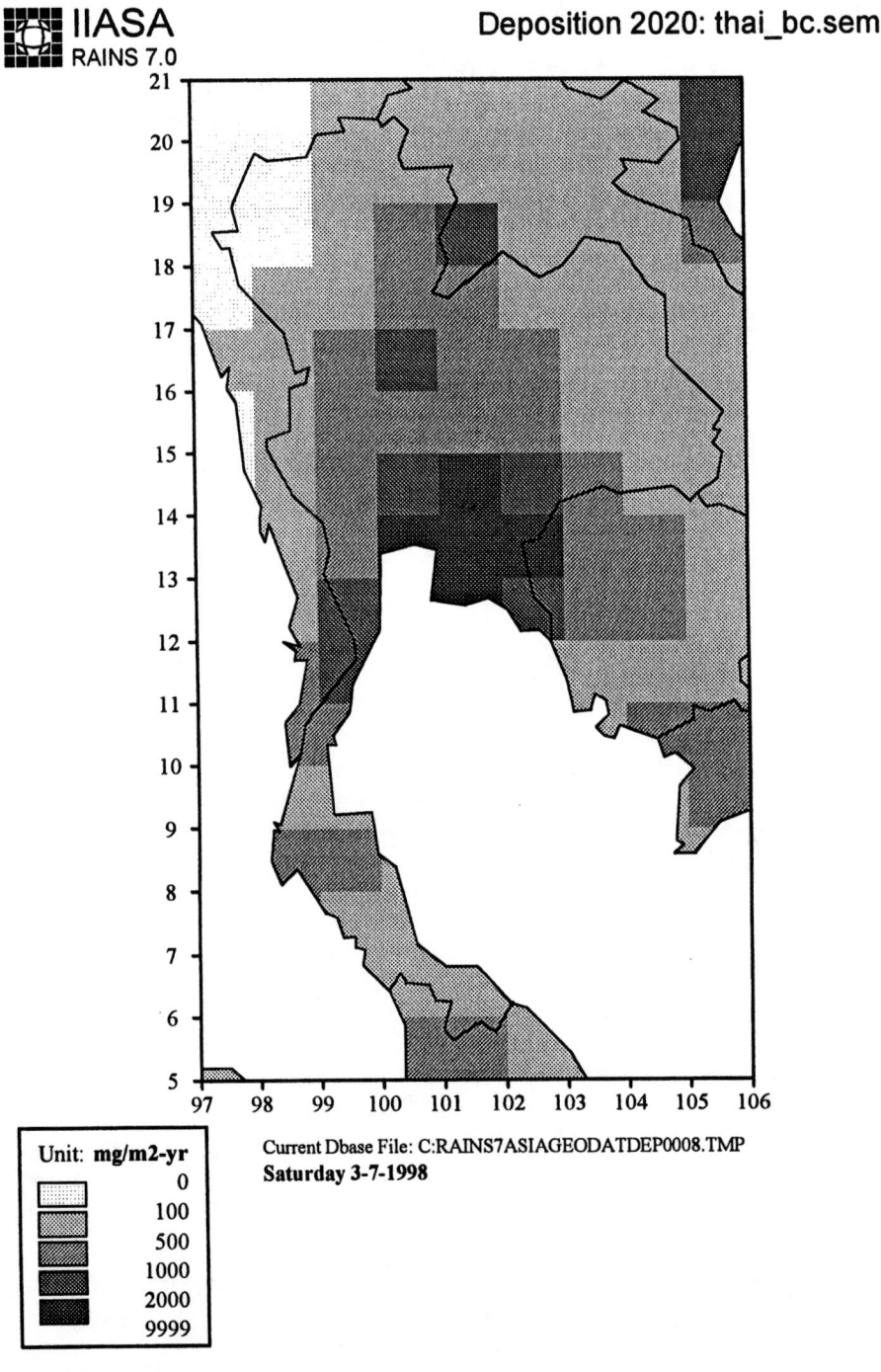

Figure 2. Sulfur Deposition for Thailand in 2020 Calculated by Rains/Asia Model.

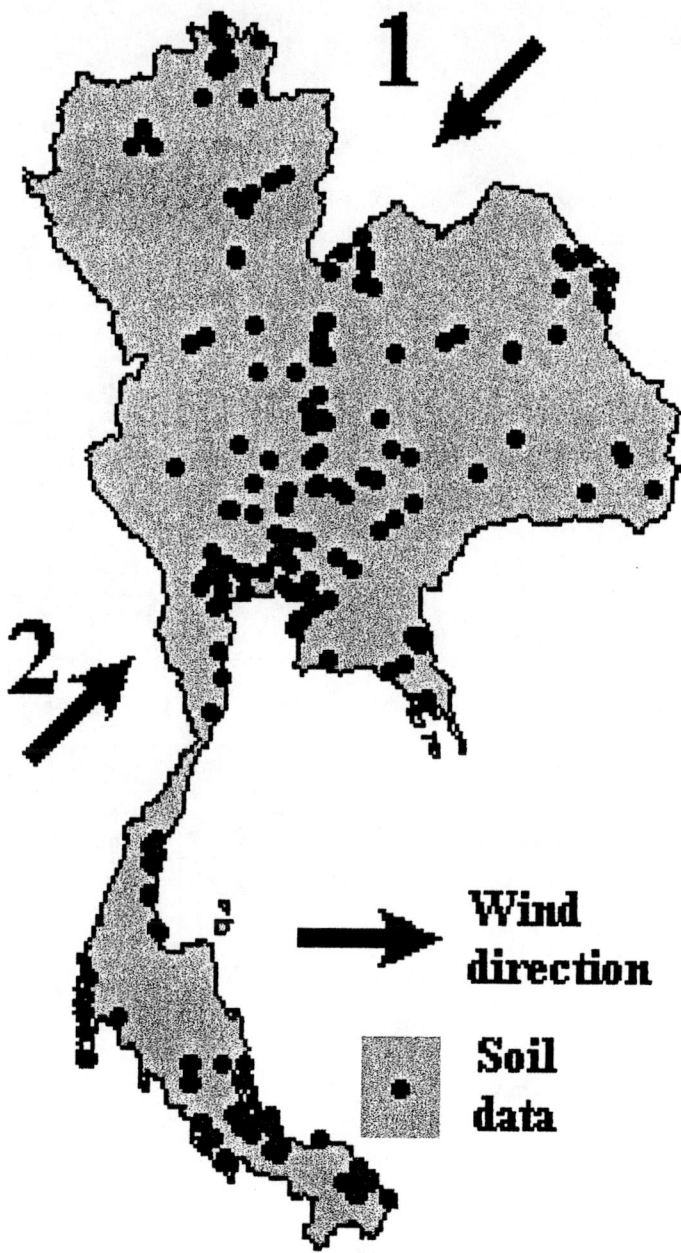

Figure 3. Map of Thailand. The filled circles depict location of sites included in DLDSIS. Arrows are used to show predominant winter (1) and summer (2) winds (monsoons).

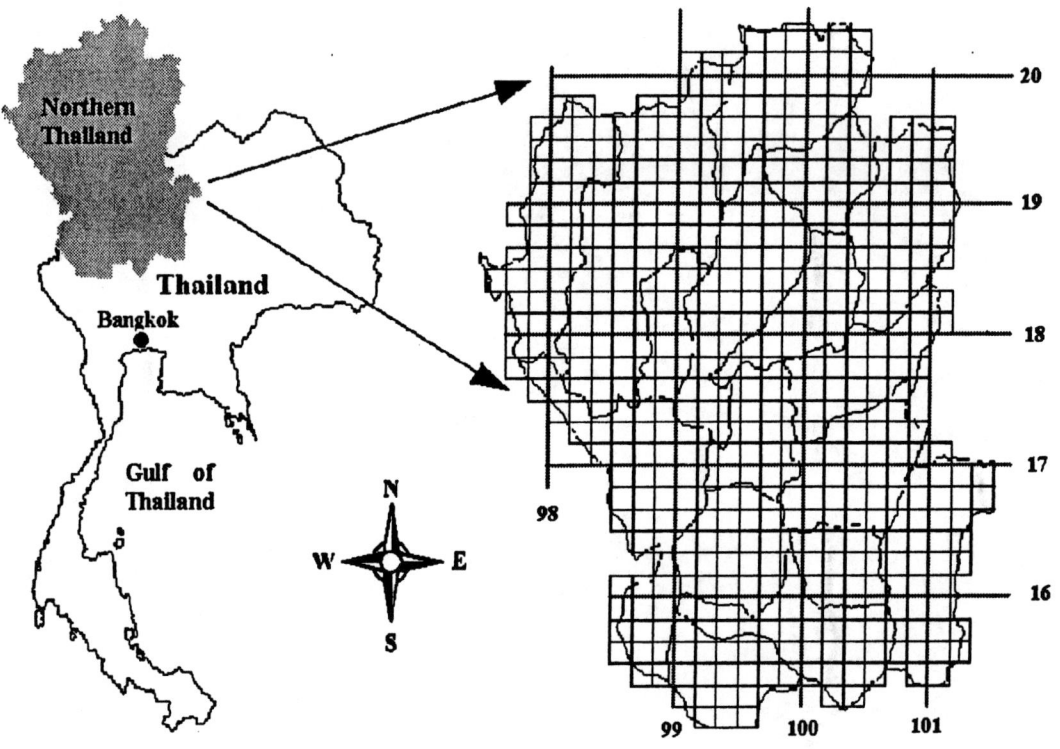

Figure 4. Map of Northern Thailand. The location of the calculation grid is on the right side. The digits along right edge mean Northern latitude. The digits along bottom edge mean the Eastern longitude.

So, for year 2020 the largest depositions in Thailand are calculated in the vicinity of Bangkok and areas directly east of the urban area. Maximum sulfur depositions in this area now reach 6.3 g S/m^2-yr, with several grid areas receiving in excess of 2.0 g S/m^2-yr. It is also significant to note the appearance of a new area of high levels of sulfur deposition in the Southern Peninsula region near city Krabi. Calculated sulfur depositions in parts of this region have increased ten-fold from less than 0.2 g S/m^2-yr in 1990 to 0.8 g S/m^2-yr. This increase is almost entirely connected with the enhancement of emissions of coal- and heavy oil-fired IPP power plants that are planning to develop in the region. For comparison, sulfur deposition monitored in the highly polluted and damaged areas of Central and Eastern Europe peaked at about 15 g S/m^2-year in the end of 1980s.

In general, based on the RAINS baseline scenario for the entire Asia region, background deposition in Thailand from sources outside Thailand is relatively insignificant throughout the country. Even in the year 2020, background deposition ranges from only about 0.003 to 0.2 g S/m^2-yr, amounting to only a fraction (usually less than 10%) of the total deposition calculated for each grid cell.

To estimate the influence of the given acidic depositions on the various ecosystems there are different approaches connected with either the experimental assessment of alterations in chemical and biological properties of ecosystem components or model calculation of ecosystem sensitivity. These methods are being used for many years successfully in Europe and North America and recently are also implemented in Asia.

Thus, this article is aimed to assess the effects of acidic deposition on Thailand's ecosystems by using the experimental and model approaches.

2. GENERAL DESCRIPTION OF THE STUDY AREA

The recent implementation of the RAIN-ASIA model is related to the effects of acidic deposition for whole Thailand. But taking into account that Northern part of the country has influenced by much more higher level of acidic deposition, the main attention in the given study was given to this region (Figure 1), where the most important source of SO$_2$ emission is Mae Moh power complex (see the right part of the Figure 4).

2.1. SOIL AND CLIMATE DATABASE FOR THAILAND

The study area is located in the northern part of Thailand (Figure 4). This area is bound between 15 ° to 20.5° North latitude and 97.5° to 101.5° East longitude. Concurrent with the variability in the topography and vegetation, which have a direct influence on soil formation, a large variation in soils can also be observed. The vegetation types from Subtropical Wet Forest to Tropical Dry Forest are very typical in the region. Analysis of the meteorological and climatological data (1956-1990) clearly shows that Northern Thailand is characterized by a humid tropical maritime climate with extremely wet summers and dry winters prevails in the area. The average temperature pattern for more than 30 years registered in Lampang province of Northern Thailand is shown in figure 5 (upper part). Annual mean temperature ranges from 24° C in Chiang Mai to 27° C in Tak. The monthly temperature variation between the warmest month (April) and the coldest month (December or January) is about 7-10° C only.

Annual precipitation generally varies from 1050 mm to 1400 mm, though it is higher in certain areas. The dry season, from November to April, is associated with southeast and north winds. The dominant wind directions for different seasons are shown in figure 3 by black arrows. Number 1 in this figure corresponds to winter wind (mostly from north and northeast), and number 2 - to summer wind (mostly from south). During some months of the year, the precipitation corresponds to less than 1 % of the annual amount of rain. In most parts of Northern Thailand, at least 2 to 3 "dry" months occur each year. The rainy season, from May to October, driven by the southwest monsoons, provides 90 percent of the annual precipitation. For most part of the year, the dominant high altitude wind directions are from the south and southwest. Meso-scale transport is characterized by low wind speed, increasing the pollutant residence time to about 24 hours. The average precipitation pattern for more than 30 years registered from year 1956 in Lampang province of Northern Thailand is shown in Figure 5 (bottom part).

The distribution of zones with different ratio between precipitation and evapotranspiration in Thailand is shown in Figure 6. Here the darker the shading, the greater the difference between precipitation and evapotranspiration. A zone with number 1 corresponds to water deficit areas (evapotranspiration is higher than precipitation for the whole year). Number 2 shows the zones where precipitation is higher, but the difference does not exceed 500 mm. Number 3 reflects the zones with relatively high level of precipitation (difference is more than 500 mm). Such pattern influences on a sensitivity of ecosystems to acidic deposition significantly (see for explanation below).

Soil development reflects the prevailing climate. The strong evapo-transpirative demand coupled with generally sparse precipitation (over much of the year) results in an "Ustic" moisture regime. Rain-fed agricultural production on upland sites is strongly limited by soil moisture deficit during the dry period. Soils with high moisture retention tend to be more productive (barring other limiting factors) than gravelly or sandy soils with poor retention. Local soils fall into two broad groups. Upland soils developed on dissected erosion surfaces, and old alluvial fans and terraces. Alluvial soils have developed on recent alluvium, associated with active sedimentation of rivers or streams.

Upland soils are variable, with characteristics reflecting the surficial geology and moisture regime. Soils developed on old, well-drained surfaces tend to be Ultisols (e.g. Paleustults). Under the Thailand Soil Classification they are described as Red Yellow Podzolic Soils, or Reddish Brown Lateritic Soils. These are highly leached, with relatively little silica and relatively abundant hydrous oxides of aluminum and iron. They tend to be moderately acidic (pH range: 5.0 to 6.2 in H_2O), with low base saturation (less than 35%), low cation exchange capacity, but high anion exchange capacity. As a result, they tend to fix phosphorus and have a high sulfur absorption capacity.

Average monthly temperature

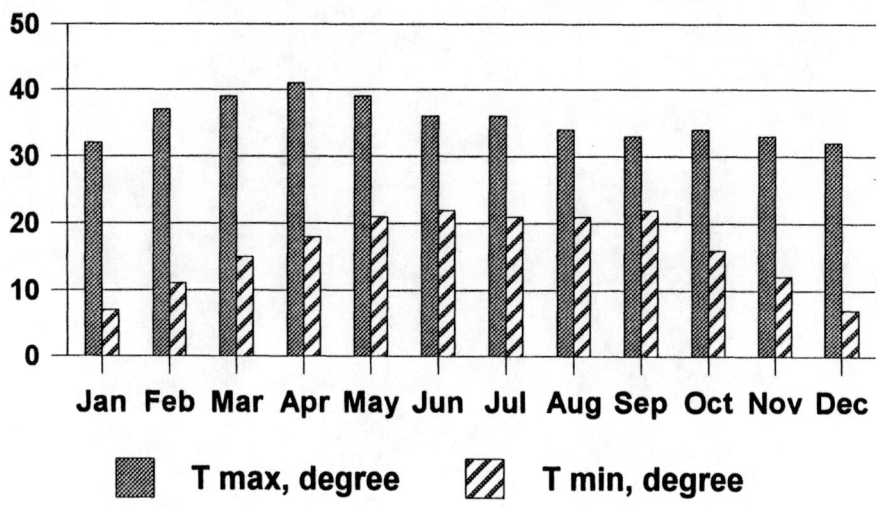

Average monthly precipitation

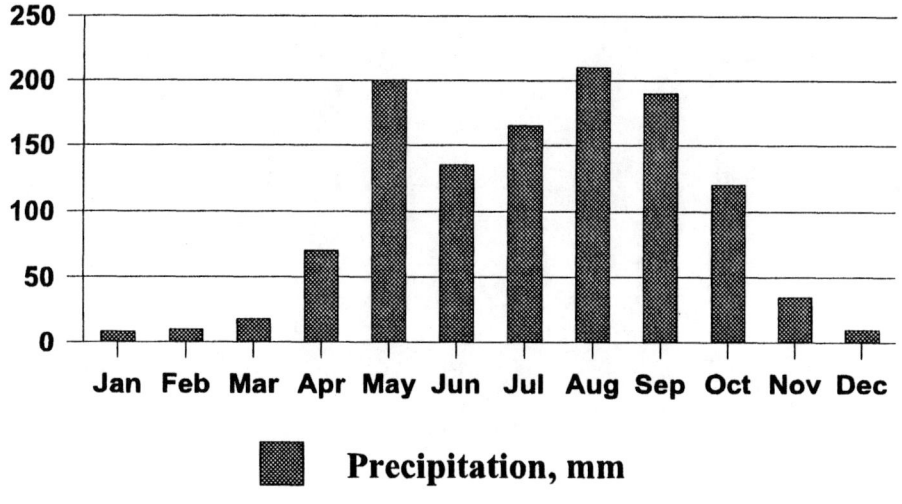

Figure 5. Average monthly temperature and precipitation for Lampang province, Northern Thailand (1956-1990).

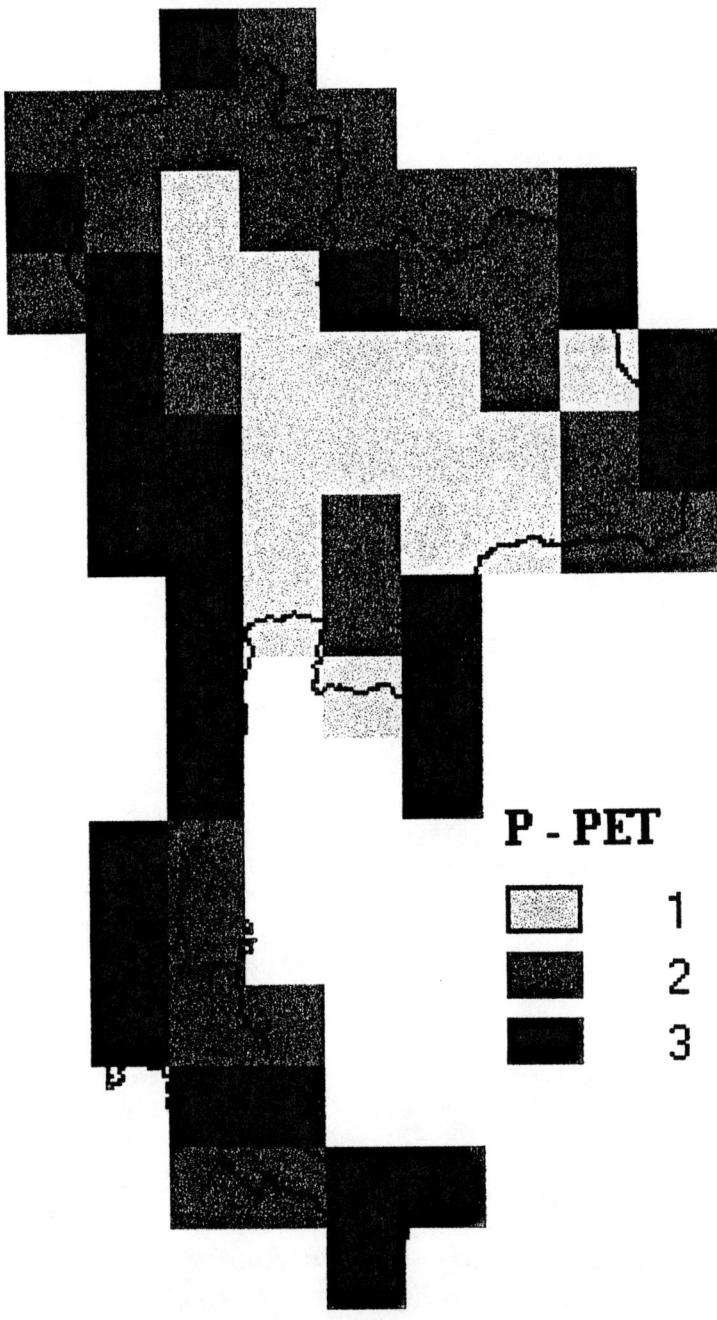

Figure 6. Distribution of zones in Thailand with different ratio between precipitation and evapotranspiration.

Table 5. Fragment of Department of Land Development Soil Information System description.

soil series number: T 319   country: THAILAND   soil description   DLDSIS 05/22/96

CLASSIFICATION	FAO/UNESCO,1974: (Tent. class.)
	USDA,1975: Fine, montmorrilonitic, isohyperthermic
Diagnostic horizons:	
(other) Diagn. criteria:	
Local classification: Samut Songkhram series : Sso	
LOCATION	: Amphoe Muang,Samut Songkhram Province Code SW 54/1
	: Latitude: 13 23 25 N Longitude: 95 56 53 E
	Altitude: 1.5 (m.a.s.l.)
AUTHOR(S) - DATE (mm.yy) : W.Van der kevie - 0.79	
GENERAL LANDFORM : tidal flat Topography: flat or almost flat	
PHYSIOGRAPHIC UNIT : tidal swamps	
SLOPE Gradient/aspect/form: %	
POSITION OF SITE :	
MICRO RELIEF Kind:	
SURFACE CHAR. Rockoutcrops: none Stoniness: none	
Cracking: nil Sealing: nil	
Salt: moderate Alkali: moderate	
SLOPE PROCESSES Soil erosion: Aggradation: nil	
PARENT MATERIAL I : marine sediments Derived from: Texture:	
Remarks: marine deposits	
WATER TABLE Depth(cm): 50 Kind: groundwater table	
DRAINAGE : imperfectly	
PERMEABILITY : slow	
FLOODING frequency: Run off: rapid	
MOISTURE CONDITIONS PROFILE :	
LAND USE : arable farming, crops, see remarks,	
VEGETATION Structure:	
Landuse/vegetation remarks: coconut,palms, palm sugar	
CLIMATE koppen : Aw Soil Moisture Regime: aquic	
no station linked to this monolith	

PROFILE DESCRIPTION

A 0- 18cm 10.0YR 3.0/3.0 moist; silty clay loam; moderate medium subangular blocky structure; firm sticky plastic; many medium roots and many coarse roots; pH(field): 8 ; gradual smooth boundary to

BA 18- 46cm 10.0YR 4.0/4.0 moist; silty clay; weak coarse subangular blocky structure and fine subangular blocky structure; sticky very plastic; few fine (7.5YR 4.0/4.0) and few fine (5.0Y 5.0/2.0) mottles; many medium roots and few fine roots; pH(field): 8 ; gradual smooth boundary to

Bg 46- 80cm 2.5Y 5.0/2.0 moist; silty clay; weak medium subangular blocky structure; many medium distinct (7.5YR 4.0/4.0) mottles; many very fine roots and common fine roots; pH(field): 8 ;
Cg1 80-115cm 5.0Y 5.0/1.0 moist; silty clay; weak coarse subangular blocky structure; sticky plastic; pH(field): 8.5;

Cg2 115-170cm 5.0GY 4.0/1.0 moist; silty clay loam; ; pH(field): 8.5;

REMARKS:
They are deep, moderately alkaline soils and are characterized by a brown or dark brown clay A horizon which overlies a greyish brown or live grey B horizon. This in turn overlies a reduced, greenish grey clay C horizon with its upper boundary at some depth below 50 cm. but within 125 cm. of the soil surface. The A

horizon may have faint mottles, whereas the B horizon has many brown or olive brown mottles. The soil has low n-values below 50 cm. and salinity is moderately high.

soil series number: T 319 analytical data <missing value = -1> DLDSIS: 05/22/96

no	top	bot mm	2> 1000	2000 1000	1000 500	500 250	250 100	100 50	tot sand	silt	clay <2	disp	bulk dens	pf- 0.0	1.0	1.5	2.0	2.3	2.7	3.4	4.2	spec surf
1	0	18	0	0	0	0	0	0	8.00	52.50	39.50	0	0.00	0	0	0	0	0	0	0	0	0
2	18	46	0	0	0	0	0	0	2.50	45.00	52.50	0	0.00	0	0	0	0	0	0	0	0	0
3	46	80	0	0	0	0	0	0	3.50	42.00	54.50	0	0.00	0	0	0	0	0	0	0	0	0
4	80	115	0	0	0	0	0	0	5.00	54.50	40.50	0	0.00	0	0	0	0	0	0	0	0	0
5	115	170	0	0	0	0	0	0	4.00	60.00	36.00	0	0.00	0	0	0	0	0	0	0	0	0

No.	pH H2O	pH KCl	Ca-CO3 %	ORG-C %	MAT. N %	EXCH Ca	CAT. Mg	K	Na	sum	Ex-trac Acid meq	Sum B+A	CEC soil	CEC clay	P	K	BASE SAT %	EC 2.5 mS/cm
1	7.20	6.80	1.65	2.30	0.200	17.60	10.70	1.70	9.50	39.50	4.20	43.70	34.10	86.30	53.90	685.00	90.00	0.00
2	8.00	7.00	1.50	2.44	0.210	12.40	12.10	1.20	11.50	38.20	3.50	41.70	34.50	65.70	14.20	465.00	92.00	500.00
3	8.00	7.20	3.15	0.98	0.080	22.50	15.00	1.20	12.30	51.00	2.00	53.00	33.70	61.80	12.40	450.00	96.00	480.00
4	5.60	5.00	1.20	3.06	0.260	12.20	16.50	1.80	10.00	40.50	11.80	52.30	30.70	75.80	26.20	715.00	77.00	600.00
5	7.00	6.80	4.95	3.44	0.300	26.00	17.30	1.91	12.50	57.70	5.10	62.80	29.10	80.80	80.60	700.00	92.00	89.00

The Ultisols have a low inherent fertility, and as such do not have a high capability for agriculture. Many show unfavorable texture, being gravelly or sandy. At best, these soils are moderately suited to upland crops, such as pineapple, sugarcane and tobacco. They may be suitable for pasture. Upland soils developed on more surfaces or soils developed in poorly drained sites in which leaching has not been so pronounced, are mostly Alfisols. These, by definition, are less leached, and have a higher base saturation (greater than 35%) than the Ultisols. However, soil acidity is more variable than that for Ultisols (pH range: 4.5 to 7.5). Some Alfisols have developed from parent materials rich in carbonates, such as limestone.

Alfisols have a higher inherent fertility than Ultisols, but their capability for agriculture may be limited by other factors. Many Alfisols are not well suited for upland crops, fruit trees and pasture due to drainage limitations. They may, however, be well suited to paddy. Soils derived from calcareous parent materials may have important fertility limitations due to the low availability of phosphorus, iron, manganese, zinc and copper.

In 1989, the Soil Survey and Classification Division, Department of Land Development, introduced a soil information system for Thailand –DLDSIS (Vearasilp and Songsawad, 1991). It contains each of the national soil series typifying pedon with a total of 244 soil series units in the database, including profile descriptions and analytical data. Each entry consists of a site description, a horizon description and analytical data (Table 5). The locations of soil profile data are shown in Figure 3 as shaded circulars. One can see that for the western and the northeastern parts of the country the data sets are limited.

2.2. ECOLOGICAL RESOURCES

The study area contains both natural and plantation forests. Total forested area was about 45.5% in 1991. Undisturbed natural forests consist of two main types, namely Mixed Deciduous and Dry Dipterocarp. Due to extensive industrial and agricultural development, as well as the establishment of teak plantations, virtually no undisturbed forest remains in the central part of study area. Undisturbed Mixed Deciduous Forest is still present on steeper slopes in the hills on medium textured soils at elevations from 50 to 600 m. This forest type is not dominated by any one family or genus, but contains numerous species of deciduous broad leaf tree. Teak (Tectona grandis) is the most abundant species, which dominating the stand in some places. Other Mixed Deciduous Forest species are Pterocarpus macrocarpus, Barringtonia acutangula, Canarium kerrii, Terminalia tomentosa, Spondias pinnata, Lagerstroemia calyculata, Xylia kerrii. Stands of Mixed Deciduous Forest may reach 35 to 40 m in height. The understory is dense with numerous seedlings of overstory trees, shrubs and bamboo.

The Dry Dipterocarp Forest is in many places associated with well-leached lateritic soils (e.g. Oxic Paleudults) at lower valley positions. This forest is dominated by trees of the Dipterocarpaceae, or Dipterocarp family, in particular those of the genera Dipterocarpus and Shorea. These forest stands are of a lower stature than the Mixed Deciduous Forest, with canopy heights ranging from 20 to 30 m. Shifting cultivation, practiced extensively in the area, has resulted in secondary forest, dominated by fast-growing pioneer trees, shrubs, or herbs and grass. Bamboo is well represent in secondary vegetation. Secondary vegetation also occupies land disturbed through other means, such as mining activity and agricultural land clearing.

Aquatic biota is very diverse throughout the region. A number of the fish species, particularly Tilapia spp. and Cyprinus spp. are stocked to provide fishery resources to the local population. In addition, the recent creation of reservoirs in the region has likely increased fisheries yield in the region due to increasing biological productivity and habitat for aquatic biota. The inhabitants of the study area use a variety of aquatic biota, including macrophytes and shellfish and benthos (Chiang Mai University, 1990). According country statistics, the national average for fish consumption in Thailand is approximately 25 kg/person/year. Therefore, the hazardous influence of acidic deposition on water ecosystems is very crucial and it has been investigated by Shchultz Int. (1991).

Wildlife habitat has undergone extensive changes in the northern Thailand in the last few decades as a result of extensive shifting cultivation. Secondary vegetation is of poor quality for many species. These types of vegetation offer insufficient forage and structural diversity required supporting a diverse fauna. For example, barking deer (Muntiacus muntjak), formerly common, are now rarely observed. Of the mammal species recorded in the Mae Moh valley near big power complex, only rats (Rattus spp.) and bats (Rhinolophus spp.) are common.

2.3. MAIN OCCURRING AND POTENTIAL IMPACTS

The results of the environmental screening conducted by the working group of Schultz Int. (1991) have focused on the following impact hypothesis:

- effects of atmospheric emissions and fugitive dust on agriculture, timber, and wildlife production, and, of course, human health,
- effects of changes in surface water hydrology and water quality on drinking water and aquatic resources.

Observation of vegetation on the slopes of Doi Pha Tup and Doi Pha Chi (the limestone ridge running north-south immediately to the east of Mae Moh Power Complex Units 4-13) suggests that significant changes have occurred recently. In one patch of forest northwest of the stack cluster numerous standing dead trees can be seen (estimated density: 10 to 15 dead trees per ha). This mortality seems large in comparison with other forested slopes.

In another patch, approximately 3 km north by northwest of the stack cluster, most of the overstory trees have died, leaving a luxuriant growth of bamboo. It is probable that the bamboo formed the understory and responded vigorously to increased light levels when the deciduous forest was killed. While it is not certain that stack emissions are responsible for the death of these trees, given that the area affected corresponds to an area of very high predicted SO_2 concentrations, it is likely that the emissions at least contributed to the problem.

The study by KBN (1994) indicates that although operation of the additional units of Mae Moh Power Plant will not exacerbate SO_2 concentrations in the Mae Moh valley, there will be areas in the mountains between the Mae Wang and Mae Moh valleys where ground level SO_2 concentrations will exceed NEB (National Environment Board) limits. The air quality evaluation conducted by Schultz Int.(1991) predicts concentrations of SO_2 on the floors of both valley (below elevation 400 m) at less than NEB limits for the annual average (100 micro gram/cub.m) and maximum 24 hour periods (300 micro gram/cub.m). However, at elevations above 400 m the largest SO_2 concentrations for annual average and maximum 24 hour period were 4,000 micro gram/cub.m and 18,000 micro gram/cub.m respectively.

Most of the soils in the immediate vicinity of Mae Moh Power complex are Ultisolic in nature, and contain abundant hydrous oxides of aluminum and iron. These can be expected to have a high sulfate adsorption capacity. Further to south of this complex, there are fairly extensive areas of Alfisols (Haplustalfs and Paleustalfs) which have intermediate sulfate adsorption capacities. Many of these soils have a high pH and will therefore not effectively retain sulfate.

Saturation of sulfate adsorption capacities may result in acidification of soil water solutions and leaching of cations. The depression of soil pH will depend on the buffering capacity of the soil. Soils with relatively high cation exchange capacities and/or high organic matter contents (e.g. Mollisols) will resist acidification. Other soils (e.g. Ultisols) will have less buffering resistance. Acidification of these soils may promote leaching and at the same time reduce nutrient retention, because their cation exchange capacity, being pH-dependent, will diminish.

Based on the information summarized above, the soils in the region have been rated for sulfur adsorption capacity and buffering capacity (Schultz Int., 1991). It is predicted that Wet soils on Alluvium will be at most risk for eventual acidification due to sulfur addition. Within that group, soils with high sand and low organic matter contents will be most susceptible. It is also notable that soils on hill slopes, which will probably be exposed to the highest concentrations of atmospheric sulfur, have a relatively low risk of acidification.

There were not many reports and studies on the effect of sulfur dioxide to the environment in Thailand until 1992 when the incident of power plant emission at Mae Moh began. Many villagers close to Mae Moh area were suffer from the serious amount of sulfur dioxide emission when one electrostatic precipitator was broken down and the thermal inversions have kept the pollutant emission near the ground level (Wibulswas and Towprayoon, 1993). There were also evidence shown that crop were damaged by both wet and dry deposition. The tropical forest surrounding the area, although no visual significant was observed but the forest soil and its ecosystem was directly effected by these depositions. The anticipated deposition shift area in 2020 indicates that the expansion of ecological diverse impact by sulfur dioxide will be occurred almost over the whole country except the southern peninsular. Therefore, the experimental study of chemical and nutrient change in the Mae Moh forest soil as well as the change of soil microorganisms was carried out using the technique of soil column applied with simulated acid rain. This study was aimed to recognize the primary change of the direct impact of acid deposition in the ecosystem where soil is the first direct acceptor.

3. METHODOLOGY

3.1. EXPERIMENTAL PART

Soil column

The forest soil was taken at the level of 25 centimeter underground from Mae Moh district, Lampang province which located in the Northern part of Thailand (see figure 4). The soil samples were packed in the acrylic column of 15 cm diameter with 50 cm high for microbiological change observation and 6.5 cm diameter with 25 cm high for chemical change study.

Acid rain application

Simulation of acid rain using sulfuric acid at various pH was applied to the soil columns with certain quantity and soil samples were taken to analyzed at the appropriate time interval..

Soil analyses

Exchangeable acidity, exchange aluminum, iron and exchange hydrogen ion were measure with potassium chloride method. Cation exchange capacity (CEC) was done by ammonium saturation method. Exchange calcium and exchange magnesium were detected by atomic absorption spectrophotometer while exchange potassium and exchange sodium were measured by flame photometer.

Microbial count

Numerical method of bacterial, fungi, actinomyces and total count were done by plate count technique with soil extract agar as cultivated medium. Nitrogen fixing bacteria in term of *Azotobacter* was counted with azotobacter medium (Page, 1982). Ammonifying bacteria and nitrifying bacteria were numerated using Most Probable Number method with nutrient broth and nitrite oxidizer medium, respectively, as the media.

3.2. CALCULATION AND MAPPING PART

3.2.1. General approaches and uncertainties

At present, there are several approaches to calculating the ecosystem sensitivity to acidic deposition and corresponding CL values. The most known one is the Steady State Mass Balance model (SSMB) including a quantitative assessment of the most relevant parameters of pollutant turnover within specific area.

In Europe the CL values were computed with SSMB method mainly for forest soils and surface waters. For Asia, an improvement to the European exercise was made by including 31 ecosystem types in the CL assessment (Hettelingh *et al*, 1995). The computation of CL is based on plant response criteria and soil stability criteria. A critical molar ratio of the concentration of base cations to aluminum in soil solution, $(BC/Al)_{crit}$, for each plant species is used as indicator of plant response. Using these critical BC/Al ratios in the SSMB method it is possible to compute maximum allowable acidifying deposition, i.e. the CL. The soil stability criterion is introduced to avoid that acid deposition leads to Al leaching in excess of Al produced by weathering and other process, e.g. in high precipitation areas. CL values for each ecosystem are computed by taking the minimum of these both criteria. In fact through broad literature review the above criteria are among the most extensive ones which have considered all the important factors related to acidic deposition-ecosystem responses.

However, there are some difficulties for using these criteria in Thailand. First, overall a considerable amount of information is required. The detailed weather, soil, geological and vegetation data are needed. In some research sites, most of the information is available, but in most areas many of the required data do not exist. Second, the relationship between base cation to aluminum and changes in plant growth, which is used to define the $(BC/Al)_{crit}$ in the equations, was determined by experiments done in areas outside of Thailand. Although plants of the same species might respond similarly, laboratory studies should be taken. With the lack of the local $(BC/Al)_{crit}$ data, the published results from other parts of Asia as well as outside of Asia could be used temporary but with caution. Third, the methods for the estimation of a soil weathering rate under acid deposition input are needed to be verified for local conditions.

The idea to assign CL to ecosystem sensitivity classes was used in the Stockholm Environment Institute (SEI) method (Cambridge et al., 1995). Sensitivity of ecosystem is determined from an assessment of the factors that indicate potential ecosystem response to acidic deposition. The selection of the environmental controlling variables has been shown that ecosystems susceptible to acid deposition can be successfully mapped on the basis of three parameters: soil buffering ability, land use, and humidity index. The output is intended to be used to indicate ecosystems at risk to acidic deposition. The main advantage of SEI method is it uses a consistent methodology that has been modified in order to tailor it to the available world-wide databases. Unfortunately, due to the data used and the criteria employed the sensitivity of areas can be exaggerated or underestimated. In order for results to be applied for scientists and policy makers, the uncertainties require additional investigation.

Taking into account the specification and diversity of ecosystems in Thailand and low degree of information support for calculation of various parameters according steady-state mass balance model for critical loads, we used a combination of both the above mentioned approaches. First of all, it is a need to assess sensitivity of different ecosystems to acidic deposition, e.g. according to SEI method, and then to calculate CL for each class of sensitivity using expert assessment and available monitoring data. This approach requires numerous cartographic materials, for example, maps of soil cover, vegetation, meteorology conditions, land use, etc. The voluminous data include maps and tables that are temporally and geographically distributed. Realization of such approach requires the handling of huge volumes of data and extensive spatial analysis including expert assessments. The using of expert-modelling GIS is proven to be useful here. The similar approach has been realized recently for Asian part of Russia (Bashkin *et al*, 1996) and for Thailand (Kozlov *et al*, 1997).

Actually there are several reasons for uncertainty here. As illustrated by Figure 7, the aggregation problem is one of the sources for uncertainty to be appeared.

Aggregation problems

Map before aggregation

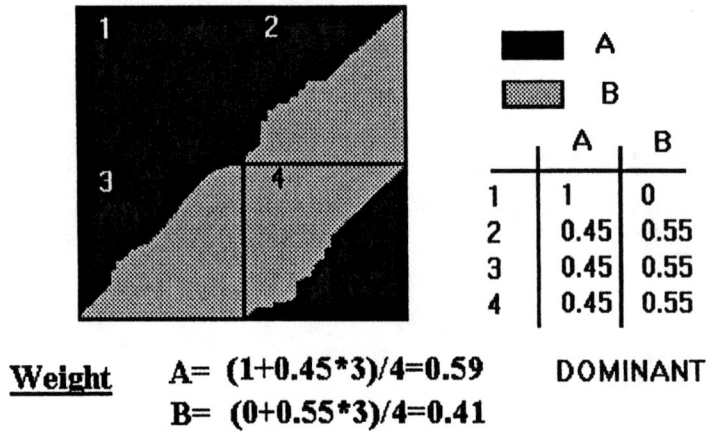

Weight A= (1+0.45*3)/4=0.59
B= (0+0.55*3)/4=0.41   DOMINANT

Map after aggregation

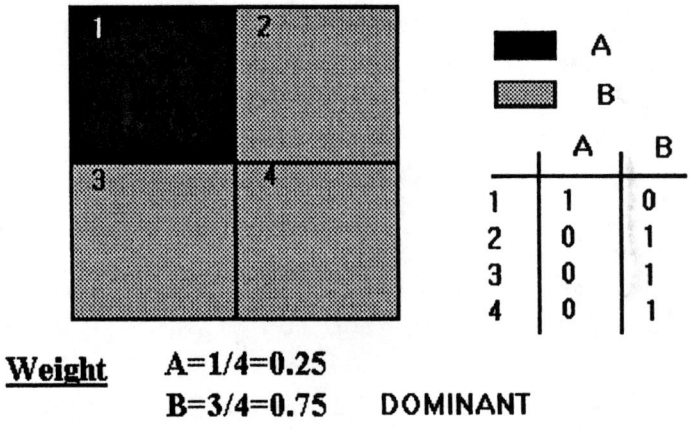

Weight A=1/4=0.25
B=3/4=0.75   DOMINANT

Figure 7. Explanation of Aggregation Uncertainty for Arbitrary Map.

The need to aggregate data arises many times during an assessment of ecosystem sensitivity or calculation CL. Usually we have various maps and other relevant data (for instance, soil or weather information etc.) in different scale and resolution. These data should be aggregated for the given scale of research. The area-weighted method is one of the commonly used. Let us imagine the situation shown in Figure 7. Map of some feature looks like upper map in this figure. We use very simple combination of only two features – A and B. Somebody can see easily that square 1 covered by feature A fully (weight is 1). For all other squares from 2 to 4 the corresponding weights for feature A are 0.45. If we calculate dominant feature for this case we get the results as follows

Weight A = (1+0.45*3)/4=0.59 Weight B= (0+0.55*3)/4=0.41

The conclusion here is feature A is dominant because 0.59 > 0.41

As concerning the second feature B, the picture is opposite. Let us aggregate the information to leave only dominant feature for each square. The result is shown in the second map (bottom part of Figure 7). The aggregated weights can be seen in the corresponding Table 1n right side of bottom picture.

If we calculate dominant feature for this case we get the results as follows

Weight A = (1+0*3)/4=0.25 Weight B= (0+1*3)/4=0.75

The conclusion here is feature B is dominant because 0.75 > 0.25

As one can see we get controversial results after using of are-weighted method. We used this simple example only to show how the uncertainties might be appeared in the case of aggregation. There are a number of sources for uncertainty in input data, calculation of some coefficients etc. during using of described methodology.

3.2.2. NEW METHODOLOGY OF SEI

Recently the revised methodology of SEI (1997) has been distributed. The approach was simplified to assess sensitivity of ecosystem by using mainly buffering ability of the soil. Such new methodology was examined for whole Thailand and someone can find results latter in this paper.

Before using revised approach we explain the idea of buffering ability which lies behind the assessment procedure. As very well known (Cambridge *et al.*, 1995) the sensitivity of terrestrial ecosystems to acidic deposition is due to a combination of the response to short- and long-term processes. If the soil is well buffered then the short-term processes are mostly irrelevant. In less well-buffered systems the concentration and seasonal impacts will be more important. Soil buffering is considered to have overriding importance in determining sensitivity and the seasonal and ion concentration effects modify this picture. Therefore, the revised SEI methodology (1997) is proposed only two main soil features to assess the sensitivity of terrestrial ecosystems to acidic deposition in tropical regions: i)soil buffering properties, ii)dilution of protons and toxic ions in very wet environments.

The most important buffering mechanism in soils is the weathering of soil minerals. In addition there is cation exchange. Mineral weathering can be considered to be rate limited and capacity unlimited (the buffering rate determines the degree of buffering; the content of soil minerals in medium-term time scales can be considered very large in most cases). Cation exchange reactions are considered to be rate unlimited but capacity limited (exchange reactions are rapid; the capacity of base cations on the exchange site is limited).

The two soil parameters which are generally measured and considered to be indicative of soil buffering are base saturation (BS, %) and cation exchange capacity (CEC, meq/100g). Base saturation is the result of rate-based reactions – either as the input or leaching rate of base cations. Inputs are derived from mineral weathering, transfer from a remote site by atmospheric deposition or transfer in ground water. Leaching rates are either due to water flux down the soil profile or presence/absence of an impermeable soil layer. Cation exchange capacity is determined by the number of negative charge sites for cation adsorption in the soil (dependent on the amount of organic matter) and the amount and type of clay mineral. CEC is also modified by pH.

The higher the base saturation, the greater is the buffering ability. This is due to the implication of high input rate (or low leaching rate) of base cations which will maintain base saturation under the influence of acidifying deposition (Figure 8). High cation exchange capacity indicates lower sensitivity as this entails greater capacity to resist short-term fluctuations in ionic concentrations and as it indicates, together with base saturation, the total exchangeable bases in the soil available for buffering by cation exchange.

In order to calculate the soil buffering, the base saturation and cation exchange capacity are linked and values for buffering determined over the rooting zone. There is obvious uncertainty in rooting depths for different vegetation types, in particular, as no information concerning the vegetation type is included in this assessment. The soil buffering class (1 – very low buffering, to 5 – very well buffered) is derived from a matrix of BS and CEC (bottom part, Figure 8). The long-term buffering associated with rate based buffering processes is given priority. In soils with base saturation >80% the dominant buffering mechanism is likely to be carbonate weathering. Therefore all soils with such BS values are allocated to the highest buffering ability category 5 (see Figure 8).

Using this methodology the buffering score for each soil type has been calculated using the soil type database (DLDSIS) developed by Soil Survey and Classification Division, Department of Land Development of Thailand. During our assessment we operate with the grid area which usually contents several soil types with the different buffering score. It is therefore more accurate to calculate an average score for each grid than for the dominant soil type. In this methodology the buffering score for a soil type is determined and then a score for the grid is calculated using area-weighted mean scores (and then rounding the values). The final scores will vary from 1 up to 5.

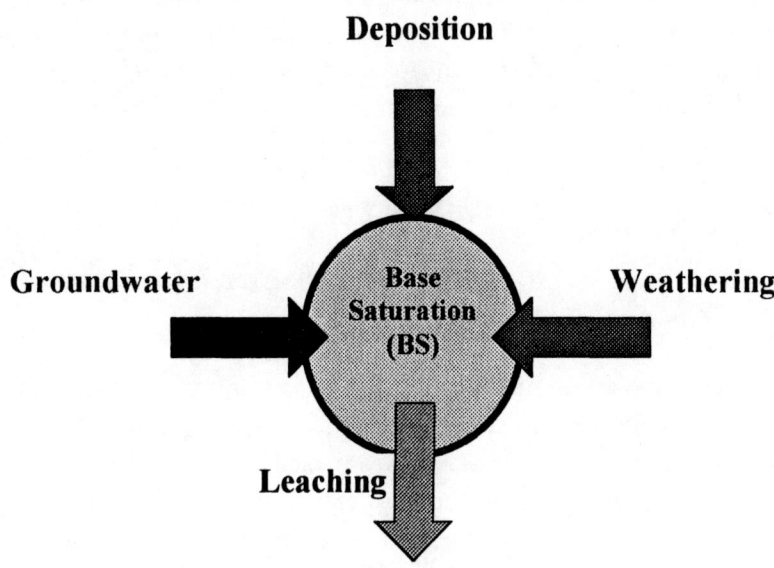

Figure 8. Elements of revised SEI methodology for assessment sensitivity of ecosystem.

In very wet systems without a long dry season (c.f. Wet Rain Forest in Thailand) the sensitivity will be reduced by one class (i.e. where annual difference between precipitation (P) and potential evapotranspiration (PE) more than 500 mm and where P is not less than PE for more than 3 months). This is related to the decrease in proton concentration in wet environments. This means that the leaching of cations from leaves will be reduced. In addition the impact of aluminum will be reduced as the concentration of aluminum will be reduced. It is still considered that the long-term acidification – i.e. the reduction in base saturation over time is the most important for the ecosystem. Therefore the dilution may modify this picture and should be analyzed additionally.

4. Results

4.1. Calculation and Mapping of Ecosystem Sensitivity to Acidic Deposition

For realization of the above mentioned methodology, the area of research was subdivided on 528 uniform cells with the size of 10 minutes along longitude and latitude (Figure 4, right part). All data need for assessment was picked up for each cell. According SEI methodology calculation and mapping for three environmental controlling variables have been done. Precipitation maps was generated from weather stations data and potential evapotranspiration was calculated using one of the most cited Thornthwaite's formula. Then, the results were classified according to moisture regime (Figure 9, A). These regimes were joint by 3 classes (humid and very humid, moist sub-humid and arid).

The soil data layer is based on the national soil map, which was digitized. The predominant soil type was taken into consideration for each initial cell. There are about 240 soil series according Thailand taxonomy or 89 soil names according USDA classification. The main problem here is there are a number of areas with so named "slope complex" for soil type. This type reflects mainly upland and mountain soils. On the final sensitivity soil map (Figure 10, B) white denotes such areas (0 at the map legend). Usually these areas are most sensitive to acidic deposition as well as areas with low buffering ability (1 and 2 on the Figure 9, B). Only a few zones have medium (3) and high buffering ability (4).

For the estimation of ecosystem sensitivity, each land-use type was assigned to one of four classes depending on the information associated with a certain vegetation type and land-use practice (Figure 10, A). According to the recommendation of the SEI, the land-use expected to be very sensitive to acidic depositions includes the Tropical Rain Forest and Mixed Evergreen Forest (1 in figure 10, A). The other areas distributed from sensitive-intermediate (2 on figure 10, A) through intermediate-insensitive (3 in Figure 10, A) up to insensitive (4 in Figure 10, A).

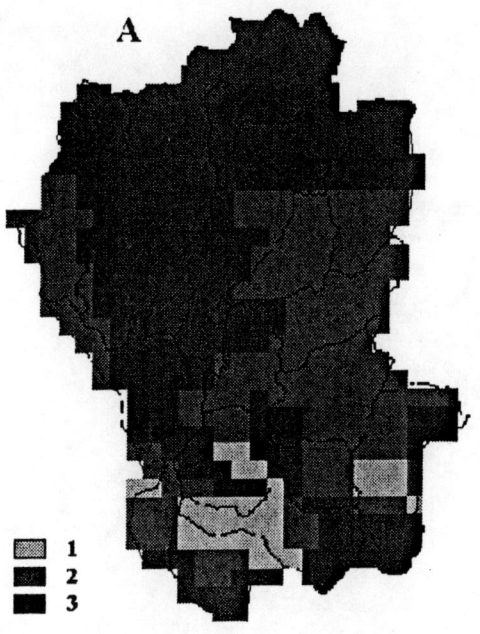

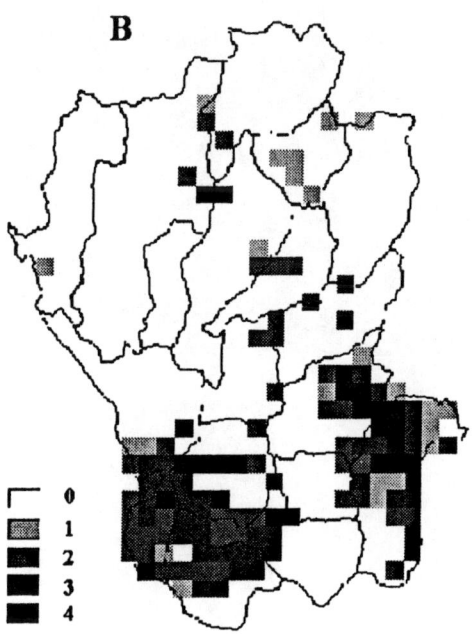

Figure 9. Sensitivity of ecosystems of Northern Thailand to acidic deposition derived from moisture regimes (A) and soil types (B). For legend description see text of the paper.

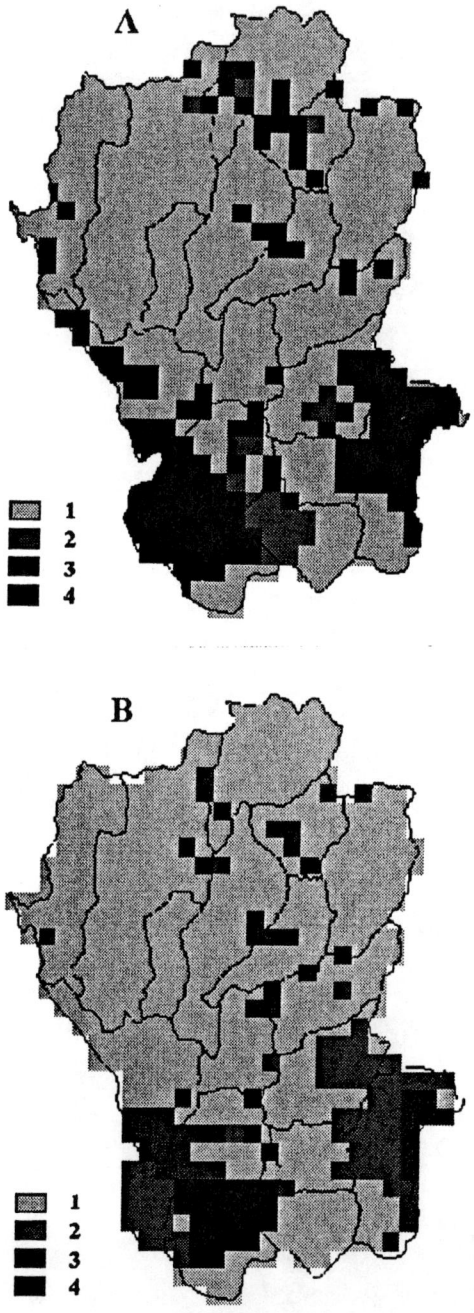

Figure 10. Sensitivity of ecosystems of Northern Thailand to acidic deposition derived from land use (A). Critical Loads for Northern Thailand (B). For legend description see text of the paper.

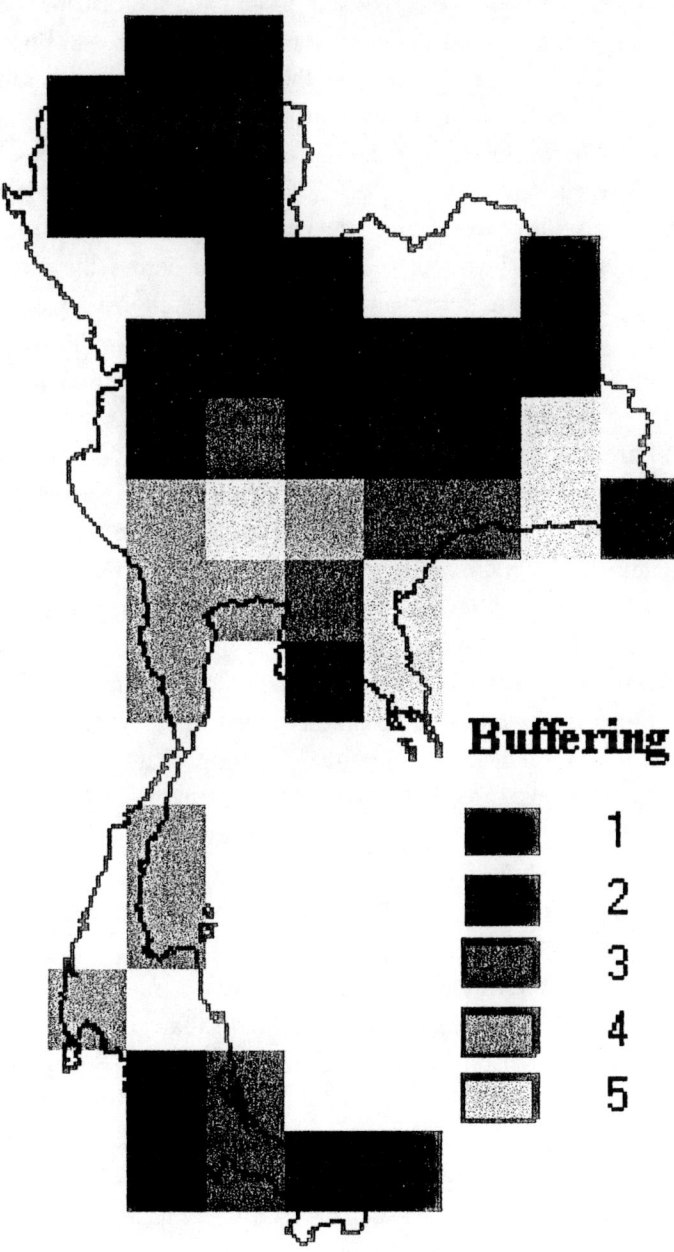

Figure 11. Buffering ability calculation for soils in Thailand according to the revised methodology of SEI.

The preliminary calculation of CL values for Northern Thailand mapped in figure 10, B. The legend means as follows 1 - less than 200 eq/ha/year, 2 - from 200 up to 400, 3 from 400 up to 700, and 4 -higher than 700 eq/ha/year. Practically the distribution of CL reflects the particular features of the soil sensitivity (compare Figure 9, B and Figure 10, B). One can see that calculated CL values across most of the study area are rather low. To assess the influence of acid rain on ecosystems of Northern Thailand sulfur depositions were calculated by using RAINS/ASIA model. The study area experiences modest excess deposition, usually between 375 and 750 eq/ha/year, although even this level of excess deposition is still typically three to six times the calculated CL. As a result of both the high sensitivity of ecosystems and levels of excess deposition across the Northern Thailand, more then 75% of the ecosystems across about 50% of the study area are at significant risk from acid deposition. Of course, these calculations should be validated, because some parameters were taken from literature or assessed by experts without sufficient field experiments.

The results of buffering ability calculation according to the revised methodology of SEI are shown in Figure 11 for the whole country. As you can see this result reflects again the main conclusion from the previous research. The most sensitive zones in Thailand (low buffering ability) concentrate in the northern and northeastern part of the country (see more darker area in the Figure 11). Several cells having soils of high sensitivity to acidic deposition can be found in the southern part too.

Potential environmental impacts in the given methodology due to sulfur deposition are accomplished by comparing calculated deposition to the critical loads. As mentioned above Critical Loads (CL) represent the maximum deposition that will not cause chemical changes leading to harmful effects on ecosystem structure and function. Detailed description of the estimation of critical loads is given elsewhere (Hettelingh et al., 1995).

Areas on the map with lower critical loads are more sensitive to damage from acid deposition. It can be seen from our calculations that most of Thailand is very sensitive to acid deposition. In fact, Thailand has some of the most sensitive areas with respect to acid deposition in all of Asia. Much of the eastern and northern sections of the country have critical loads of about 0.2 to 0.6 g S/m^2-yr. Areas of the far northwest and the southern peninsula have slightly higher critical loads in the range of 0.4 to 0.8 g S/m^2-yr.

Figures 12 and 13 show the excess deposition (exceedance) above the critical loads for the baseline emissions scenario in Thailand for the years 1990 and 2020 respectively. The legend of these maps means as follows 1 – negative exceedance, 2 – exceedance from 0 up to 0.5 g S/m^2-yr, 3 - from 0.5 up to 1, 4 – between 1 and 2 and 5 - higher than 2 g S/m^2-yr. The exceedances are calculated by subtracting the critical load from the average annual deposition for each grid. It should be mentioned that results of recent study are coincided in principle with the previous research conducted by Rains/Asia model. Already in 1990, two areas of Thailand show significant levels of deposition above the critical load. The first area is in the proximity of the Mae Moh power station in northern Thailand. Excess deposition in this area ranges from about 0.2 to 4.2 g S/m^2-yr with the greatest exceedances nearby the power station. The Bangkok region and areas to the east of Bangkok also experience a measure of excess deposition in the base year. Exceedances in this region range from 0.2 to 2.0 g S/m^2-yr.

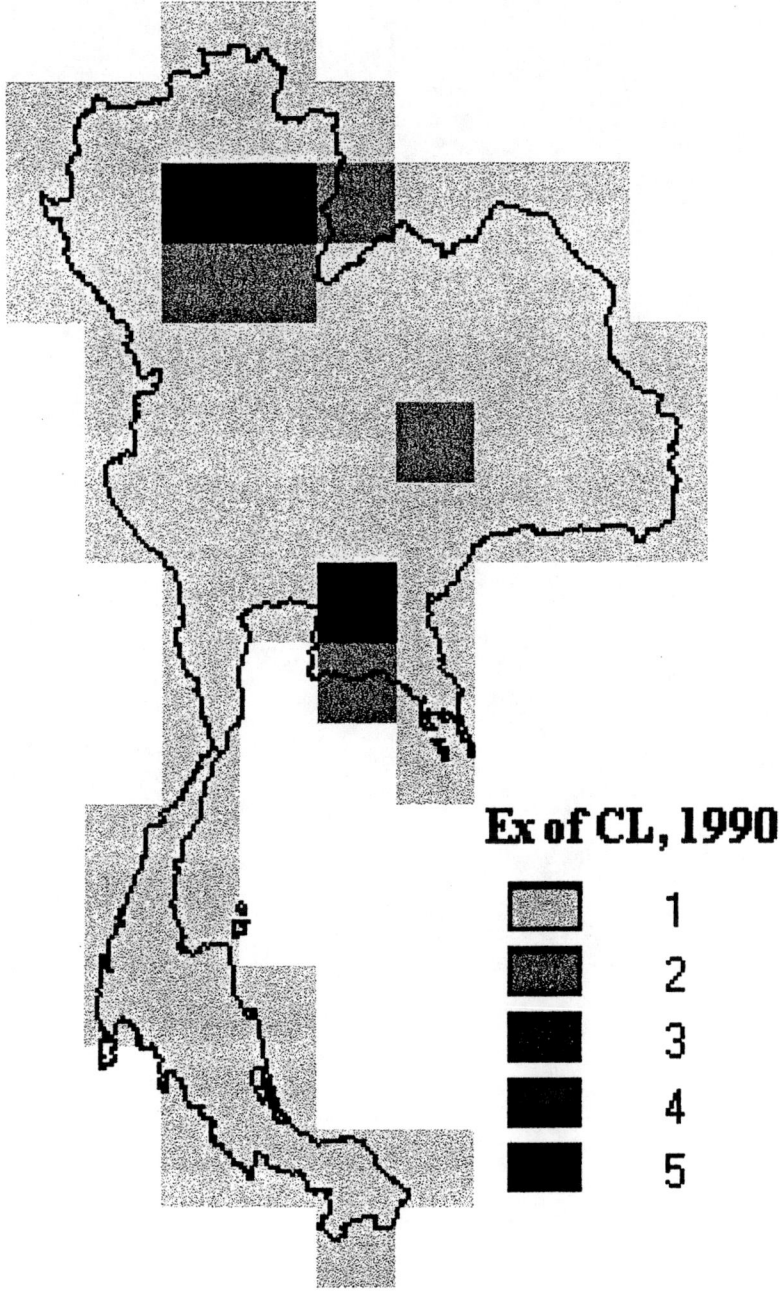

Figure 12. Exceedance of Critical Loads for baseline scenario in Thailand, 1990. For legend description see text of paper.

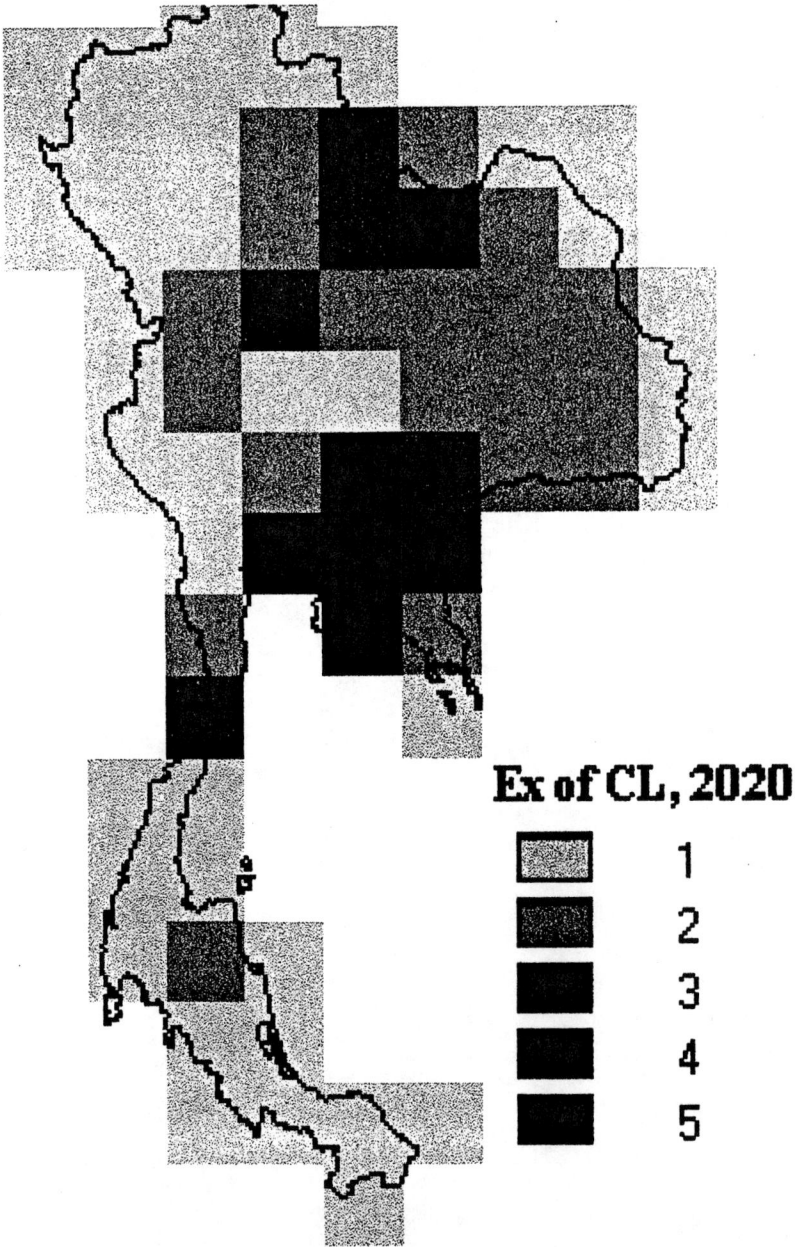

Figure 13. Exceedance of Critical Loads for baseline scenario in Thailand, 2020. For legend description see text of paper.

The environmental picture for the year 2020 is much different. In figure 13 it is apparent that, with the exception of the northwestern corner, deposition across most of Thailand will exceed the critical loads. Excess depositions range from about 0.2 to 5.6 g S/m^2-yr. The greatest exceedances occur in the east and southeast of Bangkok, where values from 1.0 to 5.6 g S/ m^2-yr cover the entire area. The remainder of the country experiences modest excess depositions under this scenario, usually up to 0.5 g S/m^2-yr although even this level of excess deposition is still typically several times the critical load. As a result of both the high sensitivity of ecosystems and levels of excess deposition across the country, 75% or more of the ecosystems across about 55% of the area of Thailand will be at significant risk from acid deposition in the year 2020.

Although the current state of scientific knowledge does not yet allow one to draw conclusions about the environmental damage implied with such excess deposition, the fact that sulfur deposition will be more than ten times above the sustainable levels in several areas may give reason for serious concern. Sulfur deposition will likely cause significant changes in the soil chemistry in several areas of the country, affecting growing conditions for many natural ecosystems and agricultural crops. If no additional countermeasures are taken, a degradation of the environmental quality in these affected areas has to be anticipated (see the experimental results below).

The transboundary effects of emissions on the whole Thailand seem to be relatively limited over the period studied according RAINS/ASIA model calculation. Background deposition of sulfur in Thailand due to emission from neighboring countries is insignificant when compared with the deposition due to sources from within Thailand. Transboundary effects due to emission from China only begin to affect Thailand in the future.

In the present study, uncertainties are difficult to evaluate. Major uncertainties may occur with space and time interpolation. The weather station network density in Northern Thailand is only 0.09/1000 km^2. This value is significantly lower than the generally recommended network density (e.g. 0.75/1000 km^2 to 5/1000 km^2 for precipitation climatology). The existing low weather station network density may also affect the interpolation of precipitation events. The larger error of deposition is associated with rainfall events with an intensity of less than 2 mm/3 hrs, which represent 10-20 % of the periods with precipitation during the rainy season.

4.2. EXPERIMENTAL RESULTS

4.2.1. Soil properties

The studied soil sample was taken from the forest area, that is close to the main lignite power plant station in Mae Moh district, Northern part of Thailand. The soil (clay texture with organic content of 1.33 %) was classified in the order of Zonal soils, lateritic soils of forested warm-temperate (suborder), Reddish brown lateritic soil (Grate soil groups), Ultisols and loamy paleustults. This soil group was found mainly in tropical soil (Kaewruenrom, 1990) where silicate clay was mainly belong to kaolinite with iron and aluminum oxide and hydroxide. The soil properties and composition are illustrated in table 6.

Table 6. The analytical results of the studied soil sample.

Parameters	The analytical result
Soil Texture Sand 39% Silk 16% Clay 45%	Clay
Bulk Density	1.58 g/cm^3
Soil Organic Matter	1.33 %
Soil moisture content	13.33 %
Soil pH	5.05
% $CaCO_3$	0.20 %
Exchange Acidity (EA)	2.9514 meq/100 g soil
Exchange Hydrogen Ion (Exch. H^+)	2.0826 meq/100 g soil
Total Aluminium	9.44 %
Exchange Aluminum Ion (Exch. Al^{3+})	0.8688 meq/100 g soil
Total Iron	7.285 %
% Active Fe (Fe^{2+}, Fe^{3+})	2.40 %
Cation Exchange Capacity (CEC)	11.4931 meq/100 g soil

Total Exchange Base (TEB)
  Exch. Ca^{2+} 2.8397 meq/100 g soil
  Exch. Mg^{2+} 2.8940 meq/100 g soil
  Exch. K^+ 0.4340 meq/100 g soil
  Exch. Na^+ 0.3000 meq/100 g soil
% Base Saturation (%BS)
	6.4677 meq/100 g soil
	Total Calcium 0.11 %
	Total Magnesium 0.915 %
	Total Potassium 2.96 %
	Total Sodium 0.155 %
	56.27 %

SO_4^{2-}	0.5959 meq/100 g soil

4.2.2. Impact of acidity on tropic soil

Simulated acid rains with H^+ concentration of 0.001-100 meq were applied to 100 g of soil in order to study the H^+ adsorption in the reddish brown lateritic soil. It was found that the H^+ buffering capability of the soil without pH change was limited at 0.1 meq/100gm soil. The higher concentration caused gradual decrease in the soil pH as shown in Figure 14. Exchange aluminum ion was increased when the addition of H^+ was higher than 2 meq/100g soil. Potassium was markedly decreased when H^+ was above 10 meq/100g soil. The other metal and nutrient were not significantly influenced by acid rain. From this study, change of chemical parameters in tropical soil can be conceptualized according to the H^+ concentration: < 0.1 meq/100 g soil; > 0.1 meq/100 g soil; and >10 meq/100 g soil as shown in Figure 15.

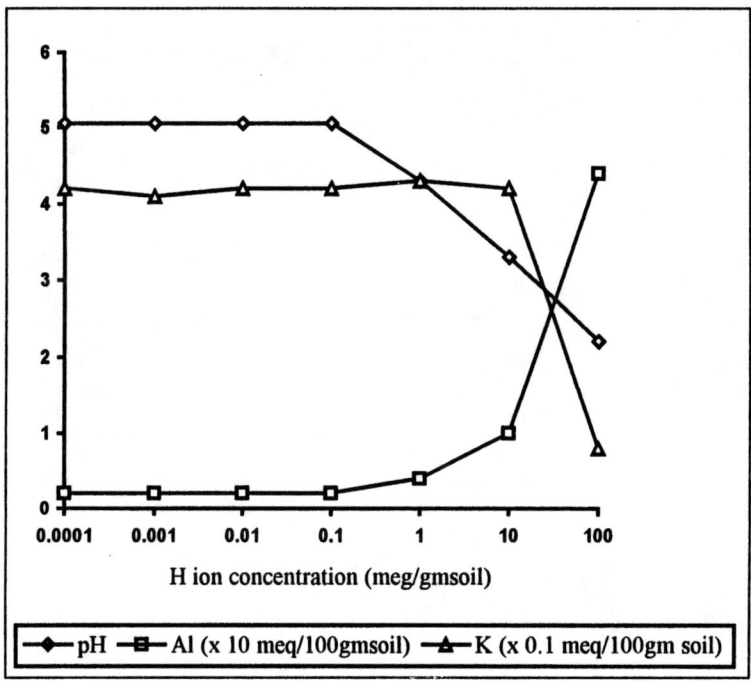

Figure 14. Change in pH, aluminum and potassium during acid rain application to the forest soil.

The study of the effect of simulated acid rain in the soil column (with the highest H^+ application at 1 meq/100gm soil) indicated unchanging in soil pH when the H^+ addition was lower than 0.2 meq/100gm soil. The decrease of pH was found only in the upper part of the column. However the leachate collected from 40 to 100 days of experiment, while the stable soil pH was observed, slightly changed to lower pH values.

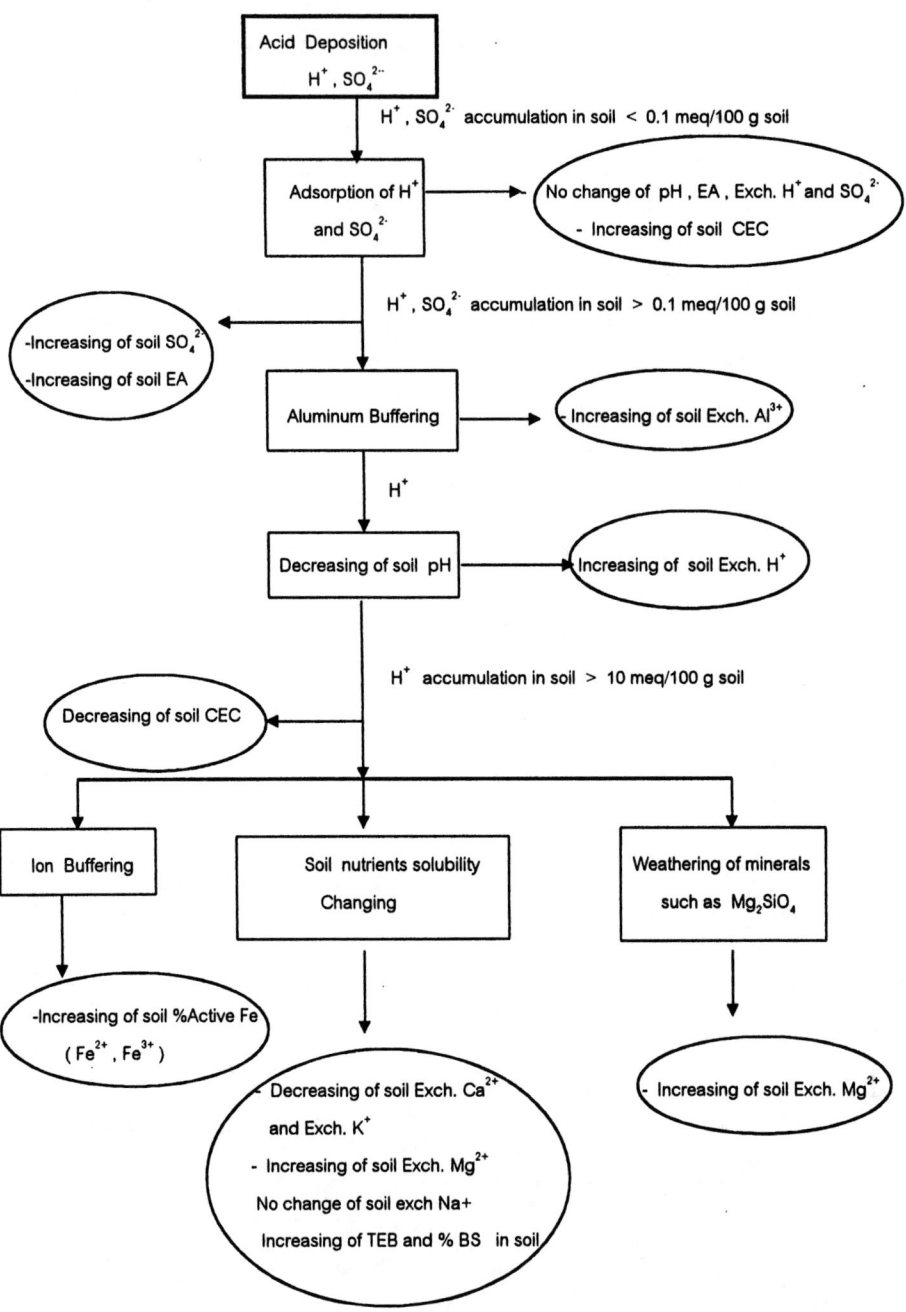

Figure 15. Conceptual diagram derived form the preliminary study showing the effect of acid deposition resulting from sulfur dioxide emission to the atmosphere through soil chemistry parameters.

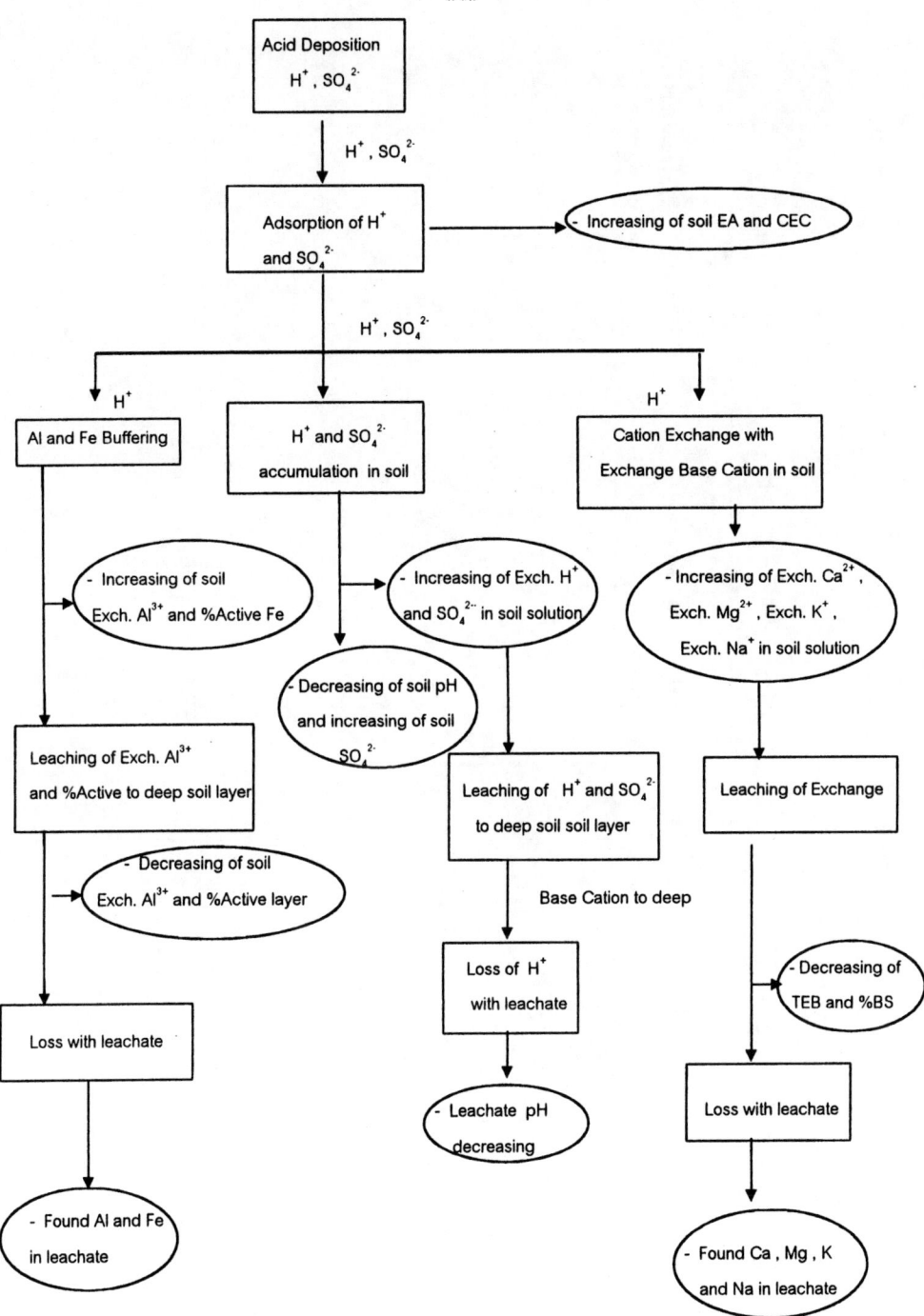

Figure 16. Conceptual diagram shows the effect of acid deposition resulting from sulfur dioxide emission to the mobility of the metal and nutrients in soil which was conducted in simulated natural condition.

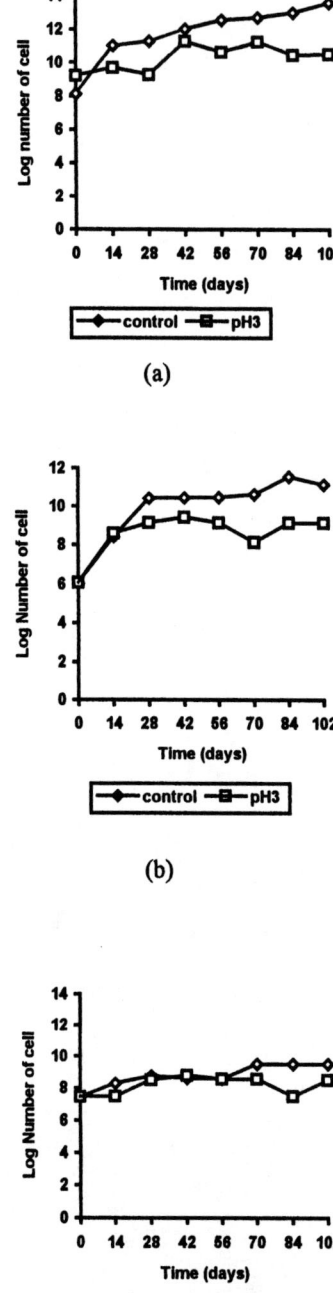

Figure 17. The microbial change effected by simulated acid rain application at the level of 6 cm from the surface soil. (a) Bacteria, (b) Actinomyces, (c) Ammonifying bacteria.

The mobility of metals (aluminum and active iron) and exchange nutrients (calcium, magnesium, potassium and sodium) in the upper layer (5 cm) of the soil was more effected by the high concentration of hydrogen ion than in the deeper layer (15 cm). The exchange aluminum, exchange active iron and exchange calcium were consequentially increased with the quantity of water and H^+ added. These phenomena lead to the reduction of total exchange base and base saturation and the increment of exchange acidity and cation exchange capacity in the soil. The exchange aluminum and iron were also found in the leachate of simulated acid rain with pH 2 (H^+ concentration of 1 meq/100gm soil) and their concentrations were higher that in variants with simulated acid rains of pH 3, 4, 5 and the control at pH 6. The exchange base cations (Ca, Mg, K) in soil were also increased in the soil solution and therefore released to leachate after 40 day. The mobility of these cations was also varied by the amount of water applied to the soil column. At the same pH of acid application, the higher metal mobility was observed when high volume of water was added. The conceptual diagram of the mobility of Al, Fe, Ca, Mg, K, as well as H^+ migrate to the next soil layer are shown in Figure 16.

4.2.3. Impact on Soil Microorganisms

The daily application of seventy milliliters of the simulated acid rain of pH 5, 4 and 3 were treated to the soil column and distilled water at pH 6 was used as control. The appropriate amount of soil at the depth level of 6, 18, 38 cm from top surface of the column was taken to analyze the microbial change. As for control, the high numerical number of the total microorganisms in the forest soil increased to 10^{12} cell/g soil during 20 days of acid application which reflect to the soil fertility. The obvious impacts were observed at the column with the highest acidic application (pH 4 and 3). The total amount of bacteria, actinomycetes and ammonifying bacteria were clearly reduced in comparison to the control, particularly at the soil level of 6 cm where high contents of hydrogen ions were accumulated as shown in Figure 17.

The moisture content due to the simulated acid rain addition showed the effect to soil microorganisms at different depth. High moisture content at the lower part of the column caused less oxygen transfer to the ecosystem while high accumulation of hydrogen ion play the important role to the upper part of the soil. In case of bacteria and actinomyces, the numerical number at the upper layer was highly effected by the acid accumulation when compare to the deeper layer. However there were evidence that the number of these two microbial groups in the lower part of the column would be highly effected if the addition of acid rain was continued. The influence of depth was opposite on the number of ammonifying bacteria which may be due to the oxygen depletion and the moisture in the soil that obstructed the ammonifying activity.

Although the increasing rate of the fungi was observed during the early acid rain application and expressed the constant growth rate after 20 days, there was no significant change in the numerical study of fungi. This may be due to the resistant of some fungi to the higher acid condition than other microorganisms as reported by Lettl (1990). It is noted that in this experiment, there was no pH change even with the addition of the simulated acid rain at pH 3. The number of nitrogen fixing bacteria and nitrification bacteria in the forest soil was very small and no difference in the numerical study was observed.

5. Conclusions

1. Assessment the effects of acidic deposition in Northern Thailand has been carried out by using two steps GIS-based procedure. Firstly, the sensitivity mapping has been conducted for Thailand conditions. On the second stage, the sensitivity of Thailand ecosystems to acidic deposition has been described by means of CL. Application of this methodology across Northern Thailand shows that CL are rather low for most part of the study area. As a result of both the high sensitivity of ecosystems and levels of excess more then 75% of the ecosystems across about 50% of the study area are at significant risk from acidic deposition. These results coincide with the assessment for the whole country using revised SEI methodology. Sulfur deposition will likely cause significant changes in the soil chemistry in several areas of Thailand, affecting growing conditions for many natural ecosystems and agricultural crops. If no additional countermeasures are taken, a degradation of the environmental quality in these affected areas has to be anticipated. In order for results to be applied for policy makers uncertainties require additional investigation.

2. The input of <1 meq/100g soil to the forest soil in the northern part of Thailand, with the organic content of 1.33 percent, has no changes in pH value due to existing hydrogen buffering capacity. Simulated acid rain at pH 2 led to Al leaching from this soil naturally enriched by aluminum and iron, although there was no significant change of pH in the soil. The acid depositions caused also the reduction of bacteria, actinomyces and ammonifying microbes in the upper part of the soil. These could lead to low nutrient cycling rate in the ecosystem as these microorganisms play the significant role in organic matter decomposition in the soil. Due to certain buffering ability of soils to acid depositions the forest ecosystem can be sustainable in short-time scale, however some chemical and biological changes can occur in soil even in this period. It will gradually decrease the ecosystem sustainability to acid deposition.

REFERENCES

Bashkin, V.N., M.Ya., Kozlov and A.Yu., Abramychev (1996). The application of EM GIS to Quantitative Assessment and Mapping of Acidification Loading in Ecosystems of the Asian Part of the Russian Federation. *Asian-Pacific Remote Sensing and GIS Journal*, Vol. 8, No.2, 73-80.

Cambridge, H.M., M.J., Chadwick, J.C.I., Kuylenstierna et al.(1995). *The Sensitivity of Ecosystems to Acidic Deposition*, Stockholm Environmental Institute, 25 p.

Chiang Mai University (1990). *Draft Final report for EIA of Mae Moh Water Supply Project - Kew Kor Mah Reservoir - Vol. II Main Report, Vol. III Environmental Impact Mitigation Plan, Vol. IV Appendices*, June 1990.

DEDP (1995): *Thailand Energy Situation*. Department of Energy Development and Promotion, Ministry of Science, Technology and Environment, Bangkok, 28p.

Environmental Impact Assessment for Mae Moh Mine and Power Plant Expansion Project(1991). *Final Report by Schultz Int.*, November 1991.

Evaluation of Alternative SO_2 Control Strategies for Mae Moh units 1 to 13 and Mae Kham AFBC Units 1 and 2 (1994). *Report of KBN Engineering and Applied Sciences, Inc.*, Bangkok, November 1994.

Hettelingh, J.-P., Sverdrup, H. and Zhao,D. (1995) Deriving Critical Loads for Asia. *Water, Air and Soil Pollution*, Vol.85, 2565-2570.

Kato, N. and Akimoto, H. (1991) *Anthropogenic Emission of SO_2 and NO_x in Asia*. Third Annual Conference on Acid Rain and Emission in Asia Bangkok, Thailand, 18-21 November 1991.

Khummongkol, P., Pairintra, R. and Santisirisomboon, J. (1996) Energy Industrial Processes in Final Report *"Thailand's National Greenhouse Gas Inventory 1990"* submitted to U.S. Country Studies Program by Office of Environmental Policy and Planning, The Royal Thai Government, Complied by Thailand Environment Institute.

Kozlov, M.Ya. and Towprayoon, S. (1995) Situation and need of critical loads assessment in Thailand. Global Analysis, Interpretation and Modelling: *First Science Conference*, Garmisch-Partenkirchen, September 25-29, H-16.

Kozlov, M.Ya. and Towprayoon, S. (1997) GIS-based method to indicate the ecosystems at risk to acidic deposition in Northern Thailand. *Asian-Pacific Remote Sensing and GIS Journal*, Vol. 9, No.2, 69-73.

Kozlov, M.Ya., Towprayoon, S. and Sirikarnjanawong, S. (1997) Application of Critical Load Methodology for Assessment the Effects of Acidic Deposition in Northern Thailand, *Proceedings of the International Workshop on Monitoring and Prediction of Acid Rain*, September 29-October 1, Seoul, Korea, 141-146.

Lettl, A.(1990) Influence of Industrial Sulfur dioxide Emission on Microorganisms of Forest Soils, *Ekologia-Cssr*, Vol. 9, No3, pp. 315-330.

NEPO (1997) *Sulfur Emission Impacts and Mitigation Alternatives in Thailand*. National Energy Policy Office, Bangkok, 118p.

Nilsson, J. and Grennfelt, P. (1988) Critical loads for sulfur and nitrogen. Proceedings of the Worshop at Stockloster, Sweden on 19-24 March, *Miljo report*, 15, Nordic Council of Ministers, Kobenhavn, 418 p.

Page, A.L. (1982) *Method of Soil Analysis, Part 2, 2nd ed.*, Madison, American Society of Agronomy, pp. 781-1079.

Kaewruenrom, A. (1990) *Soil of Thailand: Properties, distribution and use*, Division of Soil Science Faculty of Agricuture Kasetsart University, pp. 99-475.

SEI (1997) Chapter 2.5. Ecosystem Acidification. In: J.C.I Kuyelenstierna, S. Cinderby, H. Cambridge and W.K. Hicks (1998) *Regional Air Pollution in Developing Countries, Background Document for Policy Dialogue*, Bangkok, March, 1998. Edited: Johan Kuylenstierna and Kevin Hicks.

Vearasilp, T. and Songsawad, K. (1991) *Soil information system for Thailand*. Department of Land Development, Bangkok, 253p.

Wibulswas, P and S. Towprayoon. (1993) Sulfur dioxide emission from power station and oil refineries in Thailand. *The Proceeding of the International Conference on regional Environment and Climate Change in East Asia*, November 30- December 3, Taiwan, 275-277.

Chapter 11
ACID DEPOSITION AND FOREST ECOSYSTEM SENSITIVITY IN TAIWAN

LIN TENG-CHIU

National Changhua University, Changhua, Taiwan, tclin@cc.ncue.edu.tw

CONTENT

Abstract
1. Introduction
    1.1 Air pollution and forest health
    1.2 Atmospheric N and S deposition in forest ecosystems
    1.3 Critical load
    1.4 Temperate versus subtropical forests
2. Acid deposition in Taiwan
    2.1 Physical setting of Taiwan
    2.2 SO_x and NO_x emission
    2.3 Acid deposition in Taiwan
    2.4 Source receptor relationship
3. Ecosystem sensitivity assessment – an application of RAINS-ASIA IMPACT module
    3.1 Sensitivity classification for Taiwan ecosystems
    3.2 Critical load calculation
    3.3 Evaluated forest ecosystems
    3.4 Input data
    3.5 Discussion of Ecosystem Sensitivity to Acidic Deposition
4. Conclusions
References

ABSTRACT

On a basis of monitoring data, approximately 70% of the precipitation monitored in Taiwan is considered acid rain (exhibiting a pH lower than 5.6). I has been shown that acid deposition shows large spatial variation with northern Taiwan exhibiting the highest acid deposition in comparison with southeastern part. Dry deposition is least important in northern and northeastern Taiwan as these areas have the high annual precipitation with an even distribution. Dry deposition becomes proportionally more important in areas with low annual precipitation and high variability, such as southern Taiwan. Using RAINS-ASIA impact module the ecosystem sensitivity to acid deposition has been assessed and critical load of S to six forest ecosystems in Taiwan have been calculated. Results indicate that forest ecosystems in Taiwan are very sensitive to acid deposition due to their low soil pH (<5.5). Lowland subtropical forest ecosystems in Taiwan have low or moderate low critical load for S suggesting their sensitivity to acidity loading. The critical loads are exceeded for many forest ecosystems at present level of sulfur deposition that lead to hazardous consequences once the current buffering capacity will be depleted.

1. INTRODUCTION

1.1. AIR POLLUTION AND FOREST HEALTH

There has been increasing concern about the declining health of many forest ecosystems and the potential impact that this decline may have on local and global ecosystems. Acid deposition is considered a primary cause of this decline. Several pathways by which acid deposition could damage forests have been proposed: 1) soil acidification, cation depletion, and Al toxicity (van Breemen et al. 1982; Ulrich and Pankrath 1983; Schulze 1989; Likens and Bormann 1995); 2) direct and indirect influence of acidic pollutants on the photosynthetic capacity of the trees (Unworth and Ormrod 1982; Eamus and Murray 1991; Cape, 1993); 3) excess nitrogen deposition and subsequent soil acidification or physiological injury (Nihlgard 1985; Aber et al. 1989) 4) biodiversity changes due to eutrophication (Bobbink *et al.* 1997). Field studies and greenhouse experiments prove that acid deposition can affect forests in many ways; through soil microorganisms (Termorshuizen 1990; Esher et al. 1991; Vosatka et al. 1991; Wallander and Nylund 1992), soil chemistry (van Breemen et al. 1982; Ulrich and Pankrath 1983; Fasth et al. 1991; Matzner and Prenzel 1991), through nutrient cycles (McColl and Bush 1978; Shepard et al. 1989; Kress et al. 1990; Duckworth and Cresser 1991), and through plant physiology (Baker and Hunt 1986; Eamus and Murray 1991; Haines et al. 1985). The diversity of actual and potential acid deposition influences on forests emphasizes the need to further explore the relationship between acid deposition and forest ecosystem health.

Some researchers have questioned the significance of air pollution as a factor influencing forest health (Johnson 1983; Johnson and Taylor 1989; Fernandez 1989; Rehfuess 1991). Those questioning the importance of air pollution have used two basic lines of reasoning: one, that some forests experiencing heavy pollutant loading do not show prominent symptoms of decline, and secondly, that some declining forests receive little pollutant loading. However, such observations do not reduce the importance of studying acid deposition pollution on forest health, but instead emphasize the complexity of the subject.

Forest decline is the product of interactions between trees and external factors, both anthropogenic and natural. Some forests may be more resistant to the effects of acid deposition (Houston and Stairs 1973; Karnosky 1977; DeHayes and Hawley 1992; Soares et al. 1995) and therefore, forest decline may not always be observed in areas experiencing heavy acid deposition. The effects of acid deposition are likely to be additive and it may take decades before symptoms of decline are observed (Bormann 1985). Because forests are complex, long-lived ecosystems, they may respond slowly and subtly to stress (Pitelka and Raynal 1989). The failure to find prominent growth decline in areas undergoing heavy acid deposition may be primarily due to prolonged response time, especially for species more resistant to air pollution. In addition, the failure to find a clear relationship between acid deposition and forest decline in many situations may result from type II statistical errors, due to the "low power" of most field studies. Because changes in forest health are often confounded by uncontrolled variables, e.g. land-use history and plant community differences, field studies often have low power (Peterman 1990).

1.2. ATMOSPHERIC N AND S DEPOSITION IN FOREST ECOSYSTEMS

The fate and influence of high levels of atmospheric deposition of nitrogen and sulfur on forest ecosystems are the focus of many studies of acidic deposition. Although both elements are essential nutrients, very few terrestrial ecosystems encounter sulfur deficiency, while most are nitrogen limited to some degree (Brady 1990; Vitousek and Haworth 1991). The fate and influence of nitrogen and sulfur deposited on a forest ecosystem are very different (Hutchinson et al. 1972; Richter et al. 1983; Boring et al. 1988; Friedland et al. 1991; Marschner et al. 1991; Gebauer et al. 1994). Atmospheric input of sulfur is accumulated as inorganic sulfate, and are not used for growth, in contrast to nitrogen, which is used for growth (Waring and Pitman 1985; Kenk and Fischer 1988; Miller and Miller 1988; Oren et al. 1989; van Dijk et al. 1990; Rodenkirchen 1991; Nilsson and Wiklund 1992; Gundersen and Bashkin, 1994). Both the sulfate accumulated and the nitrogen used by trees for growth, increase system demand for cations, thereby potentially inducing nutrient imbalances (Skeffington and Wilson 1988; Schulze 1989; Tietema and Verstraten 1991; Aber 1992; Kaiser et al., 1993).

1.3. CRITICAL LOAD

The critical load concept is receiving much attention as a measure for developing national emission abatement strategies. It provides a receptor-based approach to control emissions of pollutants (Hornung et al. 1995). Although the determination of the critical load is complicated by the fact that it depends upon the pollutant of concern and the target ecosystems under observation, the concept is relatively simple. Critical load can be generally regarded as the highest load that will not cause harmful effects on the receptors under concern (Nilsson 1986; Bull 1992; Hettlingh et al. 1992; Langan and Hornung 1992; Bashkin, 1997). It is possible to calculate the critical load for any given pollutant. As a result, for acid deposition, several approaches have been developed to calculate critical loads for the two principal groups of pollutants, namely sulfur and nitrogen compounds (Nilsson and Grennfelt 1988; Schulze et al. 1989; Sverdrup and de Vries 1993). One research group has been investigating the emission and transport of pollutants and the calculation and mapping of critical load. They have extended their mapping efforts to Asia and have developed a RAINS-ASIA model (Carmichael et al. 1993). In contrast to the effective reduction of acidic deposition in Europe and North America through the adoption of strict pollution abatement measures, acidic deposition throughout Asia is likely to become increasingly severe due to the rapid pace of development. For example, acidic deposition is suspected to have played a role in forest health decline in China (Zhao and Seip 1991; Liu and Li 1991). The RAINS-ASIA project group has been working intensively to develop an assessment model for acid deposition in Asia. A subgroup of the RAINS-ASIA team has focused on establishing sensitivity criteria as well as computing and mapping critical loads to better assess the impact of acidic deposition on Asian ecosystems. Since the research is on a regional scale, these criteria and methodologies may not be suitable for local application as the latter requires a much higher resolution to be of practical use.

1.4. TEMPERATE VERSUS SUBTROPICAL FORESTS

The effect of acid deposition on forest ecosystems is site specific. For example, forests can act as a source or sink for atmospheric ammonia depending on concentrations above the forest (Langford and Fehsenfeld 1992) and on plant nitrogen status (Lin and Hamburg 1992). The mechanisms through which nutrient imbalances impact plant growth in temperate forests might not be operational in subtropical forests. Loss of hardening ability resulting from excess nitrogen input is one of the main mechanisms for forest damage in temperate forests (Soikkeli and Karenlampi 1984; Aber et al. 1989; Nashimoto and Kohno 1989). Yet, in lowland subtropical forests frost is uncommon. Because nutrient cycles of subtropical forests differ from those of temperate forests (Bruijnzeel 1991; Vitousek and Sandford 1986), impacts of acidic deposition on nutrient imbalances cannot be easily extrapolated from temperate forests to subtropical forests. In fact, our understanding of nutrient cycling in subtropical forests is very limited (Bruijnzeel 1991).

Although studies have substantially improved our knowledge of the mechanisms by which acid pollutants interact with plants and soils, and, in turn, affect forest health, studies are disproportionally focused on the temperate forest ecosystems of the northern hemisphere. This is partly because these regions have a longer history of industrialization and associated acid deposition than do currently developing regions. However, many subtropical regions, such as Taiwan, now experience very high levels of acid deposition. If acid deposition is detrimental to forest health, it is reasonable to expect that subtropical forests do indeed face an immediate threat. In fact, the usually low base saturation of many subtropical forest soils, especially humid subtropical forests, may make them more susceptible to detrimental pollution effects than temperate forests.

2. ACID DEPOSITION IN TAIWAN

2.1. PHYSICAL SETTING OF TAIWAN

Taiwan is an oblong-shaped island 120-200 km off the southeast coast of Mainland China (Fig. 1). It is 394 km long and up to 140 km wide with area of 36,000 km^2. The topography of Taiwan is characterized by the Central Mountain Range, a north-south trending ranges with that has over 200 peaks greater than 3,000m. The foothills are generally several hundred meters high, but may extend up to 1000 meters (Hsieh and Shen 1994). The Central Mountain Range bisects the island from north to south. It has profound influences on Taiwan's natural and cultural history.

The climate of Taiwan is subtropical and lies within the Pacific Monsoon region. The tropic of cancer passes through the south part of the island. However, due to the great altitudinal gradient, climate varies from subtropical to arctic along the mountain elevation gradient. The Central Mountain Range also affects rainfall patterns. Northern and eastern Taiwan receive heavy rains throughout the year, while the southwest has a winter dry season and high summer rains. The prevailing direction of the wind from late spring to early fall is southwest. The wind comes off the ocean and does not pass over any larger landmasses prior to arriving at Taiwan. Because Taiwan is directly in the path of Pacific typhoons, there are frequent summer typhoons (twice a year on average for the last 100 years). The typhoons, originating from oceanic air masses, typically approach the island from the east and lose strength as they cross the Central Mountain Range. From late fall to early spring, the prevailing winds are out of the northeast, with the air masses originating from Mongolia. The wind passes through part of the industrialized southwest of mainland China before arriving at Taiwan. The east coast of mainland China is characterized by heavy acid deposition and high emission levels of NOx and SOx and high atmospheric deposition capacity (Wang et al. 1993). Several studies indicate the eastward transport of air pollutants from the east coast of mainland China (Gao et al. 1993) particularly during the winter and early spring when the westerly wind are strong (Uematsu et al. 1993). The area of Taiwan most directly affected by the winter winds is the northeast, including

particularly between December and February when the strong northeast monsoon dominates the weather system.

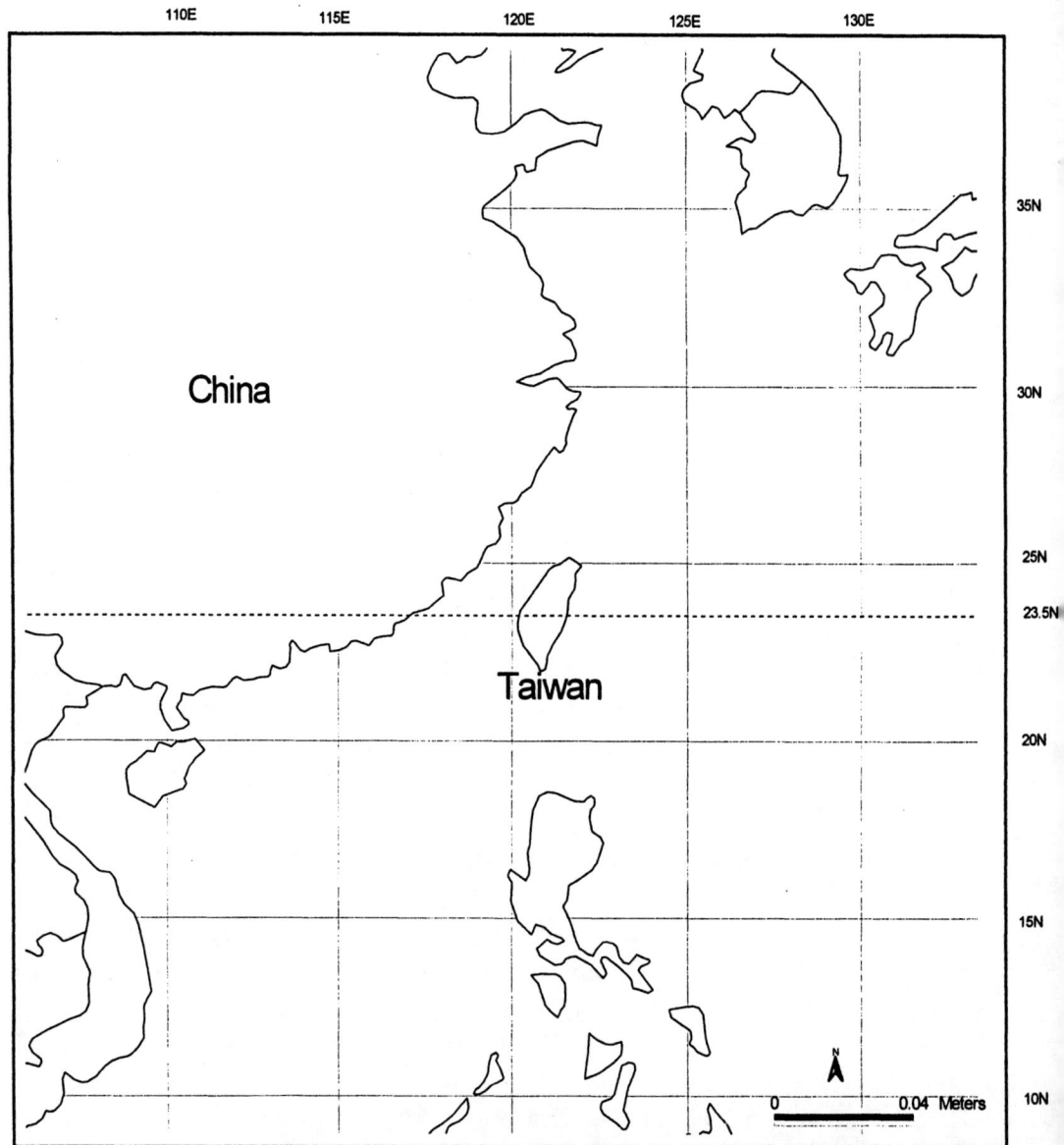

Figure 1. Location of Taiwan.

2.2. SO$_x$ AND NO$_x$ EMISSION

Like many Asian countries, recent economic development of Taiwan has been characterized by an increase in population density, increased energy demands, rapid urbanization and industrialization and as a result, increasing environmental degradation. Air pollution is particularly prominent among the diverse environmental problems that Taiwan faces. Annual emission of SO$_x$ was as high as 583,000 tons in 1991, decreased to 443,000 tons by 1995 (EPA, 1996) and is projected to further decrease to 335.000 tons by 2000 (Liu et al. 1996). However due to fast growing in moto-vehicles emissions of NO$_x$ slightly increased from 599,000 ton in 1991 to 664,000 tons in 1995 and is projected to reach 749,000 tons by 2000 (Liu et al. 1996). Given the small land area of Taiwan (36,000 km^2) the emission of SO$_x$ and NO$_x$ on a per hectare basis is very high. The patchy distribution of industrial development in Taiwan produces a mosaic of divergent ecosystems, which vary in distance from local pollution sources. The eastern half of Taiwan, with its narrow flood plain, is much less developed than the western half, which has been heavily industrialized over the past two decades. As a result, acid deposition exhibits substantial spatial differentiation.

Due to the decrease in SO$_x$ emissions and the increase in NO$_x$, the latter may prove the more important form of acid deposition in the future. If nitrogen deposition exceeds the plant growth requirement, ecosystems may become nitrogen saturated (Aber *et al.* 1989; Gundersen and Bashkin, 1994). Studies have indicated that nitrogen saturation may lead to plant nutrient imbalance, delay plant hardening as well as change the ecosystem nutrient budget. Several studies conclude that nitrogen deposition is closely related to forest decline in Europe and North America (Aber et al. 1989; Skeffing and Wilson 1988; Schulze 1989; Aber 1992) .

2.3. ACID DEPOSITION IN TAIWAN

(1) Acid deposition monitoring network

Long-term, systematic monitoring of acid deposition in Taiwan is limited to two principal networks, the Environmental Protection Agency (EPA) and the Taiwan Ecological Research Network (TERN). A nationwide network funded by the EPA has been monitoring acid deposition at 10 sites since April 1990. An additional 2 sites were added to the network in 1993 (Chen et al. 1996). The EPA monitoring network is principally located in the heavily populated coastal areas. The TERN network, funded by National Science Council, has been monitoring acid deposition at three lowland forest sites. Researchers from the Taiwan Forestry Research Institute monitor acid deposition at five lowland mountain sites and one high elevation mountain site (King et al. 1994). One of these is also part of the TERN monitoring network. Other monitoring efforts have typically been limited to a much shorter monitoring time scale. Several universities also maintain their own monitoring stations close to campus. This lack of acid deposition monitoring continuity makes cross-site comparisons difficult due to the lack of congruent data sets.

(2) Patterns of Acid Deposition in Taiwan

A synthesis analysis of 20 precipitation monitoring stations by Chen et al. (1996) provides spatial patterns of pH, sulfate deposition and nitrate deposition around the island (Fig. 2). Approximately 70% of the monitored rainfall events have a pH lower than 5.6 and can be classified as acid rain. The fact that the annual depositions are based on averages of different time spans should be noted. The annual deposition of sulfate (130 kg/ha) and nitrate (50 kg/ha) in northern Taiwan greatly exceeds the maximum sulfate deposition (43 kg/ha) reported by the National Acid Precipitation Assessment Program of the United States (NAPAP 1990).

Much of the spatial pattern in acid deposition around Taiwan can be explained by the spatial difference in the degree of urbanization and industrialization as indicated by the distribution map of population and industrial parks (Fig 2). Northern Taiwan has the highest annual sulfate and nitrate deposition whereas eastern Taiwan has lowest nitrate and sulfate deposition. This pattern coincides with that of population and industrial park density.

It is also important to note that dry deposition was not included or was only partially included in the calculation of annual sulfate and nitrate deposition as the deposition map was made base upon either wet-only or bulk precipitation. Northern and northeastern Taiwan exhibit high annual precipitation that is distributed rather evenly throughout the year. In contrast, southwestern Taiwan has less total precipitation and irregular distribution with a long dry winter. As precipitation can effectively remove pollutants in the atmosphere, dry deposition is less significant in northern and northeastern Taiwan than in southern Taiwan. Analyzing data from the EPA acid deposition monitoring system, Yei et al. (1996) indicate that dry deposition contributes about 33% and 24% of nitrate and sulfate deposition in northern Taiwan, respectively. However, in southern Taiwan, the contribution of dry deposition was as high as 70% for both sulfate and nitrate deposition. An independent study by Lin et al. (1996) using Na-ration methods (Gosz 1980) estimated that the contribution of dry deposition to sulfate and nitrate deposition was 21% in northeastern Taiwan. Northern Taiwan has patterns of precipitation similar to that of northeastern Taiwan but with slightly less rainfall. The congruence of the figures utilizing different analysis methodologies strongly supports a 20-30% dry deposition estimate for northern Taiwan.

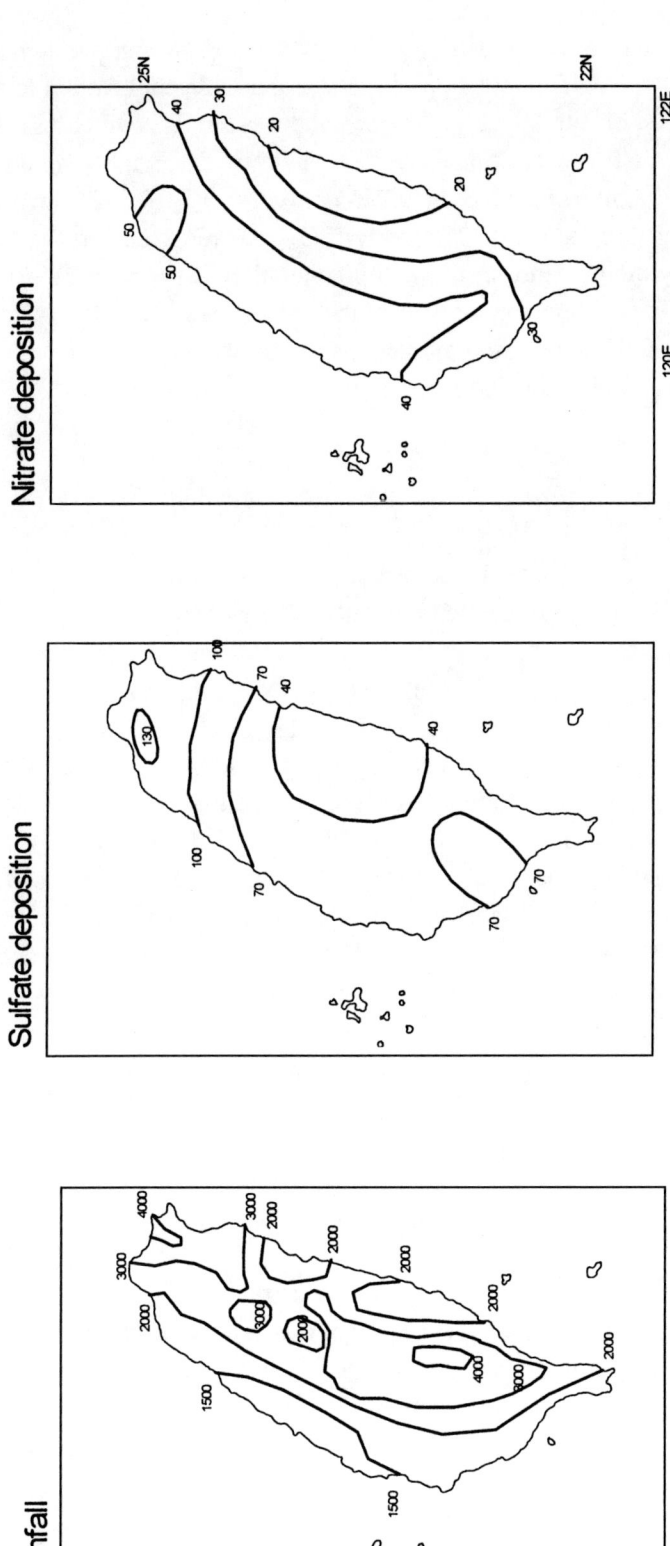

Figure 2. The isopleths of annual rainfall (mm), sulfate and nitrate deposition (kg/ha/yr) in Taiwan adapted from Chen et al. 1996.

Spatial variation in precipitation at least partly explains why although southwestern and northern Taiwan exhibit comparable population densities and industrialization levels, nevertheless the wet deposition of nitrate and sulfate of southwestern Taiwan is less than 2/3 of that of the north. In fact, EPA monitoring data (1993) indicate that Kaohsiung, in the south, showed higher concentrations of both SO_2 (0.025 ppm) and NO_2 (0.032 ppm) than that of Taipei in the north (0.017 ppm and 0.025 ppm for SO_2 and NO_2, respectively). The higher annual and temporally more evenly distributed precipitation of northern Taiwan is potentially more effective at scavenging materials from the atmosphere, thereby producing higher deposition levels of both sulfate and nitrate. More of these pollutants remain airborne in the south due to its relative paucity of precipitation.

2.4. SOURCE RECEPTOR RELATIONSHIP

Having documented the high level of acidic deposition in industrialized areas, there is a growing interest to identify the sources of acidic pollutants. Studies suggest that both local emissions and long-range pollutant transport contribute to the high levels of acid deposition in Taiwan. Yet, researchers disagree as to the significance of long-range transport. A case study of atmospheric deposition between May 13 and May 17, 1987 by Chang and Jeng (1996) indicates that long-range transport contributed 31% and 24% for the wet deposition of atmospheric sulfur and nitrogen deposition. For dry deposition the corresponding percentages for sulfur and nitrogen are 6% and 29% and for total deposition, 27% and 25%. This is contrast to the study by Zhau at al. (1996) which suggests that more than 95% of the sulfur and nitrogen deposition in Taiwan originated from local emission sources. A study by Wang et al. (1996) also postulates that although there was acidic pollutant output from long-range sources, acid deposition in East Asia is primarily from local emission sources. In the study described below, I provide details of a case study examining the contribution of local and long-range transport to total acid deposition at a forest ecosystem in northeastern Taiwan.

(1) A case study on partitioning short and long distance pollution sources

Over the past twenty years Taiwan has undergone an industrial transformation that has resulted in a patchwork of industrial development, agricultural land and lightly managed forests. In combination with the mixing of oceanic and local air masses, this diverse land use pattern produces temporally and spatially complex pollution patterns. However, the relative isolation of the island and the consistency of seasonal wind trajectories in combination with the land-use patterns, permits the separation of local and regional influences on pollutant loading. A detailed analysis of the precipitation of individual storms from two sites, one in a forest in northeastern Taiwan and the second 35 km north in Taipei, was used to identify the source of deposited pollutants at the forest site.

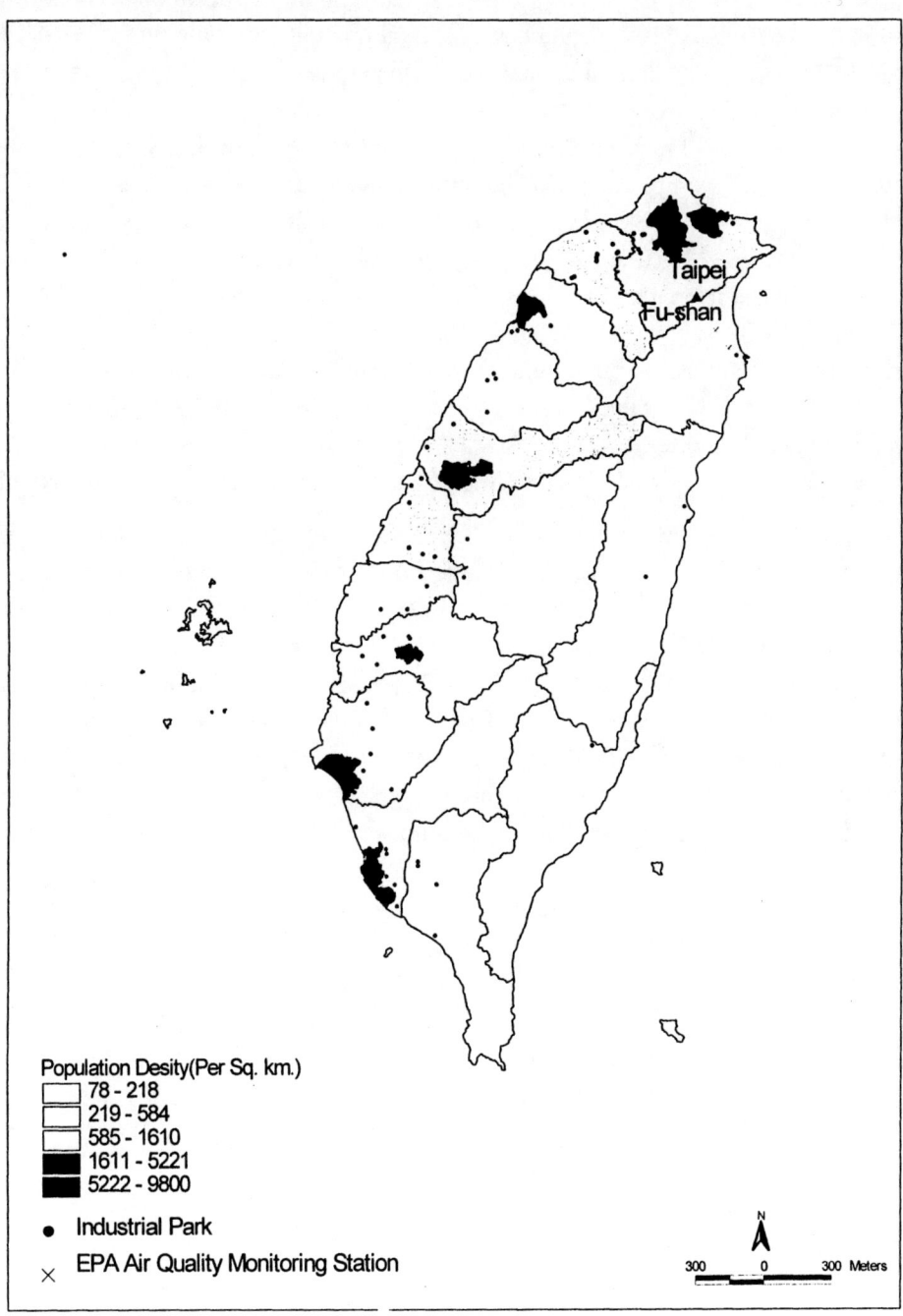

Figure 3. Map of Taiwan indicating population density and locations of industrial parks, Fu-shan Experimental Forest, and Taipei.

The forest site is a 37 ha watershed at the Fu-shan Experimental Forest in northeastern Taiwan (24° 34" N, 121° 34' E), 18 km from the coast. The moist sub-tropical forest varies in elevation from 670 m to 1100 m, and has an annual precipitation ranging from 2900 mm to 4800mm (1993 to 1997), and an annual mean temperature of 20 °C (with a range from 14°C in January to 28 °C in August).

The Taipei site is 16 km from the northwest coast (Fig. 2) and is a city of 4 million people, with an annual precipitation of 2500 mm, and an annual mean temperature of 22°C (15 °C in January, 29°C in August). Since 1980, event based precipitation has been collected and analyzed[*]. Between August 14, 1993 and August 13, 1994, all event based precipitation samples from the two sites were collected and analyzed using the same protocol (Lin et al. 1997).

Annual fluxes of sulfate and inorganic nitrogen were 250 and 279 meq/m^2/yr in Taipei and 168 and 121 meq/m^2/yr at the forest (Lin et al. 1994). These fluxes are high relative to those measured at polluted temperate forest sites, e.g. 162 and 64 meq/m^2/yr for sulfate and inorganic nitrogen, in the United States (Johnson and Lindberg 1989) and 144 and 143 meq/m^2/yr in Germany (Matzner 1989). The amount of sulfate deposited in Taipei originating from sea salt aerosols was 18% in the summer (June to August) and 10% in the winter (December to February) (Keene et al. 1986; McDowell et al. 1990; Bridgman 1992)(Table 1). At the Fu-shan forest, it was 15% in the summer and 10% in the winter.

Annual emissions of SO_x and NO_x within Taiwan in 1993 were 460000 Mg/yr and 410000 Mg/yr, respectively (EPA 1993). Of these emissions, three quarters of the SO_x and one third of the NO_x came from the generation of electricity. An additional 50% of the NO_x resulted from automobile emissions. Taipei has a large number of local emission sources as it is surrounded by industrial parks. This is in contrast to the relative isolation of Fu-shan from large local emission sources.

In the summer (June to August), the chemical composition of precipitation (concentration and flux) was much higher in the city than the forest site (Table 2). For aerosols of anthropogenic origins which can be transported long distances (e.g., nitrate and sulfate, Tanner 1990), the concentrations at the forest site were approximately 30% of those found in the city. For calcium and magnesium which can not be easily transported for long distances, (Tanner 1990) the loading in the forest site was only 14% (Ca) and 24% (Mg) of those found in the city.

The sampling device is situated on the top of the Forestry Research Institute a four floor building local in the Taipei Botanical Garden in central Taipei. Precipitation is collected using a 20 cm-diameter stainless steel bucket.

[*] The ratios are calculated from the seasalt corrected flux data in table 1.

Table 1. Mean concentartions (meq/L) and fluxes (meq/m2/period) during summer (June to August) and winter (December to February) of ions in Fu-Shan Forest and Taipei, Taiwan. Calculated from event based precipitation samples.

Ions	Summer				Winter			
	Concentration		Flux		Concentration		Flux	
	Taipei	Fu-shan	Taipei	Fu-shan	Taipei	Fu-shan	Taipei	Fu-shan
H_2O, mm			1233	1542			307	724
H^+	11.8	10.6	14.5	16.4	28.3	39.6	8.7	28.7
Na^+	144.4	35.9	178.0	55.3	193.9	68.4	59.5	49.5
K^+	8.9	8.6	11.0	13.3	19.9	6.5	6.1	4.7
Ca^{2+}	95.6	14.1	117.9	21.7	130.9	56.1	40.2	40.6
Mg^{2+}	49.3	11.9	60.8	18.3	61.9	21.4	18.9	15.5
NH_4^+	56.5	16.8	69.6	25.9	147.9	18.5	45.4	13.4
Cl^-	189.9	63.1	233.0	97.3	236.8	97.5	72.7	70.6
NO_3^-	33.6	9.3	41.4	14.4	87.6	28.0	26.9	20.3
SO_4^{2-}	97.0	28.6	119.6	44.1	238.1	86.2	73.1	62.4
Nss-K	5.8	7.9	7.2	12.1	15.6	5.0	4.8	3.6
Nss-Ca	89.4	12.5	110.2	19.3	122.5	53.0	37.6	38.4
Nss-Mg	46.0	11.1	56.7	17.1	57.0	19.8	17.5	14.3
Nss-Cl	20.5	21.0	25.3	32.4	10.8	17.7	3.3	12.8
Nss-SO4	79.6	24.3	98.1	37.4	214.7	77.9	65.9	56.4

Non-seasalt (Nss) and seasalt fractions of potassium, calcium, magnesium, chloride and sulfate was calculated using sodium as the seasalt tracer. This calculation was based on the assumption that sodium originated solely from seasalts, and its ratios to other ions not chage during aerosol farmation and transport (Keene et al. 1986)

Table 2. Mean concentrations (meq/L) and fluxes (meq/m^2/period) during the summer (June to August) and winter (December to February) of ions in Fu-shan Forest and Taipei, Taiwan. Calculated from event based precipitation samples for.

	Summer				Winter			
	Concentration		Flux		Concentration		Flux	
	Taipei	Fu-shan	Taipei	Fu-shan	Taipei	Fu-shan	Taipei	Fu-shan
H_2O (mm)			1233	1542			307	724
H^+	11.8	10.6	14.5	16.4	28.3	39.6	8.7	28.7
Na^+	144.4	35.9	178.0	55.3	193.8	68.4	59.5	49.5
K^+	8.9	8.6	11.0	13.3	19.9	6.5	6.1	4.7
Ca^{2+}	95.6	14.1	117.9	21.7	130.9	56.1	40.2	40.6
Mg^{2+}	49.3	11.9	60.8	18.3	61.6	21.4	18.9	15.5
NH_4^+	56.5	16.8	69.6	25.9	147.9	18.5	45.4	13.4
Cl^-	189.0	63.1	233.0	97.3	236.8	97.5	72.7	70.6
NO_3^-	33.6	9.3	41.4	14.4	87.6	28.0	26.9	20.3
SO_4^{2-}	97.0	28.6	119.6	44.1	238.1	86.2	73.1	62.4
nss*-K^+	5.8	7.9	7.2	12.1	15.6	5.0	4.8	3.6
nss-Ca^{2+}	89.4	12.5	110.2	19.3	122.5	53.0	37.6	38.4
nss-Mg^{2+}	46.0	11.1	56.7	17.1	57.0	19.8	17.5	14.3
nss-Cl^-	20.5	21.0	25.3	32.4	10.8	17.7	3.3	12.8
nss-SO_4^{2-}	79.6	24.3	98.1	37.4	214.7	77.9	65.9	56.4

*Non-seasalt (nss) and seasalt fractions of potassium, calcium, magnesium, chloride and sulfate was calculated using sodium as the seasalt tracer. This calculation was based on the assumption that sodium originates solely from sea salt aerosols, and its ratios to other ions did not change during aerosol formation and transport (Keene et al. 1986).

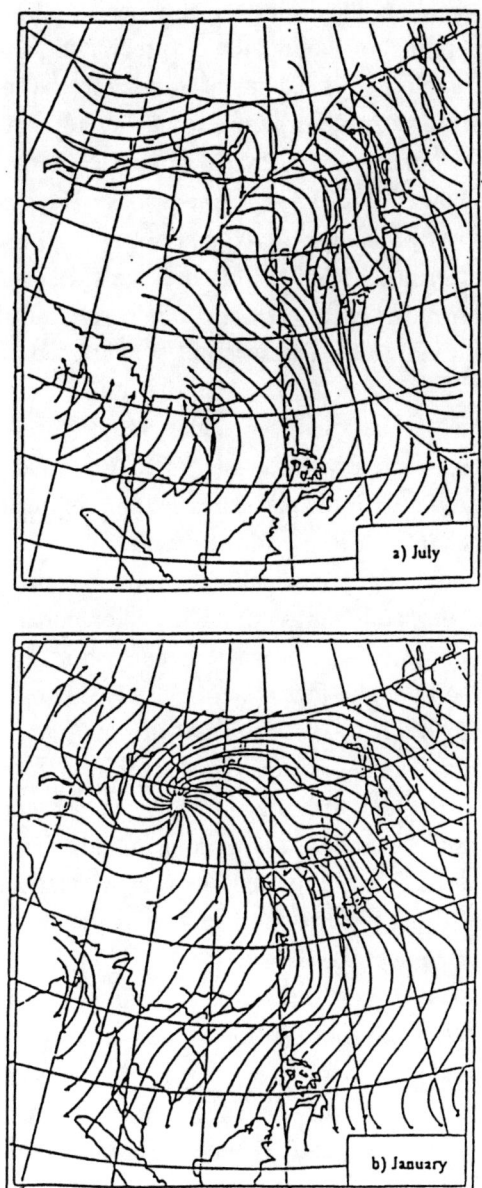

Figure 4. Average a) July and b) January surface current movement in East Asia. Adapted from Chi (1987).

The averaged surface current movement for July (Fig. 4a) clearly shows that the prevailing direction of air masses is southwest. The summer air masses come off the ocean without passing over any large landmass, except the western portion of the Taiwan, prior to arriving at the forest. Thus, anthropogenically derived chemicals, such as ammonium, nitrate and sulfate must have local origins during the summer months. For sites sharing the same pollution sources, the ratios of pollutants in precipitation should be constant, particularly for sites that are close by. Chemical ratios of the major anthropogenic pollutants between the two sites are very similar during the summer but not the winter; ammonium : nitrate : sulfate-summer 0.71 : 0.35 : 1 (Taipei) and 0.69 : 0.33 : 1 (Fu-shan); winter 0.62 : 0.37 : 1 (Taipei) and 0.21 : 0.32 : 1 (Fu-shan)[*]. The agreement of chemical ratios in the summer for anthropogenically derived and not for soil aerosols[**] is consistent with the weather pattern and therefore indicates that the two sites share pollution sources. The possibility that the agreement in chemical ratios could result from local transport of pollutants from the east coast plain to the forest is not supported by the summer weather pattern.

Multivariate factor analysis, which has been widely used to separate chemicals into groups sharing the same sources (McDowell et al. 1990; Brigeman 1992; Abbas et al. 1993), indicates that in the summer two factors explained approximately three quarters of the variation of precipitation chemistry in the forest (Table 3)[*]. The first factor has high loading on both soil aerosols and anthropogenically derived ions and the second factor has high loading on soil aerosols, Ca and Mg, alone. The factor pattern seen at Fu-shan is different from the patterns found in heavily polluted areas such as Taipei (Table 3) where the anthropogenic factor and soil aerosol factor can be easily distinguished (McDowell et al. 1990; Bridgman 1992; Abbs et al. 1993). Transport of both anthropogenically derived pollutants and soil aerosols from Taipei explains the sharing of a common factor between the two types of ions and is consistent with the summer weather wind pattern. Transport from Taipei (Factor 1) is the major source for anthropogenically derived pollutants to Fu-shan whereas both local transport from Taipei and on-site sources are important for soil aerosols (Factor 1 and Factor 2); a pattern consistent with the relationship of chemical ratios between Taipei and Fu-shan described above.

In the winter, concentrations of most chemicals were higher at both sites relative to those of the summer (Table 2). This observation is consistent with the EPA national monitoring data that documented higher pollutant emission in the winter than the summer. As in the summer, the winter concentrations of most chemicals are higher in Taipei than at Fu-shan, however the differences were greatly reduced (Table 2). This is primarily due to the fact that the increase in chemical concentrations was greater at Fu-shan than in Taipei.

---

[*] The ratio for soil aerosols, calcium to magnesium, in precipitation was 1.9 in Taipei and 1.5 in Fu-shan

[*] The factor analysis are performed on seasalt corrected precipitation chemistry data.

[*] Although in table 1 the concentration of nitrate and sulfate differ considerably in the winter, part of the difference could be attributed to the dilution effect of the much higher rainfall at Fu-shan. The dilution effect can be excluded by looking only at storms brought similar rainfall qualities to Taipei and Fu-shan.

This trend persists despite lower winter emissions in the east coast area of Fu-shan relative to those seen in the summer. Thus, local transport from the east coast can not explain the higher concentration enrichment of pollutants at Fu-shan relative to Taipei. Nor can local transport of pollutants from northwestern Taiwan, including Taipei, to the forest site explain the greater increase in concentrations of sulfate and nitrate at Fu-shan relative to Taipei.

Average January surface current movement (Fig. 4b) indicates that in the winter the prevailing air masses moving over Taiwan are from the northeast and pass over the industrialized east coast of China before arriving to Taiwan. The northeast coast of China is characterized by high emissions of NO_x and SO_x (Zhao and Sun 1986; Carmichael et al. 1993; Wang et al. 1993). Air parcel trajectory analysis has shown long-range transport of ozone from mainland China to Taiwan occurs in winter (Liu et al. 1993). Using black carbon as a tracer of air movement, long-range transport from the east coast of China to the North Pacific Ocean was evident (Parunge et al. 1994). Several studies indicate considerable transport of sulfate and nitrate from the east coast of China to the North Pacific Ocean particularly during the winter when the prevailing flow provides a transport path off the east coast (Prospero et al. 1985; Rapporteur et al. 1990; Uematsu et al. 1993; Gao et al. 1993). Northeastern Taiwan is most directly affected by the winter winds, particularly between December and February when the strong northeast monsoon dominates the weather system.

If long-range transport of nitrate and sulfate from China is important during winter months, the concentration of nitrate and sulfate should be similar at the two sites (Taipei and Fu-shan) during precipitation events caused by cold fronts from the northeast. Two consecutive storms on January 20 and 25, 1994 brought similar quantities of precipitation to the two sites and concentrations of sulfate and nitrate were similar while all other ions except hydrogen were higher in Taipei (Table 4)[*]. This is a very different pattern from that seen during the summer when concentrations of sulfate and nitrate were always much higher in Taipei than at Fu-shan. This pattern of winter and summer precipitation clearly shows the importance of long-range transport on the deposition of pollutants to the Fu-shan forest.

In the winter, two factors explain approximately 80% of the variation of precipitation chemistry at Fu-shan (Table 3). The first factor has high loadings on anthropogenically derived chemicals and the second factor has high loadings on soil aerosols. The lack of a common loadings of anthropogenically derived chemicals and soil aerosols suggests that the two types of chemicals had different sources in the precipitation at Fu-shan. If the major emission sources impacting Fu-shan were local, anthropogenically derived pollutants and soil aerosols should share high factor loadings as are seen in Taipei in the winter (Table 3). Thus, the winter pattern of factor analysis suggests that long-range transport is important for anthropogenically derived pollutants deposited at Fu-shan.

Table 3. Varimax rotated factor analysis for summer (June to August) and winter (December to February) precipitation chemistry, after correction for sea salt. Factor components with a values of > 0.5 are bold-faced to emphasize their relative importance. Only factors with an eigenvalue greater than 1 are included. Calcium, magnesium and to a lesser extent potassium ions in precipitation originate from soils whereas ammonium, nitrate and sulfate are typically from anthropogenic sources.

		H^+	K^+	Ca^{2+}	Mg^{2+}	NH_4^+	NO_3^-	SO_4^{2-}
Summer	Taipei							
	Factor 1	-0.17	**0.90**	**0.88**	**0.95**	0.39	0.35	0.33
	Factor 2	**0.76**	0.12	0.42	0.11	**0.81**	**0.67**	**0.80**
	Fu-shan							
	Factor 1	**0.82**	**0.60**	**0.67**	-0.21	**0.87**	**0.76**	**0.97**
	Factor 2	-0.01	0.24	**0.66**	**0.97**	-0.11	-0.14	0.05
Winter	Taipei*							
	Factor 1		**0.90**	**0.76**	**0.92**	0.22	**0.62**	**0.74**
	Fu-shan							
	Factor 1	**0.96**	0.49	0.08	**0.75**	**0.61**	**0.93**	**0.93**
	Factor 2	0.15	**0.51**	**0.96**	0.30	**0.69**	0.38	0.22

* H^+ is not correlated with any other ions.

To estimate the contribution of long-range transport on acidic deposition at the forest site, we need to separate it from the contribution from local sources. Although ammonium can be transported long distances, it is not emitted in sufficient quantities to act as an important long range pollutant. Almost all ammonium found in the atmosphere originates from the conversion of emitted ammonia (Asman and van Jaarsveld 1992; Fangmeier et al. 1994). In spite of the fact that ammonia can be rapidly converted to ammonium aerosol (Asman and Janssen 1987; Allen et al. 1988), the conversion ratio remains low as more than 50% of the emitted ammonia is deposited less than 4 km from the source. As a result, the deposition of ammonium aerosols principally originates from local sources (Ersiman et al. 1987; Asman and van Jaarsveld 1992). While the industrialized east coast of China represents the mainland's largest emission source of nitrate and sulfate, the major source of ammonia is from the inland agricultural centers. (Pong and Yu 1989). As a result, one would predict the contribution of long- range transport of ammonia should be insignificant.

In the summer the concentration of ammonium, nitrate and sulfate in the precipitation at Fu-shan was about 30% of that measured in Taipei (Table 2). This consistent relationship suggests that when local sources dominate the deposition of these ions, their concentrations in precipitation at the forest site should be a consistent proportion of those measured in Taipei. We calculated the importance of local pollution sources on sulfate and nitrate deposition at the Fu-shan forest site with the following two assumptions 1) long-range transport of ammonium is minimal and 2) in the absence of long-range transport, the ratios of ammonium, nitrate and sulfate between the two sites should be similar in the winter and in the summer*. In the winter the concentration of ammonium in the forest site was 12.5% of that found in Taipei. Thus the contribution from local sources on winter nitrate and sulfate deposition at Fu-shan forest represents, 7.9 meq/m^2 and 19.2 meq/m^2, or 39% and 34% of winter input to the forest. The high pollutant input annualized contribution of long-range transport from China would be 23% and 22% of the flux of nitrate and sulfate to the Fu-shan forest. This high pollutant input from long-range transport superimposed on the already high local sulfur and nitrogen input will certainly exacerbate the detrimental effects (e.g., soil acidification and soil cation depletion) of acid deposition on forest ecosystems.

These estimates represent minimum contributions of long-range transport to the lightly developed northeast of Taiwan because 1) long-range transport during spring and fall which is part of the northeast monsoon season was not considered; and 2) the assumption of no long-range transport of ammonium to both sites is overly conservative. The forest site will have an even greater probability of receiving long range ammonium deposition given that Fu-shan is more directly affected by northeast monsoon than is Taipei.

3. ECOSYSTEM SENSITIVITY ASSESSMENT—AN APPLICATION OF RAINS-ASIA IMPACT MODULE

The possible linkage of acidic deposition to forest decline has been intensively studied in Europe and North America. As the detrimental effects of acidic deposition on forest ecosystem become more evident, researchers have begun to address concerns regarding the sensitivity of ecosystems to acid deposition and the maximum safe acid deposition, also known as the critical load.

The sensitivity of an ecosystem to the adverse effects of acidic deposition relies upon the ecosystem's capacity to buffer acidic input by means of weathering processes as well as upon the vegetation tolerance to acidic deposition. Buffering is considered to play a greater role in controlling response than is tolerance because different ecosystems exhibit a greater range in buffering rates than tolerance to acidic conditions. An understanding of certain ecosystem features related to the buffering capacity is required for sensitivity assessment. Because the vegetation present at a site must fit the prevailing site factors including acid deposition, vegetation type could serve as an indicator of the acid neutralizing ability of the soils and the tolerance of plants to acid deposition.

The calculation ignores that some of the sulfate and nitrate deposited in Taipei might come from long-range transport and should be subtracted from the calculation of local contribution.

Table 4. Volume weighted mean chemical concentrations (meq/L) of rainfall events occurring between Jan 20 and 26, 1994.

	Taipei	Fu-shan
H_2O(mm)	31.0	28.0
H^+	6.9	100.0
Na^+	217.3	108.9
K^+	13.3	7.9
Ca^{2+}	128.5	47.0
Mg^{2+}	62.6	31.9
NH_4^+	127.1	21.8
Cl^-	330.7	258.9
NO_3^-	106.4	86.5
SO_4^{2-}	215.8	227.9

The critical load concept is receiving much attention as a measure for developing national emission abatement strategies in Europe as it provides a receptor-based approach to control pollutant emissions (Posch et al. 1993, 1995; Hornung et al. 1995). For acid deposition, the focus has been on two groups of pollutants, namely sulfur and nitrogen compounds. Here I use the RAINS-ASIA impact module mentioned above to examine the sensitivity of ecosystems, and to compare current acid deposition and the critical load of S of forest ecosystems in Taiwan.

Forest ecosystems cover approximately 52% of the area of Taiwan and thus represent the main ecosystem type. Forest ecosystems are critical for the maintenance of biodiversity and for soil and water conservation. Therefore, protecting forest ecosystems are the key to sustainable development of Taiwan. Even though there is little direct documentation of the detrimental effects of acidic deposition on the ecosystem level, studies have reported canopy and soil cation leaching, phenomenon which are typically associated with forest decline due to heavy acidic deposition (Kang 1992; Huang 1994; Lin 1995). As widespread pollutant loading has been documented at levels far exceeding that of the United States and Europe, the threat to Taiwanese forests should not be underestimated. Therefore, developing and/or adapting models to estimate the critical load and assess the impact of acidic deposition on Taiwanese ecosystems is crucial to identify and protect susceptible ecosystems. The RAINS-ASIA Impact module has thus been applied to evaluate the sensitivity of forest ecosystems and compute critical load of S to forest ecosystems in Taiwan.

Table 5. Classification of ecosystem sensitivity according to vegetation types used in RAINS-ASIA project and the modifications suggested in this study. Adapted from RAINS-ASIA project.

Class	Ecosystem Type	
	Determined by RAINS-ASIA Group	After modifications
Ecosystems of low sensitivity		
6	desert; semi-desert; irrigated/paddy land	paddy land
5	thorn woods; semi-arid woods	
4	dry tropical/subtropical forests; savanna; agricultural land	agricultural land
3	cool scrub; grassland; temperate broadleaf woodland; tropical Montana forests	humid subtropical broad-leaf forest
2	conifer forest/ mixed forest; tibetan cold moor; bog; mire; wet tropical forest	grass; conifer forest/mixed forest
1	tundra; main Taiga; northern and southern Taiga	tundra
Ecosystems of high sensitivity		

3.1. SENSITIVITY CLASSIFICATION FOR TAIWAN ECOSYSTEMS

One of the main purposes of determining ecosystem sensitivity is to identify ecosystems which are potentially susceptible to acidic deposition. This would help to set abatement measures to protect sensitive and important ecosystems before the onset of the detrimental effects of acidic deposition. The RAINS-ASIA Impact Module used vegetation type and soil buffering rate to determine ecosystem sensitivity levels. According to the proposed approach, ecosystems in Asia were divided into 6 sensitivity classes (Table 5). One of the six classes, the thorn woods, are not present in Taiwan. Humid subtropical broad-leaf forests are not listed in the classification system. The RAINS-ASIA Impact module uses 6 factors to classify vegetation sensitivity. The 3 most important factors are soil buffering ability (which is largely affected by soil pH), soil moisture and nutrient circulation/rooting depth. Soils with lower buffering ability, higher moisture and slow nutrient cycling rates are more sensitive to acid deposition. In comparison with montane coniferous forests, Taiwanese humid broad-leaf forest, which mostly distributed in lowland areas, have lower soil moisture, similarly low buffering ability (Kang 1992; Kang et al. 1993; Huang 1994), and faster nutrient cycling rates. Therefore, they should be less sensitive to acid deposition than conifer forests. (Table 5). Due to Taiwan's mountainous geomorphology, the majority of forested areas are distributed on steep slopes and as a result the trees tend to have shallow roots. The shallow roots combined with the high soil moisture, make the soils particularly susceptible to cation leaching. Cation leaching and the already low buffering capacity together combine to make Taiwanese forest ecosystems very sensitive to acid deposition.

3.2. CRITICAL LOAD CALCULATION

RAINS-ASIA Impact module uses a steady-state mass balance approach to assess the critical load of acidic input that would not cause long-term harmful effects on the target ecosystems. It assumed that an ecosystem is subject to risk of damage whenever the ratio of base cation and aluminum is below a critical value. Sulfur and nitrogen compounds are the two major sources of acidic input. Although including N in the module is possible, the calculation based on S is better developed. Nevertheless the S calculations still includes parameters that are often not locally available. Therefore, only S is considered in our application of the RAINS-ASIA Impact module given the fact that even for the simplified, S-only, module the required information is incomplete.

Two criteria were developed by the RAINS-ASIA project group to calculate critical loads for sulfate; a plant response criteria and a soil stability criteria. The minimum value of the two criteria is defined as the critical load for a given ecosystem. The critical load computation from the two criteria use the following two equations:

(1) Plant response criteria

$$CL = ANC_w + \{1.5 \frac{X_{(Ca+Mg+K)}ANC_w + BC_d - BC_u}{(BC/Al)_{crit} K_{gibb}}\}^{1/3} Q^{2/3}$$

$$+ \{1.5 \frac{X_{(Ca+Mg+K)}ANC_w + BC_d - BC_u}{(BC/Al)_{crit}}\}$$

where CL = critical load (eq ha^{-1} yr^{-1})
ANC$_w$ = alkalinity produced by weathering (eq ha^{-1} yr^{-1})
X$_{(Ca+Mg+K)}$ = fraction of weathering as Ca, Mg, and K (eq ha^{-1} yr^{-1})
BC$_d$ = base cation deposition (eq ha^{-1} yr^{-1})
BC$_u$ = base cation uptake by plant root (eq ha^{-1} yr^{-1})
Q = runoff (mm yr^{-1})
(BC/Al)$_{cr}$ = critical molar base cation to aluminum ratio
K$_{gibb}$ = gibbsite equilibrium constant

(2) Soil stability criteria

$$CL = 3ANC_w + (2\frac{ANC_w}{K_{gibb}})^{1/3} Q^{2/3}$$

Figure 5. Locations of six forest ecosystems studied for the critical load of S.

The plant response criterion requires information of critical molar base cation to aluminum ratio, $(BC/AL)_{cr}$, for each plant species included in the analysis (Sverdrup and Warvinge 1988; Shindo et al. 1994). However, due to the high diversity of tree species, over 100, in the mixed subtropical forest under study and the lack of $(BC/Al)_{cr}$ laboratory data for each tree species, only the soil stability criteria is used in our study. Moreover, no single value could be used for all tree species and there is also no suitable average value for all tree species.

3.3. EVALUATED FOREST ECOSYSTEMS

Six forest ecosystems were evaluated in this study (Figure 2 and Table 6). Geographically, they include forests from north to south, from east to west and from low to high elevation in Taiwan. They also include forests of varying distance from major industrial centers. Therefore, the range of evaluated forests represent the range and diversity of forest ecosystems under the threat of acidic deposition over the entire island of Taiwan.

Table 6. Data input for the calculation of critical loads (eq/ha.yr), the calculated critical loads and current deposition rates for S at six forest ecosystems in Taiwn.

Site	Fu-shan	Pin-lin	Lien-hua-chi	Pi-lu-chi	San-pan	Tai-ma-li
ANC_w (eq/ha.yr)	70	70	140	70	70	240
Q^1 (mm/yr)	2000	1600	1200	1300	2000	1250
K_{gibb}	150	150	150	300	150	150
CL (eq/ha.yr)	365	343	560	300	365	890
DR^2 (eq/ha.yr)	2800	2700	1400	650	1750	800

[1] Q: runoff, the difference between rainfall and evaportranspiration
[2] DR: deposition rate

3.4. INPUT DATA

Input data as derived from the following description are summarized in Table 7.

Weathering Rates(ANC_w): Weathering rates were estimated based on the acidity of parent material, soil texture, soil temperatue and evapotranspiration rate (De Vries 1989) using a decision tree (Sverdrup and de Vries 1993). Parent material, soil acidity, and soil texture were derived from a soil survey and expert opinions (King unpublished data). According to the decision tree, a soil with or without literization is the first factor affecting weathering rate. There is no prominent literization for soils of the 6 forest ecosystems under study and probably not for most forest soils of Taiwan. Therefore, the initial estimate of weathering is based upon soil acidity. Soil acidity is divided into 5 catagories, lower than 5.0, 5.0~6.2, 6.2~7.2 and above 7.2. The corresponding initial estimates of weathering rate are 35, 150, 350, and 650 meq/m^2.yr from low pH to high pH. The study of sulfate absorption and desorption examined the basic chemistry for 10 forest soils in Taiwan (Hung 1994). The results indicated that almost all the forest soils have pH lower than 5.0. Some of the forest ecosystems evaluated here were included I that analysis and some are near the forests that was analyzed. Therefore, all the forest ecosystems except the one in southeastern Taiwan have soil pH between 5.0-5.5.The third factor for estimating weathering rates is weather or not the soil is a lithosol. None of the forest soils evaluated is lithosol such that no correction is necessary. The fourth factor influencing weathering rates is soil temperature. For soils with temperature always greater than 0°C, no correction is necessary as is the case for the 6 forests under study. The fifth factor influencing weathering rates is soil texture. Soil texture is divided into 5 classes, coarse, coarse/medium, medium, medium/fine, fine and the corresponding multiplying factor is 0.25, 0.5, 1, 2, and 4, respectively. Finer soils have greater multiplying factor and therefore greater weathering rates. The 6 forest ecosystems under evaluation has soil texture either in the medium/fine class or in the fine class. The final factor affecting weathering rate is moisture regime. For soils with precipitation to evaportranspiration ratio of 0.5, increasing soil moisture will increase weathering rate. When the ratio reach 2.0 the correction multiplying factor is 1.0 and further increase in soil moisture will not increase weathering rate.

*Runoff (Q):*Runoff was calculated as the difference between the annual precipitation and the annual evapotranspiration.

Gibbsite equilibrium constant (K_{gibb}): The value of the gibbsite coefficient varies with soil depth and organic content of the soil. Typically, a pure organic soil such as pit, will be give the value $K_{gibb} = 3$. A soil consisting of pure mineral materials without any organic matter is given the value $K_{gibb} = 500$. For subtropical wet forests and conifer forest, the given value is $K_{gibb} = 150$ and 300, respectively.

3.5. DISCUSSION OF ECOSYSTEM SENSITIVITY TO ACIDIC DEPOSITION

Critical load of S for 5 of the 6 forest ecosystems is at the second lowest critical load level, 200-500 eq/ha.yr in Asia according to the RAINS-ASIA regional critical load map, suggesting that the 5 forest ecosystems are very sensitive to acid deposition (Table 7). The 4 ecosystems also receive levels of S deposition much higher than the critical load. The other 2 forest ecosystems have higher critical loads and are thus able to tolerate higher levels of acid deposition. Yet, they also receive S deposition at levels near or higher than the critical loads (Table 6). Local studies on soil buffering capacity to acid deposition (Hung et al. 1992) indicate that soils with a pH between 4.5 and 6.5 are very sensitive to acid deposition. All the 6 forest ecosystems and most forest ecosystems in Taiwan have soil pH within that range. This further illustrates the potential threat that acid deposition presents to forest ecosystems in Taiwan. Although the forests may appear healthy, immediate and drastic decline in forest health is possible once the current buffering capacity of the system is depleted.

Ecosystem sensitivity classification based upon vegetation type suggests that conifer forest ecosystems are more sensitive to acid deposition than humid subtropical hardwood forest ecosystems. The critical loads calculated from soil stability criteria indicate that forest ecosystems in northeastern Taiwan have lower critical loads of sulfur. Therefore, applying models based upon regional estimates show low critical load values which are highly exceeded under present deposition rates.

Many forest soils in Taiwan are able to buffer acid deposition through sulfate adsorption and cation exchange. However, most of them have very low buffering capacity and moderate pH (4.5-6.5) and are, therefore, very vulnerable to acid deposition. Although very acidic soils with pH under 4.5 have a higher ability to buffer acid deposition, they do so only with serious consequences. Studies indicate that for soils with pH lower than 4.5, the main buffering processes is through the release of aluminum ions (Hung et al. 1992; Liu et al. 1992; Liu and Chen 1996). Since aluminum ions are toxic to both plant roots and aquatic organisms, the release of aluminum is of great threat to forest ecosystems. Studies suggest that continuous exposure to acid deposition for ten years would greatly increase the release of Cu, Zn and Cd due to decreased soil pH in red soils (Liu and Chen 1996). Long-term study regarding the impact of acid deposition on forest ecosystems at Hubbard Brook suggests that the loss of cations, especially calcium, is probably the most significant repercussion of acid deposition on forest ecosystem (Nilsson and Wiklund 1992; Likens 1996). Studies in Europe and North America have unambiguously illustrated that cation leaching from the soils may lead to nutrient imbalances and thereby affect forest health. Further research concerning early detection of acid deposition induced damage is also needed. More importantly, effective pollution control measures must be enforced to protect forest ecosystems and the biodiversity and intrinsic qualities (soil and water conservation) associated with them.

4. CONCLUSIONS

Approximately 70% of the precipitation monitored in Taiwan is considered acid rain (exhibiting a pH lower than 5.6).

Acid deposition shows large spatial variation with northern Taiwan exhibiting the highest acid deposition and southeastern Taiwan showing the lowest deposition. The spatial pattern of acid deposition is correlated with the degree of urbanization and industrialization.

The contribution of dry deposition to total acid deposition shows great spatial variation and is governed by precipitation patterns. Dry deposition is least important in northern and northeastern Taiwan as these areas have the high annual precipitation with an even distribution. Dry deposition becomes proportionally more important in areas with low annual precipitation and high variability, such as southern Taiwan.

Both local and regional emission sources contribute to the high acid deposition in Taiwan, although the significance of long-range transport is of some controversy. My study using seasonal patterns of atmospheric deposition, multivariate factor analysis of precipitation chemistry and weather patterns indicates that long-range transport contribute at least 20% of S and inorganic N deposition to forest ecosystems in northeastern Taiwan.

A RAINS-ASIA impact module was used to assess ecosystem sensitivity to acid deposition and to calculate critical load of S to six forest ecosystems in Taiwan. Results indicate that forest ecosystems in Taiwan are very sensitive to acid deposition due to their low soil pH (<5.5).

Lowland subtropical forest ecosystems in Taiwan have low or moderate low critical load for S (320-740 eq/ha/yr) suggesting that they are vulnerable to acid deposition. At present many forest ecosystems are receiving levels of acid deposition far exceeding their critical loads. Although these forest ecosystems appear healthy, there may be sudden detrimental change once the current buffering capacity is depleted.

Cation leaching both from the forest canopy and forest soils was observed in some forest ecosystems. Continuous exposure to high levels of acid deposition should lead the forest to nutrient imbalance and thereby undermine forest health.

REFERENCES

Abbas M. Z. M., Bruns R. E., Scarminio I. S., and Ferreira J. R. 1993. A multivariate statistical analysis of the composition of rainwater near Cubatao, SP, Brazil. *Environmental Pollution* 79: 225-233.

Aber J. D. 1992. Nitrogen cycling and nitrogen saturation in temperate forest ecosystems. *Trends in Ecology and Evolution* 7: 220-224.

Aber J. D., Nadelhoffer, K. J., Steudler, P., and Melillo, J. M. 1989. Nitrogen saturation in northern forest ecosystems. *BioScience* 39: 378-386.

Allen A. G., Harrison R. M.and Wake M. T. 1988. A meso-scale study of the behavior of atmospheric ammonia and ammonium. *Atmospheric Environment* 22: 1347-1354.

Asman W. A. H. and Janssen A. J. 1987. A long-range transport model for ammonia and ammonium for Europe. *Atmospheric Environment* 21: 2099-2119.

Asman W.A.H. and van Jaarsveld H. A. 1992. A variable-resolution transport model applied for NHx in Europe. *Environment Pollution* 26A: 445-464.

Baker E. A. and Hunt G. M. 1986. Erosion of waxes from leaf surfaces by simulated rain. *The New Phytologist* 102: 161-173.

Bashkin V.N. 1997. The critical load concept for emission abatement strategy in Europe: a review. *Environmental Conservation*: 24(1) 5-13.

Bobbink R., Hornung M. and Roelofs J.G.M. 1997. Empirical nitrogen critical loads for natural and semi-natural ecosystems. UBA, *Manual on Methodologies and Criteria for Mapping Critical Levels/Loads and Geographical Areas Where They are Exceeded*, Berlin, Annex III.

Boring L. R., Swank W. T., Waide J. B., and Henderson G. S. 1988. Sources, fates, and impacts of nitrogen inputs to terrestrial ecosystem: review and synthesis. *Biogeochemistry* 6: 119-159.

Bormann F. H. 1985. Air pollution and forests: an ecosystem perspective. *BioScience* 35: 434-441.

Brady N. C. 1990. *The Nature and Properties of Soils*. 10th edition. Macmillan Publishing Company, New York.

Bridgman H. A. 1992. Evaluating rainwater contamination and sources in southeast Australia using factor analysis. *Atmospheric Environment* 26A: 2501-2412.

Bruijnzeel L. A. 1991. Nutrient input-output budgets of tropical forest ecosystems: a review. *Journal of Tropical Ecology* 7: 1-24.

Bull K. R. 1992. An introduction to critical loads. *Environmental Pollution* 77: 173-176.

Cape J.N. 1993. Direct damage to vegetation caused by acid rain and polluted cloud: definition of critical levels for forest trees. *Environmental Pollution* 82: 167-180.

Carmichael G. R., Amann M., Azimi A., Hbattcharya S. C., Shrestha R. M., Chadwick M., Kuylenstierna J., Cinderby S., Foell W., Green C., Hettelingh J. P., Hordijk L., Shah J., Singh M., Streets D., Bhatti N., and Zhao D. 1993. Acid rain and emissions reduction in Asia. *Proceedings of the International Conference on Regional Environment and Climate Changes in East Asia*. pp. 328-333.

Chang K.-H and Jeng F.-T. 1996. A case study on the influences of long-range transport to Taiwan's acid deposition. In the *Proceedings of International Conference on Acid Deposition in East Asia*, May 28-30. 1996. Taipei, Taiwan. p. 50-69.

Chen C.-S., Lin N.-H, Peng C.-M. and Jeng F.-T. 1996. Acidic deposition on Taiwan and associated precipitation systems. In the *Proceedings of International Conference on Acid Deposition in East Asia*, May 28-30. 1996. Taipei, Taiwan. p. 124-132

DeHayes D. H. and Hawley G. J. 1992. Genetic implications in the decline of red spruce. *Water, Air, and Soil Pollution* 62: 233-248.

De Vries W. 1991. Methodologies for the assessment and mapping the critical loads and of the impact of abatement strategy on forest soils. *The Winand Staring Centre for Integrated Land, Soil and Water Research*, Report 46, Wageningen, The Netherlands, 109pp.

Duckworth C. M. S. and Cresser M. S. 1991. Factors influencing nitrogen retention in forest soils. *Environmental Pollution* 72: 1-21.

Eamus D. and Murray M. 1991. Photosynthetic and stomatal conductance responses of Norway spruce and beech to ozone, acid mist and frost - a conceptual model. *Environmental Pollution* 72: 23-44.

EPA 1993. *Annual Report of Environmental Protection*. Environmental Protection Agency, Taipei, Taiwan.

EPA 1994. *Annual Report of Environmental Protection*. Environmental Protection Agency, Taipei, Taiwan.

EPA 1996. *Environmental Information of Taiwan*., Republic of China, EPA, Taipei, Taiwan.

Ersiman J. W., Vermetten A. W. M., Pinksterboer E. F., Asman W. A. H., Waikers-Ypelaan A. and Slanina J. 1987. Atmospheric ammonia: distribution, equilibrium with aerosols and conversion rate to ammonium. In *Ammonia and Acidification*. Symposium of the EURASAP, Bilthoven, The Netherlands, 13-15 April 1987, ed. W. A. H. Asman & H. S. M .A. Dideren. RIVM/TNO, Bilthoven, pp. 59-72.

Esher R. J., Marx D. H., Ursic S. J., Baker R. L., Brown L. R., and Coleman D. C. 1991. Simulated acid rain effects on fine roots, ecotomycorrhizae, microorganisms, and invertebrates in pine forests of the southern Untied States. *Air, Water, and Soil Pollution* 60: 269-278.

Fangmeier A., Hadwiger-Fangmeier A., van der Eerden L. and Jager H.-J. 1994. Effects of atmospheric ammonia on vegetation-a review. *Environmental Pollution* 86: 43-82.

Fasth W. J., David M. B., and Vance G. F. 1991. Sulfate retention and cation leaching of forest soils in response to acid additions. *Canadian Journal of Forest Research* 21: 32-41.

Fernandez I. J. 1989. Effects of acidic precipitation on soil productivity. In Adriano D. C. and Johnson A. H. eds. *Acidic Precipitation*: Volume II. Biological and Ecological Effects. Springer-Verlag, New York. pp. 61-84.

Friedland A. J., Miller, E. K., Battles J. J. and Thorne J. F. 1991. Nitrogen deposition, distribution and cycling in a subalpine spruce-fir forest in the Adirondacks, New York, USA. *Biogeochemistry* 14: 31-55.

Gao Y. R., Arimoto R. and Duce R. A. 1993. Atmospheric aerosols over eastern Asia and their impact on the regional atmospheric environment. *Proceedings of the International Conference on Regional Environment and Climate Change in East Asia*. pp. 78-82.

Gebauer G., Giesemann A., Schulze E. -D., and Jager H.-J. 1994. Isotope rations and concentrations of sulfur and nitrogen in needles and soils of *Picea abies* stands as influenced by atmospheric deposition of sulfur and nitrogen compounds. *Plant and Soil* 164: 267-281.

Gosz J. R. 1980. Nutrient budget studies for forests along an elevational gradient in New Mexico. *Ecology* 61: 515-521.

Gundersen P and BashkinV. 1994. Nitrogen cycling, ch.9. *Biogeochemistry of small catchments*, J.Wiley and Sons, New York, 253-277

Haines B. L., Jernstedt J. A., and Neufeld H. S. 1985. Direct foliar effects of simulated acid rain II. Leaf surface characteristics. *The New Phytologist* 99:407-416.

Hettelingh J. -P., Gardner R. H. and Hordijk L. 1992. A statistical approach to the regional use of critical loads. *Environmental Pollution* 77: 177-183.

Hornung M., K.R. Bull K. R., Cresser M., Hall J., Langan S. J., Loveland P. and Smith C. 1995. An empirical map of critical loads of acidity for soils in Great Britain. *Environmental Pollution* 90: 301-310.

Houston D. B. and Stairs G. R. 1973. Genetic control of sulfur dioxide and ozone tolerance in eastern white pine. *Forest Science* 19: 267-271.

Hsieh C. E. and Shen C. E. 1994. *Flora of Taiwan Introductory Series and Plates of Vegetation*. Editorial Committee of the Flora of Taiwan.

Hutchinson G. L., Millington R. J., and Peters D. B. 1972. Atmospheric ammonia absorption by plant leaves. *Science* 175: 771-772.

Huang J. T. 1994. *Sulfate Adsorption and Desorption of Forest Soils in Taiwan*. Master Thesis, National Taiwan University, Taipei Taiwan. 121 pp.

Hung J. J., Liu W. C., Chang C. H., Tzeng J. S. and Li S. W. 1992. *An evaluation of buffering potential of selected Taiwan agricultural soils for acid deposition*. Report of Taiwan Sugar Research Institute 138: 11-19.

Johnson A. H. 1983. Red spruce decline in the northeastern U. S.: hypotheses regarding the role of acid rain. *Journal of the Air Pollution Control Association* 33(11):1049-1054.

Johnson D. W. and Taylor G. E. 1989. Role of air pollution in forest decline in eastern North America. *Water, Air, and Soil pollution* 48: 21-43.

Kaiser W., Dittrich A. and Heber U. 1993. Sulfate concentrations in Norway spruce needles in relation to atmospheric SO_2: a comparison of threes from various forests in Germany with trees fumigated with SO_2 in growth chambers. *Tree Physiology* 12: 1-13.

Kang M. H., Lin T. C., Wang M. K., and King H. B. 19993. The effects of simulated acid rain on cation leaching of an Ultisol in Central Taiwan. *Proceedings of the Workshop on Effects of Air Pollution and Agrometerology on Crop Production in Taiwan*, C. C. Tu and C. M. Yang (eds) p. 227-241.

Kang M. H. 1992. Cation Leaching From Lien-Hua-Chi Watershed Forest Soils Treated by Simulaated Acid Rain Solutions. Master Thesis, National Taiwan University, Taipei Taiwan. 103 pp.

Karnosky D. F. 1977. Evidence for genetic control of response to sulfur dioxide and ozone in *Populus tremuloides*. *Canadian Journal of Forest Research* 7: 437-440.

Keene W. C., Pszenny A. P., Galloway J. N. and Hawley M. E. 1986. Sea-salt corrections and interpretation of constituent ratios in marine precipitation. *Journal of Geophysical Research* 91: 6647-6658.

Kenk G. and Fischer H. 1988. Evidence from nitrogen fertilization in the forests of Germany. *Environmental Pollution* 54: 199-218.

Kessler C. J., Porter T. H., Firth D., Sager T. W., and Hemphill M. W. 1992. Factor analysis of trends in Texas acidic deposition. *Atmospheric Environment* 26A: 1137-1146.

King H. B., Liou C. B., Lin T. C. and Wang, L. J. 1994. Chemistry of precipitation, throughfall, stemflow, soil solution and streamwater of six forest sites in Taiwan. *Proceedings of International Symposium on Biodiversity and Terrestrial Ecosystems*.

Kress M. W., Baker R., and Ursic S. J. 1990. Chemistry response of two forested watersheds to acid atmospheric deposition. *Water Resources Bulletin* 26: 747-755.

Langan S. J. and Hornung M. 1992. An application and review of the critical load concept to the soils of northern England. *Environmental Pollution* 77: 205-210.

Langford A.O. and Fehsenfeld F.C. 1992. Natural vegetation as a sourse or sink for atmospheric ammonia: a case study. *Science*, 225: 581-583.

Likens G. E. and Bormann F. H. 1995. *Biogeochemistry Of A Forested Ecosystem.* 2nd edition, Springer-Verlag New York, Inc. New York, USA.159 pp.

Lin T. C. 1995. *Atmospheric Deposition and Forest Canopy Processes in a Subtropical Rain Forest in Southeast Asia-Taiwan,* Doctorial Dissertation, University of Kansas, Lawrence, Kansas, USA. 124 pp.

Lin T-C. and Hamburg S. P. 1992. *Nutrient leaching and retention in an ecotonal forest in eastern Kansas.* Supplement to Bulletin of the Ecological Society of America 73: 250. (Abstract).

Lin T.-C., King H.-B., Hsia Y.-J and Liou C.-B. 1996a. Wet and dry deposition in a subtropical broad-leaf forest of northeastern Taiwan. In *the Proceedings of International Conference on Acid Deposition in East Asia,* May 28-30. 1996. Taipei, Taiwan. p. 98-106.

Lin T. C. King H. B., Wang L. J., Hsia Y. J., and Hettelingh J. P. 1996. Evaluating the Rains-Asia Impact Module for the use in Taiwan. In the *Proceedings of International Conference on Acid Deposition in East Asia,* May 28-30. 1996. Taipei, Taiwan. p. 428-438.

Lin T. C., Hamburg S. P., King H. B., and Hsia Y. J. 1997. Spatial variability of throughfall in a subtropical rain forest in Taiwan. *Journal of Environmental Quality* 26:172-180.

Liu C. M., Young C. Y. and Su W. C. 1996. Scenarios of SO_2 and NO_x emission in Taiwan. In the *Proceedings of International Conference on Acid Deposition in East Asia,* May 28-30. 1996. Taipei, Taiwan. p. 409-426.

Liu J.-C and Chen Z.-S. 1996. Effects of atmospheric acid rain on soil properties of Taiwan - A reivew. In the *Proceedings of International Conference on Acid Deposition in East Asia,* May 28-30. 1996. Taipei, Taiwan. p. 537-549.

Liu Q. and Li C. 1991. Simulating study on the effect of acid precipitation on forest soil weathering. *Journal of Environmental Science* (China) 3: 61-70.

Marschner H., Haussling M., and George E. 1991. Ammonium and nitrate uptake rates and rhizosphere pH in non-mycorrhizal roots of Norway spruce [*Picea abies* (L.) Karst.]. *Trees* 5: 14-21.

Matzner E. 1989. Acidic deposition; case study Soiling. in Adriano D. C. and M. Havas. (eds) *Acidic Precipitation I case studies.* pp. 39-84. Springer-Verlag, New York.

Matzner E. and Prenzel J. 1991. Acid deposition in the German Solling area: effects on soil solution chemistry and Al mobilization. *Water, Air, and Soil Pollution* 60: 221-234.

McColl J. G. and Bush D. S. 1978. Precipitation and throughfall chemistry in the San Francisco Bay area. *Journal of Environmental Quality* 7: 352-357.

McDowell W. H., Sanchez C. G., Asbury C. E. and Perez C. R. R. 1990. Influence of sea salt aerosols and long range transport on precipitation chemistry at El Verde, Puerto Rico. *Atmospheric Environment* 24A: 2813-2821.

Miller H. G. and Miller J. D. 1988. Response to heavy nitrogen application in fertilizer experiments in British forests. *Environmental Pollution* 54:219-231.

NAPAP (National Acid Precipitation Assessment Program) 1990. *Integrated Assessment Report.* US Government Printing Office, Washington, D.C. 1991.

Nashimoto M. and Kohno Y. 1989. Current status of forest decline and its research activities in Europe. *Denryoku Chuo Kenkyusho Hokoku (u89015)*: I-IV, 1-35. (In Japanese, with English abstract).

Nihlgard, B. 1985. The ammonium hypothesis- an additional explanation to the forest dieback in Europe. *AMBIO* 14: 2-8.

Nilsson J. 1986. Critical loads for nitrogen and sulfur. *Miljorapport 1986:11*, Nordic Council of Ministers, Copenhagen.

Nilsson J. and Grennfelt P. Critical loads for sulfur and nitrogen, *Report 1988:15*, Nordic Council of Ministers, Copenhagen, Denmark, 1988.

Nilsson L. -O. and Wiklund K. 1992. Influence of nutrient and water stress on Norway spruce production in south Sweden - the role of air pollutants. *Plant and Soils* 147: 251-265.

Oren R., Werk K. S., Meyer J., and Schulze E. -D. 1989. Potentials and limitations of field studies on forest decline associated with anthropogenic pollution. In Schulze E. -D, Lange O. L., and Oren R. eds. *Forest Decline and Air Pollution: a study of spruce (Picea abies) on acid soils*. Springer-Verlag, New York. pp. 23-36.

Parunge F., Nagamoto C., Zhao M.Y., Hansen A. D. A. and Harris J. 1994. Aeolian transport of aerosol black carbon from China to the Ocean. *Atmospheric Environment* 28: 3251-3260.

Peterman R. M. 1990. The importance of reporting statistical power: the forest decline and acidic deposition example. *Ecology* 71: 2024-2027.

Pitelka L. F. and Raynal D. J. 1989. Forest decline and acidic deposition. Ecology 70: 2-10.

Pong S. and Yu Q. T. 1989. Measurement of gaseous ammonia in air in China. *Journal of Environmental Science* (in Chinese)

Posch M., Hettelingh J. -P., Sverdrup H. U., Bull K. and de Vries W. 1993. Guidelines for the computation and mapping of critical loads and exceedances of sulfur and nitrogen in Europe. In: R.J. Downing, J-P. Hettlingh and P.M.A. de Smet (Eds.). *Calculating and Mapping of Critical Loads in Europe*, Status Report, Coordination Center for Effects (RIVM) Bilthoven, The Netherlands, pp. 25-38.

Posh M. de Smet P.M.A., Hettelingh J. -P., and Downing R.J. 1995. *Calculation and mapping of critical thresholds in Europe*: Status Report Coordination Center for Effects (RIVM) Bilthoven, the Netherland.

Prospero J. M., Savoie D. L., Nees R. T., Duce R. A. and Merrill J. 1985. Particulate sulfate and nitrate in the boundary layer over the North Pacific Ocean. *Journal of Geophysical Research.* 90 (10): 586-596.

Rapporteur H. L. I., Galloway J. N., Moody J. L., Eliassen A., Ryaboshapko A. G., Fisher B. E. A., Savoie D., Gorzelska K., Whelpdale D. M. and Hastie D. R. 1990. The long-rang transport of sulfur and nitrogen compounds. In A. H. Knap (ed.) *The Long- Range Atmospheric Transport of Natural and Contaminant Substances,* 231-257. Kluwer Academic Publishers. The Netherlands.

Rehfuess K. -E. 1991. Review of forest decline research activities and results in the Federal Republic of Germany. *Journal of Environmental Science and Health A* 26: 415-445.

Richter D. D., Johnson D. W., and Todd D. E. 1983. Atmospheric sulfur deposition, neutralization, and ion leaching in two deciduous forest ecosystems. *Journal of Environmental Quality* 12: 263-270.

Rodenkirchen H. 1991. Effects of acidic precipitation, fertilization and liming on the ground vegetation in coniferous forests of southern Germany. *Air, Water, and Soil Pollution* 60: 279-294.

Schulze E. -D. 1989. Air pollution and forest decline in a spruce (*Picea abies*) forest. *Science* 244: 776-783.

Schulze E. -D., de Vires W., Hauhs M., Rosen K., Rasmussen L., Tamm C.-O., and Nilsson J. 1989. Critical loads for nitrogen deposition on forest ecosystems. *Water, Air, and Soil Pollution* 48: 457-461.

Shepard J. P., Mitchell. M. J., Scott Y.M., Zhang Y. M., and Raynal, D. J. 1989. Measurements of wet and dry deposition in a northern hardwood forest. *Water, Air, and Soil Pollution* 48:225-238.

Shindo J., Bregt A.K., and Hakamata T. 1994. *Application and evaluation of critical load methods for acid deposition in Gunma, Japan.*

Skeffington R. A. and Wilson E. J. 1988. Excess nitrogen deposition: issues for consideration. *Environmental Pollution* 54:159-184.

Soares A., Ming Y. J., and Pearson J. 1995. Physiological indicators and susceptibility of plants to acidifying atmospheric pollution: a multivariate approach. *Environmental Pollution* 87: 159-166.

Soikkeli S. and Karenlampi L. 1984. The effects of nitrogen fertilization on the ultrastructure of mesophyll cells of conifer needles in northern Finland. *European Journal of Forest Pathology* 14: 129-136.

Sverdrup H. and de Vries W. 1993. Calculating critical loads for acidity with the simple mass balance method. *Water, Air and Soil Pollution* 72: 143-162.

Sverdrup H. and Warfvinge P. 1988. Weathering of primary minerals in the natural soil environmental in relation to a chemical weathering model. *Water, Air and Soil Pollution* 38: 387-408.

Tanner R. L. 1990. Sources of acids, bases and their precursors in the atmosphere. in Lindberg S. E., A. L. Page and S. A. Norton. (eds) *Acidic Precipitation III sources, deposition and canopy interactions.* pp. 1-20. Springer-Verlag, New York.

Termorshuizen A. J. 1990. Decline of carpophores of mycorrhizal fungi in stands of *Pinus Sylvestris*. Ph. D. Thesis. University of Agricultural Science, Uppsala, Sweden.

Tietema A. and Verstraten J. M. 1991. Nitrogen cycling in an acid forest ecosystem in the Netherlands under increased atmospheric nitrogen input: the nitrogen budget and the effect of nitrogen transformations on the proton budget. *Biogeochemistry* 15: 21-46.

Uematsu M., Komai N., Medvedev A. N. and Anikiev V. V. 1993. Long-scale transport of pollution aerosol over and deposition to the east coast of Asia. *Proceedings of the International Conference on Regional Environment and Climate Change in East Asia.* pp. 74-77.

Ulrich B. and Pankrath D. 1983. *Effects of Accumulation of Air Pollutants in Forest Ecosystems*. D. Reidel, Dordrecht.

Unworth, M. H. and Ormrod, D. P. 1982. *Effects of Gaseous Air Pollution on Agriculture and Horticulture*. Butterworth Scientific, London.

van Breemen N., Burrough P. A., Velthorst E. J., van Dobben H. F., de Wit T., Ridder T. B., and Reijnders H. F. R. 1982. Soil acidification from atmospheric ammonium sulfate in forest canopy throughfall. *Nature* 299: 548-550.

van Dijk H. F. G., de Louw M. H. J., Roelofs J. G. M., and Verburgh J. J. 1990. Impact of artificial, ammonium-enriched rainwater on soils and young coniferous trees in a greenhouse II. Effects on the trees. *Environmental Pollution* 63: 41-60.

Vitousek P. M. and Howarth R. W. 1991. Nitrogen limitation on land and in the sea: How can it occur? *Biogeochemistry* 13:87-115.

Vitousek P. M. and Sandford R. L. Jr. 1986. Nutrient cycling in moist tropical forest. *Annual Review of Systematics and Ecology* 17: 137-167.

Vosatka M., Cudlin P., and Mejstrik V. 1991. VAM population in relation to grass invasion associated with forest decline. *Environmental Pollution* 73: 263-270.

Wallander H. and Nylund J. -E. 1992. Effects of excess nitrogen and phosphorus starvation on the extramatrical mycelium of ectomycorrhizas of *Pinus sylvestris* L. *New Phytologist* 120: 495-503.

Wang Z., Huang M., Xu H., Zhou L., Gao H. and Cheng X. 1996. Studies on budget, balance and transboundary flux of sulfur in East Asia. In *Proceedings of International Conferenec on Acid Deposition in East Asia*, May 28-30, 1996. Taipei, Taiwan, 550-556.

Wang W. X., Zhang Q. Shi Q., and Hong S. 1993. Study on factors related to acidity of rain water in China. *Proceedings of the International Conference on Regional Environment and Climate Change in East Asia*. pp. 231-235.

Waring R. H. and Pitman G. B. 1985. Modifying lodge pole pine stands to change susceptibility to mountain pine beetle attack. *Ecology* 66: 889-897.

Yei C.-J., We Y. L, and Jen F. T. 1996. *Seasonal patterns of acid deposition in Taiwan*.

Zhao D. and Seip H. M. 1991. Assessing effects of deposition in southwestern China using the MAGIC model. *Water, Air, and Soil Pollution* 60: 83-97.

Zhao D. and Sun B. 1986. Air pollution and acid rain in China. *Ambio* 15(1): 2-5.

Zhao L., Huang M., Xu H., Wang Z. and Gao H. 1996. The study of sulfur deposition and transport between Taiwan and the continent of China. In the *Proceedings of International Conference on Acid Deposition in East Asia*, May 28-30. 1996. Taipei, Taiwan. p. 529-536.

CONCLUSIONS

The problem of acid rains has been known from the middle of 19th century. However for the last two decades the heated debates on air pollution control have been assumed both national and international dimensions. With the establishment and quantification of the long-range transport of air pollutants in the mid-1970s, air pollution and acid rain have been seen as an international problem requiring further research on the impact of air pollutants on the environment.

The first agreement to control transboundary air pollution has been made with the signing of the United Nation Economic Commission for Europe (UN/ECE) Convention on Long-Range Transboundary Air Pollution (LRTAP) by 32 European countries plus Canada and USA in 1979.

It should be stressed that in this agreement, the agreed control measures were not based on any concrete scientific assessment of the reduction required to protect the environment from the harmful influence of these emissions. However, since 1988 a new concept has been agreed both nationally and internationally to try to improve this situation by optimizing cost and benefits. This is the critical loads/critical levels approach and it has been adapted as the basis of modern emission abatement strategy for many the countries in Europe, parties of LRTAP Convention.

Furthermore, there is an evidence that long-range transboundary air pollution is not limited to the geographical boundaries. It is spread world widely. In order to estimate quantitatively the harmful impact of transboundary air pollution on the ecosystems in the vast regions of Eurasian continent, the scientific investigations involving various countries are requiring. Consequently these scientific investigations should include the assessment of acidification loading at the ecosystems, monitoring of the acid rain, experimental assessment of terrestrial ecosystem response to acid input, calculation and mapping of critical loads as the indicators of ecosystem sensitivity to acid deposition in different non-European regions, especially in East Asia.

Accordingly, this book presents the results of research teams from various East Asian countries concerning on different aspects of emission, depositions, acidification loading on the ecosystems, from transboundary air pollution to local influences of acidity on various ecosystem components in East Asia.

At present the problem of transboundary pollution in the vast Eurasian continent is attracted the public attention in many countries. For instance, both local and regional emission sources contribute to the high acid deposition in Taiwan, although the significance of long-range transport is of some controversy. The local study using seasonal patterns of atmospheric deposition, multivariate factor analysis of precipitation chemistry and weather patterns indicates that long-range transport contribute at least 20% of total S and inorganic N deposition to forest ecosystems in northeastern Taiwan.

One of the first steps in discovering the acidification as a large-scale environment damage was the Programme of Long-Range Transport of Atmospheric Pollution executed in Europe in 1970s. Since than the investigations have been primarily concentrated in two main areas; field measurements and model simulations. Several studies have shown that oxidised sulfur in the atmosphere can be transported to hundreds and thousands of kilometres and powerful emission sources such as power plants can produce a considerable amount of loading to far-located environmental systems.

The first chemical species that has been used for the investigation of long-range transport and transformation in the atmosphere was sulfur oxides. Since this chemical species has linear characters in physical and chemical processes in the atmosphere, the first models that has been developed to investigate the long-range transport and chemical transformation was a trajectory model with a puff that moves with the environmental wind field.approach. This puff model had algorithm, dealing with both the chemical transformation and wet and dry deposition processes.

More sophisticated and precise but more computing time required models are Eulerian models. They are based on algorithms for numerical solution of differential equations describing both chemical and physical processes affecting the pollution distribution.

Several numerical simulations have shown that in Northern Hemisphere there are three self-polluted regions - Europe, East Asia and North America. However, the effects of transboundary influence in these regions are detectable on the vast areas between them. It is worthwhile to note that the prevailing transport direction in Eurasia is from west to east. Consequently, the European pollution sources can significantly impact on Central Asia, Asian territory of Russia, and some CIS countries. In the meantime, Chinese and Russian emitters also remarkabely affect on the Japanese and Korean ecosystems.

The main question is the precision of the obtained results. In industrial regions it is often determined by the quality of the emission data, while in remote areas the quality of the model and algorithms take the first priority. The most direct way to check the accuracy of the models is to compare the model results with available measurement data, but it raises several problems. First, long-range models have relatively coarse resolutions (at least several tens of kilometres) while the measurement data are of point character. Therefore spatial representativeness of the measured data becomes crucial. Second, the difficulty is a long-term effect of acidification, which requires a characteristic time scale of several years. Consequently, it implies that measurements should also be temporally representative (to have regular character and similar features for a long period of time). To resolve these main difficulties several methodologies of the monitoring station evaluation and comparison of their results with model outputs were developed and shown in this book.

Among different problems devoted to the modeling of atmospheric deposition of acidifying compounds of sulfur and nitrogen, the most complicated ones are connected with the assessment of dry deposition of SO_2 and NO_x.

In order to calculated the dry deposition of SO_2 and NO_x over South Korea synoptic meteorological conditions are categorized into several similar types for five years from 1989 to 1993. The surface observation data from 64 stations are averaged hourly for the non-precipitating days to get diurnally varying micrometeorological data that are required for the estimation of the dry deposition velocity. With the use of the monitored SO_2 and NO_2 concentrations at 31 sites in South Korea, the estimated deposition velocity and the supplement of some empirical constants used in the Regional Acid Deposition Model (RADM), the dry deposition fluxes of SO_2 and NO_2 are estimated using the inferential method over South Korea. The estimated annual mean dry deposition flux over South Korea is about 8.07 t km^{-2} yr^{-1} (81 kg ha^{-1}) for SO_2 and 3.8 t km^{-2} yr^{-1} (38 kg ha^{-1}) for NO_2 which are corresponding to 45% and 40% of the annual total emission rate respectively. However, in the eastern parts of Korea except for the regions where high emission sources are located the dry deposition fluxes exceed the emission rates. This suggests a significant horizontal transport of pollutants from west to east due to prevailing westerly winds and local circulations.

The large attention in this volume is given to the experimental monitoring of precipitation. At present acid rain is observed regularly at many sites in East Asia, especially in China, South Korea, Hong Kong, Taiwan and Japan. Lower pHs tend to be observed in industrialized countries and regions while higher pH values tend to be observed in areas with little or no industrial activities.

For instance, the Environmental Protection Administration (EPA) of Taiwan has established twelve monitoring stations of acid deposition in Taiwan since 1990. Based on the database in the monitoring system, the main compositions of rain water were the sulfate, chloride, sodium and ammonium ions, followed by nitrate and calcium ions. The concentration ratio of sulfate to nitrate of acid deposition in Taiwan is about 5. The mean annual total deposition of sulfate ion from rain water in northern Taiwan is higher than 100 kg/ha/yr which is more than two times compared with that in the eastern USA. Monitoring results also indicates that about 70 kg/ha/yr of sulfate is deposited in southern Taiwan. The mean annual total deposition of nitrate ion from rain water in northern and western Taiwan is 40 to 60 kg/ha/yr, and < 30 kg/ha/yr in eastern and southern Taiwan. The mean annual total deposition of hydrogen ion from rain water is highest in northern Taiwan, about 1 kg/ha/yr, and < 0.4 kg/ha/yr in other regions of Taiwan. The mean annual total deposition of ammonium ion from rain water is highest in southwestern Taiwan, ranged from 20 to 30 kg/ha/yr, and < 10 kg/ha/yr in eastern and southern Taiwan.

It has been concluded that Japan has the most sophisticated and longest precipitation chemistry and air quality monitoring program in East Asia. Also, Japan is a world leader in air quality legislation, management and control. In many areas of air quality management it is on a par with the industrialized countries of Europe and North America, and in some areas, notably air pollution control technologies, it is more advanced.

With the exception of Japan, other countries in East and South-East Asia have started relatively recently monitoring of acidic precipitation and air quality. Monitoring programs vary from being relatively comprehensive (e.g. South Korea, Hong Kong, China) to rudimentary (e.g. Indonesia, Vietnam). Some countries have yet to set up air quality and acid rain monitoring programs of any kind (e.g. Myanmar, Laos, Cambodia). General air quality management, legislation, and control measures vary between countries of East Asia in the same proportion as air quality and acid precipitation monitoring.

The quality of data produced by many of the monitoring networks in the region, with the possible exception of Japan, leaves much to be desired. There is an absence of quality control procedures in many of the networks. Much improvement could be achieved by the introduction of QA/QC (quality analysis/quality certificate) schemes in acid rain and air quality monitoring networks. The use of certified reference materials (CRM) in rainwater analysis is to be encouraged and the setting up of international monitoring programs with a centralized QC/QA scheme would be desirable. This would greatly improve the quality of the data generated and allow for more meaningful comparisons to be made between different countries. Obviously, many of the earlier conclusions will have to be revised when more reliable data become available. The proposed international acid rain survey in East Asia would help by standardizing methods and procedures.

Many of the poorer countries in East and South-East Asia lack the technical expertise and instrumentation to set up adequate acid rain and air quality monitoring networks. Clearly, there is a need for international assistance in terms of providing training and financial resources to these countries.

One should be mentioned that it is difficult to make reliable comparisons in pH and other measurements reported in the literature, in the absence of a uniform methodology strictly adhered to by all parties involved. Measurements of pH are especially sensitive to a variety of factors including: sampling bottle material, sampling period (event, daily, weekly, or monthly), type of sample (bulk or wet-only), and especially the time of pH analysis. The pH is not a static parameter, but is continually changing, during and after sampling, due to chemical reactions in the sample.

In the recent years, the haze phenomenon in South-East Asia is a major hot air pollution problem and its impact on the chemistry of precipitation is still unknown. It has been speculated that forest fires could raise the acidity of precipitation. While it is known that forest fires produce acid precursor gases such as SO_2, NO and NO_2, the non-carbonaceous matter of particles produced by these fires contains Ca and K, which give rise to alkalinity in rainwater. There is a need for a research project to study both the emissions from the forest fires in South-East Asia and their impact on precipitation chemistry and rainfall acidity.

From the measurements available so far it can be concluded that acid rain is coming to be a major problem in East Asia. In many industrially developed and new developed countries such as Japan, China, Taiwan, South Korea, Thailand etc the values of pH <5 are encountered at many sites, and they are representing more that 50% of monitoring rain events in regional scale. However, in some developing countries of South-East Asia (Myanmar, Laos, Cambodia) most rainwater pH measurements tend to be around 5.6, the pH of "natural" rainwater and the acid rain precipitation are mainly due to localized industrial pollution. There is some evidence that pH values below 5 at unpolluted sites may be due to the contribution of weak organic acids, such as formic and acetic acids.

It should be stressed that most of the precipitation networks in East and South-East Asia do not monitor weak organic acids, such as formic and acetic acids, in rainwater. At many tropical sites in Africa and South America it has been demonstrated that organic acids can contribute between 40 and 80 % of the total acidity. In view of this it is important that precipitation surveys in East and South-East Asia begin routine monitoring of organic acids at the earliest stage.

The influence of acid rain on the environment is related to the various properties of different ecosystems. This influence varies depending upon physic-chemical characteristics of soil, vegetation type, stemflow and throughfall interactions of rain waters with canopy of different botanic species.

It is well known in Japan that soils close to the stems of Japanese cedar (*Cryptomeria japonica*) trees are strongly acidic. This is partly due to the leaching of hydrogen ions from the stems. Since soil acidification due to acidic deposition has not been observed in Japan, investigation of soils and soil solutions close to stems may be worthwhile to predict the situation that may occur after acidification due to acidic deposition. In a Japanese cedar forest in Gunma Prefecture, Japan, soil solutions have been collected by ceramic porous cups at a depth of 10 cm. Soil solutions far from stems (> 100 cm) are slightly acidic (pH ~ 5.8) and contain 1.0 µM of total Al in average. The speciation of Al, using cation exchange chromatography with fluorescence detection of the Al-lumogallion complex, shows that nearly 100 % of the Al consist of species with a charge less than or equal to +1 (possibly, organically chelated Al). The molar ratios of BC (= Ca^{2+} + Mg^{2+} + K^+) to total Al are extremely high, ranging from 66 to 1050. In contrast, soil solutions close to a stem (10 cm) are markedly acidic (pH ~4.5) and contain 47 µM of total Al in average. Furthermore, more than 55 % of the Al are in the form of Al^{3+}. The BC/T-Al ratios in winter decline to as low as 2, which is close to 1 that is known to be used as critical value for starting soil acidification and appearance of harmful free aluminum ions.

During last decades the declines of several tree species and forests in the different areas of Japan are noted, as well as European forests are suffering. However, they may be linked with different causes: forest succession dynamics, diseases and pests, meteorological extreme conditions, air pollutants, acidic deposition, and so on. Exposure experiments have been conducted to assess cause-effects relationships with acidic deposition and forest decline. Simulated acid rain or ozone in combination with sulfur dioxide was exposed to sixteen potted-tree species for 3 growing seasons. Rain acidity below pH 4.0 could induce deleterious effects on some broad-leaved trees, however, coniferous trees do not show any significant growth reduction. Sulfur dioxide and/or ozone induced complicated differential growth responses than the wet deposition did. Some indicated additive harmful effects of sulfur dioxide. Others showed synergistic adverse effects of ozone and sulfur dioxide. Experimental results suggest that ambient level of ozone (on of the precursor of acid deposition connected with nitrogen species in troposphere) may take an important role to reduce tree vitality, since ozone induced chronic changes in carbon allocation will be accelerated by increased nitrogen input.

Assuming that fossil fuel energy consumption will continue at the current rate in the China, sulfur emission will be double at the year of 2010. Without any countermeasures for reduction of sulfur dioxide emission, it may induce possible adverse direct effects on the natural vegetation due to increasing concentration of sulfur dioxide rather than due to increasing wet deposition of sulfate. Exposure experiments suggest that differential sensitivity of plants to primary gaseous pollutants and its critical level will be a more important factor to explain forest decline rather than the soil acidification stress alone associated with wet acid deposition.

It is known that the geochemical and biogeoichemical mobility and migration of majority of heavy metals are increasing with the decreasing soil and water pH values that are occurring due to acid deposition. So, it is of great scientific and public concern to study the influence of atmospheric acidity loading on the accumulation of hazardous metals in agricultural food production.

Representative two rural soils in Taiwan were selected to compare the extractable concentrations of heavy metals (Cd, Cu, Pb, and Zn) in soils and total concentration of heavy metals in the brown rice and vegetables. These two soils were treated with different grades of concentration of heavy metals and then simultaneously treated with artificial acid rain or without acid rain (treated with top water) by pot experiments in 1996. The Lung-tang area (affected by acid rains) in northern Taiwan and Lung-luan-tang area (unaffected by acid rains) in southern Taiwan were selected as the field sampling sites of different vegetable species for studying the effects of acid deposition on the biological accumulation of heavy metals (Cd, Cu, Pb, and Zn) in the brown rice and in the leaves of 19 vegetables sampled from 1996 to 1997. The results from pot experiments indicate that the treatment of artificial acid rain in two selected soils significantly increases the biological accumulation of studied heavy metals in the brown rice and leaves of pickled cabbage. The results from the field samples indicated that the ratios of relative concentration of Cd, Zn, and Cu, except for Pb, in nineteen vegetable species sampled from the acid rain affected region are almost higher than 1, or higher than 3, compared to the crops grown in acid rain non-affected area. These results suggest that the biological accumulation of Cd, Cu, and Zn, expect for Pb, in the leaves of vegetables was affected by the acid rain, and the rating of effectiveness on the phyto-availability of heavy metals caused by acid deposition follows the trend: Cd>Zn>Cu>>Pb.

The experimental data obtained in various countries of East Asia allow us to consider the applicability of methodology of critical loads related to an assessment of ecosystem sensitivity to acid rains. The critical load (CL) and Environmental Risk Assessment (ERA) approaches were used for the evaluation of ecosystem sustainability to acid deposition in the East Asia. The calculations of critical loads for the assessment of the sensitivity of the ecosystem to acidic deposition have been made using biogeochemical approaches including the intensity of biogeochemical cycling and period of active temperature duration. On the basis of these coefficients the soil-biogeochemical regionalization is carried out for the whole area of East Asia and the values of critical loads for acid-forming compounds are calculated using modified steady-state mass balance (SSMB) equations. In the northeastern ecosystems of the Asian part of Russia these values of critical loads for N, CL(N), and S, CL(S), compounds are shown to be less than in Europe due to many peculiarities of climate regime and biogeochemical cycling of elements. The minimum values of both CL(N) and CL(S) are <50 eq/ha/yr. and the maximum ones are >300 eq/ha/yr. being at least a few times less than corresponding European ecosystems. The ERA estimates show the maximum significance of such endpoints as N content in plant issues and surface water for many North-Eastern Asia ecosystems. For the South-Eastern ecosystems of the northern part of Thailand the minimum values are < 200 eq/ha/yr. and maximum values - >700 eq/ha/yr. and the minimum rank is related to more than 75% of the studied area. The exceedances of critical levels for various atmospheric acid forming pollutants and their precursors (O_3, SO_2, NO_x) are shown also for different natural and agricultural ecosystems of South Korea.

The experimental results obtained by Chinese scientists have obviously shown that the areas suffering from acid rain in China have extended northwards from the south of Yangtze River in 1986 to the whole East China at present. The statistical results from the Acid Rain Survey in 82 cities from 1991 to 1995 indicate that the annual average pH value of the precipitation was lower than pH 5.6 in nearly half of these cities or in 87% of the southern cities which are located in the south to the Qingling Mountain and Huaihe River, and the lowest even reached pH 3.52 in Changsha, Hunan province. In addition, the frequency of acid rain was very high (higher than 60%) in one fourth of these cities. Up to date, acid deposition appears quite severe in wide areas southern to the Changjiang (Yangtze) River and eastern to the Qingzang (Tibet) Plateau, covering 12 provinces/autonomous regions, with the formation of four core zones, namely Chongqing-Guiyang area in the southwest, Changsha-Nanchang area in the south, the southeast coastal area and the area near Qingdao in Shandong province.

The chemical composition of acid rain in China is generally different from that in Europe, with the lower pH value and the higher sulfate, calcium and ammonium concentrations. Another difference is that the concentration of calcium relative to sulfate is very high in China, while the nitrate concentration relative to other components is low. In some cases, the fluoride concentration in precipitation appears also high in China, owing possibly to the combustion of coal with high fluoride content. Besides, the alkaline fly ash and soil dust build the capacity for acid neutralizing during washout.

Based on the mineralogy controlling weathering and soil development, sensitivity of ecosystem to acid deposition is assessed with the comprehensive consideration on the effect of temperature, soil texture, land use and precipitation. The results show that the most sensitive area to acid deposition in China is podzolic soil zone in the Northeast, then followed by latosol, dark brown forest soil and black soil zones. The less sensitive area is ferralsol and yellow-brown earth zone in the Southeast, and the least sensitive areas are mainly referred to as xerosol zone in the Northwest, alpine soil zone in the Tibet Plateau, and dark loessial soil and chernozem zone in central China. These regional different soil sensitivities to acid deposition can be attributed to the differences in temperature, humidity and soil texture.

It has been shown that the assessment of ecosystem sensitivity to acidic loading depends strongly on the calculation of chemical weathering of soil base cations due to input of proton with depositions. There are two aprroches for calculating this weathering. The use of dynamic modeling approach has been identified as essential for the predicting the time taken for the ecosystems to reach equilibrium between present and potencial input of acidity. These models are much more complicate than so called simplified steady-state mass balance (SSMB) equitions. The first word indicates that the description of the biogeochemical processes involved is simplified, which is necessary when considering the large-scale application to East Asian ecosystems. The second word of the SSMB acronym indicates that only steady-state conditions are taken into account, and this leads to considerable simplification.

Accordingly, in Japan the models were developed to evaluate soil acidification caused by acidic deposition and ecosystem sensitivity to it in different spatio-temporal scales. For the prediction of catchment scale acidification, a dynamic model was made. The model took rapid chemical reactions into consideration as the quasi-equilibrium processes and processes such as chemical weathering, nutrient uptake, nitrification *etc.* as element flux at a constant rate to the soil system. Application of this model to the soil acidification experiments using simulated acid rain showed that changes in soil chemistry could be well expressed by the model and the acid loaded into the soil was neutralized in the top 10 cm horizon and soil acidification occurred there. Against the urgent acidification, main buffering mechanisms were cation exchange and dissolution of Al hydroxides whose relative contributions differed according to soil characteristics and its acidification stage.

For larger scale estimation, especially for the country scale, a steady state mass balance model was to be employed. Magnitudes of acid neutralizing capacity and acid production due to chemical weathering, base cation and nitrogen uptakes and base cation deposition were estimated for total Japanese forest ecosystems based on existent data bases of geology, soil, vegetation etc., some measured data on soil properties and parameters derived from literature. It has been additionally stressed that the mineral weathering rate has the most significant effect on the neutralizing capacity of ecosystems. An indicator was proposed to evaluate the possible ecosystem impact by acid deposition based on the steady state model under the condition of current or predicted acid deposition rates.

As it has been already mentioned, approximately 70% of the precipitation monitored in Taiwan is considered acid rain (exhibiting a pH lower than 5.6). Acid deposition shows large spatial variations with the highest acid deposition northern Taiwan and the lowest deposition in southeastern Taiwan. The spatial pattern of acid deposition is correlated with the degree of urbanization and industrialization. The contribution of dry deposition to total acid deposition shows great spatial variations with precipitation patterns. Dry deposition is least and evenly distributed in northern and northeastern Taiwan where annual precipitation amount is high while it becomes larger with high variability in southern Taiwan where annual precipitation amount is low. A RAINS-ASIA impact module is used to assess ecosystem sensitivity to acid deposition and to calculate critical load of sulfur to six forest ecosystems in Taiwan. Results indicate that forest ecosystems in Taiwan are very sensitive to acid deposition due to their low soil pH (<5.5). Lowland subtropical forest ecosystems in Taiwan have low or moderate low critical loads for S suggesting that they are vulnerable to acid deposition. Yet, many forest ecosystems are exposed to acid deposition far exceeding their critical loads. Although these forest ecosystems appear healthy, there may be a sudden detrimental change once the current buffering capacity is depleted.

Cation leaching both from the forest canopy and forest soils is observed in some forest ecosystems. Continuous exposure to high levels of acid deposition can lead the forest to be in nutrient imbalance and thereby undermines forest health.

The input of <1 meq/100g soil to the forest soil in the northern part of Thailand, with the organic content of 1.33 percent, has no changes in pH value due to existing hydrogen buffering capacity. Simulated acid rain at pH 2 led to Al leaching from this soil naturally enriched by aluminum and iron, although there is no significant change of pH in the soil. The acid depositions cause also the reduction of bacteria, actinomyces and ammonifying microbes in the upper part of the soil. These could lead to low nutrient cycling rate in the ecosystem as these microorganisms play the significant role in organic matter decomposition in the soil. Due to certain buffering ability of soils to acid depositions the forest ecosystem can be sustainable in short-time scale, however some negative chemical and biological changes occur in soil. This will gradually decrease the ecosystem sustainability to acidic loading.

The study carried out in Thailand aimed to provide the results of sensitivity assessment in term of maps for the study area, and to determine and identify the sensitive receptors and locations where abatement strategies would be implemented to reduce environmental impacts of acidification on forest and agriculture. New data have been picked up for Thailand ecosystems, and a revised methodology has been applied. Sensitivity of ecosystems to acidic deposition has been carried out by using two steps procedure. Firstly, the sensitivity mapping according to revised methodology of Stockholm Environment Institute has been conducted for Thailand conditions. The purpose of such sensitivity mapping is to define the distribution of ecosystems with the same relative level of reactions to the given rate of acidic deposition. On the second stage the sensitivity of Thailand ecosystems to acidic deposition has been described by means of Critical Loads (CL) and exceedances, using modified Steady-State Mass Balance model with simplified expert-modelling approach. The maps for CL of sulfur derived from this methodology have been overlaid with current (or projected future) deposition maps in order to show areas where the CL of sulfur is (or will be) exceeded. In order to manipulate with the numerous maps and data a geographical information system (GIS) has been used.

These data support the above mentioned point on the high sensitivity of Thai ecosystems to acidity input.

Consequently, on the basis of results shown in various chapters of this book one can conclude:

- there exists clearly transboundary air pollution problems in East Asia;
- the acidity of precipitation has been increased during the latest decades in many developed and developing countries;
- the harmful single and synergetic effects of acidity, SO_2, NO_x and O_3 are experimentally shown for various natural and agricultural ecosystems in East Asia;
- the critical loads concept is shown to be a good guide for assessing ecosystem sensitivity to acid deposition in many East Asian countries;
- the calculated critical loads are exceeded at present acidity loading for many ecosystems in East Asian countries (Russia, China, Thailand, Taiwan, Japan, South Korea);
- for the emission abatement strategy in the whole East Asian region the critical load concept has to be implemented by Environmental Risk Assessment (ERA) procedure in the stage of management and investments. For instance, in order to control acid rain and sulfur dioxide pollution, China should designate the Acid Rain Control Areas and the Sulfur Dioxide Pollution Control Areas where acid rain or serious SO_2 pollution could occur or might occur. Thermal power plants and enterprises within these areas must take measures to abate SO_2 emissions;
- it is difficult to make predictions about the future incidence of acid rain in East and South-East Asia due to the unpredictable economic situation. Many countries in the region are experiencing a slowing down of their industrial development due to the recent currency crisis, which could in turn slow down pollutant emissions. However, the result could lead to more polluted environment due to poorly pollution controlled industries because of lacking the financial resources. If the economic growth recovers to the rates experienced during the late 1980s and early 1990s, we could expect to see a greater incidence of acid rain over a wider region. ERA scenarios based on the calculations and mapping of critical loads should be applied to local, national and international scale.
- there is a clear evidence of the necessity to enforce the international cooperation in East Asia in order to develop joint emission abatement strategy. The expansion of UN/ECE Long-Range Transboundary Air Pollution Convention to the whole Eurasian continent may serve as a key role in solving the ploblems of hazardous influence of acidic deposition on ecosystem health.

INDEX

A

acetic acids, 96, 116, 117, 417
Acid Mist, 153
Acid Rain Frequency, 273, 274
acid rain monitoring, 95, 96, 103, 114, 116, 277, 416
advection scheme, 5, 8, 9, 10, 23, 44, 47, 48
aerodynamic resistance, 49, 54, 55
Aggregation Uncertainty, 358
air pollutants, 1, 44, 45, 47, 106, 114, 144, 145, 149, 168, 185, 223, 231, 238, 259, 383, 410, 413, 417
albuminoid nitrogen, 110
alkaline fly ash, 268, 419
Aluminum leaching, 139, 222
Aluminum speciation, 125
ambient air, 148, 149, 168, 170, 180, 338
analysis, factor, 394, 395, 396, 405, 406, 414
arable land, 234, 300
Arctic Deserts, 241, 243
Armenia, 39
ASEAN, 117, 118, 119, 120, 121, 122
atmospheric emissions, 354
atmospheric pollutants, 7, 44, 51, 337
Australia, 100, 102, 222, 406
Austria, 39, 45
Azerbaijan, 39

B

BC/Al ratios, 125, 126, 141, 356
Beijing, 97, 102, 179, 267, 269, 272, 282, 283, 284, 310, 311
Belarus, 39
Belgium, 39
biogeochemical cycling, 230, 232, 234, 238, 240, 241, 242, 243, 245, 246, 247, 248, 249, 251, 254, 260, 419
Black Forests, 145
Black Triangle Region, 145, 178, 180, 181, 182
Boreal Taiga Forest, 241, 245
Bosnia, 39
Brazil, 405
Broad-leaved trees, 143, 154
Broad-Leaved Trees, 155, 159
Brook C.G.D., 404
Brown Rice, 189, 205, 215
Brunei, 2, 95, 99, 110, 111, 118, 120, 337
Brunei Darussalam, 95, 99, 110, 118, 120
Bulgaria, 39

C

cadmium, 224
calcareous soils, 102
California, 101, 224
Cambodia, 95, 96, 114, 116, 121, 337, 416, 417
cation, 126, 129, 136, 138, 142, 201, 203, 234, 245, 296, 313, 314, 316, 319, 320, 323, 328, 330, 331, 334, 348, 355, 357, 360, 375, 380, 397, 398, 399, 400, 402, 404, 407, 408, 417, 420
cations and anions, Determination of, 125
Charnock formula, 29
Chemical transformations, 5
Chestnut soil, 298, 299, 301
Chinese brown coal, 1
Chongqing, 97, 179, 269, 270, 276, 282, 283, 285, 286, 287, 290, 302, 303, 310, 311
Commonwealth of Independent States, 6, 414
Coniferous trees, 143, 154

Coniferous Trees, 155, 158, 161, 163
copper, 224, 352
critical load (CL), 163, 230, 418
Critical loads of sulfur, 250, 253
crop production, 153
Czech Republic, 39, 145, 180

D

defoliation, 155, 156, 157, 158, 171, 177
Denmark, 39, 262, 410
Desert-Steppe and Desert Ecosystems, 247
Device Precision, 14
dry deposition velocity, 11, 29, 30, 36, 46, 49, 50, 51, 52, 77, 79, 415

E

East China, 59, 268, 276, 419
economic development, 238, 385
economic growth, 116, 238, 422
ecosystem sensitivity, 1, 2, 230, 231, 235, 238, 240, 255, 256, 260, 313, 315, 324, 335, 336, 347, 356, 357, 359, 362, 379, 380, 399, 405, 413, 418, 420, 421, 422
Ecosystem Sensitivity Assessment, 397
education, 46
elution time, 130, 132
e-mail, 5, 229
emission controls, 342
Emission estimates, Asian region, 5
Emission Intensity, 20
emission reduction, 7, 37, 182, 235, 238, 239, 240, 308, 342
environmental protection, 232, 304, 306
Environmental Quality Act, 108
Environmental risk assessment, 229
Environmental Risk Assessment (ERA), 230, 418, 422
Estonia, 39
Eulerian model, 5, 6, 10, 21, 43, 46, 414
Eurasia, 5, 6, 14, 18, 21, 35, 37, 42, 43, 414
Europe, 1, 6, 9, 13, 17, 21, 33, 34, 35, 38, 42, 43, 44, 45, 47, 48, 96, 100, 115, 140, 141, 142, 145, 146, 159, 168, 172, 173, 178, 180, 182, 186, 187, 188, 225, 230, 232, 241, 251, 254, 260, 262, 263, 268, 278, 279, 293, 302, 315, 333, 337, 338, 347, 356, 382, 385, 397, 398, 404, 405, 410, 413, 414, 416, 419
Europe, Eastern, 347
eutrophication, 37, 232, 233, 234, 236, 238, 380

evapotranspiration, 177, 245, 247, 249, 250, 332, 348, 350, 362, 402, 403
exchange rates, 182
Experimental forest, 125

F

Factors Affecting Tree Growth, 146
Far East, 246, 251, 254
Finland, 39, 186, 411
Florida, 222, 223
flue gas desulfurization, 101, 307, 308
fluoride, 126, 129, 130, 131, 133, 136, 268, 279, 419
Forest Decline, 145, 333, 410
Forest decline in Japan, 143
Forest Meadow Steppe, 246
Forest Soil, 262, 269, 295, 297, 377, 408
fossil fuel, 1, 144, 146, 180, 192, 336, 337, 340, 418
fossil fuels, 180, 192, 336, 340
France, 39
Fujian, 273, 274, 275, 276, 277, 278, 279, 284, 286, 291, 292, 309, 311

G

Gabbro, 295
gasoline, 340
Georgia, 39
Germany, 39, 90, 145, 180, 184, 263, 390, 408, 410, 411
Gibbsite equilibrium constant (K_{gibb}), 403
Great Britain, 39, 334, 407
Greece, 39
ground-ozone levels, 264
growth factor, 174
Growth Response, 158, 172
Guangdong, 273, 277, 284, 285, 287, 290, 291, 292, 293, 302, 309, 310
Guangzhou, 179, 270, 282, 292
Guiyang, 97, 179, 269, 282, 285, 286, 287, 293, 302
Guizhou, 102, 273, 277, 284, 285, 287, 291, 292, 293, 303

H

Hazard identification, 235
heavy metals, 189, 190, 191, 192, 193, 203, 205, 206, 207, 208, 209, 210, 211, 219, 220, 221, 222, 223, 224, 225, 232, 418

Hong Kong, 95, 96, 112, 115, 116, 117, 120, 121, 415, 416
hPa maps, 60, 61, 66, 67, 69, 70
Hubei province, 278, 285
Hungary, 39

I

India, 1, 231
Indonesia, 95, 96, 98, 103, 104, 105, 106, 116, 117, 119, 120, 121, 337, 416
industrial sector, 340
industrialization, 1, 103, 106, 115, 383, 385, 386, 388, 404, 420
Inner Mongolia, 277, 285, 301
Iran, 39
Iraq, 39
Ireland, 39
irrigation, 170
Israel, 39
Italy, v, 39

J

Jakarta, 104, 105, 106
Japan, 2, 34, 39, 45, 51, 52, 95, 96, 99, 100, 101, 115, 116, 118, 119, 120, 121, 125, 126, 128, 132, 136, 143, 144, 145, 147, 149, 150, 151, 159, 166, 168, 173, 174, 176, 178, 179, 180, 181, 182, 183, 184, 186, 246, 261, 262, 263, 278, 279, 313, 314, 317, 324, 328, 329, 330, 332, 333, 334, 411, 415, 416, 417, 420, 422
Jiangsu, 273, 274, 275, 278, 279, 284, 291, 292, 311
Jordan, 39

K

Kansas, 409
Kazakhstan, 39, 251
Kentucky, 153
Kerosene, 337, 339
kidney, 211, 214, 215, 216, 217, 218, 220, 289
Korea, 1, 2, 39, 46, 49, 50, 51, 52, 53, 59, 65, 68, 71, 74, 86, 89, 90, 95, 96, 113, 115, 116, 118, 121, 178, 179, 229, 230, 246, 249, 255, 257, 258, 260, 261, 262, 264, 265, 377, 415, 416, 417, 419, 422

L

Lagransian model, 5, 13

Latvia, 39
Lebanon, 39
legislation, 96, 101, 114, 115, 116, 416
Levels, 405
lime, 308
Lithuania, 39
local government, 146, 148, 307

M

Macedonia, 39
Malaysia, 95, 98, 108, 109, 110, 117, 118, 119, 337
manganese, 192, 224, 352
mass media, 145, 147
mesophyll, 56, 57, 411
Mexico, 407
Microbial count, 356
Mineral weathering rate (BC_{we}), 324
Missouri, 221
Modelling, 44, 45, 262, 333, 334, 377
Mongolia, 38, 39, 102, 277, 285, 301, 383
Montana, 399
Myanmar, 95, 96, 115, 116, 118, 337, 416, 417

N

National Acid Precipitation Assessment Program (NAPAP), 145, 186, 386, 409
natural gas, 342
natural resources, 238
needle chlorosis, 177
Netherlands, 39, 142, 180, 262, 333, 406, 407, 410, 411
Nevada, 187
New Mexico, 407
New York, 44, 90, 183, 187, 223, 225, 334, 406, 407, 409, 410, 411
nitrates, 11, 22, 29, 32, 34
Nitrogen chemistry transformations, 11
nitrogen oxides, 1, 11, 19, 36, 146, 147, 148, 180, 192, 231
Nitrogen uptake (N_{gu}), 330
Non-seasalt (nss), 392
North Korea, 39
Norway, 39, 45, 46, 47, 145, 192, 406, 408, 409, 410
Numerical models for evaluation of atmospheric pollution, 5

O

OECD, 7, 44, 46
one-layer Gaussian dynamic profile, 27
Open-Top Chamber (OTC), 168
Oxidised sulphur, 11
oxygen depletion, 375
Ozone, 2, 63, 143, 166, 171, 172, 173, 176, 177, 181, 183

P

Pacific, 38, 39, 114, 120, 122, 383, 395, 410
Paddy soil areas, 301
particulate matter, 147, 149
perennial ryegrass, 153, 154
permafrost zones, 240
Philippines, 95, 106, 121, 337
photosynthesis, 225
Podzolic soil, 298, 300, 302, 309
Poland, 39, 145, 180
Pollutant Exposures, 170
population densities, 388
population density, 385, 389
Portugal, 39
Pot Experiments, 189, 203
Potassium, 370, 371
poverty, 304
Precipitation Chemistry, 278, 280, 283
public interest, 146
Puerto Rico, 409

Q

Q-mode clustering, 51

R

raw materials, 307
Regional Acid Deposition Model (RADM), 50, 51, 59, 62, 63, 415
Rhizosphere, 161
Risk characterization, 236
Risk management, 236
Romania, 39
Russia, 2, 5, 6, 34, 38, 39, 43, 100, 229, 230, 231, 246, 249, 252, 253, 254, 261, 357, 414, 419, 422

S

Sakhalin, 39, 252

sandstone, 203, 206, 208, 209, 221
Scale leaf brown necrosis, 156
Schist, 295
Seoul, 2, 49, 68, 74, 79, 86, 113, 120, 121, 179, 229, 261, 262, 377
shale alluvial soil, 204, 206, 208, 209, 221
Shandong, 268, 273, 274, 275, 277, 278, 279, 284, 419
Shanghai, 269, 282, 286
shellfish, 353
Siberia, 102, 246, 251, 254
Sichuan, 272, 273, 284, 285, 286, 287, 290, 291, 292, 293, 301, 303, 308
simulated acid rain (SAR), 150, 153, 155, 176, 178
Singapore, 95, 107, 111, 118, 120, 122, 337
Slovakia, 39
Soil acidity, 201, 403
soil erosion, 238, 351
soil mineralogy, 294, 295, 296
Soil moisture content, 370
Soil Organic Matter, 370
Soil pH, 285, 317, 370
Soil properties, 335, 369
soil solution samples, 125, 126, 128, 130
soil texture, 201, 251, 268, 294, 295, 296, 297, 298, 310, 402, 419
Soil Texture, 370
soil-canopy energy budget model, 56
solar radiation, 57, 151, 168, 178, 309
Southeast Asia, 35, 38, 43, 120, 409
South-East Asia, 1, 95, 96, 97, 100, 101, 116, 117, 231, 248, 416, 417, 422
SO_x and NO_x Emission, 385
Soybean plants, 153
Spain, 39
Sri Lanka, 247
Steppe Ecosystems, 245, 246, 247
stress factors, 146
Subboreal Forest, 242, 245
sugar maples, 145
sulfur dioxide emission, 144, 168, 337, 340, 355, 372, 373, 418
surface and canopy resistance, 49
surface water, 222, 230, 232, 233, 234, 250, 255, 256, 260, 267, 292, 293, 309, 354, 356, 419
surface water hydrology, 354
Sweden, 39, 181, 186, 187, 192, 223, 262, 294, 296, 333, 334, 378, 410, 411
Switzerland, 39, 181, 188
Synoptic meteorological classification, 49, 59
Syria, 39

T

Taiga Meadow Steppe, 245
Taiwan, 2, 95, 96, 99, 115, 119, 178, 179, 181, 189, 190, 192, 193, 194, 195, 196, 197, 198, 200, 201, 202, 203, 204, 211, 212, 213, 219, 221, 222, 223, 224, 225, 301, 378, 379, 380, 383, 384, 385, 386, 387, 388, 389, 390, 391, 392, 394, 395, 397, 398, 399, 402, 403, 404, 405, 406, 407, 408, 409, 412, 414, 415, 417, 418, 420, 422
Tajikistan, 39
Thai lignite, 1
Thailand, 2, 1, 2, 95, 98, 99, 111, 112, 117, 118, 119, 120, 121, 122, 230, 231, 249, 259, 260, 261, 262, 335, 336, 337, 338, 340, 341, 342, 343, 344, 345, 346, 347, 348, 349, 350, 352, 353, 355, 357, 359, 360, 362, 363, 364, 365, 366, 367, 368, 369, 376, 377, 378, 417, 419, 421, 422
Tianjin, 98, 272, 284
Tibet, 248, 268, 269, 272, 277, 286, 287, 301, 308, 309, 419, 420
Trace Elements, 189, 190, 203, 204, 211, 215
transboundary air pollution, 1, 188, 230, 413, 414, 422
transportation, 269, 278, 304, 338
tree growth, 143, 187
Tropical Monsoon Forest, 247
Tropical Wet Forest, 242, 248
Tundra Ecosystems, 243, 245
Turkey, 39

U

U.S. Environmental Protection Agency, (EPA), 190, 193, 194, 204, 211, 213, 222, 223, 224, 235, 263, 385, 386, 388, 390, 394, 406, 415

Ukraine, 39, 48
UNESCO, 351
United Nations, 223
United States, 118, 145, 184, 201, 224, 386, 390, 398
urbanization, 1, 385, 386, 404, 420
USSR, 46, 121
Uzbekistan, 39

V

Vertical structure of model, 5
vertical winds, 12
Vietnam, 95, 96, 98, 103, 116, 117, 119, 121, 416
volcanic ash, 128, 170, 325, 328, 332

W

water conservation, 398, 404
water supply, 258
Weathering Rates(ANC$_w$), 402
Western Europe, 44
wet deposition, 30, 45, 51, 102, 119, 143, 144, 147, 149, 150, 153, 182, 191, 331, 388, 418
World Bank, 231, 234, 260, 263

X

Xerofitic Savanna, 247
Xiamen, 179, 270, 271, 276, 286

Y

Yugoslavia, 39

Z

Zhejiang, 273, 274, 275, 279, 290, 291, 292, 309, 311
Zinc, 191